Advances in Wastewater Research

Series Editor

Maulin P. Shah, Environmental Microbiology Consultant, Ankleshwar, Gujarat, India

The Advances in Wastewater Research highlights the prospective future development of wastewater treatment techniques, research and field applications. The series also focuses on the limitations and challenges associated with some generally accepted wastewater strategies and evaluates the possible applications of these corrective strategies to eliminate toxic pollutants from the environment through integrated technologies in wastewater treatment. The series covers the research areas including, but not limited to, toxic pollutants removal from wastewater, methods and procedures, materials, management of wastewater, valorisation possibilities of waste resulting from the removal of toxic pollutants from wastewater, wastewater recycling, environmental impact, and wastewater policies. The series contribute to remove, reduce, or neutralize water contaminants that threaten human health, and ecosystem productivity and integrity. The series includes peer-reviewed monographs, edited volumes, textbooks, and conference proceedings.

Ashootosh Mandpe · Devaanshi Jagwani ·
Maulin P. Shah
Editors

Nanotech Solutions for Industrial Wastewater

Springer

Editors
Ashootosh Mandpe
Department of Civil Engineering
Indian Institute of Technology Indore
Indore, Madhya Pradesh, India

Devaanshi Jagwani
Institute of Engineering and Science
IPS Academy
Indore, Madhya Pradesh, India

Maulin P. Shah
Environmental Microbiology Consultant
Microbiology Lab
Ankleshwar, Gujarat, India

ISSN 3004-9725 ISSN 3004-9733 (electronic)
Advances in Wastewater Research
ISBN 978-981-96-6878-6 ISBN 978-981-96-6879-3 (eBook)
https://doi.org/10.1007/978-981-96-6879-3

This Springer imprint is published by the registered company Springer Nature Singapore Pte Ltd.
The registered company address is: 152 Beach Road, #21-01/04 Gateway East, Singapore 189721, Singapore

Preface

Nanotechnology is a revolutionary field with innovative solutions for the most pressing ecological challenges in the future. Its numerous applications make one prominent field stand out, such as nanomaterials which play a crucial role in treating industrial wastewater since it has the potential for changing water pollution outcomes, rational use of natural resources, and sustainable development. This book is a comprehensive exploration of nanomaterials and their critical role in the treatment of industrial wastewater, catering to the needs of researchers, academics, industry professionals, and policymakers alike.

The book is divided into twelve chapters dealing with various critical features of nanotechnology in wastewater treatment. The introduction is done to Nanomaterials and Industrial Wastewater Treatment, giving the reader a background on the topic. Other subsequent chapters cover the Types of Nanomaterials, their Synthesis and Characterization, and Mechanisms that help them to react with pollutants.

Emerging pollutants in industrial wastewater pose unique challenges, and the application of nanomaterials for their remediation is discussed in detail. The book also addresses the integration of nanomaterials in advanced treatment processes, exploring their effectiveness and adaptability. As with any cutting-edge technology, it is crucial to consider the Environmental Implications and Safety Considerations of nanotechnology, ensuring that its benefits are not outweighed by unintended consequences.

Furthermore, the book has covered Technological Challenges and Future Prospects as well as the Economic and Societal Aspects of introducing nanotechnology in wastewater treatment. The practical applications of a dedicated case study on nano filters are elaborated along with real-world success. The discussion ends with Metal Oxide Nanocomposites for industrial wastewater remediation and the conclusion with the outlook of nanotechnology in this domain.

This book is the result of collaborative efforts by esteemed researchers and professionals from various disciplines and institutions worldwide. Each chapter has been carefully crafted to provide an in-depth understanding of the subject while addressing current challenges, technological advancements, and future possibilities.

Chapter 1: Introduction to Nanomaterials and Industrial Wastewater Treatment

Authored by: Vijay Kumar Baradiya, Amit Vishwakarma, Lalitesh Sinha

This chapter introduces the basic concepts of nanomaterials and their relevance in industrial wastewater treatment. It highlights the escalating global concerns regarding water pollution and the growing demand for advanced technologies to combat industrial effluents. The chapter discusses the unique properties of nanomaterials that make them suitable for water remediation and sets the stage for the subsequent chapters by providing an overview of their applications, challenges, and potential in achieving sustainable water management.

Chapter 2: Types of Nanomaterials in Wastewater Treatment

Authored by: Ashootosh Mandpe, Mohd Nasim, Harish Kumar Dwivedi

This chapter groups various nanomaterials applied in wastewater treatment, including carbon-based nanomaterials, metal oxides, quantum dots, and polymeric nanostructures. The authors describe their specific functions in adsorbing, degrading, or removing contaminants, from heavy metals to organic pollutants. Comparative analyses of their efficiency, cost-effectiveness, and environmental impact provide readers with a clear understanding of how to select suitable nanomaterials for specific applications.

Chapter 3: Synthesis and Characterization of Nanomaterials

Authored by: İlayda Bersu Kul, Gamze Tüzün, Burak Tüzün

This chapter looks into methods of making nanomaterials through chemical, physical, and biological approaches. Controlling the size, morphology, and surface properties of nanomaterials is considered crucial to improving their performance in wastewater treatment. Specific characterization techniques covering scanning electron microscopy (SEM), X-ray diffraction (XRD), and Fourier-transform infrared spectroscopy (FTIR) are provided for readers, including information on the structure and functional properties of nanomaterials.

Chapter 4: Mechanisms of Nanomaterials in Wastewater Treatment

Authored by: Radha Rani Mehra, Asmita Sharma, Devaanshi Jagwani

This chapter discusses the physicochemical mechanisms by which nanomaterials interact with wastewater pollutants. Some of the discussed topics include adsorption, photocatalysis, oxidation-reduction reactions, and membrane filtration. The authors explain how these mechanisms enhance the removal of heavy metals, dyes, pharmaceuticals, and other emerging contaminants. Case studies and experimental findings illustrate the efficiency of nanomaterials in treating complex industrial effluents.

Chapter 5: Application of Nanomaterials for Remediation of Emerging Pollutants in Industrial Wastewater

Authored by: Mudita Nagpal, Nidhi Sharma, Tanuja Nautiyal

Emerging pollutants, pharmaceuticals, pesticides, and personal care products present significant risks for ecosystems and human health. This chapter addresses the nanomaterial challenge in the remediation of the threats emerging from such pollution, focusing on their ability to degrade or remove such pollutants. Industrial applications are reported together with an estimation of their efficiency and limitations.

Chapter 6: Nanomaterials in Advanced Treatment Processes

Authored by: Doyel Bhattacharya, Shubhajit Halder, Rohit Maind

This chapter describes the integration of nanomaterials into advanced wastewater treatment processes such as advanced oxidation, membrane filtration, and catalytic degradation. The authors discuss the synergies between conventional and nanotechnology-based methods, pointing out their combined potential to achieve higher removal efficiencies and improved treatment outcomes in industrial applications.

Chapter 7: Environmental Implications and Safety Considerations of Nanotechnology in Wastewater Treatment

Authored by: Subramaniam Karuppannan, Lalitesh Sinha

It is very interesting that nanotechnology has tremendous potential, but at the same time it has environmental and safety implications also. This chapter assesses the potential hazards due to nanoparticle release into the environment, their bioaccumulation and toxicity. Measures to control the risks are also discussed in the form of regulatory frameworks and green synthesis methods to ensure sustainable use of nanomaterials.

Chapter 8: Technological Challenges and Future Prospects

Authored by: Nimmy Srivastava, Jayeeta Chattopadhyay, Ashish Ch. Singh

This chapter discusses the technological and logistic challenges in industrial wastewater treatment involving nanotechnology. Scalability, cost, energy, and long-term stability are evaluated critically. Future research directions and emerging trends in nanotechnology that can be used to overcome these challenges and maximize the potential of this technology in water remediation are discussed.

Chapter 9: Case Study: Nanofilters and Their Applications in Water and Wastewater Treatment

Authored by: Vivek Yadav, Rahil Mathakia, Pritesh Bhatt, Amit Kumar Sharma, Arun K. Soni, Viralkumar B. Mandaliya

This chapter is a practical case study that discusses the development and application of a nano filter for water and wastewater treatment. The authors reported the conception, fabrication, and field implementation of the nano filter with a major focus on its performance in the removal of targeted pollutants. The authors further discuss the possibilities of scalability and economics in industrial applications

Chapter 10: Economic and Social Aspects

Authored by: Jayeeta Chattopadhyay, Nimmy Srivastava, Tara Sankar Pathak

This chapter deals with the economic and social effects of nanotechnology use in wastewater treatment. Among topics covered are cost-benefit analysis, public perception, policy implication, and the potential to generate employment. In this chapter, the authors elaborate on how nanotechnology may assist in reaching the sustainable development goals, focusing on solving the water quality issues while ensuring an economic and social balance.

Chapter 11: Metal Oxide Nanocomposites for Industrial Wastewater Remediation

Authored by: Alaa El Din Mahmoud, Ghada Ashour

This chapter focuses on the role of metal oxide nanocomposites in industrial wastewater remediation. Authors discussed their synthesis, properties, and applications focusing on their efficacy to remove heavy metals, organic pollutants, and other contaminants. Experimental studies and real-world applications have been included to prove versatility and prospects of large-scale deployment.

Chapter 12: Conclusion and Future Outlook

Authored by: Ashootosh Mandpe, Devaanshi Jagwani, Maulin P. Shah

The final chapter sums up the major findings of the book by focusing on the transformative role that nanomaterials play in industrial wastewater treatment. The authors note the advancements in this area and provide further directions for future research to bridge remaining gaps. The authors provide recommendations on sustainable practices and policy interventions that will help ensure long-term success and safe application of nanotechnology in water management.

This book explores and presents numerous innovative ideas and strategies centered on the application of nanomaterials for industrial wastewater treatment. With the rapid industrialization and urbanization taking place globally, managing wastewater effectively has become a critical challenge. Nanomaterials, with their unique physicochemical properties, have emerged as game-changing tools in addressing the complexities of wastewater treatment. This book is an effort to deepen the understanding of how nanomaterials can revolutionize waste management practices, offering more efficient and sustainable solutions to mitigate the environmental and health impacts of industrial pollutants.

As against technical explorations, this book goes into the salient aspects of technology transfer and capacity building with a focus toward sustainability in the context of waste management. A critical approach by sharing how to effectively pass knowledge and expertise and build competency aims at capacitating professionals and organizations as well as stakeholders, helping them utilize the nanomaterial-based method of approaches significantly towards enhancing treatments.

The book also talks about the governance factor, which plays a very crucial role in highlighting the need for strong policy frameworks and regulatory measures to protect the environment from bad waste management practices. It underlines the fact that there should be strong policies to promote responsible use of nanomaterials and guide industries and agencies toward sound environmental practices. The book addresses these governance aspects, thereby providing a holistic view of industrial wastewater management and bridging the gap between technological innovations and policy interventions.

Designed as a resource for users, this book is an informative source for anyone interested in managing industrial waste effectively, especially industrial scientists, engineers, and students; as well as practicing professionals in all fields of the environment, the sciences, or sustainable waste valorization. Further, it helps policymakers, conservators, or NGOs concerned with environmental protection to gain insights into the advancement of technologies and their sustainable frameworks towards developing circular economic pathways and integrated practices of waste management.

The editors are deeply grateful to the contributors who have willingly shared their expertise, insights, and knowledge and enriched the content with their academic and professional excellence. Their cooperative efforts, patience, and commitment in compiling this resource are much appreciated.

Special thanks go to the Springer team for their great support in all the stages of the publication process. The constant guidance, assistance, and dedication they provided were fundamental to bringing this work to completion and making it available to the international community to promote knowledge sharing in the fast-evolving field of nanomaterials for industrial wastewater treatment.

It is our hope that this book will inspire innovative research, foster collaborations, and contribute to the development of sustainable and efficient solutions for industrial wastewater treatment. We extend our gratitude to all contributors and reviewers for their invaluable insights and dedication, without which this book would not have been possible.

This book aspires to be a highly influential resource in building further research, practice, and policy in the field, one that promotes both a more sustainable and environmentally secure future.

Aspiring to serve as a cornerstone for advancing research, practice, and policy, fostering sustainability and environmental security.

Indore, India — Ashootosh Mandpe
Indore, India — Devaanshi Jagwani
Ankleshwar, India — Maulin P. Shah

Contents

Chapter 1
Introduction to Nanomaterials and Industrial Wastewater Treatment

Vijay Kumar Baradiya, Amit Vishwakarma, and Lalitesh Sinha

Abstract Industrial effluent is a multifaceted environmental problem in terms of the presence of contaminants, such as heavy metals, organic chemicals, pharmaceuticals, and microorganisms. Traditional treatment schemes are not often effective for eliminating such contaminants. Nanotechnology, with its new and highly efficient nanomaterials, holds much hope. This chapter presents the basic principles of nanomaterials and their distinct characteristics like high surface area, reactivity, and adjustable functionality, which render them efficient agents in wastewater treatment. Different nanomaterials like metal oxides, carbon nanomaterials, and functionalized nanoparticles are presented along with their modes of action like adsorption, catalysis, and antimicrobial activity. The chapter also delves into the technical challenges involved in upscaling nanomaterial-based treatment systems, such as problems associated with synthesis, cost, regeneration, and environmental safety. The overview paves the way for more in-depth treatment of the topics in following chapters, highlighting nanotechnology's potential to transform sustainable industrial wastewater treatment.

Keywords Nanomaterials · Industrial wastewater · Adsorption · Photocatalysis · Heavy metals · Environmental remediation · Surface area · Antimicrobial activity · Water treatment

V. K. Baradiya · L. Sinha (✉)
Civil Engineering Department, IPS Academy, Institute of Engineering and Science, Indore, Madhya Pradesh, India
e-mail: laliteshsinha@ipsacademy.org

V. K. Baradiya
e-mail: vijaybaradiya@ipsacademy.org

A. Vishwakarma
Department of Civil Engineering, UIT, RGPV, Bhopal, India
e-mail: amitvishi@rgpv.ac.in

A. Mandpe et al. (eds.), *Nanotech Solutions for Industrial Wastewater*, Advances in Wastewater Research, https://doi.org/10.1007/978-981-96-6879-3_1

1.1 Introduction

Water forms one of the most valuable natural resources existing on Earth, which is crucial to sustain life for a human being as well as other living organisms. It is not only about having it for day-to-day drinking purposes and for agricultural purposes but also many ecosystems are maintained by water. However, the high rate of urbanization and industrialization has increased the demand for water and caused not only a depletion of available freshwater resources but also some type of global crisis related to this condition, namely water scarcity (Shevah 2015). The challenge it poses to economic development and human well-being is mainly severe in already existing areas with deficient water availability, such as those suggested by the World Health Organization (2019).

In this regard, water pollution has also emerged as a major environmental concern in the global platform, and among these, the contaminant of major concern is heavy metals. Heavy metals like lead, mercury, cadmium, and arsenic enter into the aquatic systems through various anthropogenic activities like industrial discharge, mining, and improper waste disposal (Pandey and Kumari 2023). In addition to the toxic effect on aquatic life, it poses serious health implications for human beings, including neurological, reproductive, and developmental disorders (Qadri and Faiq 2020). Heavy metal pollution in water systems causes ecological disturbances and puts the quality of safe drinking water at risk, necessitating efficient and sustainable solutions for its treatment.

These must be handled creatively, or rather innovatively, by understanding the traditional water treatment methods and taking those a step beyond. Nanotechnology holds great promise in enhancing the efficiency of the process of purification of water, especially the removal of hazardous heavy metals and contaminants (Ajith et al. 2021).

Nanotechnology has promise not only in this area of environmental remediation but also in making revolutions within other sectors, such as manufacturing, electronics, and health care, respectively. Moore (2006) explains that nanomaterials, characterized by 1–100 nm, are materials with characteristics that are not achievable in their counterparts in large quantities. Coetzee et al. (2020) note that these nanomaterials exhibit enhanced reactivity, strength, and electrical conductivity. These characteristics can constitute the basis for creating very efficient and targeted treatment systems.

Many nanomaterials have been investigated and studied in literatures for wastewater treatment applications. Among them, carbon nanotubes have gained promises due to their high ability of adsorption and mechanical strength. These unique features characterize carbon nanotubes as ideal materials for filtration and pollutant capturing (Mauter and Elimelech 2008; Upadhyayula et al. 2009). Like carbon nanotubes, TiO_2 nanomaterials have attracted researchers and scholars in photocatalytic applications due to their stability, low price, and ability to cause strong oxidation (Khan et al. 2002; Shankar et al. 2004).

1.2 Brief Overview of Nanomaterials and Their Unique Properties

Nanomaterials are considered those materials that have at least one dimension in the nanoscale range thus ranging between 1 and 100 nm. By having minute size, such materials have unique physical, chemical, and mechanical properties compared to the bulk counterparts. Its prominent properties, such as high surface area-to-volume ratio, quantum effects, and increased reactivity, and mean nanomaterials are extremely desirable and useful in a wide variety of applications in electronics, medicine, energy, and environmental remediation (Asha and Narain 2020; Khan et al. 2019a, b, c, d).

1.2.1 High Surface Area-To-Volume Ratio

Perhaps the greatest characteristic of nanomaterials is their large surface area in relation to their volume. This surface has increased reactivity, making nanomaterials useful in applications such as improving catalytic activity in chemical reactions, delivery mechanisms for drugs, and pollutant absorption (Biswas and Wu 2005). For instance, due to their larger surface area, nanoparticles have adsorbed more molecules on their surface, thus highly effective in environmental cleaning and water purification processes (Khan et al. 2019a, b, c, d).

1.2.2 Quantum Effects

In the nanodomain, materials exhibit quantum mechanical properties, which are used in determining electrical, optical, and magnetic behavior. Quantum effects allow for changes in electronic band structure, which then also grant new optical properties such as fluorescence and higher conductivity. Quantum dots emit light and have properties that can be varied to different wavelengths, making it useful in advanced imaging and display technologies (Khan et al. 2019a, b, c, d).

1.2.3 Improved Mechanical Properties

Nanomaterials because of their smaller size dimensions often display more excellent mechanical properties than their respective counterpart in the bulk material. Carbon nanotubes for instance, as well as graphene, exhibit exceptional strength and flexibility with much higher tensile strength than that of steel by Iijima (1991) and Geim and Novoselov (2007). These nanomaterials find extensive utility in materials reinforcement in structures and developed lightweight, yet strong composites for the aerospace and automotive industries.

1.2.4 Chemical Reactivity and Catalytic Activity

The high surface energy, along with the existence of active sites, is among the reasons the nanomaterials are more chemically reactive and more catalytically active. These are metal nanoparticles—including gold and platinum—which are highly used as catalysts in chemical reactions. Mostly, their application can be observed in the automotive and chemical industries because activation energy can easily be lowered, and rate enhancing processes can be easily facilitated (Astruc 2008). In addition, metal oxide NPs including TiO_2 have been proven to be used in photocatalysis for environmental use, specifically for air and water purification (Khan et al. 2002).

1.2.5 Magnetism Properties

Some nanomaterials, like iron oxide nanoparticles, are ferromagnetic, even though the properties do not manifest in the bulk. These materials show a characteristic called superparamagnetic, so they respond to magnetic fields from the exterior but do not retain magnetism when the field itself is removed (Lu et al. 2007). It is for this reason that these materials find use for a number of biotechnological applications: targeted drug delivery, magnetic resonance imaging contrast agents, and magnetic hyperthermia for the treatment of cancer (Silva et al. 2012).

1.2.6 Biocompatibility and Functionalization

Nanomaterials easily interact with biological systems; hence, they are a good candidate to be applied to a medical application. The functionalization of nanomaterial with biocompatible coatings or ligands enables targeted delivery of therapeutic agents, thereby minimizing side effects and improving efficiency in treatment (Yetısgin et al. 2020). For example, silica-based nanoparticles may be functionalized to carry drugs or imaging agents straight to cancer cells, thereby optimizing cancer therapies (Ang et al. 2014).

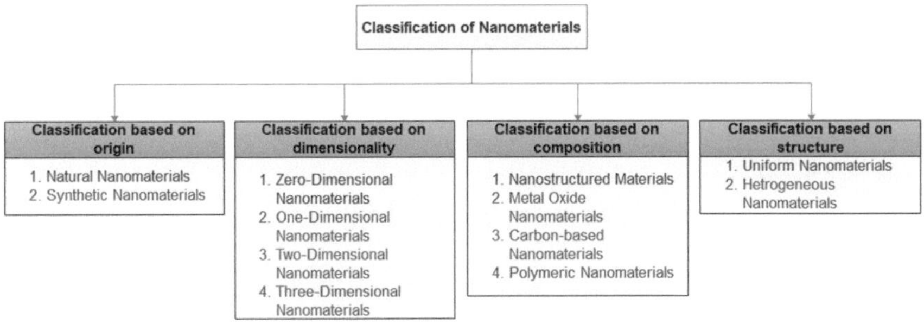

Fig. 1.1 Classification of nanomaterials

1.3 Definition and Classification of Nanomaterials

1.3.1 Definition of Nanomaterials

Nanomaterials are materials that contain structural components at the nanoscale, which specifically deals with sizes within 1–100 nm (nm). Here, materials have specific physical and chemical properties at this dimension, significantly different from those found in the bulk. Such properties go on to be associated with quantum effects, an increase in the surface area-to-volume ratio, and the impact of the dimensions of the material on the nanoscale on its behavior. Any type of substance, including metals, oxides, polymers, and composites, can be used to make up nanomaterials, which can exist in nanoparticles, nanotubes, nanowires, and nanosheets..

1.3.2 Classification of Nanomaterials

Nanomaterials can be classified based on several criteria such as origin, dimensionality, composition, and structure. Some of the major classification categories are shown in Fig. 1.1.

1.3.3 Classification Based on Origin

1. **Natural Nanomaterials**: These are naturally present in the environment, including minerals such as volcanic ash and some minerals or biologically derived materials like proteins and Zhai cellulose (Yarrow 2015).

2. **Synthetic Nanomaterials**: These include man-made materials generated by multiple chemical, physical, or biological processes. Examples include carbon nanotubes, metal nanoparticles, and engineered nanoclays (Dolez 2015).

1.3.4 Classification Based on Dimensionality

1. **Zero-Dimensional (0D) Nanomaterials**: These are nanoparticles and quantum dots, where all of the dimensions are nanometer-sized. They possess unique optical and electronic characteristics due to quantum confinement (Khan et al. 2019a, b, c, d).
2. **One-Dimensional (1D) Nanomaterials**: Their material has one dimension at the nanoscale range. These include nanowires, nanorods, and nanotubes. Such materials are usually characterized by significant aspect ratios; they find applications in electronics and nanocomposites (Zhai et al. 2010; Xia et al. 2003).
3. **Two-Dimensional Nanomaterials**: These are materials that have two dimensions and their features in the nanoscale range, such as graphene, graphene oxide, and transition metal dichalcogenides. They are characterized by exceptional mechanical and electrical properties (Geim and Novoselov 2007).
4. **3D Nanomaterials**: These are nanostructures that exist three-dimensionally in the nanoscale. These include nanostructured coatings, porous materials, as well as nanocomposites. They can serve to multiply a large number of nanomaterials with required properties (Li and Yu 2024).

1.3.5 Classification Based on Composition

1. **Nanostructured Materials**: These are typical examples of metal nanomaterials. This includes nanomaterials of metals, such as gold, silver, and platinum, mostly applied in catalysis, biomedical fields, and electronics (Perez-Juste et al. 2005).
2. **Metal Oxide Nanomaterials**: Here falls metal oxides, titanium dioxide (TiO_2), zinc oxide (ZnO), and iron oxide (Fe_2O_3) which have been more commonly used in photocatalysis, sensors, and solar cells (Khan et al. 2002).
3. **Carbon-based Nanomaterials**: This group includes carbon nanotubes, graphene, and fullerenes. They have shown to be very strong in strength, conductivity, and have uniquely electronic properties (Iijima 1991).
4. **Polymeric Nanomaterials**: It involves polymers at the nanoscale, including dendrimers, nanogels, and nanocomposites. It is primarily encountered in drug delivery and packaging applications (Dristant et al. 2023).

1.3.6 Classification Based on Structure

1. **Uniform Nanomaterials**: They possess a uniform composition throughout; for example, spherical nanoparticles that consist of only one metal.
2. **Heterogeneous Nanomaterials**: Such materials consist of distinct phases or components, such as core–shell nanoparticles, where a core material is coated by another shell material (Li et al. 2019).

1.3.7 Introduction to the Concept of Nanomaterials

Nanomaterials refer to those materials that contain at least one dimension within the nanoscale, as typically between 1 and 100 nm (nm). At such a scale, materials are characterized by properties that are distinctly different from those found in their macroscopic counterparts. It ranges from medicine, electronics, energy, environmental science, and many more fields of use.

1.3.7.1 Properties of Nanomaterials

Due to the tiny size of nanomaterials, it has a high surface-area-to-volume ratio that makes them highly reactive and strong. Quantum effects in nanomaterials can also be attributed to their electronic, optical, and magnetic properties, and thus, these are exploited for a special application. For instance, gold nanoparticles seem to change color depending on the size, but this property is exploited in imaging and sensing applications (Alivisatos 2004).

1.3.7.2 Types of Nanomaterials

There are several classes including nanoparticles; these refer to the materials at the nanoscale and are used in drug delivery, catalysts, or sensors. Others include nanotubes refer to cylindrical structures of great strength and electrical properties and are mostly used in nanocomposites or electronics. Others are nanosheets that are two-dimensional materials such as graphene, of great conductivity or mechanical properties used in energy storage and flexible electronics.

1.3.7.3 Synthesis Methods

Several synthesis techniques are involved in making nanomaterials. This includes the top-down approach such as lithography and milling and the bottom-up approach, including chemical vapor deposition, as well as sol–gel processes. Such synthesis techniques allow for fine control of size, shape, and functionalization critical for tailoring of materials to specific applications (Zhao et al. 2015a, b).

1.3.7.4 Applications of Nanomaterials

The nanomaterials, due to their special properties, have a huge space in the applications field. Because they help in facilitating medicinal imaging, targeted delivery of drugs through nanomaterials has been observed. In environmental sciences, nanomaterials are applied to remove pollutants and treat water. Their application in electronics, transistors, sensors, etc., indicates the overall potential of the application industry.

1.3.8 Definition Based on Size (Nanoscale Dimensions)

The term "nanomaterials" is defined by their size, usually those materials with nanoscale dimensions, which are usually measured to range from 1 to 100 nm in size. At this scale, the materials exhibit properties, both physical and chemical, that distinguish them uniquely from other counterparts, and in some areas, they can be used for several applications in different disciplines (National Nanotechnology Initiative Strategic Plan 2021).

1.3.8.1 Nanoscale Dimensions

The nanoscale is characterized by dimensions that lie on the order of 10^−9 m. A nanometer is large enough to be measured but about 100,000 times smaller than the diameter of a human hair and about 1/1000th the size of a red blood cell. This scale is significant because materials at the nanoscale often exhibit quantum mechanical effects that alter properties such as electrical conductivity, optical absorption, and reactivity (Roco 2005).

1.3.8.2 Special Properties

These materials have been characterized at the small scale to possess a relatively high area-to-volume ratio, thereby enhancing their reactivity and interaction with their environment. This has endowed them with strength, electrical properties, and optical properties, which have created niches for nanomaterials to be used in medicine, electronics, and environmental remediation (Cao 2015).

1.3.8.3 Classes of Nanomaterials

Different groups of nanomaterials can be differentiated based on dimensionality:

1. Zero-dimensional (0D): Nanoparticles having spherical morphology and having no length, width, or height at the nanoscale. Examples include quantum dots.
2. One-dimensional: Such nanowires and nanotubes have nanoscale extent in terms of length but are remarkably larger in the other two dimensions. Examples include carbon nanotubes.
3. Two-dimensional: Such nanosheets include graphene and transition metal dichalcogenides, which have nanoscale thickness while being extensive in the other two dimensions. Example includes graphene oxide.
4. Three-dimensional (3D): Nanostructured materials in which nanoscale features are retained throughout the entire three-dimensional volume, such as nanocomposites (Huang et al. 2013).

The immense importance that these materials have due to size at the nanoscale will allow for prospective applications in medicine, such as drug delivery systems; catalytic activities in chemical reactions; and the overall improvement in electronic device functionality (Khan et al. 2019a, b, c, d). Therefore, it is important to understand and realize the role of size in determining the behavior and properties of nanomaterials for them to be applied effectively in technology as well as in research.

1.3.9 Categories of Nanomaterials

These variations in properties and potential applications classify nanomaterials by dimensionality, structure, and composition. The most commonly used classifications include nanoparticle, nanotube, nanowire, and nanosheet—which have separate characteristics—making them useful in medicine, electronics, and environmental science (Roco 2005).

1. **Nanoparticles**: These nanomaterials are of 0D. All three dimensions of the particles lie in the range of the nanoscale, and all their dimensions vary between 1 and 100 nm in size. They are spherical, cubic, and irregular. Due to high surface area and reactivity, nanoparticles are extensively used in drug delivery systems, imaging, and as catalysts (Alivisatos 2004).
2. **Nanotubes**: They are one-dimensional 1D nanomaterials with a cylindrical configuration; carbon nanotubes (CNT) is the well-known compound featuring mechanical, electrical, and thermal properties to exceptional levels. They are used in composite materials and sensors and nanodevices (Iijima 1991).
3. **Nanowires**: Similar to nanotubes, nanowires are one-dimensional structures but more likely to be of elongated shape and prepared from a very broad range of materials, including metals, semiconductors, and oxides. Nanowires can be potentially very useful for electronics and photonics as well as energy storage

applications, because their electrical properties can be very unique (Wang et al. 2006).

4. **Nanosheets**: They are 2D nanomaterials possessing a nanoscale thickness and extend laterally. The most significant nanosheet is graphene, a single layer of carbon atoms structured in a hexagonal lattice. Nanosheets exhibit very exceptional electrical, thermal, and mechanical properties, thus being amenable to applications in flexible electronics, energy storage, and sensors (Geim and Novoselov 2007).
5. **Nanocomposites**: Nanoscale fillers, nanoparticles, or nanotubes distributed in a polymer, metal, or ceramic matrix. Nanocomposites are known to have mechanical, thermal, and barrier properties enhanced from the baseline material. They are used in packaging, coatings, and structural applications (Thostenson and Chou 2001).

Others include quantum dots, which are semiconductor nanodots that display quantum mechanical behavior, and dendrimers, a class of branched macromolecules at nanoscale dimensions. These both have well-defined applications in biotechnology and electronics, among other disciplines (Rogers et al. 2009).

1.4 Characteristics of Nanomaterials

Nanomaterials possess distinctive properties that set them apart from bulk materials, primarily due to their nanoscale dimensions and high surface area-to-volume ratios. Figure 1.2 shows an overview of the nanomaterial characteristics.

1.4.1 Nanoscale Properties

Nanoscale materials possess properties greatly differing from those of materials at the size of bulk. These comprise physical, chemical, and biological properties in a range that falls between 1 and 100 nm. Optimization of usage in medicine, electronics, environment, and other fields is possible only by knowing the properties of nanoscale (Zhang et al. 2020a, b, c, d, e).

(a) **High Surface Area-to-Volume Ratio**: The most prominent feature that defines nanoscale materials is their very high surface area-to-volume ratio. This characteristic enhances the reactivity of nanoscale materials and their interaction with other materials, which increases their activity in catalysis, drug delivery, and environment remediation applications (Guerra et al. 2018).

(b) **Quantum Effects**: Quantum mechanical effects become pronounced at the nanoscale and determine electronic, optical, and magnetic properties of the material. For instance, semiconductor nanoparticles called quantum dots have size-dependent optical properties, so their emission color depends on their size

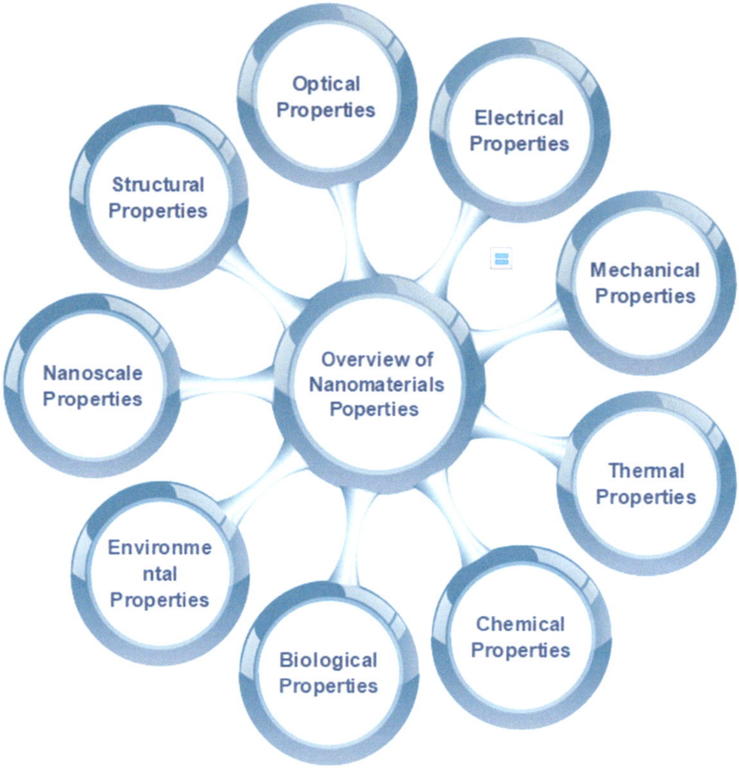

Fig. 1.2 Overview of the nanomaterial characteristics

(Nirmal and Brus 1999). Such properties find their application in bioimaging, laser technologies, and displays.

(c) **Mechanical Strength**: In general, nanomaterials exhibit greater mechanical strength than bulk material. For example, carbon nanotubes possess tensile strength and elasticity which are highly superior and therefore appropriate for filling composite materials as well as for developing lightweight high-strength structures (Ajayan et al. 2000).

(d) **Thermal Properties**: Nanomaterials can also exhibit varying thermal conductivity and heat capacity. For instance, materials at the nanoscale may exhibit superior thermal conductivity in contrast to the bulk situation, which is advantageous for applications such as thermal management of electronics (Pop et al. 2006). Conversely, some nanomaterials are quite suited as thermal insulators, which is useful for energy-efficient systems.

(e) **Optical Properties**: The optical behavior of nanoscale materials can be quite different from their bulk counterparts and often come along with some phenomena, for example, surface plasmon resonance. Colloidal metal nanoparticles were discovered to be colored when conduction electrons oscillate collectively in the presence of light (Liu et al. 2007). It is widely used in many sensing and imaging applications.

(f) **Chemical Reactivity**: Nanoscale materials have more reactivity in chemical terms due to their more extensive surface area and active sites. They consequently act as effective catalysts in the catalysis of chemicals as well as in the degradation of pollutants in environmental applications (Khan et al. 2019a, b, c, d).

1.4.2 Structural Properties

The structural properties of nanomaterials are considerable and define the behavior, performance, and suitability of nanomaterials for various applications. The dimensions of nanoscale, morphology, and arrangement significantly differ from those of their bulk counterparts. Thus, understanding the structural properties of nanomaterials is necessary to exploit their unique characteristics in fields such as electronics, materials science, and biomedicine (Zhao et al. 2015a, b).

(a) **Crystallinity**: Nanomaterials show impact of crystallinity toward mechanical and optical properties. These are nanocrystalline materials, rich in grain-boundary density which are stiffer and harder compared to their coarse-grained counterpart. This is due to the Hall–Petch effect which has shown that reduced grain sizes cause an increase in force to move dislocations.

(b) **Morphology**: The shape and size of nanomaterials greatly affect their properties and performance. For example, spherical nanoparticles can be rod-like or plate-like but have different reactivity and interact with light. Morphology can thus be tailored during synthesis, for instance, in drug delivery as the shape affects its mechanism of cellular uptake and distribution (Huang et al. 2016).

(c) **Porosity**: Nanomaterials tend to have high porosity, which is accompanied by enhanced surface area, and enables them to perform adsorption and catalysis. In the case of mesoporous silica nanoparticles, the material is designed with uniform pore sizes, which helps the material to ensure a controlled drug release when it comes to pharmaceutical applications (Zhao et al. 2013). Porosity comes into play as an essential characteristic in environmental application, such as in pollutant adsorption and filtration.

(d) **Surface Structure**: The surface of nanomaterials, as a rule, has defects, roughness, and functional groups. These are the factors that control their chemical reactivity and the process of interaction with other materials. Thus, nanoparticles modified by functional groups provide enhanced stability and biocompatibility, which make them useful for biomedical applications (Khan et al. 2019a, b, c, d).

(e) **Nanostructured Materials**: Nanomaterials can be realized in several forms such as thin films or coatings but more typically as composites. The nanocomposite has been engineered with unique structure to realize targeted functionalities. For example, an appropriate embedding of nanoparticles in a polymer matrix leads to materials with improved mechanical properties and thermal conductivities over the latter (Thostenson and Chou 2001).
(f) **Self-assembly**: Many nanomaterials can undergo self-assembly wherein they spontaneously arrange themselves in structured arrangements. This property is exploited in the complex nanostructures to be made for applications in drug delivery, sensors, and electronics. Self-assembly allows the construction of certain materials with specific functionalities and hierarchical structures.

Therefore, structural properties like crystallinity, morphology, porosity, and surface structure and the ability to form nanostructured materials by way of self-assembly critically affect the characteristics of these nanomaterials and their possible applications in different types of fields.

1.4.3 Optical Properties

The optical properties determine the behavior of nanomaterials and their applications in areas such as photonics, sensing, and biomedical imaging. These properties will generally depend on factors such as the nanoscale dimensions, shape, composition, and arrangement of nanomaterials, which have a tremendous potential to differ from those of their bulk counterparts. Thus, understanding the optical properties of nanomaterials appears quite fundamental when aiming to optimize their use in more advanced technologies (Khan et al. 2019a, b, c, d).

(a) **Size-dependent Optical Properties**: The principal feature of nanomaterials, as well as quantum dots, which are semiconducting nanoparticles, lies in the size-dependent optical properties. Quantum confinement effects make the emission wavelength of quantum dots dependent on size and thus adjustable for the whole visible spectrum (Nirmal and Brus 1999). This is exploited in various applications from displays to biological labeling.
(b) **SPR**: Metal nanoparticles, particularly gold and silver display SPR, where conduction electrons oscillate upon incidence by light. This is associated with an enhancement of absorption and scattering properties by light, hence colorful appearance along with strong optical signals. SPR has been widely applied in biosensing, imaging, and photothermal therapy applications.
(c) **Nonlinear Optical Properties**: Many of the nanomaterials have nonlinear optical properties which come into play at high light intensities. Nonlinear effects are responsible for the manifestation of phenomena such as second-harmonic generation, two-photon absorption, and self-focusing. Such properties are very important in applications related to optical switching, telecommunications, and some advanced imaging techniques (Baker et al. 2014).

(d) **Light Scattering**: The light scattering behavior by nanomaterials depends on their size, shape, and surface roughness. For their sizes on the nanoscale, new or interesting ways of scattering light are then possible to provide support for applications in sensors and imaging. It has already been known that scattered light is enhanced by gold nanoparticles from nearby molecules that SERS applications can thus be found useful. At times, nanomaterials exhibit optical absorption and transmission properties that are higher than those exhibited by the bulk material, and this has tremendous applications in photovoltaics and photothermal therapy. This increased absorption is attributed to a surface area highly increased at the nanoscale and a distinct electronic structure (Khan et al. 2019a, b, c, d).

(e) **Photonic Applications**: With optical properties, nanomaterials have found applications in laser devices, LEDs, and optical filters. The nanomaterials can thus be engineered to incorporate photonic bandgap materials that control the transport of light to maximize the photonic-device function (Song et al. 2010).

1.4.4 Electrical Properties

The understanding of the electrical properties of nanomaterials forms the basis for understanding their behavior and performance in electronics, energy storage, sensors, as well as many more applications. These properties are sometimes quite different from those of the bulk material, especially because of quantum effects, high surface area-to-volume ratios, and nanoscale dimensions. Such research into electrical properties could also have potential applications in the development of highly advanced technologies in nanotechnology, semiconductor devices, as well as renewable energy.

(a) **Conductivity**: Nanomaterials can have superior conductivity compared to their bulk counterparts. For example, carbon nanotubes exhibit exceptional electrical conductivity due to their one-dimensional nature and high aspect ratio and are of utility in conductive composites, flexible electronics, and interconnects (Ajayan et al. 2000). Most of the nanomaterials have semiconductor properties. Quantum dots and nanowires are very apt examples of such a nanomaterial. The size and shape dependence in semiconductor materials influences the size-dependent electronic properties of materials. In fact, it has led to the use of nanomaterials in optoelectronics, including LEDs and solar cells (Murray et al. 2001).

(b) **Charge Transport Mechanisms**: Due to surface states, defects, and quantum confinement effects, the charge transport mechanisms in nanomaterials are different from those in a bulk material. Depending upon the type of material and environmental conditions, charge transport in nanoparticles occurs by tunneling, hopping, or thermal activation (Baker et al. 2014). Knowing these mechanisms will help optimize the nanomaterials in electronic devices. The dielectric properties of nanomaterials are controlled by the influence of the permittivity and loss, hence purely dealing with their nanoscale dimensions and morphology. In particular, such nanoscale dielectric materials may have high permittivity and low loss, which may suitable the application in capacitors, insulators, and various high-frequency devices (Wang et al. 2016).

(c) **Thermoelectric Properties**: Some of the nanomaterials can exhibit thermoelectric properties, which enable them to transform temperature gradients into electrical voltage. Therefore, the efficiency of thermoelectric materials is enhanced at the nanoscale level by reduced thermal conductivity and increased electrical conductivity, which provides scopes for waste heat recovery and cooling (Gao et al. 2017).

(d) **Supercapacitors and Batteries**: With the advancement of nanomaterials, energy storage devices such as supercapacitors and batteries can be enhanced. Nanostructured materials have a high surface area and high porosity, which will improve the charge storage capacity as well as allow for faster rates of charge/discharge. As a result, it delivers energy as well as power density (Zheng et al. 2018).

1.4.5 Mechanical Properties

The mechanical properties of nanomaterials would be the primary consideration to define their properties and suitability for a variety of applications, mainly in material science, structural engineering, and nanotechnology areas. Because the particular microstructure, elevated surface area/volume ratio and nanoscale dimensions were so far reported for nanomaterials, the latter often had diffusable character in comparison with its bulk counterparts. Presently, knowledge of mechanical properties of nanomaterials plays an essential role for the optimization of use in advanced applications.

(a) **Strength and Hardness**: Many nanomaterials have been shown to possess greater strength and hardness compared with the bulk materials. The fine grain-size nanocrystalline materials have higher mechanical strength through the Hall–Petch effect, in which smaller grains inhibit the dislocation movement. They have advantages in applications demanding a strength-to-weight ratio such as aerospace and automotive components (Wang et al. 2015).

(b) **Elastic Modulus**: The nanoscale structure can drastically modify the elastic modulus, which is a measure of the stiffness of the material. Carbon nanotubes (CNTs), as a typical example, show an extremely high elasticity modulus and are considered excellent reinforcement fillers for composite materials. The elasticity of CNTs has been attributed to both unique atomic arrangements and strong covalent bonds (Ajayan et al. 2000).

(c) **Tensile and Compressive Strength**: The tensile and compressive strength of nanomaterials highly depends on its chemical composition and structure. The tensile strength of nanofibers and nanocomposites is often greater due to their larger aspect ratios along with an ideal mechanism of stress transfer (Yuan et al. 2014). The compressive strength of nanomaterials will also be dependent on porosity and microstructural defects.

(d) **Ductility and Toughness**: While nanomaterials possess high strength, they are likely to be lacking in ductility, which is the material's ability to deform under stress, and toughness, a measure of energy that causes fracture. However, some nanostructured materials, generally nanocrystalline metals, can provide good strength with suitable ductility through proper processing routes (Liu et al. 2016).

(e) **Fatigue Resistance**: Nanomaterials also exhibit better resistance to fatigue—the weakening of a material under cyclic loading and unloading—an important consideration in structural applications. Nanostructuring can enhance the resistance to fatigue since it reduces the size of defects and dislocations that cause failure under cyclic loading (Wang et al. 2015).

(f) **Impact Resistance**: Impact loads are again essential for applications in protective equipment and impact-resistant structures. At times, such impacts can become the reason behind a material failure, and hence, this remains an important aspect within the study of nanomaterials. Nanocomposite material comprised of nanoparticles dispersed in a polymer matrix features improved impact resistance. Nanoparticles provide energy-dissipating mechanisms (Raghavan et al. 2014).

1.4.6 Thermal Properties

The thermal properties of nanomaterials behave and perform in many applications, especially in electronics, energy storage, and thermal management systems. Due to the special structure and nanoscale size, nanomaterials indicate considerably different thermal properties compared with the corresponding bulk materials. Hence, studying such properties is indispensable for the optimization of nanomaterial design and functionality in advanced technologies (Zhou et al. 2015).

(a) **Thermal Conductivity**: In nanomaterials, the structural and composition-dependent thermal conductivity can be either enhanced or diminished compared with the value of the bulk material. For instance, carbon nanotubes (CNTs) and graphene are some examples which are recently reported to possess excellent values of thermal conductivities and are capable for heat dissipation and thermal management applications (Balandin et al. 2008). This material is very valuable due to the ability of intense phonon transport and specific lattice structure.

(b) **Thermal Stability**: Most nanomaterials have sufficient thermal stability with respect to performance in high temperature applications; a number of the metal oxides and nanocomposites can be exposed to relatively higher temperatures without their degradation. Nanoscale TiO_2 has high thermal stability; it is used for catalysis and as a pigment in coating applications.

(c) **Heat Capacity**: Nanomaterials have varied heat capacity than the bulk. Heat capacity depends on the temperature as increased temperatures lead to reduced dimensions and surface effect consequences. Higher thermal energy storage properties at lower temperatures make nanoscale material apply in specific uses. Storage of thermal energy is required in applications that require thermal insulation and energy preservation.

(d) **Thermal Expansion**: The coefficient of thermal expansion in nanomaterials is significantly influenced by their sizes and geometries. For instance, nanoparticles will lead to a degree of decreased thermal expansion in nanocomposites, which thereby limits the polymer chains' mobility within it. Dimension stability of nanomaterial-based applications depends largely on understanding this behavior of expansion due to the need for tight tolerances in the application area.

(e) **Phase Change Materials (PCMs)**: Some nanomaterials may also be phase change materials. Several PCMs absorb or release heat depending on the phase change. To advance the nanoparticle incorporation within the PCMs to enrich its thermal properties, potentially even to enhance heat storage capacity and the thermal conductivity of this system, it can be utilized within thermal management systems and energy storage applications (Zheng et al. 2015).

(f) **Thermal Interfaces**: Nanomaterial thermal interface characteristics are important for addressing heat transfer issues in electronic devices. Inclusion of nanomaterials within the interfaces will improve the thermal conductivity and reduce the thermal resistance, and it consequently enhances the thermal performance of the electronic systems (Feng et al. 2018). This characteristic ensures the reliability of electronic components along with a longer lifespan.

1.4.7 Chemical Properties

Chemical properties define reactivity, stability, and reaction of nanomaterials with other substances. These are often very different from those of bulk material due to increased surface area-to-volume ratio and unique electronic characteristics at the nanoscale. Such knowledge of the chemical properties of nanomaterials is important to optimize their applications in catalysis, drug delivery, and environmental remediation, etc. (Gao et al. 2015).

(a) **Reactivity**: Nanomaterials present greater chemical reactivity compared to most of their corresponding bulk counterparts since they have a large surface area. The high reactivity is added to the reason for the high application of nanocatalysts in many catalytic reactions carried out more efficiently than their counterparts catalyzed by larger particles. For example, noble metal nanoparticles of gold and platinum have been utilized in catalytic applications owing to their high reactivity and selectivity.

(b) **Stability**: The stability of nanomaterials is one of the major concerns in relation to the performance and potential lifetime in diverse applications. Different changes such as in size and shape, and surface modifications are observed to vary the stability of nanomaterials under different conditions. For instance, stabilizing nanoparticles within biological systems that improves their effectiveness as drug delivery agents by using surface coatings or functionalization (Zhang et al. 2016a, b).

(c) **Surface Chemistry**: In this regard, surface chemistry serves as the foundation for the interaction of nanomaterials with other substance. Functional groups and other chemical surface modifications play crucial roles in changing adsorption, reactivity, and the overall behavior of a nanomaterial in chemical reactions. Tailoring the surface chemistry of nanomaterials allows for the design of specific functionalities, such as enhancing adsorption capabilities for environmental remediation (Liu et al. 2017).

(d) **Acid–Base Properties**: Acid–base properties of nanoparticles, hence, are likely to be drastically different from their bulk analogues due to their high surface energy and reactivity. Even the surface charge of nanoparticles can be pH dependent; this has an impact on the behavior of nanoparticles as a whole in aqueous solutions. Therefore, these properties become quite important in drug delivery applications where there may be a pH-sensitive mechanism of release (Khan et al. 2019a, b, c, d).

(e) **Oxidation and Reduction**: In general, nanomaterials have very unusual oxidation and reduction properties, which are tapped into applications as diversified as in sensors and catalysis. For instance, the electronic properties of metal nanoparticles can be precisely controlled by oxidation or reduction; thus, their great potential in catalysis is enhanced (Yuan et al. 2018). Controlling the oxidation states of nanomaterials is considered to modify their chemical behavior.

(f) **Corrosion Resistance**: Nanomaterials can gain enhanced corrosion resistance as compared to the respective macroscale counterpart due to their specific microstructure and surface characteristics. Coatings and composites at the nanoscale offer more surface protection against corrosion than other types and can thus be used in hostile environments (Zhao et al. 2020). With corrosion-resistant nanomaterials, several systems would live longer.

1.4.8 Biological Properties

It has been established that the biological properties of nanomaterials are among the most important properties when it comes to interactions between nanomaterials and living organisms; therefore, it finds applicability in many areas, including medicine, drug delivery, diagnostics, and environmental health. Owing to the fact that nanomaterials have unique size, shape, and surface characteristics, different biological behavior can be exhibited compared to bulk materials. Therefore, knowledge of these properties is important for the safe and effective use of nanomaterials in biomedical and environmental applications (Harris et al. 2018).

(a) **Biocompatibility**: This is the ability of the material to react with the living systems without developing any significant immunoresponse. In general, gold nanoparticles as well as silica nanoparticles have excellent biocompatibility that makes them nanocarriers for the delivery and imaging agents (Patel et al.

2013). Meanwhile, biocompatibility is dependent on surface properties and modification for the material, thereby underlining the design importance.

(b) **Cellular Uptake**: Nanomaterials can be well internalized in cells, a property that one can utilize toward targeted drug delivery and therapy. The mechanism of uptake by the cells often depends on the size, shape, and surface chemistry of the nanomaterials. For example, smaller nanosize particulates are readily more efficiently internalized by cells through processes mediated by endocytosis, whereas larger particles are seen remaining on the cell surface. These mechanisms are important in optimizing drug delivery systems.

(c) **Toxicity**: Toxicity is a potential problem for the nanomaterials because they are of very small size and have large surface areas, which may result in unforeseen biological interactions. Nanomaterials may induce cytotoxicity and genotoxicity in living organisms along with oxidative stress. Some metal nanoparticles have been reported to generate reactive oxygen species (ROS) inside the cells, hence damaging the cells (Chen et al. 2014). The toxicity of nanomaterials needs to be assessed to ensure safe use in medical as well as environmental applications.

(d) **Biodistribution**: Distribution and accumulation in biological systems—biodistribution highlights a group of variables that govern the biodistribution profiles of nanomaterials, including particle size, surface charge, and shape. For instance, positively charged nanoparticles may be accumulated in higher concentration within certain tissues like the liver and spleen as compared with their negatively charged counterparts (Sadrzadeh et al. 2016). Biodistribution should be determined for assessing efficacy and safety applications in nanomedicine.

(e) **Immunogenicity**: Describes the ability of a substance to evoke an immune response. In certain cases, immune reactions by nanoparticles may lead to inflammation or an allergic reaction. Modulating the surface of nanoparticles, it is possible either to increase or decrease their immunogenicity, thereby facilitating scientists to design agents that elicit a minimum of immune responses for a therapeutic application (Kwon et al. 2016).

(f) **Antimicrobial Activity**: A few nanomaterials have proved to be antimicrobial in nature. These are primarily metal nanoparticles of silver and copper. These nanomaterials are applied either through coatings, wound dressings, or infection control. Their action is basically due to the formation of metal ions or ROS capable of cleaving microbial cell membranes (Sondi and Salopek-Sondi 2004).

1.4.9 Environmental Properties

The environmental properties of nanomaterials are crucial for ascertaining its effect on the ecosystem, potential application in removing pollutants, and overall sustainability. As nanomaterials exhibit unique properties at the nanoscale, it interacts with matrices of the environment in a different way compared to bulk materials. The properties of these nanomaterials help assess the risk and benefit of the nanomaterials used in different applications (Baker et al. 2017a, b).

(a) **Environmental Persistence**: The persistence of nanomaterials in the environment must be evaluated to predict possible ecological effects. Nanomaterials could degrade at varying rates than their bulk counterparts. For instance, carbon-based nanomaterials are known to persist longer in both soil and aquatic ecosystems. That way, it could lead to persistent environmental accumulation over many years of exposure to the environment (Sadrzadeh et al. 2017).
(b) **Mobility and Transport**: The mobility of the nanomaterials in the environmental system controls their distribution and feasible bioavailability. The transport of the nanomaterials could be efficiently affected by various factors, such as particle size and shape, surface charge, and aggregation. A smaller nanoparticle tends to be more mobile, hence allowing higher dispersion and possible ecological exposure (McGillicuddy et al. 2019).
(c) **Interaction with Environmental Pollutants**: Nanomaterials can interact with environmental pollutants in influencing their fate and transport. For example, some nanomaterials can adsorb heavy metals and organic pollutants, thus enhancing the removal of these pollutants from contaminated environments. High surface areas and the reactivity of nanomaterials make them to be efficient adsorbents particularly for environmental remediation applications (Li et al. 2018). However, interaction between nanomaterials and pollutants can also give rise to unanticipated by-products such as toxic by-products.
(d) **Toxicity to Non-Target Organisms**: One of the main sources of concern about the utilizations of nanomaterials in environmental applications is the possibility of toxicity to non-target organisms. Nanomaterials may interfere with microbial communities, aquatic organisms, and also terrestrial plants and thus potentially disrupt ecosystems. For instance, some metal nanoparticles were considered toxic to algae and aquatic invertebrates, raising concerns regarding the ecological safety of such formulations (Gonzalez et al. 2019).
(e) **Bioaccumulation and Biomagnification**: Bioaccumulation and biomagnification of nanomaterials in food webs is one important environmental concern. Some nanomaterials could be absorbed by organisms that will be transferred through the food chain and potentially accumulate to higher concentrations in the top predators. The bioaccumulation potential of nanomaterials constitutes the basis for the ecological risk evaluation.

Nanomaterials have unfolded their applicability very rapidly in environmental remediation. The peculiar properties attributed to nanomaterials make their application capable for the extraction of contaminants from soils and aqueous mediums. This makes their involvement very promising for the decontamination of polluted sites. For example, zero-valent iron nanoparticles have been used to remediate chlorinated solvents and heavy metals in groundwater (Gonzalez et al. 2020).

1.5 Overview of Industrial Wastewater Treatment

Industrial wastewater treatment is the removal of contaminants from wastewater emitted through industrial activity before it is released into the environment. The process of treatment is still an essential component in maintaining public health, conserving natural water bodies, and ensuring compliance with environmental standards. Generally, industrial wastes generally encompass a mix of pollutants that range from heavy metals to organic substances and nutrients, thus requiring effective technologies toward treating such wastes (Kumar et al. 2020a, b).

(a) **Types of Industrial Wastewater**: According to Davis et al. (2019), industrial wastewater varies in composition from one industry type to another. Basically, manufacturing processes, food processing, pharmaceutical companies, and petrochemicals are the common sources. Most industries create wastewaters with different contaminant concentrations and thus treated differently (Davis et al. 2019). For example, textile industries may produce wastewaters containing high amounts of dyes and other chemicals, while some metal finishing operations may produce wastewaters containing heavy metals.

(b) **Treatment Methods**: Industrial wastewater is usually treated by considering the amalgamation of physical, chemical, and biological process. The methods can broadly be classified as primary, secondary, and tertiary treatments.

(c) **Primary Treatment**: This is done in the initial stage by the removal of huge suspended materials and large solids through different processes including screening, flotation, and sedimentation. In this, the total pollutant load is reduced to make it easier for further treatment (Ghosh et al. 2018a, b, c).

(d) **Secondary Treatment**: The biodegradation takes place here to break down organic matter and nutrients. Techniques utilized include the activated sludge, biofilm reactors, and even anaerobic digestion in order to improve the growth of microorganisms that break down the organic pollutant (Metcalf and Eddy 2014).

(e) **Tertiary Treatment/Third Treatment**: This process covers the removal of residue contaminates, which include nutrients, heavy metals, and pathogens. Advanced oxidation processes, membrane filtration, and chemical precipitation are mainly used for superior treatment efficiency (Mabey et al. 2019).

(f) **Emerging Technologies**: Innovative technologies in industrial wastewater treatment have been developed to enhance the treatments processes with regard to efficiency as well as sustainability. The attention given to nanomaterials, membrane bioreactors, and electrochemical methods is increasing efficiencies in removing pollutants and minimizing sludge production (Baker et al. 2017a, b). These may provide more efficient and environmentally friendly means of treating complex wastewater streams (Fig. 1.3).

(g) **Regulatory Framework**: The governmental regulatory body controls the discharge of industrial wastewater. These regulatory bodies, among other countries, have their modifications of Clean Water Acts; so, the industries have to

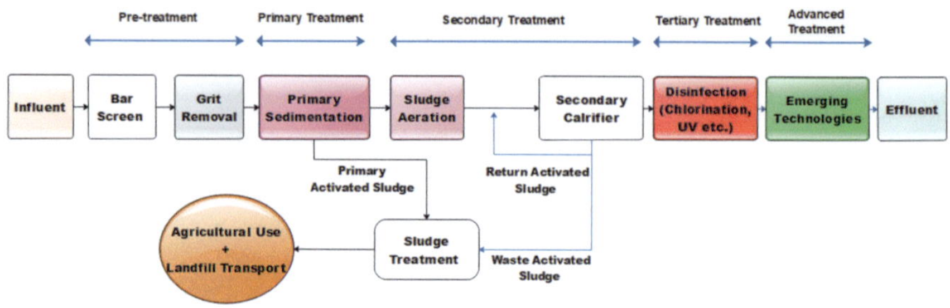

Fig. 1.3 Treatment of industrial wastewater

be very compliant with the treatment system that might be in place or implemented (Khan et al. 2019a, b, c, d). The industries must measure and report their wastewater characteristic and ensure that their effluent meets the specifications before it can be considered safe for discharge.

(h) **Challenges and Future Perspectives**: Despite the growing interest in wastewater treatment technologies, many challenges persist. Among these, perhaps the toughest ones include emerging contaminant treatment, management of complex streams, and cost-effective operations. More recently, future research has become more focused on developing efficient treatment methods that can improve resource recovery from wastewater and enhance sustainability in industrial processes (Zhang et al. 2020a, b, c, d, e).

1.6 Sources and Types of Industrial Wastewater

Industrial wastewater is typically generated by a vast array of manufacturing process and activities and tends to be compositionally diverse as well as ecologically dangerous. It is thus important to understand the sources and types of industrial wastewater in order to determine appropriate treatment measures as well as abate possible environmental risk posed by the discharge into the environment (Lema et al. 2019a, b).

- **Sources of Industrial Wastewater**: A variety of industries can produce industrial water supply. Each of these industrial wastes has unique characteristics. The most common origin includes:
- **Manufacturing Industries**: This category falls in a wide range, including the production of textiles, paper, chemicals, and metals. Water usually acts as a coolant, solvent, and transport medium in the manufacturing process and thereby generates polluted water containing different kinds of contaminants (Raja et al. 2017).
- **Food and Beverage Processing**: Wastewater generated by the food processing industry is generally voluminous and may contain organic matter, fats, oils, and greases, as well as nutrients, including nitrogen and phosphorus. Some of the

processes in this industry include meat processing, dairy manufacturing, and brewing (López et al. 2018).

- **Pharmaceuticals and Biotechnology**: Wastewater from the pharmaceutical industry contains high concentrations of active pharmaceutical ingredients (APIs), solvents, and other chemicals, which may be challenging to treat (Kummerer 2008).
- **Petrochemical and Oil Refineries**: Petrochemical manufacturing processes produce wastewater containing a mixture of hydrocarbons, heavy metals, and various organic compounds. Oil refineries too are found to produce wastewater with high concentrations of pollutants, such as phenols and sulfides (Bocca et al. 2014).
- **Classification of Industrial Wastewater:** Industrial wastewater may be classified based on characteristics and processes that are generating it, such as

  (a) **Effluent.** Such is treated or untreated wastewater discharged from industrial processes. Quality may be highly variable depending on the specific industrial activity and treatment applied (Ghosh et al. 2018a, b, c).
  (b) **Process Wastewater**. This is wastewater produced directly from manufacturing processes, containing contaminants unique to the production process. For example, textile manufacturing may have releases of wastewater containing dyes and surfactants, whereas metal finishing may lead to releases of wastewater concentrated with heavy metals (Patterson et al. 2020).
  (c) **Cooling Water:** Water is used as a coolant in the cooling system to cool machinery and for various processes and will emit wastewaters at high temperatures and possibly induce thermal pollution. In addition to this, such water might carry along with it several biocides and scaling agents that are used in such cooling systems (Baker et al. 2017a, b).
  (d) **Sanitary Wastewater:** This wastewater is generated by the restroom facilities and has sewage. Although usually not considered as industrial wastewater, it can be mixed with process water at a manufacturing site (Jiang et al. 2019).
  (e) **Storm Water**: Runoff rainfall may pollute industrial sites. Any storm water runoff shall therefore become part of industrial wastewater. Such runoff may have oils, heavy metals among other contaminants originating from the facility activities (Roesner et al. 2016).

1.7 Characteristics of Industrial Wastewater

The characteristics of industrial wastewater are extremely diverse depending on their origin, such as the following:

(a) **Physical Properties**: Examples include temperature, turbidity, and color that may indicate the suspended presence of organic matter or solids.

(b) **Chemical Properties**: Among the chemical parameters most frequently considered are pH, biochemical oxygen demand (BOD), chemical oxygen demand (COD), and the levels of nutrient-nitrogen and phosphorus (Mocé et al. 2021b).
(c) **Biological Properties**: The presence of microorganisms and pathogens and organic matter can determine the biodegradability as well as the toxicity of wastewater (Ghosh et al. 2018a, b, c).

1.8 Challenges Posed by Industrial Wastewater Pollution

Industrial wastewater pollution is considered one of the most considerable challenges facing environmental sustainability, public health, and economic development. With the scale and intensity of industrial production activities on the rise, volumes of wastewater increase, thereby involving complexity to impact the environment as well as society. Effective management strategies and implementations of advanced treatment technologies will be required to meet these challenges (Zhang et al. 2020a, b, c, d, e).

(a) **Environmental Degradation**: Industrial wastewater is always filled with a wide variety of contaminants in the form of heavy metals, organic compounds, and nutrients. The environmental effect is pollution of waters when the contaminants are released into water bodies without proper or rudimentary treatment. Heavy metals, such as lead, mercury, and cadmium, can find their way into the food chain and cause harm to aquatic life and humans through bioaccumulation, as stated by Jiang et al. (2018).
(b) **Public Health Risks**: This contamination of the water sources poses a significant risk to the health of the community using it directly for drinking and irrigation. Direct exposure of the community to industrial wastewater chemicals may lead to the respiratory symptoms, the skin problems, and the long-term exposures such as cancer. The waterborne diseases due to pathogens contained within untreated wastewater will worsen public health dilemmas.
(c) **Economic Impacts**: Contamination of water bodies would have devastating economic impacts on communities and industries. Only few livelihoods dependent upon fishery or agriculture will be lost where the sources of water are contaminated, hence increasing poverty and social inequalities. In addition, industries that do not properly manage their wastes are likely to face litigation, remediation costs, and loss of reputation, affecting profitability in the long run (Zhang et al. 2020a, b, c, d, e).
(d) **Regulatory Compliance**: The regulatory compliance of most industries in wastewater discharge is one of the key issues. Regulatory laws for wastewater treatment are becoming more stringent day by day, and thus, industries need to upgrade themselves with advanced treatment technologies and monitoring systems. For SMEs with limited resources, this can be a difficult compliance challenge and will not guarantee non-compliance with resulting penalties (Lema et al. 2019a, b).

(e) **Treatment Issues**: There are technical challenges to its treatment process because of the complexity of industrial wastewater. Depending on the types and concentrations of contaminants, industries may require unique solutions for treatment, as it is diversified in qualities of wastewater. Membrane filtration, advanced oxidation processes, and bioremediation are some of the advanced treatments required, as traditional methods are inadequate to take out all the contaminants (Ghosh et al. 2018a, b, c). Such advanced treatments are highly expensive and have a tremendous energy demand for many industries.

(f) **Resource Recovery Opportunities**: Though industrial wastewater poses huge challenges, the effluent also presents an opportunity for resource recovery. Nutrient recovery, energy generation, and water reclamation can transform the wastewater from a burdensome waste product to a resource, but often the resource recovery system requires a lot of investment with specialized expertise; it is especially difficult for industries to overcome them (Moce et al. 2021a).

1.9 Importance of Proper Treatment for Environmental Protection and Regulatory Compliance

Properly treated industrial wastewater is required to protect the environment and meets appropriate regulatory standards. Effective treatment reduces the harmful effects of pollutants on ecosystems and human beings and also helps industries achieve legal and social satisfaction. The primary causes that make effective treatment of wastewater necessary are as follows:

1.9.1 Environmental Protection

(a) **Removal of Toxic Pollutants**: The proper treatment process eliminates or degrades toxic substances like heavy metals, organic pollutants, and dangerous chemicals that would otherwise enter the bodies of natural water and ravage aquatic ecosystems (Metcalf & Eddy 2003).

(b) **Eutrophication Prevention**: Wastewater containing highly nutritional nitrates and phosphates may make water bodies experience eutrophication that will cause algal blooms and oxygen depletion in those water bodies. The nitrates and phosphates are also removed during proper treatment; hence, those harmful effects are also avoided (Crittenden et al. 2012).

(c) **Water Quality Protection**: Reuse of treated wastewater does not cause any problem or inconvenience in rivers, lakes, or oceans that protect and conserve the quality of that water resource and avoid degradation of biodiversity (Spellman 2008).

1.9.2 Health Protection

(a) **Clean Drinking Water**: Industrial effluents may lead to contamination of subterranean and surface water, hence being unsafe for human consumption. The prevention of the effluent's contaminants from reaching the environment ensures that humans do not ingest contaminated microbes or toxic chemicals, hence protecting their health (Eckenfelder 2000).

(b) **Population of Persistent Substances Reduced**: Treatment technologies also reduce persistent substances that are accumulated in the food chain; there is therefore a reduction of risk tied to consumption of contaminated food and water (Crittenden et al. 2012).

1.9.3 Compliance to Regulations

(a) **Compliance with Legal Standards**: Industrially, the businesses ought to meet the local, national, and also international standard requirements for the disposal of wastewater. Effective treatment ensures that it has limited pollutants; otherwise, it faces legal action and fines (Spellman 2008).

(b) **Protection from Industrial Disturbances**: Non-compliance results in several regulatory action measures include closure or even further prohibition of carrying out the operations. Effective treatment ensures that the business runs its operations uninterruptedly in trade (Eckenfelder 2000).

1.9.4 Economic Benefits

(a) **Long-term Cost Savings**: Inadequate wastewater treatment can contribute to environmental remediation, law liabilities, and loss of reputation in the long run (Metcalf & Eddy 2003).

(b) **Resource Recovery**: Advanced treatment processes will enable recovery of valuable resources such as water, energy, and nutrients from wastewater, enabling waste to be turned into economic assets (Crittenden et al. 2012).

1.9.5 Social and Ethical Responsibility

(a) **Corporate Social Responsibility (CSR)**: Appropriate treatment of waste is indicative of companies' concern for environmental stewardship as well as social responsibility, and by this, there is an improvement in their image before the public and the society in general (Spellman 2008).

(b) **Sustainable Development**: The process contributes positively to sustainable industrial practice. Effective waste treatment reduces adverse environmental impact and promotes responsible use of natural resources (Crittenden et al. 2012).

1.10 Fundamentals of Wastewater Treatment Processes

Wastewater treatment processes are designed to eliminate impurities in water; therefore, it can be safely released for discharge or reused. There are primary, secondary, and tertiary treatments, and the outcome ones depend on the nature of the impurities. Here is a summarized general overview of the basic stages in wastewater treatment processes.

(a) **Primary Treatment**:

Purpose: Remove massive amounts of solids and organic substance in wastewater.

Treatment Technologies:

Screening: Mechanical removal of coarse solids such as plastics, leaves, and rags (Metcalf & Eddy 2003).

Sedimentation: Enables the suspended solids in the wastewater to settle out by forming sludge, which is then removed from the bottom of the sedimentation tanks (Spellman 2008).

Outcome: It lowers the TSS and organic load of the wastewater.

(b) **Secondary Treatment**:

Objective: Biological processes involved degrade and remove dissolved organic matter and nutrients.

Processes accommodated in

Activated Sludge: Microorganisms hydrolyze, oxidize, and stabilize organic pollution in the aeration tanks into biomass and carbon dioxide (Crittenden et al. 2012).

Trickling Filters: Wastewater is sprinkled over a bed of stones or plastic media coated by a thin layer of microbial film that breaks down the organic pollutant (Eckenfelder 2000).

Lagoons and Stabilization Ponds: Natural processes including sunlight, algae, and bacteria treat the wastewater over an extended period of time (Spellman 2008).

Outcome: BOD and suspended solids significantly reduced.

(c) **Tertiary Treatment**:

Objective: Improve water quality through the elimination of remaining pollutants, including nutrients, heavy metals, and pathogens.

Process involved;

Filtration: This process relates to the removal of small particles using sand, membrane, or carbon filters (Crittenden et al. 2012).

Disinfection: This process kills pathogens or detrimental microbes using chlorine, UV light, or ozone (Metcalf and Eddy 2003).

Nutrient Removal: Such processes like biological or chemical precipitation remove excess nitrogen and phosphorus from the water, which leads to eutrophication in receiving waters (Spellman 2008).

Outcome: Production of the highest quality effluent for discharge or use.

(d) **Advanced Treatment**:

Objective: Removing pollutants like trace organics, pharmaceuticals, or heavy metals that are not adequately removed by conventional treatments.

Processes Included:

Advanced Oxidation Processes (AOPs): Addition of ozone, hydrogen peroxide, and ultraviolet light to break and oxidize such tough organic pollutants (Crittenden et al. 2012).

Ion Exchange and Adsorption: Elimination of certain ions or contaminants by resins or activated carbon (Eckenfelder 2000).

Result: This can ensure that all remaining contaminants are removed and meet stringent discharge standards or reuse requirements.

1.11 Treatment Techniques of Physical, Chemical, and Biological in Wastewater

The wastewater treatment encompasses several forms of treatment to remove impurities and purify water. These are broadly categorized into physical, chemical, and biological techniques. All these techniques emphasize types of pollution and a specific treatment technique different from each other to attain the desired degree of purification. Figure 1.4 shows various treatment techniques used in treatment of wastewater.

(a) **Physical Treatment Techniques**:

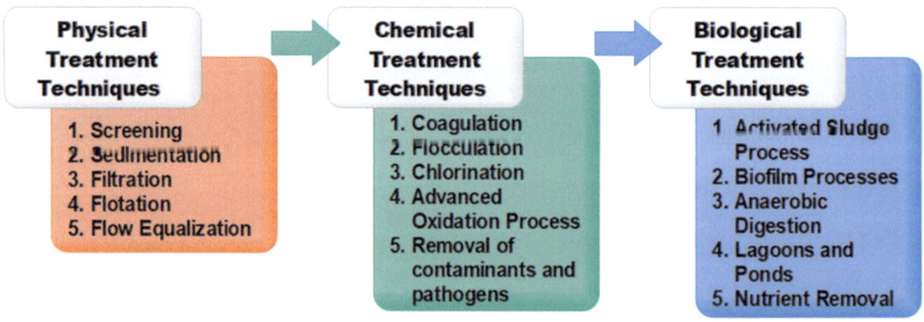

Fig. 1.4 Treatment techniques in wastewater

Objective: Mechanical treatment practices mainly focus on the removal of suspended particulate matter and suspended solids within wastewater by not changing the chemical character of pollutants.

Primary Techniques:

- **Screening**: Use mechanical screens to remove solid waste or plastic, leaves, rags, etc. before introducing them to the plant (Metcalf & Eddy 2003).
- **Sedimentation**: It is a process that enables sewerage to settle in a tank whereby the denser materials settle at the bottom, forming sludge. According to the design, it can capture as much as 90% of suspended solids (Spellman 2008).
- **Filtration**: During filtration, finer materials and particles are removed through various media such as sand, activated carbon, or membranes, thus making the water clearer and of better quality (Crittenden et al. 2012).
- **Flotation**: Air is injected into the wastewater, causing lighter particles, such as oils and grease, to float to the surface for easy removal.
- **Flow Equalization**: Regulates the flow and composition of wastewater to minimize fluctuations that could impact downstream treatment steps.
- **Chemical Treatment Methods**:

Objective: Chemical treatment methods are aimed to change the chemical nature of pollutants or make it easier to remove them through reactions.

Key Techniques

- **Coagulation and Flocculation**: Chemicals referred to as coagulants are added to wastewater, which eventually makes the suspended particles destabilize and agglomerate to settle at the bottom of the water (Eckenfelder 2000).
- **Chlorination**: This is the technique by adding chlorine in disinfection for the killing of bacteria and pathogens, and thus, ensuring effluent exiting the plant is safe and at acceptable limits before discharge (Metcalf & Eddy 2003).
- **Advanced Oxidation Processes (AOPs)**: This refers to various combinations of oxidants, for example, ozone and hydrogen peroxide, combined with UV light or catalysts to break complex organic contaminants down into simpler, nontoxic forms (Crittenden et al. 2012).
- **Effectiveness**: Chemical treatment techniques are effective in removal of dissolved contaminants and pathogens, often improving the overall treatment efficiency when used along with physical and biological techniques.
- **Biological Treatment Techniques**:

Objective: Biological treatment techniques involve the action of living organisms—mainly microorganisms—that degrade organic matter and nutrients present in the wastewater.

Key Techniques:

- **Activated Sludge Process**: The microorganisms use the organic pollutants in the aerated tanks to form the biomass and carbon dioxide; then it settled letting the treated water to be separated from the sludge (Crittenden et al. 2012).

- **Biofilm Processes**: Trickling filters and rotating biological contactors make use of mechanisms by which microorganisms grow on surfaces to break down the organic matter as wastewater over passes (Eckenfelder 2000).
- **Anaerobic Digestion**: This is the biological process of decomposing microorganisms on organic matter in the absence of oxygen so that methane and digestate can be achieved, which act as manure (Spellman 2008).
- **Lagoons and Ponds**: Natural or man-made ponds use microorganisms to gradually decompose pollutants, making them ideal for treating smaller volumes of wastewater in rural areas.
- **Nutrient Removal**: Biological processes such as nitrification and denitrification transform nitrogen compounds, while phosphorus is eliminated through biological or chemical methods to prevent water eutrophication.

1.12 Stages of Wastewater Treatment

It is a multi-stage process in which wastewater is purified to an acceptable level for its release or reuse. It can be classified into three phases: primary, secondary, and tertiary treatment. Each is focused on a different kind of pollutant and specific processes to enhance the quality of the effluent.

Primary treatment involves removing the largest solid particles and some organic matter through physical methods. Screening is the first stage in which large debris such as plastics, leaves, and rags are eliminated by using a mechanical screen to prevent interference with later stages (Metcalf & Eddy 2003). The next is sedimentation where the denser solids settle at the bottom of tanks as sludge that can be removed for further treatment or disposal (Spellman 2008). Typically, primary treatment can cut TSS by as much as 50–70% and reduce BOD by 30–40%, giving much relief to subsequent treatment phases, according to Crittenden et al. (2012).

Secondary or Biological Treatment This mainly is the biological breakdown of dissolved organic matter and nutrients. The most effective treatment processes in this stage are activated sludge process, which involves microorganisms in the aeration tanks using the organic pollutants; it is broken down into biomass and carbon dioxide, which then settles to separate the water treated from the sludge (Crittenden et al. 2012). The other method is trickling filters; here, the wastewater trickles over the media coated with a microbial film that degrades the organic material (Eckenfelder 2000). The stabilization ponds allow for the treatment of wastewater under natural means, through solar energy, algae, and bacteria in a period running for a number of months. This third step may produce an up to 85–95% removal of BOD and suspended solids from the wastewater, making its effluent quality remarkably excellent (Metcalf & Eddy 2003).

The **tertiary** step, the final one, treats the water containing the last of the remaining impurities or contaminants. These impurities include the nutrients, pathogens, etc. Filtration using either sand or membrane filters removes fine particles and impurities (Crittenden et al. 2012). Disinfection via chlorination, UV light, or ozonation eliminates pathogens to thereby make the effluent safe for discharge (Metcalf & Eddy 2003). Nutrient removal processes remove excess N and P through biological or chemical means to prevent eutrophication of the receiving water bodies (Eckenfelder 2000). Tertiary treatment has effluent of superior quality that can be released to sensitive environments and is, in fact, reused for irrigation or industrial purposes.

Overall, these successive treatment processes for wastewater significantly mitigate pollution loads, increase the security of the effluent, and thus offer flexible alternatives for reuse in different industries.

1.13 Nanomaterials in Wastewater Treatment

Nanomaterials have been a promising technology in wastewater treatment because of their unique features, specifically high surface area, reactivity, and tunable physical and chemical properties. They can remove a wide range of pollutants, from heavy metals and organic contaminants to pathogens and nutrients.

Types of Nanomaterials Used in Wastewater Treatment:

1.13.1 Carbon-Based Nanomaterials

(a) **Graphene and Carbon Nanotubes**: These have a great potential for adsorption. They are capable of removing organic pollutants, heavy metals, and dyes from water. Because they have very high surface area and their surface is functionalizable as well, they capture contaminants (Lu and Su 2010).
(b) **Activated Carbon Nanoparticles**: The application of nanoparticles in adsorption mainly targets organic pollutants because of their surface area and porosity (Ding et al. 2015).

Metal and Metal Oxide Nanoparticles:

(c) **Nanoscale Zero-Valent Iron (nZVI)**: nZVI has extensively been applied to heavy metals along with chlorinated organic compounds through oxidation–reduction reactions (Fu et al. 2014).
(d) **Titanium Dioxide (TiO_2) Nanoparticles**: It can degrade organic pollutants under UV light as utilized in photocatalytic processes. It can remove toxic compounds along with pathogens from water and thus find effective utility in removing toxins and pathogens (Chong et al. 2010).

Polymeric Nanomaterials:

(e) **Dendrimers and Nanogels**: Such materials can be designed for selective binding and elimination of contamination, such as heavy metals and organic pollutants, based on complexation and ion-exchange mechanisms (Diallo et al. 2005).

1.13.2 Removal Mechanisms of Pollutants

(a) **Adsorption**: Graphene and activated carbon-based nanomaterials present favorable conditions for the adsorption of different kinds of pollutants, such as dyes, heavy metals, and organic compounds due to their high surface areas and porous structures (Ding et al. 2015).
(b) **Catalysis and Photocatalysis**: The metal oxide nanoparticles of TiO_2 and ZnO can catalyze the oxidation of organic contaminants and pathogens, converting harmful substances into less toxic or less reactive forms upon interaction with light (Chong et al. 2010).
(c) **Redox Reactions**: nZVI and other metal nanoparticles may catalyze redox reactions that lead to depletion or transformation of toxins such as chlorinated compounds and heavy metals into less harmful species (Fu et al. 2014).

1.13.3 Advantages of Nanomaterials in Wastewater Treatment

(a) **High Efficacy**: Nanomaterials can remove various contaminants at very high efficiencies due to their high reactivity and large surface area (Lu and Su 2010).
(b) **Selective**: The use of functionalized nanomaterials can make it possible to selectively remove specific contaminants, thereby increasing treatment efficiency (Diallo et al. 2005).
(c) **Multi-Functionality**: Some nanomaterials can remove simultaneously more than one form of contaminants, including organic compounds, heavy metals, and pathogens, thus reducing the need for multiple steps for treatment (Fu et al. 2014).

1.13.4 Role of Nanomaterials in Enhancing Treatment Efficiency in Wastewater Treatment

Nanomaterials have increasingly been included in wastewater treatment owing to their special physicochemical properties including high surface area, high reactivity, and potential at the molecular level to be engineered. All these result in much better treatment efficiency compared to conventional methods by improving the removal rates of pollutants and expanding the range of contaminants that can be efficiently treated.

(a) **Enhanced Adsorption Capacity**: Especially, carbon-based nanomaterials, such as carbon nanotubes and graphene, have very high surface areas and easily tunable surface chemistry as very efficient adsorbents to a wide range of contaminants. For instance, graphene oxide was shown to display a very efficient adsorption capability for heavy metals and dyes because of the abundance of its functional groups and high surface area (Lu and Su 2010). Therefore, contaminants can be removed at much smaller concentration levels compared with traditional adsorbents.

(b) **Catalytic and Photocatalytic Activity Improve**: Photocatalytic processes involve the use of highly immense quantities of titanium dioxide nanoparticles and zinc oxide. These materials degrade organic pollutants and pathogens through light irradiation and transform them into harmless substances, such as carbon dioxide and water. High surface area and reactivity of these nanoparticles make them most suitable for the efficient degradation of the pollutants; hence, recalcitrant compounds that prove resistant to common treatment methods become most amenable for removal (Chong et al. 2010).

(c) **Efficient Redox Reactions**: The most applied nanomaterial in the treatment of wastewater is nanoscale zero-valent iron. This material exhibits some level of redox activity and enables nZVI particles to convert various species of reductive contaminant, such as chlorinated organic compounds and heavy metals, into less toxic or more removable forms. Considering such efficiency toward hard contaminants, which are otherwise difficult to remove, it has often been thought of as an excellent treatment for wastewater (Fu et al. 2014).

(d) **Selective Pollutant Removal**: Functionalized nanomaterials can be designed to selectively capture certain pollutants. Dendrimers and nanogels, for example, can be functionalized with certain functional groups toward the selective removal of heavy metals or organic pollutants. This selectivity removes the need for many treatment steps and makes the overall treatment process much more efficient (Diallo et al. 2005). Such selectivity is very handy when the industrial effluents to be treated have complex compositions of pollutants.

(e) **Multi-Functional Treatment Applications**: Some nanomaterials have multi-functional capabilities through adsorption, catalysis, and ease of separation in treated water by magnetic fields. This has made the process much easier and consequently minimizes the high energy and chemical consumption while ensuring remarkably high removal efficiencies (Qu et al. 2013).

(f) **Enhanced Disinfection and Pathogen Removal**: Due to their highly antimicrobial activity, nanomaterials like silver nanoparticles and copper oxide can be incorporated into wastewater treatment systems. That way, it becomes easier to sanitize the variety of pathogens, including bacteria, viruses, and protozoa that in turn enhances the disinfection process with minimal concentrations of conventional disinfectants and even chlorine that interfere with the production of harmful by-products (Klaine et al. 2008).

Nanomaterials will improve efficiency and remove a wider range of pollutants from wastewater, allow for selective targeting to desired contaminants, and reduce the footprint and complexity of treatment systems. The biggest challenges to realizing the full potential of nanomaterials in large-scale applications include cost, environmental impact, and scalability.

1.13.5 Benefits of Nanomaterial-Based Approaches

Due to these unique properties, nanomaterials have recently attracted much attention: they have a larger surface area, better reactivity, and selectivity, which allows them to be used in catalysis, environmental remediation, and biomedical applications. Their high surface area-to-volume ratio makes it possible to introduce a larger number of active sites per mass unit, thereby enhancing their reactivity and effectiveness in chemical reactions. This is particularly beneficial in catalysis as nanomaterials are known to increase reaction rates as well as selectivity, making the processes more efficient and sustainable (Zhang et al. 2020a, b, c, d, e). In addition, tailoring the surface properties of the nanomaterials allows for specific interactions between the target molecules to boost their selectivity toward desired reactions or processes (Chen and Wang 2021).

1.13.6 Nanomaterials Used in Wastewater Treatment

Nanomaterials have quickly gained the status of becoming an effective agent in the treatment of wastewater primarily due to their prominent physicochemical properties and high ability in removing contaminants. The major categories used include:

(a) **Metallic Nanomaterials**: It comprises nanoparticles such as silver, gold, and iron, which possess excellent antimicrobial properties and have been efficiently known for the removal of heavy metals and organic pollutants (Gupta et al. 2019).

(b) **Metal Oxide Nanomaterials**: These include such nanoparticles of titanium dioxide (TiO_2), zinc oxide (ZnO), and iron oxide. They have been used mainly based on their photocatalytic and adsorption capabilities allowing the degradation of organic pollutants as well as removal of heavy metals from contaminated water sources.

(c) **Carbon-Based Nanomaterials**: The materials include graphene oxide, carbon nanotubes, and activated carbon, which have a large surface area and high adsorption ability to be applied in the removal of organic pollutants and heavy metals from wastewater (Mishra and Ahmad 2020).

(d) **Nanocomposites**: These are artificially assembled composites of two or more different nanomaterials. Such nanocomposites aim at improving properties of the individual constituents for synergistic effects for improved efficacy in contaminant removal (Sharma et al. 2021).

1.13.7 Sensor Technologies for the Determination and Monitoring of Pollutants

Advanced sensor technologies have revolutionized the determination and monitoring of contaminants in environmental systems. Sensors nowadays provide many of the real-time, highly sensitive, and selective detection of many different types of pollutants in the air, water, and soil. Some of the most important sensor technologies include

(a) **Electrochemical Sensors**: Electrochemical sensors are efficient to identify gases such as CO, NO_x, So_x, and heavy metals, as they can provide strong sensitivity, low cost, and ease of being inserted into portable electronic devices (Xie et al. 2020). These sensors work on the basis of measuring the electrical signal resulting from the reaction of target pollutants with sensor material.
(b) **Optical Sensors**: Optic sensors, which include fluorescence-based and surface plasmon resonance sensors, in organic compounds, heavy metals, and pathogen detection, are applied. High selectivity and fast detection very quickly are possibilities for the sensors (Wang et al. 2021).
(c) **Biosensors**: These are based on biological elements including enzymes, antibodies, or DNA for detecting specific pollutants and have a high specificity and sensitivity toward organic compounds, toxins, or pathogens (Rodriguez-Mozaz et al. 2022). The use of biosensors also cuts across the analysis of water quality and trace determination of harmful substances.
(d) **Nanomaterial-Based Sensors**: Advances in nanomaterial-based sensors such as graphene, carbon nanotubes, and metal nanoparticles have enhanced detection limits and response times. These nanomaterial-based sensors are quite useful for heavy metal sensing, organic pollutant sensing, and sensing of gases (Zhang and Li 2021).

The two technologies together lead to greater observation of the environment and better pollution control with higher sensitivity and specificity along with the possibility of real-time data acquisition.

1.13.8 Challenges in Industrial Wastewater and the Need for Innovative Solutions

Industrial wastewater poses a significant threat to the environment due to its complex composition, often containing heavy metals, toxic chemicals, organic pollutants, as well as high concentrations of nutrients. Such contaminants pose extreme threats both to aquatic ecosystems and to human health unless treated before discharge (Mantzavinos and Kalogerakis 2011). Traditional treatments such as chemical coagulation, biological activities, and filtration are usually inadequate for industrial effluents with different kinds of pollutants at high concentrations. In addition, they may be energy intensive and expensive and can produce secondary pollutants (Fu and Wang 2011).

It is along this line that new initiatives in the field of wastewater treatment solutions should not only consider improving efficiency and sustainability but also flexibility as a response to heterogeneous industrial profiles of wastewater. Promising so far are technologies such as advanced oxidation processes, membrane filtration, and nanotechnology-based treatments. These new approaches facilitate higher efficiencies in pollutant removal and smaller energy consumptions with resource recovery opportunities, thus presenting attractive alternatives for modern wastewater treatment systems (Sharma et al. 2018).

1.13.9 Importance of Nanomaterials in Wastewater Treatment

The properties of nanomaterials are quite exceptional: they have high surface area, tunable pore size, and reactivity. Such characteristics bring great ease to the removal of a variety of pollutants, such as heavy metals, organic contaminants, and pathogens, in wastewater (Qu et al. 2013). Adsorption, catalysis, and filtration processes using nanomaterials are mostly far superior to conventional methods.

For example, metal oxides (e.g., TiO_2 and ZnO) and nZVI are highly applied photocatalysts and redox agents that degrade persistent organic pollutants and reduce the levels of toxic metals in the environment (Reddy et al. 2020). Carbon-based nanomaterials, including graphene oxide and carbon nanotubes, possess adsorption capabilities toward organic pollutants and can be functionalized to target particular contaminants (Mishra and Ahmad 2020). Further, nanocomposites—which involve a combination of more than one nanomaterial—exhibit synergy effects that maximize their potential performance in complex matrices of wastewater (Sharma et al. 2021).

One of the advantages of incorporating nanomaterials in wastewater treatment systems is the improvement of pollutant removal efficiency and reduction of operational costs and energy consumption. For these reasons, they have received popularity in providing sustainable and efficient wastewater management solutions.

1.14 Applications of Nanomaterials in Industrial Wastewater Treatment

Industrial wastewaters pose a serious threat to the aquatic and environmental systems due to the kinds of pollutants they contain. The potential of nanomaterials in the treatment of industrial wastewater has been quite enormous, considering that their exceptional physicochemical properties include very high surface areas, enhanced reactivity, and tunable functionalities. These materials are notably efficient in removing broad ranges of pollutants, including heavy metals, organic contaminants, and dyes, which are often encountered in industrial effluents (Gao et al. 2020).

(a) **Adsorption of Heavy Metals**: Zerovalent iron (nZVI), titanium dioxide, and graphene oxide nanomaterials are reported widely in use for adsorption and removal of heavy metals, including lead, cadmium, and arsenic. These nanomaterials have high adsorption capacities, and selectivity toward certain ions can be engineered using desirable properties.
(b) **Organic Pollutants Degradation** Photocatalytic nanomaterials are used in the degradation of organic pollutants, including pesticides and pharmaceuticals, through advanced oxidation processes. Such materials include titanium dioxide and zinc oxide that work by photogeneration of ROS, which breaks down complex organic molecules into less harmful compounds (Rajhans et al. 2019).
(c) **Membrane Filtration**: The incorporation of nanomaterials into membrane technologies increases the efficacy of filtration and reduces fouling. Nanocomposite membranes that include carbon nanotubes and metal–organic frameworks (MOFs) have a high level of permeability and selectivity, making them applicable to many industrial effluents with great loads of pollutants (Ahmad et al. 2021).
(d) **Removal of Dyes and Colorants**: Textile and dyeing industry effluent usually contains persistent colorants. Functionalized silica and metal oxide nanoparticles are used for the adsorption and degradation of such dyes; hence the hazardous effect produced by them on the environment is decreased (Katheresan et al. 2018).

1.15 Adsorption Processes Using Nanomaterials

Nanomaterial-based adsorption processes have gained immense interest in water and wastewater treatment due to their superior properties, such as high surface area and abundant active sites, along with the possibility of tuning functionalities on their surface. Nanomaterials render themselves highly efficient for removing a wide variety of contaminants, including heavy metals, organic pollutants, and dyes (Ali et al. 2019).

(a) **Heavy Metal Removal**: The nanomaterials most frequently used for the adsorption of heavy metals, including lead, mercury, and cadmium, are graphene oxide, carbon nanotubes, and magnetic nanoparticles. These large surface area and highly reactive nanomaterials have been shown to preferentially and effectively bind and remove metal ions from aqueous solutions—more often than activated carbon, the often-trusted conventional adsorbent (Fu and Wang 2011).
(b) **Organic Pollutant Adsorption**: Mesoporous silica, metal–organic frameworks (MOFs), and carbon-based materials are used for the removal of organic contaminants such as pesticides, pharmaceuticals, and industrial chemicals. These

adsorbents have high selectivity and capacity toward several organic molecules due to their variable pore sizes with available surface functionalities (Li et al. 2020).

(c) **Dye Removal**: Dyes originating from the textile and dyeing industries are one of the biggest threats to the environment. Nanomaterials such as titanium dioxide (TiO_2), zinc oxide (ZnO), and functionalized silica have been used for dye removal and degradation in wastewater treatment, which afford effective and sustainable methods for removing color (Muthusaravanan et al. 2018).

(d) **Multifunctional Nanocomposites**: Mixing different nanomaterials into a nanocomposite enhances adsorption performance through synergistic effects. For example, magnetic graphene oxide composites can be separated easily after adsorption with enhancement in overall adsorption capacity for various types of pollutants (Sharma et al. 2019).

1.16 Catalytic Degradation of Pollutants

Catalytic degradation of pollutants is a very effective way of controlling environmental pollution, especially during wastewater treatment and air purification. This method uses catalysts to accelerate chemical reactions degrading harmful substances to lesser toxic or nontoxic compounds. Various catalytic materials, such as metals, metal oxides, and nanomaterials, have been developed and utilized in order to enhance the degradation process of organic pollutants, heavy metals, and gases (Feng et al. 2020).

(a) **Metal-Based Catalysts**: Transition metals in the form of Pt, Pd, and Cu are commonly used because of their high catalytic activity. These metals can serve as good catalysts for oxidation, reduction, or hydrolysis reactions that are a part of the degradation processes of organic pollutants, such as phenol and chlorinated compounds. (Kumar et al. 2018).

(b) **Metal Oxides**: Examples of metal oxides include titanium dioxide (TiO_2) and zinc oxide (ZnO), which are most commonly used for photocatalytic degradation. TiO_2 and ZnO are semiconductor materials that produce ROS upon irradiation with UV or visible light, thus degrading organic pollutants such as dyes and pharmaceuticals into nontoxic by-products.

(c) **Nanocatalysts**: These catalytic degradations with nanomaterials have been widely used for their large surface area and unusual characteristic properties. The nanocatalysts, for example, gold nanoparticles and silver nanoparticles and even metal–organic frameworks, exhibit higher catalytic activity and specificity, thus enabling widespread applications in the environmental medium (Zhang et al. 2020a, b, c, d, e).

(d) **Enzyme Catalysis**: Another active biocatalytic approach for pollutant degradation is achieved by catalytic action of enzymes. Enzyme-catalyzed degradation of complex organic compounds is another promising approach with the catalytic action of laccases and peroxidases, which are environmentally benign compared to chemical catalysts (Singh et al. 2021).

1.17 Membrane Filtration and Separation Techniques

Membrane filtration and separation techniques find wide applications in various industries and environmental sectors mainly because they are efficient to remove pollutants in water and wastewater. The selectively permeable membranes used to separate substances based on size, charge, or chemical properties provide advantages such as low energy consumption, high efficiency, and minimal utilization of chemicals involved in the processes (Wang et al. 2020).

(a) **Micro Filtration (MF)**: In this process, the membrane pore diameters range from 0.1 to 10 μm for separating suspended solids, bacteria, and larger particles from liquids. The application areas of microfiltration include drinking water production as well as wastewater reuse (Baker and Dudley 1998).
(b) **Ultra Filtration (UF)**: Ultra filtration membranes have pores in the size range of 1–100 kDa and are very efficient in removing colloids, proteins, and macromolecules. The technique is commonly used in food and beverages industries and also in industrial effluent treatment (Khan et al. 2021).
(c) **Nano Filtration (NF)**: Membranes between ultrafiltration and reverse osmosis usually can pass monovalent ions, rejecting the divalent ions and larger organic molecules. Although NF can find varied application, it is found to be most suitable in water softening and color and organic matter removal from wastewater (Van der Bruggen and Vandecasteele 2003).
(d) **Reverse Osmosis (RO)**: This high-pressure membrane separation technique is quite capable of removing a wide range of dissolved salts, organics, and pathogens, hence suitable for desalination as well as wastewater treatment. RO membranes contain pores at the molecular level such that only water molecules pass through while retaining larger contaminants (Madaeni 1999).
(e) **Membrane Bioreactors (MBR)**: In MBR, a biological treatment process is combined with membrane filtration, therefore effectively removing organic and inorganic pollutants from the wastewater. This will imply higher removal rates of pollutants and fewer land requirements as opposed to conventional treatment systems (Liu et al. 2015).

1.18 Advanced Oxidation Processes (AOPs) Employing Nanomaterials

Advanced oxidation processes are advanced water and wastewater treatment techniques, which use strong oxidants to break down organic pollutants, pathogens, and toxic substances. The incorporation of nanomaterials in AOP has enhanced significantly the efficiency and effectiveness of contaminant removal from AOPs because nanomaterials have the specific properties, such as high surface area, increased reactivity, and improved mass transfer (Zhang et al. 2019).

(a) **Nanoparticles in AOPs**: TiO_2, ZnO, and Fe_2O_3, among others, have been reported as the most frequently used photocatalysts in AOPs. These nanoparticles, when exposed to UV and visible light, can yield reactive oxygen species in the form of hydroxyl radicals (–OH) and superoxide anions ($–O_2^-$). Thereby, these nanoparticles have been established to break down extensively a wide range of organic pollutants such as dyes, pesticides, and pharmaceuticals, according to Dhananjay et al. (2021).
(b) **Nano-Enabled Fenton Reactions**: The nanoscale iron particles increase the generation of hydroxyl radicals from the reaction between hydrogen peroxide (H_2O_2) and iron salts that take part in the Fenton reaction. The use of nZVI has been reported to increase Fenton-like processes by increasing reactive iron species that enhance high effective degradations of organic contaminants in solution (Xiao et al. 2020). Graphene oxide, carbon nanotubes, and other carbon nanomaterials also support the improvement of AOPs. These materials act as carriers of catalysts or oxidants, thereby enhancing stability and reactivity and assisting in generating ROS (Zhang et al. 2020a, b, c, d, e).

Results from the integration of AOPs with other treatment processes, such as membrane filtration or biological treatment, show promising results. The nanomaterial-based structures can be used to improve the performance of integrated systems and to enable efficient pollutant removal concurrently along with the recovery of valuable resources (Deng et al. 2019). The use of nanomaterials in AOPs presents a new aspect that characterizes water treatment, leading to improved efficiency and sustainability in the solutions proposed for complex wastewater streams.

1.19 Challenges and Future Directions

(a) **Environmental and Health Risks**: These types of nanomaterials also pose potential toxicity and environmental persistence risks to ecosystems and human health. Safe and sustainable nanomaterials are an important area of research in today's times (Klaine et al. 2008).

(b) **Scalability and Cost**: The cost of the production is high, and the use of nanomaterials in large-scale wastewater treatment plants is hard to upscale (Qu et al. 2013).
(c) **Regulatory and Safety Concerns**: Safe use and disposal guidelines along with regulatory steps are required for the use of nanomaterials in the wastewater treatment processes as they have toxic effects on aquatic species (Klaine et al. 2008).

1.20 Conclusions

Nanomaterial incorporation in industrial wastewater treatment is, thus, very promising. Efficiency, selectivity, and cost-effectiveness promise high efficiency in pollutant removal. Advanced nanoscale solutions for complex contamination treatment in wastewater have appeared through the unique surface properties, reactivity, and adaptability of nanomaterials, with reference to heavy metals, organic pollutants, and pathogens. Nanomaterial-based technologies that use nanosorbents, photocatalysts, and nanocomposites hold growing potential as supplementary alternatives or even as stand-alone solutions to traditional methods of treatment.

However, several critical challenges remain. The potential toxicity of nanomaterials poses ecological and human health risks, particularly if nanoparticles are released unintentionally into the environment during or after treatment. Moreover, the scalability of nanomaterial-based technologies from laboratory to industrial levels is limited by current cost constraints, resource availability, and the energy requirements for nanomaterial synthesis and regeneration. In addition, long-term environmental impact and degradation behaviors of nanomaterials in wastewater systems must receive adequate study and standardization in risk assessment protocols.

Future work and development should focus upon developing eco-friendly synthesis strategies that are cost-effective as well as materials that make recoveries easier and thus easily reused. It further demands the regulatory frameworks and the establishment of safety standards of the products that will facilitate their safe usage and therefore mitigate risks in health perspectives related to nanomaterials. Further, future sustainability can also be achieved as it allows the interfusion of nanotechnology into the other innovation seen through the application of the use of machine learning along with artificial intelligence in the water treatment process to ensure optimization with regards to its treatment aspects.

In summary, despite being on the edge of scientific progress, full development is realized by nanotechnology in wastewater treatment, along with innovation, and also the commitment to safety and sustainability. Nanomaterials can, in general, make a crucial difference toward filling the demand gap in terms of clean water supplies across industrial and global scale demands by overcoming present day challenges and utilizing combined endeavors of different disciplines.

References

Ahmad H, Bhat A, Mushtaq A (2021) Nanocomposite membranes in wastewater treatment: recent advances and future prospects. Environ Sci Pollut Res 28(6):6675–6690

Ajayan PM, Wang ZL, Shenoy VB (2000) Carbon nanotubes as potential long-term storage media. Adv Mater 12(15):123–128

Ajith MP, Aswathi M, Priyadarshini E, Rajamani P (2021) Recent innovations of nanotechnology in water treatment: a comprehensive review. Bioresour Technol 342:126000

Ali I, Peng C, Khan ZM (2019) Application of nanomaterials in environmental cleanup: recent trends and future prospects. Environ Sci Pollut Res 26:23472–23486

Alivisatos AP (2004) The importance of size in nanotechnology. Nature 429(6988):1–2

Ang CY, Tan SY, Zhao Y (2014) Recent advances in biocompatible nanocarriers for delivery of chemotherapeutic cargoes towards cancer therapy. Org Biomol Chem 12(27):4776–4806

Asha AB, Narain R (2020) Nanomaterials properties. In: Polymer science and nanotechnology. Elsevier, pp 343–359

Astruc D (2008) Transition-metal nanoparticles in catalysis: from historical background to the state-of-the art. Nanopart Catal 16:1–48

Baker RW, Dudley M (1998) Membrane technology and applications. McGraw-Hill

Baker RM et al (2014) Nonlinear optical properties of metal nanoparticles: a review. Opt Mater Express 4(8):1618–1631

Baker TW et al (2017) Emerging technologies for industrial wastewater treatment: a review. Environ Sci Nano 4(7):1411–1428

Baker TW et al (2017) The challenges of industrial wastewater management: a review. Environ Sci Nano 4(9):1735–1754

Balandin AA et al (2008) Superior thermal conductivity of single-layer graphene. Nano Lett 8(3):902–907

Biswas P, Wu CY (2005) Nanoparticles and the environment. J Air Waste Manag Assoc 55(6):708–746

Bocca B et al (2014) Industrial wastewater management: methods and challenges. J Hazard Mater 264:142–158

Cao G (2015) Nanostructures and nanomaterials: synthesis, properties, and applications. Cambridge University Press

Chen S et al (2014) Cytotoxicity and genotoxicity of metal nanoparticles: a review. Environ Toxicol Pharmacol 37(3):1122–1131

Chen X, Wang Y (2021) Selective catalysis using nanomaterials: design strategies and applications. Nat Nanotechnol 16:1234–1242

Chong MN, Jin B, Chow CW, Saint C (2010) Recent developments in photocatalytic water treatment technology: a review. Water Res 44(10):2997–3027

Coetzee D, Venkataraman M, Militky J, Petru M (2020) Influence of nanoparticles on thermal and electrical conductivity of composites. Polymers 12(4):742

Crittenden JC et al (2012) MWH's water treatment: principles and design, 3rd edn. Wiley, Hoboken

Davis AP et al (2019) Types of industrial wastewater: sources and treatment options. Water Environ Res 91(1):25–37

Deng H, Wang Y, Gu J (2019) Integrated advanced oxidation processes with membrane filtration: a review. Sep Purif Technol 221:202–215

Dhananjay J, Sinha R, Sahu SK (2021) A review on the use of nanomaterials in advanced oxidation processes for the removal of contaminants from wastewater. Environ Sci Pollut Res 28(25):32395–32420

Diallo MS, Christie S, Swaminathan P, Johnson Jr JH, Goddard III WA (2005) Dendrimer enhanced ultrafiltration. 1. Recovery of Cu(II) from aqueous solutions using PAMAM dendrimers with ethylenediamine core and terminal NH2 groups. Environ Sci Technol 39(5):1366–1377

Ding L, Zou B, Hu X, Wang X (2015) Application of nanomaterials for the removal of pollutants from wastewater. Prog Chem 27(1):1–12

Dolez PI (2015) Nanomaterials definitions, classifications, and applications. In: Nanoengineering. Elsevier, pp 3–40

Dristant U, Mukherjee K, Saha S, Maity D (2023) An overview of polymeric nanoparticles-based drug delivery system in cancer treatment. Technol Cancer Res Treat 22:15330338231152083. https://doi.org/10.1177/15330338231152083

Eckenfelder WW Jr (2000) Industrial water pollution control, 3rd edn. McGraw-Hill Education, New York

Feng H et al (2018) Thermal interface materials for electronic devices: a review. Mater Sci Eng R Rep 127:1–20

Feng Y, Yu Y, Zhao Y (2020) Catalytic degradation of organic pollutants: a review of recent advances and future challenges. Chem Eng J 392:123614

Fu F, Wang Q (2011) Removal of heavy metal ions from wastewaters: a review. J Environ Manag 92(3):407–418

Fu F, Dionysiou DD, Liu H (2014) The use of zero-valent iron for groundwater remediation and wastewater treatment: a review. J Hazard Mater 267:194–205

Gao J et al (2015) Chemical properties of nanomaterials: a review. Nanomaterials 5(4):2328–2357

Gao H et al (2017) Enhanced thermoelectric performance of nanomaterials. Nat Rev Mater 2(11):17049

Gao Y, Li Z, Zhang W (2020) Nanomaterials for water and wastewater treatment: applications and implications. Water Res 182:115908

Geim AK, Novoselov KS (2007) The rise of graphene. Nat Mater 6(3):183–191

Ghosh D et al (2018) Characteristics and treatment of industrial wastewater: a review. Environ Sci Pollut Res 25(22):22137–22151

Ghosh D et al (2018) Primary treatment of industrial wastewater: a review. Environ Sci Pollut Res 25(29):29000–29012

Ghosh D et al (2018) Treatment technologies for industrial wastewater: current challenges and future prospects. J Hazard Mater 344:872–889

Gonzalez MA et al (2019) Toxicity of nanomaterials to non-target organisms: implications for ecological risk assessment. Nanotoxicology 13(5):658–672

Gonzalez MA et al (2020) Application of nanomaterials for environmental remediation: a review. Sci Total Environ 703:135505

Guerra FD, Attia MF, Whitehead DC, Alexis F (2018) Nanotechnology for environmental remediation: materials and applications. Molecules 23(7):1760

Gupta S, Irshad M, Singh J (2019) Metal nanoparticles for water treatment: a green perspective. J Environ Manag 231:131–140

Harris AT et al (2018) Biological properties of nanomaterials: an overview. Nanomaterials 8(4):227

Huang X, Zeng X, Wang M (2013) Classification of nanomaterials based on dimensionality and structure. Nanotech Rev 2(1):61–76

Iijima S (1991) Helical microtubules of graphitic carbon. Nature 354(6348):56–58

Jiang G et al (2018) Environmental impacts of industrial wastewater: a global perspective. Environ Pollut 237:254–264

Jiang C et al (2019) Treatment of sanitary wastewater from industrial sources. J Clean Prod 239:118072

Katheresan V, Kansedo J, Lau SY (2018) Efficiency of various recent wastewater dye removal methods: a review. J Environ Chem Eng 6(4):4676–4697

Khan SU, Al-Shahry M, Ingler WB Jr (2002) Efficient photochemical water splitting by a chemically modified n-TiO_2. Science 297(5590):2243–2245

Khan Y et al (2018) Thermal properties of TiO_2 nanomaterials: a review. J Mater Chem a. 6(2):123–145

Khan I, Saeed K, Khan I (2019) Nanoparticles: properties, applications and toxicities. Arab J Chem 12(7):908–931

Khan Y et al (2019) Acid-base properties of nanoparticles: implications for drug delivery. J Controlled Release 303:119–136

Khan Y et al (2019) Regulatory frameworks for industrial wastewater discharge: global perspectives. Environ Policy Gov 29(6):423–436
Khan Y, Ullah N, Khan MI (2019) Nanomaterials: a review of synthesis and applications. Environ Chem Lett 17(3):1041–1051
Khan A, Ahmad M, Rahman A (2021) Overview of ultrafiltration technology: principles, applications, and recent advances. Sep Purif Technol 265:118393
Klaine SJ et al (2008) Nanomaterials in the environment: behavior, fate, bioavailability, and effects. Environ Toxicol Chem 27(9):1825–1851
Kumar P, Singh A, Sharma R (2018) Metal-based catalysts for the degradation of organic pollutants: a review. Environ Sci Pollut Res 25(20):19591–19612
Kumar A et al (2020) Industrial wastewater treatment: an overview. J Environ Manage 254:109735
Kumar R, Mondal K, Panda PK, Kaushik A, Abolhassani R, Ahuja R et al (2020) Core–shell nanostructures: perspectives towards drug delivery applications. J Mater Chem B 8(39):8992–1027
Kummerer K (2008) Pharmaceuticals in the environment: sources, fate, effects, and their sustainability. Springer, Berlin
Kwon SH et al (2016) Immunogenicity of nanomaterials: a review. Adv Drug Deliv Rev 106:178–189
Lema JM et al (2019) Industrial wastewater management: challenges and solutions. Water Res 153:64–78
Lema JM et al (2019) Sources and types of industrial wastewater: a review. Water Res 153:24–34
Li Z, Yu C (eds) (2024) Nanostructured materials, 1st edn. Elsevier, Amsterdam, 456 p. ISBN: 9780443192562
Li Y et al (2018) Interaction of nanomaterials with environmental pollutants: implications for remediation. Chem Eng J 347:839–848
Li Z, Ji S, Liu Y, Cao X, Tian S, Chen Y, Niu Z, Li Y (2019) Well-defined materials for heterogeneous catalysis: from nanoparticles to isolated single-atom sites. Chem Rev 120(2):623–682
Li Y, Zhang X, Yang X (2020) Adsorption of organic pollutants by nanomaterials: recent developments and environmental applications. Environ Sci: Nano 7(3):1027–1040
Liu Y et al (2007) Synthesis and characterization of metal nanoparticles: a review. J Mater Chem 17(17):1634–1646
Liu Y, Zhang T, Huang H (2015) Membrane bioreactors for wastewater treatment: a review of recent advances. Environ Sci Technol 49(7):4041–4055
Liu H et al (2016) Mechanical properties of nanocrystalline metals. Mater Sci Eng A 675:12–20
Liu J et al (2017) Surface chemistry and its effects on the adsorption properties of nanomaterials. Environ Sci Nano 4(8):1530–1540
López MI et al (2018) Wastewater from food processing: sources, composition, and treatment technologies. Environ Sci Technol 52(4):1843–1854
Lu C, Su F (2010) Adsorption of natural organic matter by carbon nanotubes. Sep Purif Technol 58(1):113–121
Lu AH, Salabas EE, Schüth F (2007) Magnetic nanoparticles: synthesis, protection, functionalization, and application. Angew Chem Int Ed 46(8):1222–1244
Mabey A et al (2019) Tertiary treatment of industrial wastewater: methods and challenges. Environ Technol Rev. 8(1):1–15
Madaeni SS (1999) Deterioration of membrane performance: causes and remedies. Desalination 123(3):219–231
Mantzavinos D, Kalogerakis N (2011) Treatment of industrial wastewater effluents: an overview. Chemosphere 82(3):407–415
Mauter MS, Elimelech M (2008) Environmental applications of carbon-based nanomaterials. Environ Sci Technol 42(16):5843–5849
McGillicuddy E et al (2019) Mobility of nanoparticles in environmental systems: a review. Environ Pollut 244:643–658

Metcalf & Eddy, Inc. (2003) Wastewater engineering: treatment and reuse, 4th ed. McGraw-Hill Education, New York
Metcalf E, Eddy HP (2014) Wastewater engineering: treatment and resource recovery. McGraw-Hill Education, New York
Mishra A, Ahmad R (2020) Carbon-based nanomaterials for the removal of heavy metals from water: a review. J Environ Chem Eng 8(5):104473
Moce P et al (2021a) Resource recovery from industrial wastewater: challenges and opportunities. J Clean Prod 299:126873
Mocé P et al (2021b) Understanding the characteristics of industrial wastewater: implications for treatment. Environ Eng Sci 38(6):395–407
Moore MN (2006) Do nanoparticles present ecotoxicological risks for the health of the aquatic environment? Environ Int 32(8):967–976
Murray CB et al (2001) Colloidal synthesis of nanocrystals and nanocrystal superlattices. Annu Rev Mater Res 31(1):373–418
Muthusaravanan S, Sivarajasekar N, Vivek JS (2018) Phytoremediation of heavy metals: mechanisms, methods and enhancements. Environ Chem Lett 16:1339–1359
National Nanotechnology Initiative Strategic Plan (2021, Oct 8) Alexandria (VA): National Nanotechnology Coordination Office. https://www.nano.gov/2021strategicplan
Nirmal M, Brus L (1999) Photonic properties of semiconductor nanocrystals. Nature 391(6663):1–2
Pandey S, Kumari N (2023) Impact assessment of heavy metal pollution in surface water bodies. In: Metals in water, pp 129–54.
Patel KD et al (2013) Biocompatibility of gold nanoparticles in drug delivery: a review. Pharm Res 30(8):1951–1962
Patterson JA et al (2020) Industrial wastewater treatment: best practices and technologies. J Environ Eng 146(9):04020137
Perez-Juste J, Pastoriza-Santos I, Liz-Marzan LM, Mulvaney P (2005) Gold nanorods: synthesis, characterization and applications. Coord Chem Rev 249(17–18):1870–1901
Pop E et al (2006) Thermal conductance of a carbon nanotube. Phys Rev Lett 95(9):095502
Qadri R, Faiq MA (2020) Freshwater pollution: effects on aquatic life and human health. In: Fresh water pollution dynamics and remediation, pp 15–26
Qu X, Alvarez PJ, Li Q (2013) Applications of nanotechnology in water and wastewater treatment. Water Res 47(12):3931–3946
Raghavan R et al (2014) Impact resistance of nanocomposite materials: a review. Compos Sci Technol 104:1–9
Raja A et al (2017) Industrial wastewater sources and their treatment: a review. Environ Sci Pollut Res 24(12):11070–11088
Rajhans K, Singh P, Srivastava V (2019) Photocatalytic degradation of organic pollutants using nanomaterials: a review. Environ Chem Lett 17(4):1325–1340
Reddy DHK, Lee SM, Kwon YJ (2020) A comprehensive review on the development of metal oxide nanomaterials and their application for the removal of heavy metals in wastewater. J Ind Eng Chem 83:1–19
Roco MC (2005) Environmental implications of nanotechnology: a preliminary assessment. Environ Sci Technol 39(11):4337–4345
Rodriguez-Mozaz S, Weinberg HS, Barcelo D (2022) Biosensors for water quality assessment and pollution detection. Environ Sci Technol 56(2):489–501
Roesner LA et al (2016) Stormwater runoff and its impacts on water quality: a review. Water Environ Res 88(7):580–598
Rogers JA et al (2009) Nanomaterials for electronic applications. Mater Today 12(6):22–29
Sadrzadeh A et al (2016) The effect of surface charge on the biodistribution of nanoparticles: a review. J Nanobiotechnol 14:60
Sadrzadeh A et al (2017) Persistence of nanomaterials in the environment: implications for ecological risk assessment. Environ Toxicol Chem 36(4):919–929

Shankar MV, Anandan S, Venkatachalam N, Arabindoo B, Murugesan V (2004) Novel thin-film reactor for photocatalytic degradation of pesticides in an aqueous solution. J Chem Technol Biotechnol 79(11):1279–1285
Sharma P, Singh R, Hussain S (2018) Innovative treatment technologies for industrial wastewater: an overview. Environ Sci Pollut Res 25(5):3898–3910
Sharma P, Singh R, Hussain S (2019) Nanocomposites for enhanced removal of pollutants from wastewater: recent trends and challenges. J Environ Manag 232:132–145
Sharma P, Singh R, Hussain S (2021) Nanocomposites in wastewater treatment: recent advances and challenges. Environ Sci Pollut Res 28:47013–47025
Shevah Y (2015) Water resources, water scarcity challenges, and perspectives. In: Water challenges and solutions on a global scale, pp 185–219
Silva A, Silva-Freitas É, Carvalho J, Pontes T, Araújo-Neto R, Silva K et al (2012) Magnetic particles in biotechnology: from drug targeting to tissue engineering. Adv Appl Biotechnol 1:13
Singh P, Gupta R, Sharma N (2021) Enzyme-based catalysis for the degradation of environmental pollutants: a review. Bioresour Technol Rep. 13:100634
Sondi I, Salopek-Sondi B (2004) Silver nanoparticles as antimicrobial agents: a case study. J Colloid Interface Sci 275(1):177–182
Song K et al (2010) Photonic properties of nanostructured materials: a review. Mater Sci Eng R Rep 70(2):39–54
Spellman FR (2008) Handbook of water and wastewater treatment plant operations. CRC Press, Boca Raton
Thostenson ET, Chou TW (2001) Processing-structure-property relationship in carbon nanotube/epoxy composites. Appl Phys Lett 78(6):982–984
Upadhyayula VK, Deng S, Mitchell MC, Smith GB (2009) Application of carbon nanotube technology for removal of contaminants in drinking water: a review. Sci Total Environ 408(1):1–13
Van der Bruggen B, Vandecasteele C (2003) Distillation vs. membrane filtration: theoretical and practical comparisons. Desalination 157(1–3):35–40
Wang ZL, Ouyang M, Hwang H (2006) Nanowires and nanotubes: fabrication, properties, and applications. Adv Mater 18(15):1989–2005
Wang Y et al (2015) Mechanical properties of nanostructured materials: a review. Mater Sci Eng R Rep 99:1–24
Wang Y et al (2016) Dielectric properties of nanostructured materials: a review. J Appl Phys 120(2):025103
Wang Y, Gao Y, Huang X (2020) Recent advances in membrane filtration technologies for water treatment: a review. Environ Sci: Water Res Technol 6(7):1705–1731
Wang Q, Liu G, Gao J (2021) Advances in optical sensors for pollutant detection: principles and applications. Trends Anal Chem 144:116437
World Health Organization (2019) The state of food security and nutrition in the world 2019: safeguarding against economic slowdowns and downturns, vol 2019. Food & Agriculture Org
Xia Y, Yang P, Sun Y, Wu Y, Mayers B, Gates B et al (2003) One-dimensional nanostructures: synthesis, characterization, and applications. Adv Mater 15(5):353–389
Xiao Y, Liu D, Zhang W (2020) Enhanced Fenton-like degradation of organic pollutants using nanoscale zero-valent iron: mechanisms and applications. Chemosphere 247:125923
Xie X, Zhang Z, Liu Y (2020) Electrochemical sensors for environmental monitoring: design, application, and challenges. Sens Actuators B Chem 305:127–134
Yarrow D (2015) Geology into biology: carbon, minerals, and microbes-tools to remineralize soil, sequester carbon, and restore the earth. In: Geotherapy: innovative methods of soil fertility restoration, carbon sequestration, and reversing CO_2 increase. CRC Press/Taylor and Francis Group, pp 195–223
Yetisgin AA, Cetinel S, Zuvin M, Kosar A, Kutlu O (2020) Therapeutic nanoparticles and their targeted delivery applications. Molecules 25(9):2193

Yuan W et al (2014) Nanofibers: a review on their mechanical properties and applications. J Nanomater 2014:1–14
Yuan Y et al (2018) Redox properties of metal nanoparticles: implications for catalysis. ACS Catal 8(9):8733–8747
Zhai T, Fang X, Li L, Bando Y, Golberg D (2010) One-dimensional CdS nanostructures: synthesis, properties, and applications. Nanoscale 2(2):168–187
Zhang P, Li J (2021) Nanomaterial-based sensors for detecting environmental pollutants: a review. Nano Today 37:101028
Zhang L et al (2016) Surface modification of nanoparticles for drug delivery: a review. Int J Nanomed 11:2027–2041
Zhang Z et al (2016) Cellular uptake of nanomaterials: Mechanisms and implications for drug delivery. J Controlled Release 241:225–237
Zhang X, Zhang J, Hu W (2019) Photocatalytic degradation of organic pollutants using metal oxides: mechanisms and applications. J Photochem Photobiol A Chem 377:234–250
Zhang L et al (2020) Challenges and future directions in industrial wastewater treatment. Sci Total Environ 706:135778
Zhang S, Chen G, Liu Y (2020) Advances in nanomaterials for environmental catalysis: a review. Mater Today 38:1–20
Zhang X, Zhang J, Hu W (2020) Photocatalytic degradation of organic pollutants using nanomaterials: mechanisms and applications. J Photochem Photobiol A Chem 377:234–250
Zhang Y et al (2020) Challenges posed by industrial wastewater pollution: an overview. Environ Sci Pollut Res 27(8):8230–8242
Zhang Y, Li X, Liu J (2020) Recent advances in nanomaterial-based catalysis for sustainable chemical processes. Chem Rev 120(5):2810–2878
Zhao Y et al (2013) Mesoporous silica nanoparticles: a versatile platform for drug delivery applications. J Control Release 167(1):122–132
Zhao X, Chen J, Li M (2015) Structural properties of nanomaterials: influence of nanoscale dimensions. Mater Sci Eng R Rep 99:1–23
Zhao Y, Zhang Y, Xu G (2015) Nanomaterials: synthesis and applications. Nanotech Rev. 4(5):379–392
Zhao Y et al (2020) Corrosion resistance of nanomaterials: a review. Mater Today Proc. 28:1215–1220
Zheng Y et al (2015) Phase change materials for thermal energy storage: a review. Energy Storage Mater 1:44–62
Zheng J et al (2018) Nanomaterials for energy storage applications: a review. J Mater Chem a 6(7):2653–2676
Zhou Y et al (2015) Thermal properties of nanomaterials: a review. J Nanomater 2015:1–12

Chapter 2
Types of Nanomaterials in Wastewater Treatment

Ashootosh Mandpe, Mohd Nasim, and Harish Kumar Dwivedi

Abstract The increasing scarcity of clean drinking water and pollution of freshwater sources as a result of industrialization have created a need for innovative wastewater treatment technologies. Nanomaterials have proven to be very efficient alternatives to traditional treatments because of their high surface area, reactivity, and catalytic activity. This chapter gives an in-depth overview of different nanomaterial types such as metal and metal oxide nanoparticles, carbon-based nanomaterials, polymeric nanomaterials, nanocomposites, zeolites, nanozymes, quantum dots, and nanofibers and their unique characteristics, action mechanisms, and wastewater treatment applications. The chapter further discusses their applications in practical applications such as adsorption, photocatalysis, membrane filtration, nanocatalysis, disinfection, electrochemical treatment, and magnetic separation. Although nanomaterials present enormous promise for effective, sustainable, and affordable water purification, the issues of scalability, environmental safety, and long-term sustainability need to be solved in order to see widespread practical applications.

Keywords Nanomaterials · Wastewater treatment · Metal nanoparticles · Carbon nanotubes · Graphene · Adsorption · Photocatalysis · Nanofibers · Quantum dots · Nanozymes · Zeolites · Environmental remediation · Advanced oxidation processes

A. Mandpe (✉)
Department of Civil Engineering, Indian Institute of Technology Indore, Indore, Madhya Pradesh, India
e-mail: as_mandpe@iiti.ac.in

M. Nasim · H. K. Dwivedi
Civil Engineering Department, IPS Academy, Institute of Engineering and Science, Indore, Madhya Pradesh, India
e-mail: mohdnasim@ipsacademy.org

H. K. Dwivedi
e-mail: harishdwivedi@ipsacademy.org

A. Mandpe et al. (eds.), *Nanotech Solutions for Industrial Wastewater*, Advances in Wastewater Research, https://doi.org/10.1007/978-981-96-6879-3_2

2.1 Introduction

Out of 71% water covering the earth, only 1% is available for consumption by mankind. Industrialization also led to contamination of water with heavy metals, ions, bacteria, virus, and other pollutants. Waste water occurs when water quality deteriorates due to consumption by the industry. For primary treatment; techniques available for waste water treatment are bio-sand, reverse osmosis, adsorption, flocculation, and filtration, but they don't remove the heavy metals so efficiently. Nanomaterial treatment has higher efficiency in comparison with above mentioned techniques. The efficiency of the nanomaterial is due to the nanoscale property, adsorption, high surface-to-volume ratio, catalytic activities, and reactivity. Various nanomaterial that has been used for the treatment of water includes: Silver, carbon, iron, iron oxide, and titanium dioxide nanomaterials. Colloidal form of silver nanomaterial is used as an antimicrobial agents working against naturally occurring microorganism by disrupting vital cellular component causing cell lysis and death of pathogenic bacteria. As it tends to indicate hydrogen bonding, hydrophobic interactions, and ion exchange with the wastewater pollutants, it gains a lot of interest due to being carbon-based material. Both mobile and immobile contaminates get oxidized into hydroxides by iron nanomaterials. Photocatalyst is used as iron oxide nanomaterial for removing the impurities. TiO_2 nanomaterials are very photosensitive along with being ecofriendly. They work through the process of photo degradation of contaminates. Some of the other nanomaterials used for waste water treatment are pristine cellulose nanomaterials, nano-porous carbon xerogels, nanostructured black carbon, and exfoliated graphite nano-platelets. These nanomaterials are cost-effective, ecofriendly, and time-saving solutions and are widely being used nowadays to treat the waste water.

Applications of nanomaterials in wastewater treatment include high surface area and reactivity that may adsorb contaminants, including heavy metals, organic pollutants, and pathogens. Materials such as carbon nanotubes and nanosilica are primarily used for this purpose. Nanocatalysts facilitate faster chemical reactions which may further break down pollutants in a more efficient manner. For example, metal nanoparticles such as silver or titanium dioxide can catalyze organic pollutant degradation under ultraviolet light, and nanomaterials can improve membrane technology in both permeability and selectivity. Nanocomposite membranes could filter more efficiently smaller particles and pathogens compared to traditional membranes; titanium dioxide nanoparticles might be activated by light to produce reactive oxygen species that can oxidize a vast range of organic pollutants in water; nanosensors will identify targeted pollutants at very low concentration levels, thereby creating the possibility of real-time monitoring of water quality; nanomaterials stabilize emulsions, thus preventing the re-suspension of settled solids, and enhance sedimentation efficiency in treatment plants; certain nanomaterials may favor the activity of bacteria that degrade the pollutants, thereby enhancing the effectiveness of biological treatment methods.

Nanomaterials, including those of different kinds, present the typical advantage of wide application in the purification and treatment of water. These are titanium nanotubes, zinc oxide nanoparticles, nano-enhanced filters, stimuli-responsive polymers, silica coatings, chalcogenides-based nanomaterials, many nanomaterials including adsorption, catalytic, and reactive properties, carbon nanotubes, nanocomposites, graphene, nanomembranes, nanofibers, and nanocatalysts (Dhull et al. 2024), and nanoribbon-based nanomaterials that exhibit a highly significant relevance for the removal of contaminants through heavy metals, organic pollutants, and microorganisms from wastewater. Moreover, these nanomaterials have unique physicochemical properties and a high specific surface area with strong adsorption abilities appropriate for water purification processes. However, this depends on the varied properties of the nanomaterials that can contribute to the efficiency, sustainability, and cost-effectiveness of wastewater treatment technologies toward the preservation of water and environmental health.

Nanomaterials have been attracting interest in the context of wastewater treatment due to their outstanding properties, such as high surface area, high reactivity, and possibility of functionalization for specific applications. These nanomaterials are getting increasingly integrated with various wastewater treatment technologies, which leads to increased efficiency, enhanced selectivity, and potential for sustainable water management, but still issues remain in real-world applications concerning environmental concerns, cost, and scalability. A summary overview of various types of nanomaterials utilized for wastewater treatment is offered in Table 2.1.

2.2 Nanoparticles

2.2.1 Metal Nanoparticles

These include gold (Au) (Li et al. 2019), silver (Ag) (Li et al. 2019), copper (Cu) (Chen et al. 2012), and iron (Fe) (Agrawal et al. 2021a, b) nanoparticles. Silver nanoparticles are particularly known for their antimicrobial properties, while iron nanoparticles, especially zero-valent iron (ZVI), are widely used for the reduction and removal of contaminants like heavy metals and organic pollutants (Qasem et al. 2021).

2.2.2 Metal Oxide Nanoparticles

Examples include titanium dioxide (TiO_2), zinc oxide (ZnO), and iron oxide (Fe_2O_2) (Rathore et al. 2023). These nanoparticles are used for photocatalytic degradation of organic pollutants and disinfection. TiO_2 nanoparticles, in particular, are well-known for their photocatalytic activity under UV light (Hirota et al. 2021).

Table 2.1 Overview of nanomaterials for wastewater treatment

Type of nanomaterial	Examples	Properties/ Functions	Applications	Reference
Nanoparticles	Silver (Ag) Iron (Fe) Titanium dioxide (TiO_2)	High reactivity Antimicrobial Photocatalytic	Disinfection Heavy metal removal Organic pollutant degradation	Zhang et al. (2016)
Carbon-based nanomaterials	Carbon nanotubes (CNTs) Graphene Fullerenes	High surface area Electrical conductivity Adsorptive	Adsorption of pollutants Filtration enhancement	Kumari et al. (2023)
Polymeric nanomaterials	Dendrimers Nanosponges	Functionalizable Porous structure	Removal of heavy metals Organic dyes adsorption	Sahu et al. (2023)
Nanocomposites	Metal/Metal oxide nanocomposites Magnetic nanocomposites	Enhanced reactivity Magnetic separation	Photocatalysis Contaminant recovery	Naseem and Waseem (2022)
Zeolite nanomaterials	Nanosized Zeolites	High ion exchange capacity Porous structure	Ion exchange Adsorption	Krstić (2021)
Nanozymes	Cerium oxide (CeO_2)	Enzyme-like activity Catalytic degradation	Organic pollutant degradation	Hong et al. (2022)
Quantum dots	Semiconductor Nanocrystals	Unique optical properties Photocatalytic	Detection Pollutant degradation	Abd Rani et al. (2020)
Nanoscale zero-valent iron	nZVI	High reactivity Large surface area	Contaminant reduction Organic pollutant removal	Li et al. (2024)
Nanofibers	Electrospun polymers Carbon nanofibers	High surface area-to-volume ratio	Filtration Pollutant removal	Kriklavova and Lederer (2011)

2.3 Carbon-Based Nanomaterials

2.3.1 Carbon Nanotubes (CNTs)

These cylindrical nanostructures possess high surface area, mechanical strength, and electrical conductivity. They are used for adsorption of organic pollutants, heavy metals, and for enhancing the performance of membranes in filtration systems (Devi et al. 2023).

2.3.2 Graphene and Graphene Oxide (GO)

Graphene is a single layer of carbon atoms arranged in a hexagonal lattice. It is used for adsorption of contaminants, and GO, with its oxygen-containing functional groups, has enhanced solubility and adsorption capacity, making it suitable for removing heavy metals and organic pollutant (Rana et al. 2024).

2.3.3 Fullerenes

These are spherical carbon molecules that exhibit unique properties, such as the ability to form complexes with organic and inorganic compounds, making them effective for the removal of various contaminants from water (Kausar 2024). Carbon nanomaterials are remarkable for developing this new approach of wastewater treatment. The notable properties of these include high mechanical strength and chemical stability of the CNT, an extremely large surface area leading to a high adsorption capacity, and higher electrical conductivity for membrane filtration and catalytic application. Graphene and graphene oxide are also remarkable with exceptional electrical conductivity and thermal stability. Their specific surface area is quite large, promoting adsorption; functional groups of graphene oxide also improve solubility in water and interaction with contaminants. Due to its spherical shape and high electron affinity, fullerenes can form complexes even with organic and inorganic types of pollutants. So fullerenes are very well applied in the process of photocatalysis and adsorption.

2.4 Polymeric Nanomaterials

These are highly branched, star-shaped polymers with a high density of functional groups on their surface. Dendrimers can be tailored for specific interactions with pollutants, making them effective for the removal of heavy metals, organic dyes, and

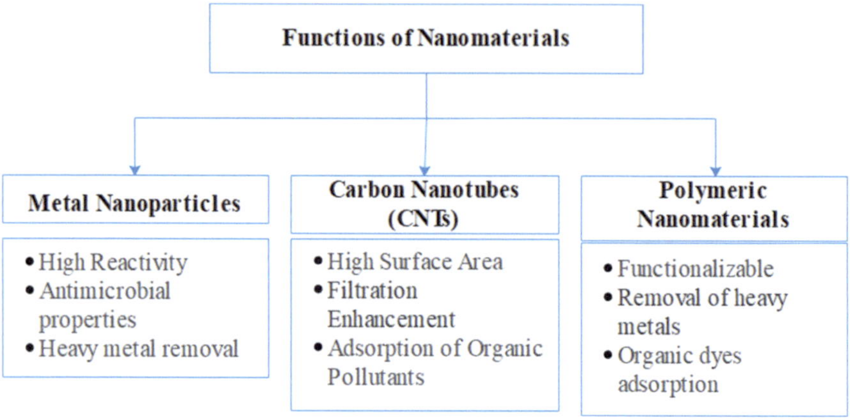

Fig. 2.1 Functions of some nanomaterials

other contaminants (Adeola and Nomngongo 2022). The kind of Polymeric Nanomaterials known as Dendrimers have a very branched, tree-like structure with multiple functional groups, providing them excellent solubility and potential as encapsulants of contaminants. Their structure can be modified according to the functional group to remove particular kinds of pollutants. Nanosponges also exhibit a sponge-like structure because of their pores, thus having large surface areas enabling them to absorb and hold pollutants within their structure. They, too, have a high loading capacity and can potentially be utilized several times (Fig. 2.1).

2.5 Nanocomposites

These combine metal or metal oxide nanoparticles with other materials (like carbon-based materials or polymers) to enhance their properties. For example, combining TiO_2 with CNTs can improve photocatalytic efficiency and facilitate the removal of a broader range of contaminants (Naseem and Durrani 2021). Nanocomposites are composites made to improve properties. Metal/metal oxide nanocomposites allow the addition of metal nanoparticles and other materials like CNTs or polymers to increase photocatalytic efficiency and remove contaminants. Magnetic nanocomposites contain magnetic nanoparticles, such as Fe_2O_3, where nanoparticles may be separated from water by using outer magnets (Chu et al. 2020). This property makes them very efficient in adsorption processes and subsequent removal of pollutants, thereby making the nanocomposites reusable.

2.6 Zeolite Nanomaterials

Zeolites are crystalline aluminosilicates with a porous structure that can trap ions and molecules. Nanosized zeolites have a larger surface area and higher reactivity, making them effective for ion exchange, adsorption, and catalysis in wastewater treatment processes (Rahman et al. 2022). Zeolite nanomaterials, including nanosized zeolites, possess a crystalline structure with pore sizes of uniform size and hence exhibit high ion exchange capacity and good potential for trapping heavy metals and other ions. It is stable and reusable and can adsorb various contaminants.

2.7 Nanozymes

Nanozymes are nanomaterials that exhibit enzyme-like activities. They can catalyze the degradation of pollutants, mimicking natural enzymes. Examples include cerium oxide (CeO_2) nanoparticles, which have been used to degrade organic pollutants in water through catalytic oxidation (Hong et al. 2022). The nanozymes, like cerium oxide (CeO_2), exhibit mimicked catalytic activity from natural enzymes. It has high redox activity that would degrade organic pollutants via catalytic oxidation and remains stable under numerous environmental conditions.

2.8 Quantum Dots

Quantum dots are semiconductor nanocrystals with unique optical and electronic properties. They can be used for detecting and removing contaminants in water through fluorescence-based sensors, and some types of quantum dots can also facilitate photocatalytic degradation of pollutants (Shivaji et al. 2023). Quantum dots are semiconductor nanocrystals whose quantum confinement effects render them uniquely optically and electronically active. Quantum dots are very sought after for applications in contaminant detection, due to their high photoluminescence efficiency apart from the potential photocatalytic degradation of pollutants under visible light.

2.9 Nanoscale Zero-Valent Iron (nZVI)

nZVI is widely used in environmental remediation, including wastewater treatment. Its high reactivity and large surface area allow it to effectively reduce and remove a variety of contaminants, including chlorinated organic compounds, heavy metals, and nitrates (Li et al. 2014). nZVI has nanoscale sizes, which give it high reactivity and a large surface area; it is effective for the reduction and immobilization of a wide range of contaminants, including the majority of heavy metals and organic pollutants.

2.10 Nanofibers

Lastly, nanofibers such as Electrospun Polymers provide better surface area-to-volume ratio along with high adsorption capacities. They can be functionalized for targeted removal of contaminants. They are mechanically flexible. Carbon nanofibers are highly thermostable along with electrical conductivity and have very high surface areas along with better enhancement in terms of adsorption, so suitable for efficient organic pollutant filtration as well as removal. Nanofibers, particularly those made from polymers or carbon, have high surface area-to-volume ratios and can be functionalized for specific pollutant removal. Electrospun nanofibers are used in filtration membranes to enhance the removal efficiency of contaminants from water (Agrawal et al. 2021a, b).

The overview of types of nanoparticles has been described in this chapter as shown in Fig. 2.2.

Table 2.1 offers a detailed overview of the properties of various nanomaterials, emphasizing their distinctive characteristics and justifying their effectiveness in wastewater treatment applications.

2.11 Properties of Nanomaterials for Wastewater Treatment

Table 2.2 summarizes the main characteristics and applications of various nanomaterials in water treatment, focusing on their selectivity and effectiveness in removing contaminants. Nanoparticles emerged as a potential material for contaminant removal in water. Silver nanoparticles (Ag) have been shown to possess high antimicrobial properties, targeting bacteria, viruses, and fungi because of their large surface area-to-volume ratio, which enhances contact with the contaminants. It has also been demonstrated by Jain et al. (2021) and Ganguly et al. (2021). Iron nanoparticles (Fe), specifically zero-valent iron (ZVI), is highly reactive and efficient in removing and reducing heavy metal and chlorinated organic compounds (Agrawal et al. 2021a, b). Titanium dioxide (TiO_2) has been reported with high photocatalytic activity in UV light for degrading and oxidizing organic pollutants into less harmful components. In addition, TiO_2 is stable and non-toxic, thus making it safe for application in the environment (Han et al. 2009).

Carbon-Based nanomaterials also plays a very critical role in wastewater treatment. The high mechanical strength, chemical stability, and large surface area make

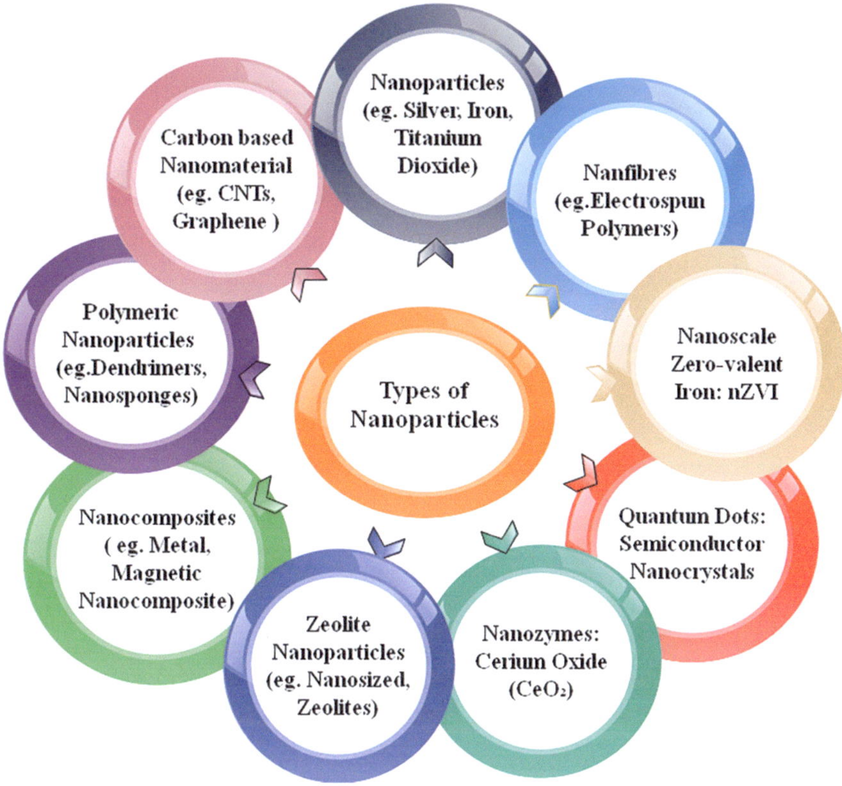

Fig. 2.2 Types of nanoparticles

carbon nanotubes work so hard to result in high adsorption capacity. Its high electrical conductivity makes it useful in membrane filtration and catalytic applications (Anzar et al. 2020; Pan et al. 2023). Graphene and graphene oxide have high electrical conductivity, high thermal stability, and large specific surface area which makes them very valuable. The functional groups of graphene oxide (GO) facilitate solubility in water and interaction with contaminants. Fullerenes have spherical structure with high electron affinity that allows them to form complexes with organic and inorganic contaminates; hence they are suited for photocatalysis and adsorption (Yao et al. 2020).

Dendrimers are Polymeric Nanomaterials with a highly branched structure and contain a high number of functional groups contributing to their high solubility and the possibility of encapsulating the pollutants. These can be functionalized for the removal of particular pollutants. Nanosponges are nanomaterials having a sponge-like porous structure, which therefore gives a large surface area. This ability gives them the scope of adsorbing and trapping the pollutant inside their structure. A huge loading capacity is provided, which may also be recyclable (Garg et al. 2024).

Table 2.2 Properties of nanomaterials for wastewater treatment

Type of nanomaterials		Key properties	References
Nanoparticles	Silver nanoparticles (Ag)	1. Strong antimicrobial properties, effective against bacteria, viruses, and fungi 2. High surface area-to-volume ratio, enhancing interaction with contaminants	Jain et al. (2021), Ganguly et al. (2021)
	Iron nanoparticles (Fe)	1. High reactivity, particularly zero-valent iron (ZVI) 2. Effective in the reduction and removal of heavy metals and chlorinated organic compounds	Agrawal et al. (2021a, b)
	Titanium dioxide (TiO_2)	1. Strong photocatalytic activity under UV light 2. Oxidation and degradation of organic pollutants into less harmful substances 3. Stability and non-toxicity, making it safe for environmental applications	Han et al. (2009)
Carbon-based nanomaterials	Carbon nanotubes (CNTs)	1. High mechanical strength and chemical stability 2. Large surface area, leading to high adsorption capacity 3. High electrical conductivity, useful for membrane filtration and catalytic applications	Anzar et al. (2020), Pan et al. (2023)

(continued)

Table 2.2 (continued)

Type of nanomaterials		Key properties	References
	Graphene and graphene oxide	1. Exceptional electrical conductivity and thermal stability 2. Large specific surface area for adsorption graphene oxide (GO) has functional groups that enhance water solubility and interaction with contaminants	Anegbe et al. (2024)
	Fullerenes	1. Unique spherical structure with high electron affinity 2. Ability to form complexes with organic and inorganic contaminants 3. Potential for use in photocatalysis and adsorption	Yao et al. (2020)
Polymeric nanomaterials	Dendrimers	1. Highly branched, tree-like structure with numerous functional groups 2. High solubility and ability to encapsulate contaminants Tailorability for specific pollutant removal through functionalization	Nyabadza et al. (2024)
	Nanosponges	1. Porous, sponge-like structure with large surface area 2. Ability to absorb and trap pollutants within their structure 3. High loading capacity and potential for multiple uses	Garg et al. (2024)

(continued)

Table 2.2 (continued)

Type of nanomaterials		Key properties	References
Nanocomposites	Metal/metal oxide nanocomposites	1. Enhanced properties by combining metal/metal oxide nanoparticles with other materials (e.g., CNTs, polymers) 2. Increased photocatalytic efficiency and contaminant removal capabilities	Kumar et al. (2022), Alhalili (2023)
	Magnetic nanocomposites	1. Incorporation of magnetic nanoparticles (e.g., Fe_2O_2) for easy separation from water using external magnets 2. Effective for adsorption and subsequent removal of pollutants, enabling reuse of nanocomposites	Vallinayagam et al. (2021), Ali et al. (2019)
Zeolite nanomaterials	Nanosized zeolites	1. Crystalline structure with uniform pore sizes 2. High ion exchange capacity for trapping heavy metals and other ions 3. Stable, reusable, and capable of adsorbing a wide range of contaminants	de Magalhães et al. (2022), Liaquat et al. (2024)
Nanozymes	Cerium oxide (CeO_2)	1. Enzyme-like catalytic activity, mimicking natural enzymes 2. High redox activity, enabling degradation of organic pollutants through catalytic oxidation 3. Stability under a range of environmental conditions	Xu and Qu (2014)

(continued)

Table 2.2 (continued)

Type of nanomaterials		Key properties	References
Quantum dots	Semiconductor nanocrystals	1. Quantum confinement effects, resulting in unique optical and electronic properties 2. High photoluminescence efficiency, making them suitable for detecting contaminants 3. Potential for photocatalytic degradation of pollutants under visible light	Mei et al. (2022)
Nanoscale zero-valent iron (nZVI)	nZVI	1. High reactivity due to the nanoscale size and large surface area 2. Effective in reducing and immobilizing a wide range of contaminants, including heavy metals and organic pollutants 3. Environmental compatibility and potential for in situ remediation	Alazaiza et al. (2022), Zafar et al. (2021)
Nanofibers	Electrospun polymers	1. High surface area-to-volume ratio, leading to high adsorption capacity 2. Can be functionalized for specific contaminant removal Mechanical flexibility and potential for incorporation into filtration membranes	Cui et al. (2020), Zhang et al. (2021)

(continued)

Table 2.2 (continued)

Type of nanomaterials		Key properties	References
	Carbon nanofibers	1. High thermal stability and electrical conductivity 2. Enhanced surface area and adsorption properties 3. Useful in filtration and removal of organic pollutants	Lazarenko et al. (2022), El-Aswar et al. (2022)

Nanocomposites improve the properties of the nanoparticles by alloying various materials. Metal/metal oxide nanocomposites enhance the efficiency of photocatalysis and contaminant removal when alloyed with CNT and polymers (Kumar et al. 2022; Alhalili 2023). Magnetic nanocomposites alloy magnetic nanoparticles, such as Fe_2O_2, making it easy to separate using exterior magnets from the water. This makes it efficient for adsorption, followed by the removal of pollutants, allowing the nanocomposites to be reused (Vallinayagam et al. 2021; Ali et al. 2019). Zeolite nanomaterials, such as Nanosized Zeolites, have a crystalline structure with pores of the same size, thus exhibiting strong ionic exchange. The materials are stable, non-destructible, and reusable; they can easily allow the adsorption of many contaminants available in nature (de Magalhães et al. 2022; Liaquat et al. 2024). Cerium oxide and nanozymes, however, show enzymic catalytic activities by mimicking the natural enzymes due to high redox activity for the catalytic oxidation of organic pollutants leading to degradation, and under several environmental conditions.

Quantum dots are semiconductor nanocrystals that have unique optical and electrical properties caused by quantum confinement effects. These dots exhibit high quantum yield in photoluminescence, making them a potential candidate for detecting contaminants, and have shown potential photocatalytic degradation of pollutants under visible light (Mei et al. 2022). This reactivity of nanoscale zero-valent iron (nZVI) is attributed to its nanoscale size and high surface area, thus allowing it to effectively reduce and immobilize various kinds of contaminants which range from metals to organic pollutants. Additionally, it is environmentally compatible and suited for in situ remediation purposes (Alazaiza et al. 2022; Zafar et al. 2021).

Electrospun Polymers nanofibers are characterized by having a very high surface area-to-volume ratio and, thus, possess a large adsorption capacity. Nanofibers can be customized for particular contaminant removal as well as mechanically flexible to incorporate them into filtration membranes (Cui et al. 2020; Zhang et al. 2021). Carbon nanofibers show good thermal stability and electrical conductivity as well as high surface area and adsorption properties, leading to increasing interest for applications such as filtration and organic pollutant removal (Lazarenko et al. 2022; El-Aswar et al. 2022).

2.12 Applications in Wastewater Treatment

Nanomaterials are increasingly being applied in the industry sector, and for many reasons related to their specific properties that include enormous surface area, reactivity, and specificity. Therefore, nanomaterials are in high demand for various advanced processes of contaminant removal. Some of the industrial wastewater treatment procedures which employ nanomaterials:

2.12.1 Adsorption

Motitswe et al. (2024) recently designed a new rGO/ZnO nanocomposite for the efficient removal of lead and cadmium from wastewater, surpassing the use of typical adsorbents. Toxic metals found in wastewater are one of the major environmental concerns. A method for effective removal of toxic metals is needed for wastewater treatment. The rGO/ZnO nanocomposite synthesized for metal adsorption showed impressive, nearly 100% efficiency for Pb(II) and Cd(II) ions. Thus, this new nanocomposite is promising in the application for industrial wastewater treatment plants, offering selective and advanced removal of lead and cadmium contaminants from the sources of water that are contaminated. The adsorption of heavy metals from wastewater by nanostructured adsorbents such as carbon nanotubes (CNTs), graphene, and metal–organic frameworks (MOFs) has elicited potential efficacy which is said to be superior compared to conventional adsorbents. Advanced materials hence have superior adsorption capacities and efficiencies owing to high surface areas combined with adjustable surface chemistry. Carbon nanotubes, graphene, and MOFs provide excellent removal capabilities for heavy metals based on their interactions at the molecular level that are strong enough to make them an attractive answer toward mitigating the problem of water pollution and ensuring efficient wastewater treatment processes. The application of nanostructured adsorbents is widely recognized to mark a significant feature of environmental remediation with sustainable and innovative approaches toward reducing heavy metal contamination of water (Al-Amrani and Onaizi 2024).

Carbon-based nanomaterials, including carbon nanotubes and graphene oxide, have shown high adsorption for heavy metals, dyes, and pharmaceuticals (Thakur et al. 2024), primarily through large surface area and highly tunable surface chemistry. These nanomaterials specifically have demonstrated improved adsorption, which is, in essence, attributed to their structural characteristics that afford a high surface-to-volume ratio, thus efficiently interacting with numerous kinds of harmful impurities in water as well as wastewater effluent streams. The adjustable surface chemistry of carbon nanotubes and graphene oxide enhances the scope of their improved adsorption possibilities, thus making them promising candidates for a variety of applications in the remediation environment to facilitate the effective and sustainable removal of pollutants from aqueous solutions. Some of the new materials with unique properties include metal oxides, such as TiO_2, Fe_3O_4 (Cai et al. 2024; Ani and Egbosiuba 2023), or zeolites (Kumari et al. 2024; Liaquat et al. 2024; Tufail et al. 2024) which are commonly found in treatment processes in industrial wastewater because they act as efficient adsorbents on any contaminants. Such high surface area materials are equipped with an ion exchange potential and structural strength to adsorb a rather varied class of pollutants such as heavy metals and organic dyes.

Zeolites are an effective group for the removal of organic and inorganic impurities from wastewater due to their microporous structure and ion exchange properties. Incorporating zeolites in polymer matrices enhances the performance of membranes with increasing selectivity and permeability to contaminants like heavy metal and dyes. It also proves versatile in dealing with different types of wastewaters, including oil-field and radioactive waste. The higher adsorption efficiencies are obtained by doping or modifying metal oxides, like TiO_2 and Fe_3O_4, nanoparticles, which are very good at removing cationic dyes. The recycled composite retains its adsorption efficiency after several cycles. Electrostatic interactions and hydrogen bonding are the predominant mechanisms in the adsorption process and play a key role in the removal of the contaminants. Despite their effectiveness, however, challenges such as regeneration and the potential for leaching of contaminants need to be addressed to make zeolites and metal oxides more sustainable and efficient in practical applications for wastewater treatment.

2.12.2 Photocatalysis

These are, in fact, newly emerging nanomaterials which, from recent research perspectives, appear to be the most valid solution for photocatalytic applications in industrial wastewater treatment as a promising way of combating the rapidly increasing scale of pollution. Their distinctive properties of large surface area and tailor-made functionalities ensure improved degradation of harmful pollutants, thus a significant feature of such a process. Below is an overview of the efficiencies of a range of nanomaterials in photocatalytic treatments for wastewater treatment. Nanomaterials like Cu_2O/CeO_2 demonstrate superior charge carrier transfer dynamics to provide photocatalytic reactions and impressive degradation rates of pollutants, such as methyl orange and tannery wastewater (Bekele et al. 2024; Mohan et al. 2024). The process's energy efficiency by using sun light as a free-of-cost, green source of energy instead of conventional fuels at high costs increases its allure (Priya and Sagadevan, 2024). A variety of nanomaterials, such as metal oxide nanocomposites like TiO_2, ZnO, and CuO, show marked photocatalytic activities, reaching almost complete degradation of dyes under visible light (Gupta et al. 2023). Chalcogenides associated with minimal band gaps have been efficient in remediating dissolved metal ions and organic pollutants in wastewater (Priya and Sagadevan 2024). These applications reach industrial contexts, where photocatalytic processes were particularly successful for treating challenging wastewater streams, including tannery effluent characterized by high toxic organic compounds and heavy metal concentration (Mohan et al. 2024). Nanomaterials also show proficiency in degrading persistent pharmaceutical contaminants (Hasan et al. 2024) while evinced versatility across diverse industrial applications. However, the potential of nanomaterials in photocatalysis poses greater promises toward challenges, such as being scalable and with good long-term stability; therefore, areas that need research and development are pertinent.

Photocatalytic nanomaterials have the capability to break down organic contaminants, dyes, pharmaceutical residues, and pathogens within wastewater through photolysis when irradiated with either UV or visible light illumination. TiO_2 nanomaterials are the most commonly used nanomaterials in this application owing to their potentiality to destructure these contaminants into harmless byproducts such as carbon dioxide and water. This environment-friendly approach makes use of light-induced nanomaterial properties for degrading pollutants into non-harmful substances, thus helping in the restoration of wastewater and conservation of water purity (Chen et al. 2020).

2.12.3 Membrane Filtration

Among the materials, nanomaterials such as nanocomposites, metal–organic frameworks (MOFs), carbon-based nanomaterials, and nanoparticles are significantly integrated into filtration membranes for an improvement in the removal efficacy of contaminants in wastewater. Because they can dramatically increase filtration efficiency while decreasing fouling, these materials can be applied in various modes of filtration technologies, including nanofiltration, reverse osmosis, and ultrafiltration (Nain et al. 2022). Nanocomposites, metal–organic frameworks (MOFs), carbon-based nanomaterials, and nanoparticles incorporated into polymeric membranes are gaining interest due to the unique attributes and functional roles in filtration practices. These nanomaterials depict improved surface areas, toughness, and chemical stability that can lead to effective extraction of the pollutant. The interaction conditions with the contaminants is essential in attaining the maximum performance criteria upon their incorporation into membranes (Khraisheh et al. 2021). Nanomaterials are primarily used in filtration membranes, which are applied to treat wastewater. Here, nanofiltration, reverse osmosis, and ultrafiltration are used to separate specific pollutants which include heavy metals, organic impurities, and microorganisms. The combination of sophisticated nanomaterials in these membranes creates a multipurpose barrier, allowing for the effective separation of unwanted substances from water (Zamel et al. 2023).

Advantages of incorporation of nanomaterials into filtration membranes has led to unprecedented advancements in terms of operational efficiency. Improvements in permeability and selectivity, which are associated with energy consumption reductions and reductions in operational costs connected with treatments, underpin the improvements in filtration efficiency. Moreover, a high, consistent level of performance over long periods is critically dependent upon the resistance of the membranes to fouling (Noah 2023).

2.12.4 Nanocatalysis

Noble metal nanoparticles such as gold, silver, and platinum with iron oxide and bimetallic nanoparticles act as a very good nanocatalyst in wastewater treatment (Vaiano et al. 2020). These nanocatalysts catalyze chemical reactions to cause the degradation of complex organic molecules and toxicity. They are very essential in industries that produce tough-treating pollutants, which are known to release toxic substances like phenols and PCBs. Because of their high surface area and reactivity, they have vast application in the pharmaceutical sector, textiles, and petrochemical sectors. Nanocatalysts function in such a way to hasten the following reactions which reduce the time factor, energy usage, and general efficiency of treatment, allowing for more sustainable wastewater processes (Mbarek et al. 2022).

Titanium dioxide, a nanocatalyst, has proved to have excellent prospects in photocatalytic processes in the cleanup of the environment. TiO_2 is a catalyst that may either degrade persistent organic compounds or microbial pathogens that are otherwise resistant in the event that they are subjected to UV light exposure instead of traditional methods. This makes nanocatalysts based on TiO_2 very effective in wastewater treatment, particularly in decomposing complicated pollutants like dyes, pharmaceuticals, and harmful microorganisms. Their capacity to capture light energy and initiate chemical reactions positions them as super in curbing environmental pollution (Nezhad et al. 2024).

Green synthesis of nanocatalysts: The use of a biological entity, in this case, a bacteria, algae, or a fungus constitutes an ecofriendly and renewable synthesis procedure as compared to the use of chemical technique (Ningthoujam et al. 2022). In addition to minimizing the environmental effects of nanoparticle synthesis, this process further enhances the stability and bio-compatibility of the resultant nanocatalysts. Green synthesis provides nanocatalysts by exploiting nature's process in such a way that it is operational in environmental applications and aligned with principles of green chemistry. In fact, nanoparticles obtained through a biological approach are more stable against agglomeration and surface features and which make them catalytically active as compared to those other routes (Roy et al. 2021).

Plant extract-mediated AgNPs have drawn increasing interest within recent times due to higher potential antibacterial activity (Kumar et al. 2021). These nanoparticles can treat wastewater well, contaminated with pathogenic microorganisms; thus, they become useful for water purification systems. AgNPs interfere with microbial growth by disrupting bacterial cell membranes and thus are considered a very potent disinfectant for water sources. The synthesis of green nanoparticles using plant extracts reduces the usage of harmful chemicals during production and gives an environmental-friendly way to synthesize nanoparticles besides increasing the efficacy of the nanospheres in antibacterial activity.

The commonly used material in the remediation of groundwater processes is zero-valent iron nanoparticles (Zafar et al. 2021). In reality, these are highly used for removing ions of heavy metals such as arsenic and lead. By passing through reductive reactions, these toxic metal ions are converted into less harmful counterparts, from which harmful ions may be prevented from being migrated into the water supply meant for drinking purposes. Complete penetration in contaminated sites by ZVI nanoparticles improves a focused remediation process. This makes ZVI an emerging remediation technology, which is worthy of further research, for the heavy metal pollution of groundwater, especially in situations where conventional methods may not be applied or seem too costly.

Hybrid nanocatalysts, Pd-TiO_2, among others, have shown dramatic enhancement in photocatalytic degradation processes. Hybrid Pd/TiO_2 nanocatalysts create synergism and hence improve the catalytic activity of such systems under visible light. This expands the photocatalysis spectrum range from UV up to visible light, hence furthering the application of these hybrid nanocatalysts in environmental applications (Obaiah et al. 2023). Materials such as these break down industrial dyes and pharmaceuticals under weaker light conditions, making them one of the best agents used in wastewater treatment and pollution remediation through other processes (Anucha et al. 2022).

However, bioinspired nanocatalysts, which incorporate catalytic properties in gold nanoparticles stabilized by plant extracts, have recently drawn much attention in this regard, since they showed great efficiency in the degradation of dyes and other environmental pollutants (Aazam et al. 2023). Such nanocatalysts not only support efficient pollution degradation but also are compatible with environmentally friendly practices by using natural materials in synthesis. Gold nanoparticles stabilized by plant-based substances improved their stability and reactivity, hence better catalytic performance. This method is aligned with the trend of using bioinspired materials for environmental protection; it provides a green, environment-friendly alternative to chemically synthesized nanomaterials.

Another promising class of nanocatalysts in this regard are MOFs. MOFs are regarded as high-performing degraders of organic pollutants through advanced oxidation processes. MOFs have a very porous structure that expands its surface area significantly for catalytic reactions (Wen et al. 2021). Such a unique property allows MOFs to achieve efficiencies over 90% in eliminating organic pollutants from water, making it highly effective in advanced wastewater treatment technologies. Thus, the versatility and high efficiency of MOFs make them prime candidates for large-scale environmental remediation applications such as the treatment of industrial effluents and purification of contaminated water bodies.

The infusion of carbon nanotubes is bringing a massive revolution into filtration technologies due to enhanced adsorption and catalytic degradation of contaminants (Peng et al. 2021). High surface area-to-volume ratio, electrical conductivity, and the ability to trap and degrade pollutants at the molecular level give CNTs as an ideal candidate for such applications. These membranes are very good at removing organic pollutants and heavy metals from water, giving a multifunctional solution to purification of water. CNT-doped membranes also exhibit benefits such as recyclability and regeneration, thereby providing an economic benefit for more extensive water treatment applications.

Graphene-based nanoadsorbents possess strong adsorption capabilities and also exhibit bactericidal properties. Such nanoadsorbents are notably useful for the removal of dyes, heavy metals, and other toxic agents from water. This two-dimensional structured graphene allows it to interact with many pollutants (Hsu et al. 2023). These pollutants get trapped on the surface of graphene and are unable to be released into the environment. Additionally, with antibacterial properties, graphene-based materials can be exploited in treating water contaminated with harmful microorganisms, making both concepts more viable for chemical and biological contamination. Magnetic nanocarriers, including nanoscale zero-valent iron (nZVI), have been studied and developed as selective remediation agents of various contaminants. These nanoparticles are easy to manipulate because of the exterior magnetic fields; they can precisely target the polluted area. Efficacy in nZVI particles cuts across different types of pollutants such as chlorinated compounds, heavy metals, and organic dyes (Venkateshaiah et al. 2022). The further recovery and reuse of the nanoparticles after remediation also contribute to making them not only a cleanup option but also sustainable and efficient in managing contaminated water and soil.

Nanocatalysts can be ideal solutions to the rapidly increasing emerging contaminants such as microplastics and pharmaceuticals, which would not easily be removed via traditionally methodical wastewater treatment (Dhull et al. 2024). Microplastics most of the time pass through conventional filtration systems due to their small size and the tendency of persistence in the environment. Pharmaceuticals, however, including some antibiotics and hormones, seem to remain chemically stable in wastewater, which causes their accumulation in water bodies. Such contaminants pose a serious environmental and health risk, such as structure-altering effects on aquatic ecosystems and antibiotic-resistance-promoting micro-ecological exposures. Nanocatalysts with high surface areas, reactivity, and ability to facilitate specific chemical reactions can be used for the degradation of such pollutants, thus making them an attractive alternative in advanced water treatment technologies.

For example, photocatalysts like titanium dioxide (TiO_2) could degrade microplastics into smaller particles with lower toxicity through oxidation reactions catalyzed by UV light (Ge et al. 2022). This could also involve nanocatalysts like silver nanoparticles (AgNPs) and iron-based materials that can degrade pharmaceutical residues by breaking complex organic molecules into simpler, non-toxic compounds. Such versatility and efficacy of nanocatalysts in treating emerging contaminants make them one of the excellent agents for modern treatment plants at wastewater treatment plants hoping to gain stricter environmental standards.

2.12.5 Electrochemical Treatment

Nanomaterial-based electrodes significantly improve electrochemical oxidation efficiency by possessing better catalytic properties, increased surface areas, and higher conductivity (Abhishek et al. 2022). These electrodes degrade the wide range of refractory organic pollutants, such as POPs, pharmaceutical residues, and industrial dyes, typically resistant to conventional wastewater treatment processes.

Electrochemical oxidation is a process in which the active species, for example, hydroxyl radicals, from the electrical oxidation attack and break the complex organic molecule. The employment of nanomaterial electrodes boosts this process because these materials are more reactive in catalytically enhancing the rates of reactions at the electrode and give better and faster degradation of pollutants (Yaqoob et al. 2020).

For example, using nanoparticles such as TiO_2, graphene, and carbon nanotubes (CNTs) (Wang et al. 2014; Huang et al. 2018) as the electrode material has been considerably explored in view of electrochemical treatment. These materials possess far better electron transfer abilities that improve the oxidation reaction efficiency. Furthermore, the increased surface area provided by the nanomaterial enhances contact between the reaction species and the pollutants and, hence, a better degradation performance.

A study investigated the use of TiO_2 nanotube-based electrodes for the electrochemical degradation of pharmaceutical residues, such as antibiotics and anti-inflammatory drugs, which are often found in wastewater (Cerro-Lopez et al. 2020; Torrinha et al. 2020). The results showed that TiO_2 nanotube electrodes enhanced the generation of reactive species and significantly increased the degradation efficiency compared to traditional electrodes. Pharmaceuticals, which are typically resistant to biological treatments, were effectively broken down into less harmful byproducts, demonstrating the potential of nanomaterial-based electrodes in advanced wastewater treatment technologies.

Graphene-based electrodes have been applied in the electrochemical oxidation of persistent organic pollutants, including polychlorinated biphenyls (PCBs) and dioxins (Sun et al. 2020). In a case study, researchers used graphene-coated electrodes to degrade POPs in industrial wastewater. The high conductivity and large surface area of graphene facilitated the generation of reactive oxygen species, which efficiently broke down these pollutants. The study reported significant removal rates, demonstrating graphene's effectiveness in treating hazardous chemicals that resist degradation through conventional processes.

Industrial dyes, such as those used in the textile industry, are known for their complex structures and high toxicity, which make them difficult to treat. In a case study, carbon nanotube (CNT)-based electrodes were employed in the electrochemical oxidation process for the degradation of synthetic dyes (Irfan et al. 2022) like methylene blue and Congo red (Nidheesh et al. 2018). The CNT electrodes demonstrated superior performance due to their high surface area and conductivity, which accelerated the breakdown of dye molecules into less toxic substances. The study highlighted the effectiveness of CNT-based electrodes in reducing both color and chemical oxygen demand (COD) in wastewater, presenting a viable solution for the textile industry's wastewater treatment challenges.

2.12.6 Magnetic Separation

Nanomaterials, especially magnetic iron oxide nanoparticles (Fe_3O_4) (Shadi et al. 2020), are thought of as materials with high efficiency for environmental remediation. The main reason behind this is their unique magnetic properties and large surface area. Functionalization of these particles with specific ligands or polymers will enable them selectively to target and bind the contaminants within water, such as heavy metals, dyes, and organic pollutants, in wastewater. Once the nanoparticles adsorb the contaminants, there is an external magnetic field that aids in elutriation of the contaminants from the water, thus making it easy to sort out contaminants quickly and thoroughly. This application does not only make the removal process of contaminants easier but allows the reuse of the nanoparticles.

Magnetic iron oxide nanoparticles, like Fe_3O_4, have exclusive characteristics including superparamagnetic properties, high surface-to-volume ratio, and ease of functionalization, which render them highly suitable for applications that are related to environmental cleanup. The ability to modify the surface of Fe_3O_4 with ligands or polymers in order to enhance binding specificity toward targeted contaminants, which include heavy metals, dyes, and organic pollutants, may enhance its utility. Nanoparticles are functionalized by incorporating various organic or inorganic coatings. The stability of such aqueous media and the selective ability of adsorption toward contaminants increases with the adjustment of the coating used for the functionalization of nanoparticles. Example: Fe_3O_4 nanoparticles, with a variety of biocompatible polymers such as polyethylene glycol or chitosan, can be coated or chemically modified using ligands that contain thiols, amines, or carboxyl groups, for interaction with pollutants (Vohl et al. 2024).

Some of the most common toxic substances which MNPs (Shukla et al. 2021) can effectively be applied to as a treatment for wastewater include heavy metals, selectively bound by MNPs, including but not limited to arsenic (As), chromium (Cr), lead (Pb), mercury (Hg), and cadmium (Cd). The binding is usually due to electrostatic interactions, complexation, or covalent bonding between metal ions and functional groups on the surface of the nanoparticle (Ojemaye et al. 2017). Many synthetic dyes are employed in various industries such as textile and paper mill industries. Such dyes, in the category of persistent environmental pollutants, can be adsorbed easily from wastewater through functionalized MNPs and separated magnetically (Darwesh et al. 2019). Organic contaminants like pesticides, pharmaceuticals, and industrial chemicals are highly damaging pollutants that can be effectively adsorbed by functionalized MNPs (Wang et al. 2018).

Water contaminated with arsenic was treated using Fe_3O_4 nanoparticles functionalized with citrate ions. The nanoparticles selectively adsorbed As(III) and As(V) species present in contaminated water. Application of the magnetic field was used to recover the arsenic contaminates along with the nanoparticles after the adsorption. High efficiency of adsorption was achieved, reducing the concentration of arsenic to the limits of WHO. The nanoparticles could be regenerated and reused many times without significant loss of efficiency. This work demonstrated that functionalized magnetic nanoparticles are a cost-effective and effective arsenic removal material for sources of drinking water contaminated by arsenic in areas with this contamination, such as Bangladesh and western Bengal in India (Bhardwaj et al. 2019).

Industrial effluent containing hexavalent chromium (Cr(VI)) was treated using Fe_3O_4 nanoparticles covered by chitosan. Chromium is a known carcinogen, and therefore, removal is critical for human health as well as the environment. Chitosan coating improved the selective binding ability of nanoparticle onto Cr(VI). Separation of nanoparticles from water was done with the help of a magnetic field. High removal efficiency more than 90% was accomplished with short contact time. Nanoparticles in the recycling were relatively easy to recover and reused many times. This method appears to be a fast, efficient, and environmentally friendly means for tackling chromium-contaminated effluents typically found within leather tanning, metal plating, and textile industries (Bahador et al. 2021).

2.12.7 *Disinfection*

Silver nanoparticles (AgNPs) are a potent antimicrobial material used in the disinfection of pathogens like bacteria, viruses, and fungi in wastewater (Bhardwaj et al. 2021). Their broad-spectrum antimicrobial activity allows them to kill or inhibit microorganisms, making them valuable for treating effluents from industries where hygiene and sterilization are critical, such as food processing and healthcare. AgNPs can be synthesized using various methods, including chemical reduction, green synthesis, and biological synthesis. Once synthesized, AgNPs can be incorporated into water treatment systems, membranes, or filters for disinfecting wastewater (Zahoor et al. 2021).

Silver nanoparticles have proven effective in removing a wide range of microbial contaminants from wastewater, including bacteria (e.g., *Escherichia coli, Salmonella, Staphylococcus aureus*), viruses (e.g., HIV, Influenza), and fungi (e.g., *Candida albicans*) (Verma and Maheshwari 2019). Their high antimicrobial efficacy, combined with their ability to remain stable in water, makes them particularly useful for treating effluents from industries that generate pathogenic wastewater.

In the food processing industry, silver nanoparticles have been successfully used for disinfection in this sector. A poultry processing facility in the United States incorporated silver nanoparticles into its wastewater treatment process to address microbial contamination issues. The treated water met the microbiological safety standards required for discharge or reuse, and the system proved highly efficient, with over 95% pathogen reduction (Tufail and Liaqat 2022).

Healthcare facilities, including hospitals, generate large volumes of wastewater contaminated with pathogens such as bacteria, viruses, and fungi. Proper treatment is crucial to prevent the spread of infections and resistant organisms into the environment. In South Korea, a major hospital faced challenges with the high microbial load in its wastewater, which included Staphylococcus aureus and Pseudomonas aeruginosa, both known for causing healthcare-associated infections (HAIs). The hospital implemented silver nanoparticle-based filtration to treat its wastewater before discharge (Selvarajan et al. 2021).

Municipal wastewater contains a wide variety of pathogens, including viruses that can cause diseases such as gastroenteritis, hepatitis, and respiratory infections. Traditional wastewater treatment methods often fall short in eliminating viruses, especially smaller, more resistant ones. In Brazil, researchers incorporated silver nanoparticles into the tertiary treatment stage to improve viral disinfection in a municipal wastewater treatment plant (Rafique et al. 2019). The silver nanoparticles interacted with viral particles, destabilizing their protein coats and inhibiting their ability to infect host cells. The viral load in the treated water was reduced by more than 90%, enhancing the overall safety of the treated wastewater and reducing the risk of viral contamination in the receiving water bodies.

2.12.8 *Advanced Oxidation Processes (AOPs)*

Advanced oxidation processes, AOPs, represent just one of the several innovative methods that have been found useful for wastewater treatment. In these, they are created to degrade and remove a broad variety of organic pollutants through the generation of highly reactive hydroxyl radicals-°OH. This radical can attack any complex organic compound and break it down into simpler, non-toxic materials. Thus, AOPs prove quite effective in dealing with the resistant or recalcitrant pollutants (Bermúdez et al. 2021).

Nanomaterials have a significant role in the enhancement of the efficiency of AOPs. The nanoscale dimension and the unique physicochemical properties of these nanomaterials make them have an effective role in the catalysis of reactive oxygen species, specifically hydroxyl radicals. Nanomaterials have a high ratio of surface volume with more active sites for chemical activity and electronic tunability with relevance to specific applications (Pandis et al. 2022).

Nanomaterial-based AOPs are advanced and sustainable methods to handle different water pollutants in wastewater. The ROS produced by nanomaterials, such as TiO_2, ZnO, and Fe_2O_3, enhances these processes and eventually brings on strong degradation of pharmaceuticals, dyes, and persistent organic pollutants. Versatility and scalability along with fewer environmental impacts make these processes viable approaches to wastewater treatment in industrial and municipality scopes (Dong et al. 2022).

The most frequent nanomaterials for AOPs are the following:

2.12.8.1 TiO_2 Nanoparticles

TiO_2 is one of the most widely investigated nanomaterials in AOPs. It has strong photocatalytic properties under UV light, making it highly effective in producing hydroxyl radicals for pollutant degradation. Moreover, the high stability and non-toxicity of TiO_2 and the availability of this material have made it highly popular in the field of wastewater treatment.

Pharmaceutical compounds, including antibiotics and hormones (Mahmoud et al. 2017), have become a major concern in recent times because they demonstrate bioactivity and are persistent in the environment. Residuals of pharmaceutical compounds in wastewater pose serious environmental and health hazards. TiO_2 nanoparticles have been applied in degrading such compounds by their illumination with UV light, which releases hydroxyl radicals capable of decomposing them into less harmful products. In the current studies, it has been proved that TiO_2 is very efficient for the degradation of a wide range of pharmaceuticals and it is most promising for the hospital and pharmaceutical industries for the treatment of wastewater.

2.12.8.2 Zinc Oxide (ZnO) Nanoparticles

Practically similar to TiO_2 photocatalytic activity but with a bandgap, slightly wider, so that its sensitivity extends to both UV and visible light. Versatility and low cost make it a substance amenable to large-scale applications, especially on degrading industrial pollution such as dyes.

Dyes manufactured in the process of the textile and industries are considered major pollutants of the effluent of wastewater which frequently produces very highly colored and toxic effluents. ZnO nanoparticles activated by UV or visible light have been used for the degradation of the dye molecules in wastewater (Aftab et al. 2022). Their catalytic properties allow for the breakdown of large dye molecules, resulting in colorless and non-harmful products. ZnO nanoparticles work efficiently in the degradation of synthetic dyes, specifically methylene blue and rhodamine.

2.12.8.3 Fe_2O_3 Nanoparticles

Fe_2O_3 nanoparticles are widely used in the removal of POPs from wastewater. They can be catalytically activated via several methods, for example, by visible light or Fenton-like reaction processes wherein the iron catalyzes the generation of hydroxyl radicals from hydrogen peroxide. Their propensity to facilitate the degradation of hard organic compounds makes it a significant advantage in treating polluted waters in the areas of heavy industrial activities.

Fe_2O_3 nanoparticles have also been utilized to treat persistent organic pollutants (POPs)—in short, harmful substances such as PCBs and PAHs (Boulkhessaim et al. 2022). Compounds are stable to decomposition in conventional treatment techniques. AOPs, apart from Fe_2O_3 nanoparticles ability to catalyze the creation of hydroxyl radicals in Fenton or photo-Fenton reactions, can degrade POPs into less dangerous substances. In these studies, pollutant concentrations had been brought to environmentally safe levels.

2.12.9 *Bio-nanotechnology Approaches*

Nanoparticles hybridization with microorganisms or enzymes is the latest technology developed for wastewater treatment (Saravanan et al. 2022). This hybrid system is integrated with the high reactivity and catalytic properties of nanomaterials along with selective and efficient degradation capabilities of biological entities. The significant improvement in micro- or enzymatic processes by nanoparticles can lead to the effective breakdown of pollutants, more so organic contaminants that resist traditional biological treatments.

Nanoparticle–microorganism and nanoparticle–enzyme hybrid systems stand as an emerging approach for the treatment of wastewater (Darwesh et al. 2019), based on the catalytic strength of nanomaterials combined with the biological efficiency of microorganisms and enzymes. Therefore, improvements in organic pollutant degradation, bioreactor performance, and sustainable and cost-effective industrial wastewater treatment approaches have been reported. The hybrid systems in real life and examples indicate application in oil refineries, textiles, as well as treatment of municipal wastewater and, hence, likely to have a higher chance of being implemented in larger numbers in the future (Al-Maqdi et al. 2021) (Fig. 2.3).

(a) **Nanoparticles Enhance Microbial and Enzyme Activity**: Nanoparticles can perform many functions when combined within microbe-based or enzyme-based systems.
(b) **Enhancing Bioavailability of Pollutants**: Nanoparticles can adsorb pollutants on their surfaces, thereby focusing these pollutants within the vicinity of microorganisms or enzymes, thus increasing chances of degradation (Chauhan et al. 2022). This is especially beneficial with hydrophobic organic compounds, which are generally not available to microorganisms in water environments.

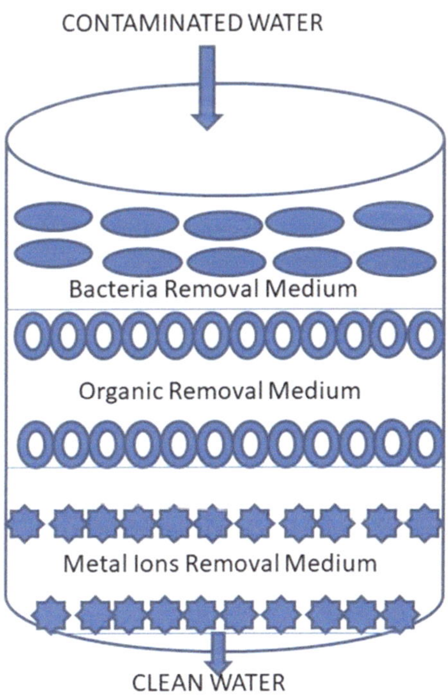

Fig. 2.3 A nanomaterial bed reactor

(c) **Electron transfer improvement**: Ample nanoparticles, such as iron oxides and carbon-based nanomaterials, facilitate the bridge of electron transfer between the microbes and pollutants. This enhances the microbial metabolic activity while there is an increase in rates of reduction and oxidation of the pollutant (Agarwal et al. 2019).
(d) **Enzyme stabilization**: The immobilization of enzymes on nanoparticles stabilizes and protects them from degradation through harsh WW environments. Multiple reuse of the biocatalysts can be achieved thus elevating the sustainability of the process (Khan et al. 2021).

Some nanoparticles possess antimicrobial properties: for instance, colloidal silver or copper (Ahmad et al. 2017). Such nanomaterials can distinguish between toxic pathogenic microorganisms responsible for pollution and beneficial microbial communities that break down organic pollutants. Hybrid systems, which combine nanoparticles with microorganisms or enzymes, appear promising for the removal of large classes of wastewater contaminants, especially organic pollutants such as hydrocarbons, phenols, dyes, and pharmaceutical residues. The different systems can be applied in various configurations that include biofilters, bioreactors, or membrane bioreactors (Naha et al. 2023).

2.12.9.1 Nanoparticle–Microorganism Systems

In these systems, the function of nanoparticles increases the accessibility of organic pollutants to microorganisms or their activity. For instance, magnetic iron oxide (Fe_3O_4) nanoparticles have been added to the anaerobic bioreactors that are used in the removal of wastewater containing various kinds of organic pollutants. This increases the rate of electron transfer involved in the microbial oxidation of organic pollutants available in the water. Thus, the efficiency of the overall bioreactor system improves. It has been proven, through experiments, that this method can increase COD and other pollutant removal rates by 30% (Hassanpourmoghadam et al. 2023).

Microorganisms such as Pseudomonas species are added to petroleum industry wastewater, which contains oil spills and hydrocarbons (Varjani 2017; Amran et al. 2022). The action of these microbes increases the metabolic activities of the water, therefore increasing the rate of electron transfer and hydrocarbon breakdown. This mechanism has been put in place in the treatment of oil refinery wastewater to enhance the remediation process of oily sludge.

2.12.9.2 Nanoparticle–Enzyme Systems

Such immobilization of enzymes onto nanomaterials would, in these systems clearly, enhance their stability and make their reuse possible for the degradation of pollutants in wastewater.

Laccase is another enzyme that has been used as a catalyst for decomposing recalcitrant organic pollutants. It has been attached to TiO_2 nanoparticles for the treatment of dye-contaminated wastewater (Li et al. 2017). It enhances the surface area of nanoparticles; therefore, it enhances enzyme loading along with interaction between laccase and dye molecules. This hybrid system was very effective for the degradation of azo dyes that are not degraded by traditional treatments. This technology has been used in real-world textile wastewater treatment plants with removal rates up to 90%.

Lipase is an enzyme that catalyzes the reaction for the degradation of fats and oils. Immobilization of lipase on ZnO nanoparticles provides a means for use in applications such as wastewater treatment processes containing high levels of oil and grease in food process industries. The new system of immobilized lipase is stable and reusable multiple times, hence a cost-effective system. On-site deployment has indicated substantial enhancement in oil and grease removal efficiency over the traditional method (Filho et al. 2019).

Magnetic iron oxide nanoparticles were used in an anaerobic bioreactor at a municipal wastewater treatment plant in China in this study, aimed at treating wastewater. Fe_3O_4 nanoparticles were added to enhance the anaerobic digestion of organic matter. This upgraded DIETing process had organisms break down complex organic pollutants into biogas that consisted of methane and carbon dioxide. The yields showed 25% enhanced biogas and COD-removing efficiency as compared to the control reactor, hence increasing efficiencies and making the process more economical (Zhang et al. 2020).

A bioreactor system was employed in the oil refinery from the Middle East by integrating the gold nanoparticles with the hydrocarbon-degrading bacteria, Pseudomonas (Kharangate-Lad and D'souza 2021). The system was aimed at degrading the petroleum hydrocarbons in the wastewater of the refinery. With the aid of gold nanoparticles, electron transfer within the bacterial cells was facilitated, thus making them work efficiently for degradation. Consequently, the bioreactor showed the removal of more than 80% of the TPH within a fraction of time compared with that required for conventional bioreactors.

TiO_2 nanoparticles, coupled with immobilized laccase, utilized for degradation of toxic azo dyes in textile wastewater in a textile treatment plant located in India (Sreedharan and Bhaskara Rao 2019). The conventional biological systems and treatment in the plant were not readily accepting due to the use of effluents loaded with dyes to cause challenges. This means that with the nanoparticle–enzyme hybrid system, the study was able to degrade over 90% of the azo dyes in the wastewater, and in doing so improved the overall quality of the water by reducing environmental hazards associated with the process of dyeing used for cloth manufacturing. Water treated in this manner was reused in the textile manufacturing process to enhance sustainability.

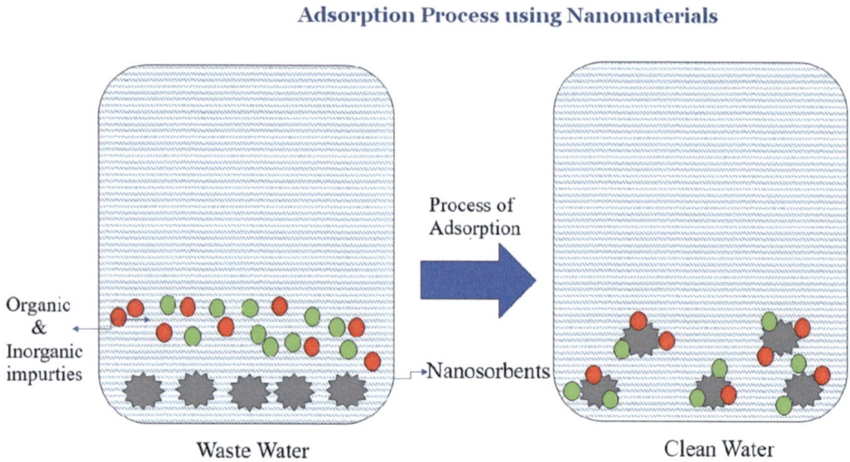

Fig. 2.4 Enhanced adsorption with nanomaterials

2.12.10 *Nanoparticle-Assisted Coagulation and Flocculation*

The application of nanoparticles can intensify both the coagulation and flocculation processes. Both processes are associated with the removal of suspended materials, colloidal particles, and organic matter from wastewater. For coagulation, nanoparticles neutralize charges on the dissolved particles or colloids in wastewater, thus allowing them to clump together (Jabbar et al. 2022). With flocculation, these aggregates grow into bigger masses or "flocs" that can be easily removed from water either by settling or by filtration (Fig. 2.4).

2.12.10.1 Improved Coagulation and Flocculation

(a) Metal Oxide Nanoparticles: Al_2O_3, Fe_2O_3, and Fe_3O_4 are the most utilized since they assist in the aggregation of suspended particles, which enhances the degree of water clarity (Mahdi 2019). The nanoparticles perform the role of bridging agents. It increases the sticking between smaller particles and larger aggregates that enable settling and removal.
(b) Clay Nanoparticles: Montmorillonite and kaolinite clay nanoparticles are highly effective because their layered structure with high ion exchange capacity affords a mechanism for the adsorption and fixation of contaminants. Their small size with a wide surface area allows them to trap effectively suspended solids and organic matter (Yang et al. 2024).
(c) Graphene and Graphene Oxide: It has been found that the electrical and surface properties of graphene, together with graphene oxide, are quite high and great materials for adsorbing organic compounds and other contaminants. They

enhance the process of coagulation through their effect of forming bridges between particles and enhance settling processes of contaminants from water (Velusamy et al. 2021).

2.13 Real-World Applications

The following examples illustrate successful applications of nanoparticles in industrial wastewater treatment processes.

(a) ***Paper Mill Wastewater***: Researches showed that metal oxide nanoparticles are so effective for suspending matters removal, organic matters like lignin, and other impurities present in paper mill wastes. For instance, coagulation study was conducted at an industrial paper mill using iron oxide nanoparticles to improve the process significantly reducing COD, turbidity, and total suspended solids. The nanoparticles increased sludge settling, which minimized the volume of sludge. The treatment process was significantly improved. Solute removal rate increased to 90% and organic contaminants were fully eliminated with subsequent improvement in the quality of effluent (Mahmoudabadi et al. 2022).

(b) ***Textile Industry Wastewater***: Graphene oxide and iron oxide nanoparticles have been used for the treatment of wastewater within the textile mills for the removal of dyes mainly azo dyes that are resistant to degradation by traditional treatments. Iron oxide nanoparticles along with graphene oxide treated dye-laden effluent from a textile manufacturing unit in one real-world application. The respective nanoparticles used as an adsorbent helped in enhancing the coagulation process and also breaking complex dye molecules for maximum efficiency of removal. Thus, as much as 95% of the dyes could be removed; COD and TDS witnessed considerable reduction. This rendered the treated wastewater in compliance with discharge regulations (da Silva et al. 2021).

(c) ***Food Processing Plants***: Effluents from food processing plants are high in organics, fats, oils, and suspended solids. In industrial-scale treatment, the clay nanoparticles, for instance montmorillonite, were dosed into the wastewater management system of a food processing industry aiming to enhance coagulation. The nanoparticles also facilitated adsorption of fats and oils along with organic matter and suspended particles which improved the flocculation process. Significant reduction of FOG content in the plant with improved removal efficiency of suspended solids and COD up to 85% that eased the subsequent treatment operations and provided better scope for recycling of treated water (Naseer et al. 2024).

(d) ***Oil Refinery Wastewater***: Iron oxide and titanium dioxide nanoparticles have been applied for oil refinery wastewater treatment, in particular for effluents with hydrocarbons and oily sludge. The most impressive industrial application is the preparation of Fe_3O_4 nanoparticles complexed with chemical coagulants to enhance further better removals of oil, suspended solids, and heavy metals from refinery wastewater. In addition to working as adsorption enhancers, nanoparticles served as a catalyst in the hydrocarbon breakdown while enhancing the process of coagulation. Oil separation improved while hydrocarbon degradation and heavy metal removal exceeded 90% oil content reduction with good water quality for reuse or safer disposal (Nawaz et al. 2022).

(e) ***Pharma Industry Effluent***: This application is particularly challenging as it contains the presence of persistent organic compounds, active pharmaceutical ingredients (APIs), and solvents. In this goal, iron oxide NPs were added to the system with enhanced organic pollutant and API removal (Malhas et al. 2023). These nanoadsorbents, through coagulation and adsorption mechanisms, helped break complex organic material structures in wastewater. Reduction of pharmaceutical-related impurities to the level of 80%, reduction of COD, and content of toxic organic matter make effluent safe for natural water bodies.

(f) ***Dairy Industry Wastewater***: Another challenge in dairy wastewater is its high organic content, fats, and proteins. It is illustrated by an example of the use of bentonite nanoparticles for the treatment of effluents from a dairy processing plant. Bentonite clay nanoparticles improve coagulation and adsorption of organic matter, fats, and proteins to enhance flocculation efficiency. The treatment achieved in excess of 80% COD, fats, and suspended solids removal leaving the effluent less difficult to handle at succeeding biological treatment processes (Lingamurthy et al. 2023) (Fig. 2.5; Table 2.3).

Nanomaterials are revolutionizing industrial wastewater treatment due to their specific characteristic properties, for example, a high surface area and reactivity. These materials could target various pollutants that include organic compounds, heavy metals, and pathogens. Enhanced efficiency drives the ability of improving coagulation, flocculation, adsorption, and catalytic processes to lead to faster pollutant removal and degradation. Nanomaterials are used in advanced water treatment techniques, like photocatalysis, membrane filtration, and disinfection. Such practices and applications are considered to be greener and more sustainable by using fewer chemicals compared to those used in the textile industry, oil refining, pharmaceuticals, food processing, and others. Such enhancement results in cleaner effluent and a more effective process for wastewater management.

2.14 Conclusion

This study demonstrates the potential of nanomaterials in industrial wastewater treatment from all of these different industries, including paper mills, textiles, food processing, oil refineries, pharmaceuticals, and dairy processing. All these sectors generate wastewater with specific and complex contaminants, which could be organic matter, dyes, suspended solids, oils, hydrocarbons, and pharmaceutical residues. Conventional treatment procedures often cannot reach the levels of efficiency required to be able to achieve environmental discharge requirements.

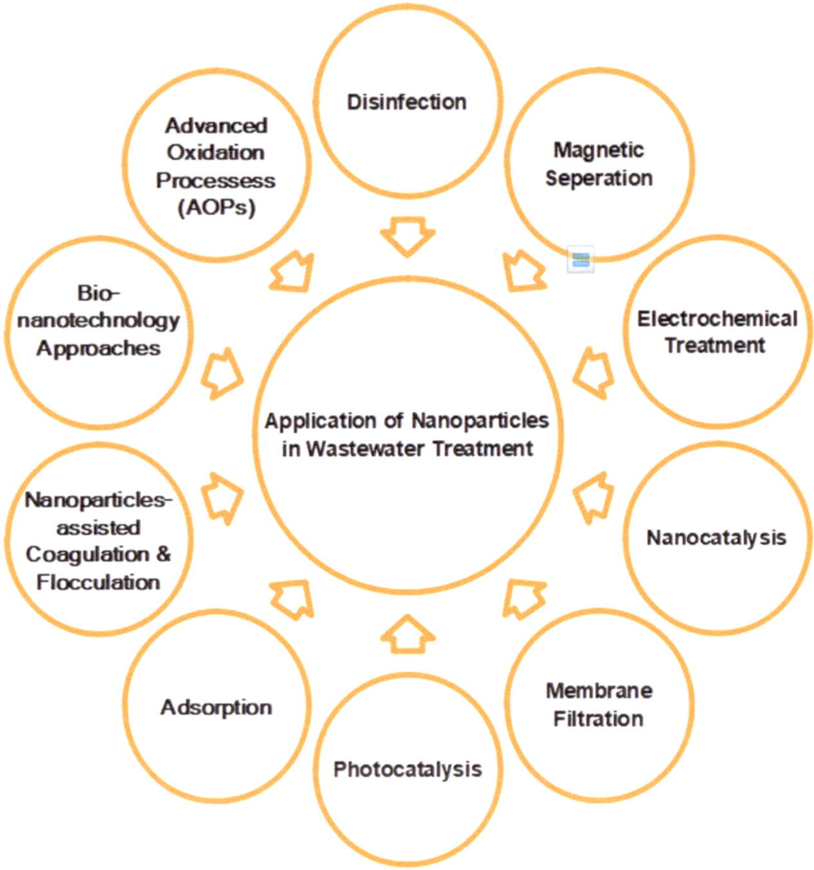

Fig. 2.5 Application of nanoparticles in wastewater treatment

However, with nanomaterial incorporation, performance has been enhanced greatly since it focuses on targeting pollutants more specifically.

For example, in paper mill treatments, iron oxide nanoparticles caused a huge improvement in coagulation processes. These improvements gave a COD and other impurities reduction by 90%. The amount of sludge produced was also reduced greatly. Graphene oxide and iron oxide nanoparticles had significant impact in the treatment of wastewater from textile industry involving hard-to-degrade azo dyes. Apart from facilitating coagulation, these nanoparticles fragmented large molecules of dyes and achieved a removal efficiency up to 95%. This demonstrated that nanomaterials could face the challenge posed by persistent pollutants, which may not respond well to traditional treatments.

The adsorption and flocculation ability of montmorillonite nanoparticles showed excellent potential for reducing FOG as well as suspended solids in food processing plants. Hence, the improvement allowed an 85% reduction in COD and FOG content

Table 2.3 Industries leveraging nanomaterials to enhance the treatment of specific contaminants in their wastewater

Industry	Nanomaterials used	Key contaminants targeted	Key benefits
Paper mills	Aluminum oxide (Al_2O_3), Iron oxides (Fe_3O_4)	Suspended solids, lignin, organic matter	Improved removal of suspended solids, enhanced sludge settling
Textile industries	Graphene oxide, Iron oxide (Fe_2O_3), TiO_2	Dyes (especially azo dyes), Organic pollutants, TSS	High dye removal efficiency, enhanced treatment of persistent dyes
Food processing plants	Clay nanoparticles (montmorillonite), Aluminum oxide	Fats, oils, organic matter, suspended solids	Effective removal of FOG and organic matter, reduced COD
Oil refineries	Iron oxide (Fe_3O_4), Titanium dioxide (TiO_2)	Hydrocarbons, oils, suspended solids, heavy metals	Accelerated hydrocarbon degradation, oil separation
Pharmaceutical industry	Iron oxide (Fe_3O_4), Aluminum oxide	Pharmaceuticals, organic compounds, active ingredients	Reduction of pharmaceutical contaminants and toxic organic matter
Dairy industry	Clay nanoparticles (bentonite), Aluminum oxide	Organic matter, fats, proteins, suspended solids	Enhanced removal of organic pollutants, improved flocculation
Chemical manufacturing	Zinc oxide (ZnO), Iron oxide (Fe_3O_4)	Heavy metals, chemical residues, organic pollutants	Efficient removal of toxic heavy metals and chemical contaminants
Mining industry	Iron oxide (Fe_3O_4), Graphene oxide	Heavy metals, sulfates, suspended solids	Removal of heavy metals, improved sedimentation of solids
Leather tanning industry	Zinc oxide (ZnO), Silver nanoparticles	Chromium, dyes, organic pollutants	Effective chromium and dye removal, reduced toxic load

that eased handling during subsequent stages of treatment and encouraged reusing the treated water. Oil refinery wastewater, which always contains hydrocarbons and oily sludge, also proved to be improved by iron oxide and titanium dioxide nanoparticles. These nanomaterials catalyzed the hydrocarbon breakdown and significantly improved the removal of suspended solids and heavy metals to over 90% in oil content.

Similarly, the iron oxide nanoparticles were used to upgrade the treatment of pharmaceutical effluents, which are notorious for their complex organic contaminants and APIs. These nanomaterials have enabled an 80% degradation of APIs and other toxic compounds in addition to lowering organic pollutant load to safer levels.

Bentonite clay nanomaterial has proven promise in wastewater, especially where the sources in the dairy industry come, with their very high amounts of organic content, fat content, and proteins content to reduce significantly COD content more than 80% accompanied by suspended solids.

The high surface area and reactivity of nanomaterials, along with their ability to act as adsorbents and catalysts, have revolutionized industrial wastewater treatment. In advanced treatment techniques such as adsorption, catalytic degradation, photocatalysis, and membrane filtration, these materials not only enhance the rate of pollutant removal but also minimize chemical usage and environmental footprint associated with the treatment processes. Such characteristics make nanomaterials major contributors toward more sustainable and efficient practices for wastewater management.

The literature further establishes scalability for various real-world industrial implementations, where the economic effects of diminished chemical usage, sludge formation, and operational costs make nanomaterials an economically efficient solution. Environmental effects are also pertinent because cleaner effluents and higher recycling possibilities dovetail with wider imperatives for sustainability.

From all this, it can be deduced that nanomaterials are an efficient and sustainable tool for dealing with the various pollutants present in industrial wastewater. Their better removal efficiency and the opportunity of wider application in most industries make them the powerful players in the future control of industrial wastewater. The results point to the significance of nanomaterials integration into conventional wastewater treatment solutions: the great implications in gaining regulatory compliance, water reuse, and environmental impact associated with industrial processes.

References

Aazam ES, Akhtar N, El-Said WA (2023) In-situ fabrication of bio-inspired nanocatalysts for enhanced oxygen reduction reaction in alkaline media. Mater Chem Phys 293:126884

Abhishek KJ, Reddy S, Acharya S, Lakshmi B, Deepak K, Naveen CS et al (2022) A review on nanomaterial-based electrodes for the electrochemical detection of chloramphenicol and furazolidone antibiotics. Anal Methods 14(34):3228–3249

Adeola AO, Nomngongo PN (2022) Advanced polymeric nanocomposites for water treatment applications: a holistic perspective. Polymers (Basel) 14:2462

Aftab S, Shabir T, Shah A, Nisar J, Shah I, Muhammad H et al (2022) Highly efficient visible light active doped ZnO photocatalysts for the treatment of wastewater contaminated with dyes and pathogens of emerging concern. Nanomaterials (Basel) 12(3):486

Agarwal P, Gupta R, Agarwal N (2019) Advances in synthesis and applications of microalgal nanoparticles for wastewater treatment. J Nanotechnol 2019(1):7392713

Agrawal S, Ranjan R, Lal B, Rahman A, Singh SP, Selvaratnam T, Nawaz T (2021a) Synthesis and water treatment applications of nanofibers by electrospinning. Processes 9(10):1779

Agrawal TA, Bogale FM, Aragaw BA (2021b) Iron-based nanoparticles in wastewater treatment: a review on synthesis methods, applications, and removal mechanisms. J Saudi Chem Soc 25(8):101280

Ahmad IZ, Ahmad A, Tabassum H, Kuddus M (2017) Applications of nanoparticles in the treatment of wastewater. In: Handbook of ecomaterials. Springer, Cham, pp 1–25

Al-Amrani WA, Onaizi SA (2024) Adsorptive removal of heavy metals from wastewater using emerging nanostructured materials: a state-of-the-art review. Sep Purif Technol 11:127018

Alazaiza MY, Albahnasawi A, Copty NK, Bashir MJ, Nassani DE, Al Maskari T et al (2022) Nanoscale zero-valent iron application for the treatment of soil, wastewater and groundwater contaminated with heavy metals: a review. Desalination Water Treat 253:194–210

Alhalili Z (2023) Metal oxides nanoparticles: general structural description, chemical, physical, and biological synthesis methods, role in pesticides and heavy metal removal through wastewater treatment. Molecules 28(7):3086

Ali I, Peng C, Naz I, Amjed MA (2019) Water purification using magnetic nanomaterials: an overview. In: Magnetic nanostructures: environmental and agricultural applications, pp 161–179

Al-Maqdi KA, Elmerhi N, Athamneh K, Bilal M, Alzamly A, Ashraf SS et al (2021) Challenges and recent advances in enzyme-mediated wastewater remediation—a review. Nanomaterials (Basel) 11(11):3124

Amran RH, Jamal MT, Pugazhendi A, Al-Harbi M, Ghandourah M, Al-Otaibi A et al (2022) Biodegradation and bioremediation of petroleum hydrocarbons in marine ecosystems by microorganisms: a review. Nat Environ Pollut Technol 21(3)

Anegbe B, Ifijen IH, Maliki M, Uwidia IE, Aigbodion AI (2024) Graphene oxide synthesis and applications in emerging contaminant removal: a comprehensive review. Environ Sci Eur 36(1):15

Ani IJ, Egbosiuba TC (2023) Metal oxides and their nanocomposites in wastewater treatment. In: Adsorption through advanced nanoscale materials. Elsevier, pp 185–203

Anucha CB, Altin I, Bacaksiz E, Stathopoulos VN (2022) Titanium dioxide (TiO_2)-based photocatalyst materials activity enhancement for contaminants of emerging concern (CECs) degradation: In the light of modification strategies. Chem Eng J Adv 10:100262

Anzar N, Hasan R, Tyagi M, Yadav N, Narang J (2020) Carbon nanotube—a review on synthesis, properties and plethora of applications in the field of biomedical science. Sens Int. 1:100003

Bahador F, Foroutan R, Esmaeili H, Ramavandi B (2021) Enhancement of the chromium removal behavior of Moringa oleifera activated carbon by chitosan and iron oxide nanoparticles from water. Carbohyd Polym 251:117085

Bekele T, Mebratie G, Girma A, Alamnie G (2024) Characterization and fabrication of p-Cu_2O/n-CeO_2 nanocomposite for the application of photocatalysis. Colloids Surf A Physicochem Eng Asp 685:133271

Bermúdez LA, Pascual JM, Martínez MDMM, Poyatos Capilla JM (2021) Effectiveness of advanced oxidation processes in wastewater treatment: state of the art. Water (Basel) 13(15):2094

Bhardwaj A, Rajput R, Misra K (2019) Status of arsenic remediation in India. In: Advances in water purification techniques. Elsevier, pp 219–258

Bhardwaj AK, Sundaram S, Yadav KK, Srivastav AL (2021) An overview of silver nanoparticles as promising materials for water disinfection. Environ Technol Innov 23:101721

Boulkhessaim S, Gacem A, Khan SH, Amari A, Yadav VK, Harharah HN et al (2022) Emerging trends in the remediation of persistent organic pollutants using nanomaterials and related processes: a review. Nanomaterials (Basel) 12(13):2148

Cai H, Zhang L, Wei J, Hou Y, Wei Y, Zhou S et al (2024) Ultra-efficient and selective adsorption of cationic dyes by Ti-doped SiO_2 functionalized hydrophilic Fe_3O_4 nanoparticles with superior structural stability. J Water Process Eng 57:104729

Cerro-Lopez M, Cisneros JJ, Méndez-Rojas MA, Castro-Pastrana LI (2020) Nanotechnologies for removal of nonsteroidal anti-inflammatory drug from wastewater. In: Non-steroidal anti-inflammatory drugs in water: emerging contaminants and ecological impact, pp 277–301

Chauhan G, González-González RB, Iqbal HM (2022) Bioremediation and decontamination potentials of metallic nanoparticles loaded nanohybrid matrices–a review. Environ Res 204:112407

Chen D, Cheng Y, Zhou N, Chen P, Wang Y, Li K et al (2020) Photocatalytic degradation of organic pollutants using TiO_2-based photocatalysts: a review. J Clean Prod:121725. https://doi.org/10.1016/j.jclepro.2020.121725
Chen Y, Wang D, Zhu X, Zheng X, Feng L (2012) Long-term effects of copper nanoparticles on wastewater biological nutrient removal and N_2O generation in the activated sludge process. Environ Sci Technol 46(22):12452–12458
Chu TD, Quach DT, Nguyen XT, Nguyen TS, Pham DT, Kim DH (2020) Synthesis and properties of magnetic-semiconductor Fe_3O_4/TiO_2 heterostructure nanocomposites for applications in wastewater treatment. J Magn 25(1):1–7
Cui J, Li F, Wang Y, Zhang Q, Ma W, Huang C (2020) Electrospun nanofiber membranes for wastewater treatment applications. Sep Purif Technol 1(250):117116
da Silva MP, de Souza ZSB, Cavalcanti JVFL, Fraga TJM, da Motta Sobrinho MA, Ghislandi MG (2021) Adsorptive and photocatalytic activity of Fe_3O_4-functionalized multilayer graphene oxide in the treatment of industrial textile wastewater. Environ Sci Pollut Res Int 28(19):23684–23698
Darwesh OM, Matter IA, Eida MF (2019) Development of peroxidase enzyme immobilized magnetic nanoparticles for bioremediation of textile wastewater dye. J Environ Chem Eng 7(1):102805
de Magalhães LF, da Silva GR, Peres AE (2022) Zeolite application in wastewater treatment. Adsorpt Sci Technol 2022:4544104
Devi MK, Yaashikaa PR, Kumar PS, Manikandan S, Oviyapriya M, Varshika V, Rangasamy G (2023) Recent advances in carbon-based nanomaterials for the treatment of toxic inorganic pollutants in wastewater. New J Chem 47(16):7655–7667
Dhull P, Saini N, Aamir M, Parveen S, Husain S (2024) Function of nanomaterials in the treatment of emerging pollutants in wastewater. In: Detection and treatment of emerging contaminants in wastewater, pp 93–112
Dong G, Chen B, Liu B, Hounjet LJ, Cao Y, Stoyanov SR et al (2022) Advanced oxidation processes in microreactors for water and wastewater treatment: development, challenges, and opportunities. Water Res 211:118047
El-Aswar EI, Ramadan H, Elkik H, Taha AG (2022) A comprehensive review on preparation, functionalization and recent applications of nanofiber membranes in wastewater treatment. J Environ Manage 301:113908
Filho DG, Silva AG, Guidini CZ (2019) Lipases: sources, immobilization methods, and industrial applications. Appl Microbiol Biotechnol 103:7399–7423
Ganguly K, Dutta SD, Patel DK, Lim KT (2021) Silver nanoparticles for wastewater treatment. In: Aquananotechnology. Elsevier, pp 385–401
Garg A, Lai WC, Chopra H, Agrawal R, Singh T, Chaudhary R, Dubey BN (2024) Nanosponge: a promising and intriguing strategy in medical and pharmaceutical Science. Heliyon
Ge J, Zhang Z, Ouyang Z, Shang M, Liu P, Li H, Guo X (2022) Photocatalytic degradation of (micro) plastics using TiO_2-based and other catalysts: properties, influencing factor, and mechanism. Environ Res 209:112729
Gupta S, Narayan H, Jain RK (2023) A review of some metal-oxide based nanocomposites for photocatalytic treatment of wastewater. Adv Nat Sci: Nanosci Nanotechnol 14(4):043003
Han F, Kambala VS, Srinivasan M, Rajarathnam D, Naidu R (2009) Tailored titanium dioxide photocatalysts for the degradation of organic dyes in wastewater treatment: a review. Appl Catal A 359(1–2):25–40
Hasan MM, Islam MD, Mondal B (2024). Nanomaterials for the removal of pollutants from pharmaceutical wastewater. In: Development in wastewater treatment research and processes. Elsevier, pp 171–193
Hassanpourmoghadam L, Goharrizi BA, Torabian A, Bouteh E, Rittmann BE (2023) Effect of Fe_3O_4 nanoparticles on anaerobic digestion of municipal wastewater sludge. Biomass Bioenerg 169:106692

Hirota M, Sugita Y, Ishijima M, Ikeda T, Saruta J, Maeda H, Ogawa T (2021) UV photocatalytic activity of titanium dioxide (TiO_2) surface contaminated with bacterial biofilm: implications for photo-restoration of osteoconductivity. Mater Today Adv 12:100182

Hong C, Meng X, He J, Fan K, Yan X (2022) Nanozyme: a promising tool from clinical diagnosis and environmental monitoring to wastewater treatment. Particuology 71:90–107

Hsu CY, Rheima AM, Mohammed MS, Kadhim MM, Mohammed SH, Abbas FH et al (2023) Application of carbon nanotubes and graphene-based nanoadsorbents in water treatment. BioNanoScience 13(4):1418–1436

Huang Y, Chen D, Hu X, Qian Y, Li D (2018) Preparation of TiO_2/carbon nanotubes/reduced graphene oxide composites with enhanced photocatalytic activity for the degradation of rhodamine B. Nanomaterials (Basel) 8(6):431

Irfan M, Shah A, Iftikhar FJ, Hayat MN, Ashiq MN, Shah I (2022) Electrochemical sensing platform based on functionalized multi-walled carbon nanotubes and metal oxide for the detection and degradation studies of orange II dye. ACS Omega 7(36):32302–32312

Jabbar KQ, Barzinjy AA, Hamad SM (2022) Iron oxide nanoparticles: preparation methods, functions, adsorption and coagulation/flocculation in wastewater treatment. Environ Nanotechnol Monit Manag 17:100661

Jain K, Patel AS, Pardhi VP, Flora SJ (2021) Nanotechnology in wastewater management: a new paradigm towards wastewater treatment. Molecules 26(6):1797

Kausar A (2024) Fullerene in water remediation nanocomposite membranes—cutting edge advancements. Charact Appl Nanomater 7(2):4945

Khan S, Naushad M, Al-Gheethi A, Iqbal J (2021) Engineered nanoparticles for removal of pollutants from wastewater: current status and future prospects of nanotechnology for remediation strategies. J Environ Chem Eng 9(5):106160

Kharangate-Lad A, D'souza NC (2021) Current approaches in bioremediation of toxic contaminants by application of microbial cells; biosurfactants and bioemulsifiers of microbial origin. In: Rhizobiont in bioremediation of hazardous waste, pp 217–263

Khraisheh M, Elhenawy S, AlMomani F, Al-Ghouti M, Hassan MK, Hameed BH (2021) Recent progress on nanomaterial-based membranes for water treatment. Membranes. Available from: https://doi.org/10.3390/membranes11120995

Kriklavova L, Lederer T (2011) A review study of nanofiber technology for wastewater treatment. In: 3rd International conference Nanocon, Czech Republic, Thomson Reuters web of knowledge, pp 106–112

Krstić V (2021) Role of zeolite adsorbent in water treatment. In: Handbook of nanomaterials for wastewater treatment, pp 417–481

Kumar S, Basumatary IB, Sudhani HP, Bajpai VK, Chen L, Shukla S, Mukherjee A (2021) Plant extract mediated silver nanoparticles and their applications as antimicrobials and in sustainable food packaging: a state-of-the-art review. Trends Food Sci Technol 112:651–666

Kumar RR, Kumar KU, Haranath D (2022) Synthesis, characterization, and applications of ZnO–TiO_2 nanocomposites. In: Nanoscale compound semiconductors and their optoelectronics applications, pp 271–314

Kumari S, Chowdhry J, Kumar M, Garg MC (2024) Zeolites in wastewater treatment: a comprehensive review on scientometric analysis, adsorption mechanisms, and future prospects. Environ Res:119782

Kumari S, Kaur M, Dhania G (2023) Carbon nanomaterials for wastewater treatment. In: Modern Nanotechnology: volume 1: Environmental sustainability and remediation. Springer International Publishing, Cham, pp 255–277

Lazarenko NS, Golovakhin VV, Shestakov AA, Lapekin NI, Bannov AG (2022) Recent advances on membranes for water purification based on carbon nanomaterials. Membranes 12(10):915

Li S, Wang W, Liu Y, Zhang WX (2014) Zero-valent iron nanoparticles (nZVI) for the treatment of smelting wastewater: a pilot-scale demonstration. Chem Eng J 254:115–123

Li G, Nandgaonkar AG, Wang Q, Zhang J, Krause WE, Wei Q et al (2017) Laccase-immobilized bacterial cellulose/TiO_2 functionalized composite membranes: evaluation for photo-and bio-catalytic dye degradation. J Membr Sci 525:89–98
Li S, Li J, Wang W, Zhang WX (2019) Recovery of gold from wastewater using nanoscale zero-valent iron. Environ Sci Nano 6(2):519–527
Li S, Li L, Zhang W (2024) Nanoscale zero-valent iron (nZVI) for heavy metal wastewater treatment: a perspective. Engineering 36:16–20
Liaquat I, Munir R, Abbasi NA, Sadia B, Muneer A, Younas F et al (2024) Exploring zeolite-based composites in adsorption and photocatalysis for toxic wastewater treatment: preparation, mechanisms, and future perspectives. Environ Pollut:123922
Lingamurthy DB, Hiregoudar S, Nidoni U, Ganachari SV, Patil VB, Hiremath V (2023) Adsorptive removal of organic pollutants from milk-processing industry effluents through chitosan-titanium dioxide nanoadsorbent-coated sand. Environ Sci Pollut Res 30(9):24101–24119
Mahdi N (2019) Synthesis and applications of discrete TiO_2 nanotubes in water treatment and superhydrophobic coatings
Mahmoud WM, Rastogi T, Kümmerer K (2017) Application of titanium dioxide nanoparticles as a photocatalyst for the removal of micropollutants such as pharmaceuticals from water. Curr Opin Green Sustain Chem 6:1–10
Mahmoudabadi ZS, Rashidi A, Maklavany DM (2022) Optimizing treatment of alcohol vinasse using a combination of advanced oxidation with porous α-Fe_2O_3 nanoparticles and coagulation-flocculation. Ecotoxicol Environ Saf 234:113354
Malhas R, El Achkar JH, Misbah B, Al RS (2023) Optimizing oil removal from oil-water emulsions using novel iron oxide magnetic nanoparticles. Water Air Soil Pollut 234(9):564
Mbarek WB, Escoda L, Saurina J, Pineda E, Alminderej FM, Khitouni M, Suñol J (2022) Nanomaterials as a sustainable choice for treating wastewater: a review. Materials 15(23):8576
Mei A, Xu Z, Wang X, Liu Y, Chen J, Fan J, Shi Q (2022) Photocatalytic materials modified with carbon quantum dots for the degradation of organic pollutants under visible light: a review. Environ Res 214:114160
Mohan TS, Soundharya R, Mathew BB, Anantharaju KS (2024) Photocatalytic nanomaterials and their application in tannery wastewater treatment. In: Development in wastewater treatment research and processes. Elsevier, pp 147–163
Motitswe MG, Badmus KO, Khotseng L (2024) Application of reduced graphene oxide-zinc oxide nanocomposite in the removal of Pb(II) and Cd(II) contaminated wastewater. Appl Nano 5(3):162–189
Naha A, Antony S, Nath S, Sharma D, Mishra A, Biju DT et al (2023) A hypothetical model of multi-layered cost-effective wastewater treatment plant integrating microbial fuel cell and nanofiltration technology: a comprehensive review on wastewater treatment and sustainable remediation. Environ Pollut 323:121274
Nain A, Sangili A, Hu SR, Chen CH, Chen YL, Chang HT (2022) Recent progress in nanomaterial-functionalized membranes for removal of pollutants. iScience:104616
Naseem T, Durrani T (2021) The role of some important metal oxide nanoparticles for wastewater and antibacterial applications: a review. Environ Chem Ecotoxicol 3:59–75
Naseem T, Waseem M (2022) A comprehensive review on the role of some important nanocomposites for antimicrobial and wastewater applications. Int J Environ Sci Technol 19(3):2221–2246
Naseer A, Younas F, Munir R, Sardar MF, Noreen S, Cui Z (2024) Clay-based nanocomposite materials used in treatment of wastewater: recent advancements, cost investigation and future perspectives. Sep Purif Rev:1–18
Nawaz R, Sahrin NT, Haider S, Ullah H, Junaid M, Akhtar MS, Khan S (2022) Photocatalytic performance of black titanium dioxide for phenolic compounds removal from oil refinery wastewater: nanoparticles vs nanowires. Appl Nanosci 12(11):3499–3515
Nezhad SA, Jaleh B, Darabi E, Dorranian D (2024) Laser-assisted preparation of TiO_2/carbon/Ag nanocomposite for degradation of organic pollutants. Materials 17(16):4118

Nidheesh PV, Zhou M, Oturan MA (2018) An overview on the removal of synthetic dyes from water by electrochemical advanced oxidation processes. Chemosphere 197:210–227
Ningthoujam R, Singh YD, Babu PJ, Tirkey A, Pradhan S, Sarma M (2022) Nanocatalyst in remediating environmental pollutants. Chem Phys Impact 4:100064
Noah NM (2023) Current status and advancement of nanomaterials within polymeric membranes for water purification. ACS Appl Nano Mater
Nyabadza A, Makhesana M, Plouze A, Kumar A, Ramirez I, Krishnamurthy S, Vazquez M, Brabazon D (2024) Advanced nanomaterials and dendrimers in water treatment and the recycling of nanomaterials: a review. J Environ Chem Eng:112643
Obaiah G, Kemparajegowda N, Gireesha J, Mylarappa M (2023) Comparative study of TiO_2 and palladium doped TiO_2 nanocatalysts for water purification under solar and ultraviolet irradiation. Chem Inorg Mater 1:100002
Ojemaye MO, Okoh OO, Okoh AI (2017) Surface modified magnetic nanoparticles as efficient adsorbents for heavy metal removal from wastewater: progress and prospects. Mater Express 7(6):439–456
Pan J, Zou L, Liao Z, Lin Z, Chen J (2023) Study on the properties of carbon nanotube (CNTs) reinforced AlSi10Mg composites fabricated by powder metallurgy. Materials 16(11):3905
Pandis PK, Kalogirou C, Kanellou E, Vaitsis C, Savvidou MG, Sourkouni G, Zorpas AA, Argirusis C (2022) Key points of advanced oxidation processes (AOPs) for wastewater, organic pollutants and pharmaceutical waste treatment: a mini review. ChemEngineering 6(1):8
Peng J, He Y, Zhou C, Su S, Lai B (2021) The carbon nanotubes-based materials and their applications for organic pollutant removal: a critical review. Chinese Chem Lett 32(5):1626–1636
Priya A, Sagadevan S (2024) Chalcogenides-based nanomaterials for contaminant removal in wastewater treatment. In: Structural and chemical features of chalcogenides. IntechOpen
Qasem NA, Mohammed RH, Lawal DU (2021) Removal of heavy metal ions from wastewater: a comprehensive and critical review. NPJ Clean Water 4(1):1–5
Rafique M, Tahir MB, Sadaf I (2019) Nanotechnology: an innovative way for wastewater treatment and purification. In: Advanced research in nanosciences for water technology, pp 95–131
Rahman RO, El-Kamash AM, Hung YT (2022) Applications of nano-zeolite in wastewater treatment: an overview. Water 14(2):137
Rana K, Kaur H, Singh N, Sithole T, Siwal SS (2024) Graphene-based materials: unravelling its impact in wastewater treatment for sustainable environments. Next Mater 3:100107
Rani U, Ng LY, Ng CY, Mahmoudi E (2020) A review of carbon quantum dots and their applications in wastewater treatment. Adv Colloid Interface Sci 278:102124
Rathore C, Yadav VK, Gacem A, AbdelRahim SK, Verma RK, Chundawat RS, Gnanamoorthy G, Yadav KK, Choudhary N, Sahoo DK, Patel A (2023) Microbial synthesis of titanium dioxide nanoparticles and their importance in wastewater treatment and antimicrobial activities: a review. Front Microbiol 14:1270245
Roy A, Elzaki A, Tirth V, Kajoak S, Osman H, Algahtani A, Islam S, Faizo NL, Khandaker MU, Islam MN, Emran TB (2021) Biological synthesis of nanocatalysts and their applications. Catalysts 11(12):1494
Sahu A, Dosi R, Kwiatkowski C, Schmal S, Poler JC (2023) Advanced polymeric nanocomposite membranes for water and wastewater treatment: a comprehensive review. Polymers 15(3):540
Saravanan A, Kumar PS, Hemavathy RV, Jeevanantham S, Jawahar MJ, Neshaanthini JP, Saravanan R (2022) A review on synthesis methods and recent applications of nanomaterial in wastewater treatment: challenges and future perspectives. Chemosphere 307:135713
Selvarajan R, Sibanda T, Pandian J, Mearns K (2021) Taxonomic and functional distribution of bacterial communities in domestic and hospital wastewater systems: implications for public and environmental health. Antibiotics 10(9):1059
Shadi AMH, Kamaruddin MA, Niza NM, Emmanuel MI, Hossain MS, Ismail N (2020) Efficient treatment of raw leachate using magnetic ore iron oxide nanoparticles Fe_2O_3 as nano adsorbents. J Water Process Eng. 38:101637

Shivaji K, Sridharan K, Kirubakaran DD, Velusamy J, Emadian SS, Krishnamurthy S, Devadoss A, Nagarajan S, Das S, Pitchaimuthu S (2023) Biofunctionalized CdS quantum dots: a case study on nanomaterial toxicity in the photocatalytic wastewater treatment process. ACS Omega 8(22):19413–19424

Shukla S, Khan R, Daverey A (2021) Synthesis and characterization of magnetic nanoparticles, and their applications in wastewater treatment: a review. Environ Technol Innov 24:101924

Sreedharan V, Bhaskara Rao KV (2019) Biodegradation of textile azo dyes. In: Nanoscience and biotechnology for environmental applications, pp 115–139

Sun B, Li Q, Zheng M, Su G, Lin S, Wu M, Li C, Wang Q, Tao Y, Dai L, Qin Y (2020) Recent advances in the removal of persistent organic pollutants (POPs) using multifunctional materials: a review. Environ Pollut 265:114908

Thakur A, Kumar A, Singh A (2024) Adsorptive removal of heavy metals, dyes, and pharmaceuticals: carbon-based nanomaterials in focus. Carbon 217:118621

Torrinha Á, Oliveira TM, Ribeiro FW, Correia AN, Lima-Neto P, Morais S (2020) Application of nanostructured carbon-based electrochemical (bio)sensors for screening of emerging pharmaceutical pollutants in waters and aquatic species: a review. Nanomaterials 10(7):1268

Tufail MS, Liaqat I (2022) Silver nanoparticles and their applications: a comprehensive review. Pure Appl Biol 11(1):315–330

Tufail MK, Ifrahim M, Rashid M, Haq IU, Asghar R, Uthappa UT, Selvaraj M, Kurkuri M (2024) Chemistry of zeolites and zeolite-based composite membranes as a cutting-edge candidate for removal of organic dyes & heavy metal ions: progress and future directions. Sep Purif Technol:128739

Vaiano V, Sannino D, Sacco O (2020) The use of nanocatalysts (and nanoparticles) for water and wastewater treatment by means of advanced oxidation processes. In: Elsevier eBooks, pp 241–264. https://doi.org/10.1016/b978-0-12-819941-1.00009-2

Vallinayagam S, Rajendran K, Lakkaboyana SK, Soontarapa K, Remya RR, Sharma VK, Kumar V, Venkateswarlu K, Koduru JR (2021) Recent developments in magnetic nanoparticles and nano-composites for wastewater treatment. J Environ Chem Eng 9(6):106553

Varjani SJ (2017) Microbial degradation of petroleum hydrocarbons. Bioresour Technol 223:277–286

Velusamy S, Roy A, Sundaram S, Kumar MT (2021) A review on heavy metal ions and containing dyes removal through graphene oxide-based adsorption strategies for textile wastewater treatment. Chem Rec 21(7):1570–1610

Venkateshaiah A, Silvestri D, Wacławek S, Ramakrishnan RK, Krawczyk K, Saravanan P, Pawlyta M, Padil VV, Černík M, Dionysiou DD (2022) A comparative study of the degradation efficiency of chlorinated organic compounds by bimetallic zero-valent iron nanoparticles. Environ Sci Water Res Technol 8(1):162–172

Verma P, Maheshwari SK (2019) Applications of silver nanoparticles in diverse sectors. Int J Nano Dimension 10(1):18–36

Vohl S, Kristl M, Stergar J (2024) Harnessing magnetic nanoparticles for the effective removal of micro- and nanoplastics: a critical review. Nanomaterials 14(14):1179

Wang C, Cao M, Wang P, Ao Y, Hou J, Qian J (2014) Preparation of graphene–carbon nanotube–TiO_2 composites with enhanced photocatalytic activity for the removal of dye and Cr (VI). Appl Catal A Gen 473:83–89

Wang T, Ai S, Zhou Y, Luo Z, Dai C, Yang Y, Zhang J, Huang H, Luo S, Luo L (2018) Adsorption of agricultural wastewater contaminated with antibiotics, pesticides and toxic metals by functionalized magnetic nanoparticles. J Environ Chem Eng 6(5):6468–6478

Wen Y, Feng M, Zhang P, Zhou HC, Sharma VK, Ma X (2021) Metal organic frameworks (MOFs) as photocatalysts for the degradation of agricultural pollutants in water. ACS ES&T Eng 1(5):804–826

Xu C, Qu X (2014) Cerium oxide nanoparticle: a remarkably versatile rare earth nanomaterial for biological applications. NPG Asia Mater 6(3)

Yang X, Zhou Y, Hu J, Zheng Q, Zhao Y, Lv G, Liao L (2024) Clay minerals and clay-based materials for heavy metals pollution control. Sci Total Environ:176193
Yao S, Yuan X, Jiang L, Xiong T, Zhang J (2020) Recent progress on fullerene-based materials: synthesis, properties, modifications, and photocatalytic applications. Materials 13(13):2924
Yaqoob AA, Parveen T, Umar K, Mohamad Ibrahim MN (2020) Role of nanomaterials in the treatment of wastewater: a review. Water 12(2):495
Zafar AM, Javed MA, Hassan AA, Mohamed MM (2021) Groundwater remediation using zero-valent iron nanoparticles (nZVI). Groundwater Sustain Dev 15:100694
Zahoor M, Nazir N, Iftikhar M, Naz S, Zekker I, Burlakovs J, Uddin F, Kamran AW, Kallistova A, Pimenov N, Ali KF (2021) A review on silver nanoparticles: classification, various methods of synthesis, and their potential roles in biomedical applications and water treatment. Water 13(16):2216
Zamel D, Khan AU, Waris A, Ebrahim A, Abd El-Sattar NEA (2023) Nanomaterials advancements for enhanced contaminant removal in wastewater treatment: nanoparticles, nanofibers, and metal-organic frameworks (MOFs). Results Chem. https://doi.org/10.1016/j.rechem.2023.101092
Zhang XF, Liu ZG, Shen W, Gurunathan S (2016) Silver nanoparticles: synthesis, characterization, properties, applications, and therapeutic approaches. Int J Mol Sci 17(9):1534
Zhang Z, Guo L, Wang Y, Zhao Y, She Z, Gao M, Guo Y (2020) Application of iron oxide (Fe_3O_4) nanoparticles during the two-stage anaerobic digestion with waste sludge: impact on the biogas production and the substrate metabolism. Renew Energy 146:2724–2735
Zhang Y, Wang F, Wang Y (2021) Recent developments of electrospun nanofibrous materials as novel adsorbents for water treatment. Mater Today Commun 27:102272

Chapter 3
Synthesis and Characterization of Nanomaterials

İlayda Bersu Kul, Gamze Tüzün, and Burak Tüzün

Abstract Nanotechnology is one of the most innovative and fast-growing fields of science and engineering today. Playing an important role in this field, nanomaterials offer unique physical, chemical, electrical, and mechanical properties for various applications. These properties of nanomaterials arise from their atomic or molecular structures, whose dimensions are on the nanometer scale. In recent years, great advances have been made in the synthesis and characterization of nanomaterials. These developments open new horizons in both basic research and industrial applications. It has been seen that many studies have been carried out on the synthesis and characterization of nanomaterials. Synthesis refers to the process of creating nanomaterials suitable for a specific purpose. Different synthesis methods allow controlling the size, shapes, compositions, and properties of materials. Various synthesis techniques, such as top-down and bottom-up approaches, allow the production of a wide range of nanomaterials. Characterization covers the methods used to study the properties and structure of synthesized nanomaterials. Various characterization methods such as microscopic, spectroscopic, and diffraction techniques help us determine the surface structure, chemical composition, crystal structure, and other important properties of nanomaterials. This information is critical for understanding and optimizing the functionality of nanomaterials.

Keywords Synthesis · Characterization · Nanomaterials · Industrial applications

İ. B. Kul
Department of Molecular Biology and Genetics, Science Faculty, Sivas Cumhuriyet University, Sivas, Turkey

G. Tüzün
Department of Chemistry, Faculty of Science, Sivas Cumhuriyet University, Sivas, Turkey

B. Tüzün (✉)
Plant and Animal Production Department, Technical Sciences Vocational School of Sivas, Sivas Cumhuriyet University, Sivas, Turkey
e-mail: theburaktuzun@yahoo.com

A. Mandpe et al. (eds.), *Nanotech Solutions for Industrial Wastewater*, Advances in Wastewater Research, https://doi.org/10.1007/978-981-96-6879-3_3

3.1 Definition and Importance of Nanomaterials

Nanoparticles are ultra-small particles at the nanoscale, usually in the size range of 1 to 100 nm (nm), where matter exhibits unique properties. This size range represents an area where quantum effects prevail and the optical, electrical, and magnetic properties of materials differ from the macroscopic scale. Magnetic nanoparticles are special nanoparticles at this nanoscale that react to external magnetic fields and exhibit magnetism properties (Khan and Hossain 2022).

Magnetic nanoparticles have a wide range of applications, from medicine to materials science, from electronics to environmental technologies. They are used in medicine, especially in targeted drug delivery systems in cancer treatment and in diagnostic methods such as magnetic resonance imaging (MRI). They play an important role in materials science, the development of high-density data storage systems, and various sensor applications. They are also used in environmental treatment technologies to remove toxic substances and in water purification systems (Tang and Lo 2013). There are several distinct kinds of metallic nanoparticles with extensive uses across various industries, such as alloys, magnetic, gold, and silver in Fig. 3.1.

The reason why magnetic nanoparticles are used in so many fields is their scalable production, their susceptibility to surface modifications, and their ability to be easily controlled by external magnetic fields. These features make them multifunctional and flexible so they can be customized for specific applications.

Nanoparticles have a wide range of uses, including biological labeling (Bruchez et al. 1998), medical treatments (Davis et al. 2008), use in various industry sectors (Stark et al. 2015), use in common products like clothing or cosmetics (El-Rafie et al. 2010), optical devices, sensor technology, and electronics (Shipway et al. 2000), and

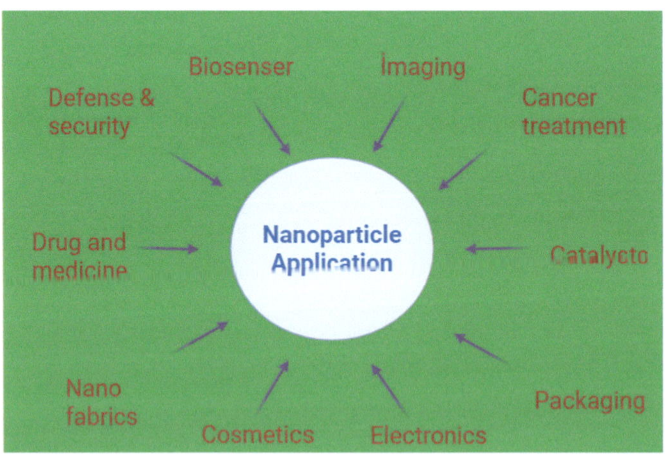

Fig. 3.1 Applications of nanoparticles

treatment of certain cancers (Shinkai et al. 1999). The unique qualities of nanoparticles, such as their ability to tolerate extreme heat and oxidation and their antimicrobial activity, have drawn interest recently (Hasan 2015).

To confirm that the manufactured particles are at the nanoscale, characterization is also crucial. The word "characterization" in material science refers to the broad, basic procedures used to investigate a material's structure and qualities. Understanding the subject matter from a scientific perspective requires this basic procedure. Characterization encompasses methods needed to investigate the microscopic structures and properties of materials, i.e., any material analysis procedure including density calculations, thermal analysis, and mechanical testing (Beek et al. 2004). The centuries-old characterization techniques are being joined by ever-evolving, more sophisticated methods. Characterization lets us ascertain the structure and content of materials as well as the success or failure of the approach. While some methods are quantitative, others are qualitative (Baxter and Aydil 2005).

This chapter will concentrate on commonly applied and efficient methods for characterizing nanoparticles in Fig. 3.2.

The pharmaceutical industry is starting to recognize the benefits of nanomedicine, as seen by the recent regulatory approval of numerous treatments based on nanoparticles (Law et al. 2005). More are in preclinical or clinical development stages of the pharmaceutical pipeline. Preclinical characterization has proven to be a rate-limiting stage in the commercialization process for a number of these nanoparticle treatments. A concerted interdisciplinary approach is frequently necessary for the

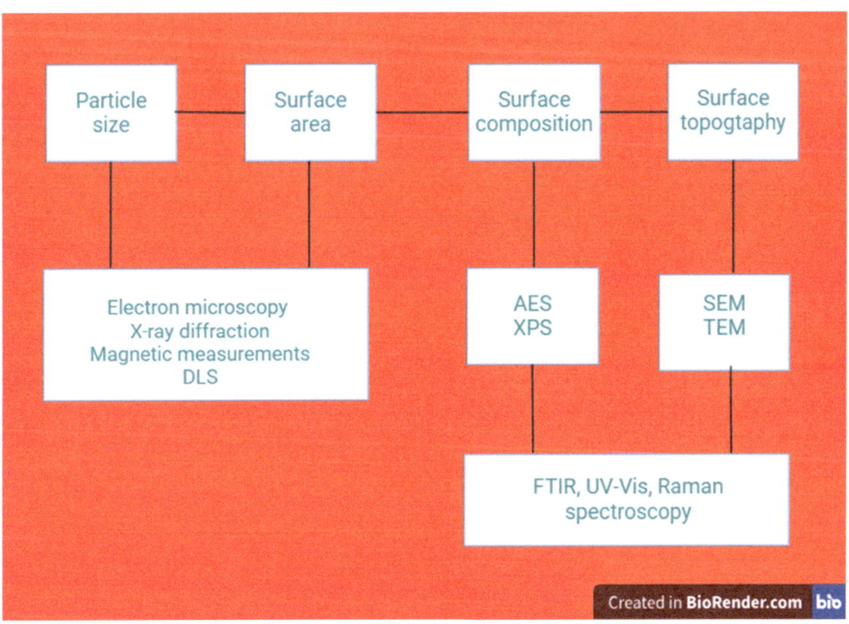

Fig. 3.2 Common characterization methods for nanoparticles

characterization of nanoparticles because they represent a broad range of materials, such as liposomes, dendrimers, nanocrystals, metal colloids, and fullerenes, and they are intended for an equally broad spectrum of clinical purposes, such as those of medical devices, vaccine formulations, carriers of diagnostic/imaging agents, or therapeutics to treat malignant, inflammatory, viral, neurodegenerative, and other disorders (Nakayama et al. 2008).

Drug developers are forced to design and validate novel methods for preclinical assessment of nanoparticle safety and toxicity because there aren't enough standard assays and methods for nanoparticle characterization. Regulatory agencies are then required to evaluate data generated from unfamiliar techniques without a history of substantial supporting scientific literature.

Nanoparticle treatments present a multitude of characterization issues, even in cases where known approaches are relevant. Measuring the molecular weight and spectral properties of soluble small-molecule medications is often sufficient for identification; if the molecule is sufficiently pure, the drug is probably functioning. But because multifunctional nanoparticles require more components to cooperate in order to function, they need to be characterized much more thoroughly.

A multipart entity, like a nanoparticle, can be meaningfully physicochemically characterized by evaluating each component separately as well as the stoichiometery, connections, and chemical stability of those connections (Austin and Klimpel 1964; Hall et al. 2007). Assessing physicochemical properties, pyrogenicity and sterility, biodistribution (absorption, distribution, metabolism, and excretion; ADME), and toxicity characterization—which encompasses both in vitro testing and in vivo animal studies—are all important steps in thoroughly characterizing a therapeutic based on nanoparticles in Fig. 3.3. A reasonable characterization cascade's levels must all be customized to be pertinent to nanomedicine. Here, we focus on a select few characterization approaches that are particularly effective for nanoparticles or that are particularly suggestive of the toxicity or efficacy of particles, rather than attempting to examine all characterization techniques applicable to nanoparticles.

These techniques for physicochemical characterization include localization in tissue by chemical composition detection using energy dispersive x-ray (EDX) spectroscopy and size characterization by transmission-electron microscopy (TEM) and dynamic light scattering (DLS).

Nanoparticles are unlike conventional pharmaceuticals in that they are multicomponent multifunctional systems. Testing for pyrogenicity and sterility as well as in vitro testing for hemolysis, thrombogenicity, complement activation, and phagocytosis-mediated immune system interactions are also covered. Lastly, we review multicomponent diagnostic labeling investigations performed in vivo. Every technique that is covered is assessed for its suitability for use in nanomedicine as well as its advantages and disadvantages.

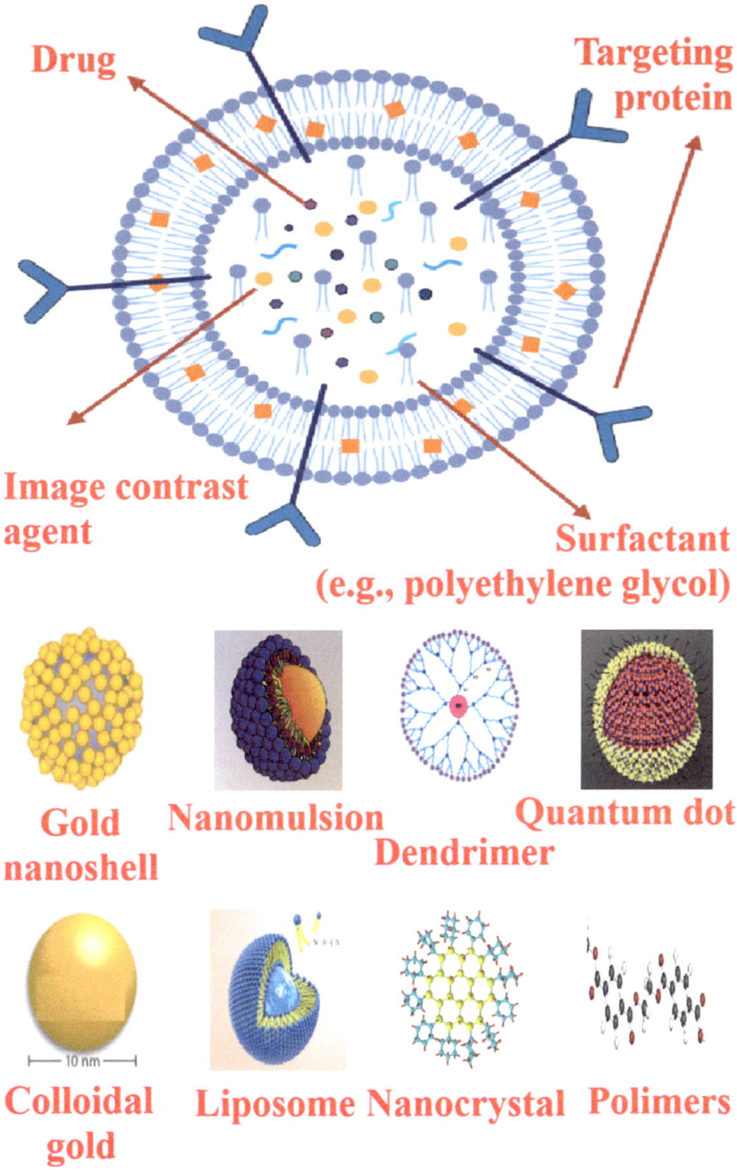

Fig. 3.3 Nanomaterials for drug delivery

3.2 Synthesis of Nanomaterials

3.2.1 *Top-Down Methods*

3.2.1.1 Mechanical Grinding

Mechanical grinding is a process using mechanical force to reduce the size of solids. It is a method generally used in industrial production processes. The grinding process occurs by crushing, shredding, or grinding the material until it reaches the desired size. This process is usually done using equipment such as grinding machines, mills, or crushers. Mechanical grinding is an effective method to reduce the size of the material without changing its physical properties. This method is generally used for pulverization or fine grinding (Bao et al. 2016).

3.2.1.2 Lithography

Lithography of magnetic nanoparticles is a technique generally used in the production of micro- and nanoscale structures. This method enables the transfer of a specific pattern onto a substrate, and this pattern is then processed through various processes, allowing the creation of magnetic nanoparticles (Suh et al. 2012). Lithography is widely used, especially in the semiconductor industry and in the field of nanotechnology, and also plays an important role for the production of magnetic nanoparticles.

The production of magnetic nanoparticles by lithographic methods allows precise control of the size, shape, and magnetic properties of these particles. This control optimizes the use of particles in specific areas such as biomedical applications (Mack 2007). For example, magnetic nanoparticles used as contrast agents for magnetic resonance imaging (MRI) must have a certain size and magnetic susceptibility.

The lithography process generally involves transferring the pattern through a mask onto a photoresist-coated substrate, using a light source (e.g., ultraviolet light). Next, the light-exposed parts of the photoresist are dissolved by the development process, and then the unwanted parts of the substrate are removed by the etching process. The remaining photoresist pattern serves as a template for depositing magnetic material onto the substrate. Finally, the photoresist is removed and the desired pattern of magnetic nanoparticles is obtained (Biju 2014).

This process provides a high level of precision and repeatability in the production of magnetic nanoparticles, making them ideal for a variety of technological applications. Lithography is a critical step in determining the properties of magnetic nanoparticles and is an important factor in the success of future applications of these particles.

3.2.1.3 Chemical Distribution

Chemical dispersion is one of the top-down approaches used in nanomaterial synthesis and aims to obtain nanostructured materials by separating a large piece of material into small components. In this method, the material is generally broken down by chemical reactions or chemical solutions. For example, a metal or semiconductor material is broken down into micro- and nanoscale components using chemical solvents or corrosive chemicals (Liu and Cohen 2014). This technique is used to create complex nanostructures, especially when combined with processes such as lithography, metal deposition, and chemical etching. For example, in the semiconductor industry, photolithography and chemical dispersion techniques are widely used to fabricate thin-film transistors and other nanoscale electronic components (Trotta and Mele 2019).

Advantages of chemical dispersion include the ability to produce complex and fine patterns, high precision, and the ability to process a wide range of materials. However, these methods often produce waste and involve working with chemical reagents, requiring environmental and health care. Therefore, appropriate safety precautions should be taken when using chemical distribution techniques and caution should be exercised in waste management (Chen and He 2001).

3.3 Bottom-Up Methods

3.3.1 Sol–Gel Method

The sol–gel method is gaining more popularity and industrial use compared to other synthesis methods due to its ability to produce high-quality nanoparticles on an industrial scale and its wide range of applications. This technique stands out with its ability to produce alloy nanoparticles in a single step by combining multiple metal or metal oxide precursors in specific proportions. In addition to being suitable for industrial scale, the sol–gel process also provides an advantage in that it can carry out the production process at lower temperatures compared to other techniques that can synthesize alloy products in a single step, such as plasma and electrochemical methods (Singh et al. 2014).

The sol–gel method allows producing highly homogeneous composite materials with high purity of 99.99% on an industrial scale. Additionally, the sol–gel method generally operates at lower temperatures ranging from 70 to 320 °C; this is a significant advantage considering that other methods operate in temperature ranges of 1400–3600 °C (Niederberger 2007). The sol–gel process works through sol-to-gel conversion, and this conversion occurs by changing the pH or adjusting the concentration of the solution. The compression process leads to the formation of the gel, an infinite three-dimensional network (Alagiri et al. 2012).

Another advantage of the sol–gel process is that nano-sized porosity can be achieved in the material. This enables the production of products with a very large surface area, which is of great benefit in industrial applications such as hydrogen storage or catalyst loading. Drying the gel provides a solid and hard structure due to volumetric shrinkage, and by controlling the drying conditions you can manipulate the porosity of the product. Various methods can be used to achieve this porosity, such as dip coating or spin coating. Thin films made by the sol–gel method have wide applications in the electronics and chemical industries (Zorkipli et al. 2016).

The sol–gel process can also be used to produce composite and nanocomposite materials. For this purpose, secondary materials are added to the porous structure. This addition is done through techniques such as chemical reactions or purification. When nanoporous materials are used in catalytic applications, catalytic efficiency increases and production costs decrease thanks to the large specific surface area. Tempering processes are another technique used to produce dense parts, but increasing the temperature can cause the grains to grow and form a coarse-grained structure (Bokov et al. 2021).

The sol–gel method is a versatile synthesis method used in various applications, from thin film production to nanocomposites, from membrane production to catalytic industries. The advantages of the method, such as high purity, large specific surface area, and ability to operate at low temperatures, make it popular in industrial and research fields. Different stages of the sol–gel process are shown in Fig. 3.4 (Nakaso et al. 2003).

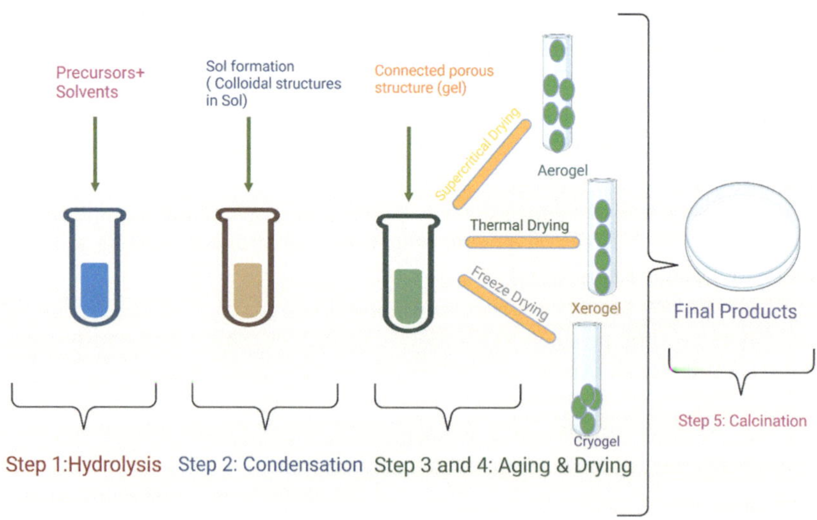

Fig. 3.4 Schematic of the different stages of the sol–gel process

3.3.2 Chemical Vapor Deposition (CVD)

Chemical vapor deposition (CVD) is the process of forming thin films or coatings by reacting or precipitating chemical components in the gas phase on the surface. This technique is used in various fields, from semiconductors to optical materials, from advanced ceramics to surface coatings.

This procedure is thought to have started when Von Guerkie created a sulfur ball for amusement that causes friction to produce sparks. Sparks flew from their rubbing against one another, forming hydrogen sulfate. Henry managed to ignite a spark in 1798 and successfully deposit carbon in hydrocarbon gas.

Using the recently developed Volta batteries, Sir Humpry Davy initiated the first chemical vapor coating experiments in 1802. The CVD technique was first developed in the 1890s in an effort to fortify bulb filaments. Aylesworht used the CVD method to form metal films in 1896 by undergoing hydrogen reduction.

While CVD was historically done in atmospherically open environments, studies started to be conducted at various pressure conditions when vacuum equipment were developed. An atmosphere comprising nitrogen or argon gas could be used to manage the CVD process and raise the density in the medium. German manufacturers began producing work tools with titanium carbide coatings using the CVD process in the 1930s. In 1948, Lander and Germer (APCVD) used the contemporary thermal atmospheric CVD technology. Theorer demonstrated the (epitaxial) coating of silicon single crystal using the CVD technique in 1961. There are two types of CVD reactors: hot wall reactors and cold wall reactors.

Only the substrate and its holder are heated in reactors with cold walls. Hot wall reactors are those in which a collection of substrates is present and the atmosphere is heated throughout. The discovery of the laser that produces local heating during CVD in 1973 led to its instant integration into all CVD systems.

LPCVD, or low pressure CVD, works at extremely low pressures (0.25 Torr), while APCVD operates at atmospheric pressure and extremely high temperatures. With the LPCVD technique, coatings are considerably better protected and have less particle pollution. In the semiconductor industry, the LPCVD coating process is utilized to create semiconductor components for the doped silicon coating process.

Very thin-film coatings are created via CVD at intermittently low pressure. One method for depositing atoms in a single layer on a surface is the intermittent CVD technique known as atomic layer coating, or ALD. The ALCVD technique enables the rapid creation of a single perspective (front) on the surface. The initial chemical vapor coating methods utilizing plasma were referred to as "luminescence discharge deposition." Later on, it was dubbed PECVD, or plasma-assisted CVD.

Von Bolton coated diamond crystals in 1911 using C_2H_2 that had been crushed in mercury vapor. In 1934, carbon-containing vapors were used to cover the walls of gas discharge tubes with inert, glassy carbonic coatings. Diamond-like carbons (DLC) were discovered by Schmellenenier in 1953 while using the plasma deposition technique to create carbon films. Different procedures (such as pyrolysis, plasma deposition, and carbon-ion beam techniques) can be used to manufacture DLC films.

Different procedures (such as pyrolysis, plasma deposition, and carbon-ion beam techniques) can be used to manufacture DLC films.

Sterling and Swann (1965) talked about the role plasmas play in CVD breakdown. The molecules are usually not totally dissociated during plasma deposition, which guarantees that the molecules' and their constituent components' spectrum orders are neutral and ionized. For instance, the Si_2H_6 arrangement is crucial in preparing the surface for absorption when employing the PECVD process to deposit from silicon to silene. On oxidized surfaces, O_3 (ozone) is more likely to be absorbed than O_2. Superfine particles can be produced by vapor phase nucleonation and plasma dissociation.

Reinberg found the radiofrequency-powered parallel-plate PECVD reactor in 1971. Lower temperatures could be used to treat CVD materials in this reactor. He was encapsulating semiconductors on glass films (phospho-silicate-glass PSG) and Si_3N_4 (nitrogen and sylene), typically at a pressure of 0.5 Torr. Optical coatings have long made use of the plasma coating method.

"Calutron" was the name of the first metal ion beam-type source developed during World War II. With magnetism, the ions generated by Calutron can be separated into isotopic masses and accelerated to up to 35 kV. Uranium-235 isotopes are separated from uranium-238 using a calutron. When negatively charged particles are thrown at vaporized uranium atoms, the evaporated atoms turn positively charged. The magnetic field then accelerates the positively charged particles, which eventually disintegrate and produce current based on their masses in the layer they have hit.

Polymerization of plasma is another CVD technique. De Wilde and Thenard made the initial discovery of this technique in 1874. Nevertheless, it wasn't put to use until the 1960s. Very thin polymer coatings were developed to prevent corrosion on food packaging after the 1960s.

The working principle of CVD is that usually gaseous precursors react chemically on a substrate under certain temperature and pressure conditions, forming a solid material (Kuo 2013).

3.3.2.1 Typical Applications of CVD

Electronics Manufacturing: CVD is widely used in the semiconductor industry to create thin-film transistors, capacitors, and other microelectronic components in integrated circuits. Materials such as silicon, silicon nitride, and silicon dioxide are produced by this process (Sundgren and Hentzell 1986).

Coating and Surface Modification: CVD is used to produce hard coatings to increase wear resistance, corrosion resistance, or heat resistance. Hard coatings such as diamond-like carbon and boron carbide are also made by CVD (Wang et al. 2003).

Optical Materials and Coatings: High-quality optical coatings, anti-reflection coatings, and other optical devices can be produced by CVD (Kudyakova et al. 2017).

Advanced Ceramics Production: Advanced ceramics and composites are produced under high temperature and pressure by CVD. Titanium nitride, aluminum nitride, and similar materials are synthesized in this way (Kamins 2012).

There are many types of CVD, such as:

Low-Pressure Chemical Vapor Deposition (LPCVD): It operates under low pressure conditions and is often used in integrated circuits (Kamins 2012).
Plasma-Assisted Chemical Vapor Deposition (PECVD): It uses plasma for the reaction of gases and makes it possible to process at lower temperatures (Silva et al. 2010).
Metal Organic Chemical Vapor Deposition (MOCVD): Materials such as semiconductors and LEDs are produced using metal organic components (Gan et al. 2020).

Advantages of CVD include its wide range of applications, the capacity to produce high-quality thin films, and the ability to process a wide variety of materials. However, CVD often requires high temperature and special equipment, and some precursors can be toxic or hazardous, so safety and environmental issues must be considered.

3.3.3 *Hydrothermal and Solvothermal Synthesis*

3.3.3.1 Hydrothermal Synthesis

Different chemical compounds and materials can be produced using the hydrothermal synthesis method. The physical and chemical processes are carried out in an aqueous solution within a closed system at temperatures above 100 °C and pressures above 1 atm.

The technique relies on water's and aqueous solutions' capacity to dilute compounds (such as certain oxides, silicates, and sulfides) that are essentially insoluble under typical circumstances; these conditions include high temperatures (500 °C) and high pressures (10–80 MPa, occasionally up to 300 MPa). The starting pH of the environment, the synthesis temperature and duration, and the system pressure are the fundamental factors of hydrothermal synthesis that control the process kinetics and the characteristics of the products generated. The procedure is done in autoclaves or sealed steel cylinders that can withstand high pressure and temperature.

A chemical synthesis technique known as hydrothermal synthesis is carried out at high pressure and temperature. The reactions happen in a solution containing water. This technique can be used to create nanoparticles and regulate their sizes (Inoue 2005).

3.3.3.2 Solvothermal Synthesis

In the solvothermal synthesis of nanoparticles, the size distribution and crystallinity of the particles can be precisely controlled. The precursor material is dissolved in a hot solvent (such as n-butyl alcohol). Reactions are carried out in an autoclave, at temperatures above the boiling point of the solvent and at available pressure. If the solvent used is water, the process is called "hydrothermal."

For example, in solvothermal processes where titanium-n-butoxide (TNB) $C_{16}H_{36}O_4Ti$ is used as a precursor and n-butyl alcohol and propanecarboxylic acid are used as solvents, different nanocrystalline TiO_2 is obtained, as seen in TEM images.

Solvothermal synthesis is a chemical synthesis technique carried out at high pressure and temperature in an organic solvent. It is possible to accurately control the size distribution and crystallinity of nanoparticles. A heated solvent, such as n-butyl alcohol, is used to dissolve the precursor material (Lai et al. 2015). Reactions are conducted in an autoclave with available pressure and at temperatures higher than the solvent's boiling point. The process is referred to as "hydrothermal" if water is the solvent (Singh et al. 2016).

3.4 Biological Methods

3.4.1 Biological Synthesis

The process of producing nanoparticles and other nanomaterials utilizing biological systems, such as microbes, plants, or animals, is referred to as biological synthesis, or biogenic techniques (Srivastava et al. 2021). These methods have the potential to be more economical, environmentally benign, and sustainable than conventional chemical and physical synthesis techniques. By reducing the use of hazardous chemicals and energy-intensive procedures, biogenic synthesis adheres to the principles of green chemistry (Pandit et al. 2022). Biological approaches involve the utilization of biochemical reactions by microorganisms or plants to change or reduce metal ions. Through enzyme reactions that take place either inside or outside of the cell, living organisms transform the metal ions they absorb from the external environment into nanoparticles in this process (Pantidos and Horsfall 2014).

Microorganisms: In biogenic synthesis, bacteria, fungi, and algae are frequently employed. These microbes can transform metal ions into nanoparticles and have a variety of metabolic processes. For instance, certain bacteria can use enzymatic mechanisms to decrease metal ions to create silver and gold nanoparticles (Shafey 2020).

Plants: Due to their natural reducing agents and bioactive components, plants are a perfect source for biological synthesis. Metal ions can be reduced by plant extracts, which will result in the formation of metal nanoparticles (Inkson 2016).

3.5 Characterization of Nanomaterials

Currently, nanotechnology is applied in nearly every industry. Because of this, it is the most significant area of study and a growing field in the twenty-first century, growing at a very quick pace. An overview of the characteristics, synthesis, and characterization of nanomaterials (NMs) is given in this working chapter. NMs have demonstrated incredibly powerful physico-chemical qualities. Because of their enormous surface area and extremely small, unique size, nano-based materials are used in practically every discipline. The generated NMs exhibit remarkable mechanical, optical, electrical, and magnetic capabilities. Both top-down and bottom-up methods are used to prepare NMs. Every process technique has benefits and drawbacks of its own. Accurate knowledge on the general characterization techniques and nanomaterial synthesis processes is crucial.

3.6 Microscopic Techniques

Characterization of nanomaterials includes a set of techniques used to study the properties of materials such as size, shape, surface properties, composition, purity, stability (Tang and Yang 2017). These techniques are important for a better understanding of nanomaterials. Here are some microscopic techniques frequently used in the characterization of nanomaterials:

Scanning Electron Microscope (SEM): SEM is used to examine surface morphology. The high-energy electron beam is focused on the surface of the sample and reflected electrons or secondary electrons are collected. In this way, the surface details of the sample can be displayed (Tang and Yang 2017).
Transmission Electron Microscope (TEM): TEM is used to examine the internal structure of the sample. It focuses on thin sections or nanoparticles. The electron beam passes through the sample and is collected by the detector. In this way, detailed images at the atomic level can be obtained (Bracewell 1989).

3.7 Spectroscopic Techniques

Fourier Transform Infrared Spectroscopy (FTIR): FTIR is used to analyze the chemical composition of the material. Infrared rays measure the molecular vibrations of the sample and thus chemical groups can be identified (Long 1977).
Raman Spectroscopy: Raman spectroscopy is used to study the molecular vibrations of the material. The laser beam is focused on the surface of the sample and the scattered light spectrum is analyzed. In this way, information about the chemical structure of the material can be obtained (Mulvaney and Keating 2010).

X-Ray Photoelectron Spectroscopy (XPS): X-Ray photoelectron spectroscopy (XPS) is a critical tool for studying the surface composition and chemical properties of magnetic nanoparticles. Magnetic nanoparticles are widely used in biomedical applications such as medical imaging, targeted drug delivery, hyperthermia therapy, as well as in industrial fields such as magnetic storage and catalysts (Long 1977). These nanoparticles are usually made of magnetic materials such as iron oxide, and their surfaces are functionalized with polymers, biomolecules, or other chemical components to make them compatible with biological systems (Kudelski 2008).

XPS can analyze what elements are present on the surface of magnetic nanoparticles and their chemical bonds. This is important information for understanding the interaction and functionality of nanoparticles with biological environments. XPS also controls the quality and homogeneity of surface coatings, ensuring that nanoparticles can be used safely in biological systems. Determining the Fe^{+2} and Fe^{+3} ratios, especially in iron oxide nanoparticles, provides information about the surface oxidation state, and this information is critical in the evaluation of magnetic properties (Epp 2016).

Besides this, XPS can detect possible contaminants or foreign elements in magnetic nanoparticles, which is essential to ensure the purity and performance of nanoparticles. Thus, XPS serves as an important tool for the synthesis, characterization, and quality control of magnetic nanoparticles, enabling the safe and effective preparation of these materials for biomedical and industrial applications.

3.8 Diffraction Techniques

Diffraction techniques are powerful analytical methods used to study the atomic and molecular structure of materials. These techniques reveal the internal structure of a material by analyzing the propagation and interference patterns of light or other types of waves (e.g., X-rays, neutrons, or electrons) transmitted or reflected by a material. The main diffraction techniques are:

X-Ray Diffraction (XRD): X-Ray diffraction (XRD) method plays an important role. A general approach that includes the steps of starting from sample preparation, performing XRD measurement, analyzing the obtained data and interpreting the results is as follows: First, magnetic nanoparticles are spread or pressed on a clean and dust-free surface, thus preparing the sample. The sample is then appropriately placed in the XRD device and the measurement parameters are set. The resulting XRD patterns are compared with the characteristic peaks of magnetic materials. Information about the crystal structure and phase composition can be obtained from the positions of the peaks, and information about the crystal size can be obtained from the peak widths. If necessary, more detailed structural analysis methods can be used. Finally, XRD results are evaluated together with other characterization techniques to examine the relationship between the crystal properties and magnetic performances of magnetic nanoparticles. This holistic analysis is important for detailed characterization of magnetic nanoparticles (Hoo et al. 2008).

Electron Diffraction: Electron diffraction is a technique used especially for the analysis of thin films, surfaces and nanomaterials. High-energy electrons interact with the sample, creating specific patterns. These patterns are formed as a result of the interaction of electrons at intervals between atomic planes and can be examined with Transmission Electron Microscopy (TEM) or Scanning Electron Microscopy (SEM).

Neutron Diffraction: Neutron diffraction is an important method, especially in the analysis of magnetic materials and hydrogen-containing compounds. Neutrons interact with atomic nuclei and magnetic moments, thus obtaining information about atomic positions and magnetic structure. This technique is performed in large research facilities, reactors or spallation sources.

3.9 Other Characterization Methods

Dynamic Light Scattering (DSL): Dynamic light scattering (DLS) is also an important technique in the characterization of magnetic nanoparticles. DLS provides information about the hydrodynamic diameter distribution by measuring the Brownian motion of nanoparticles (Khan 2008). Characterization of magnetic nanoparticles with DLS is carried out as follows: First of all, magnetic nanoparticles are suspended in a liquid medium (usually water or a suitable solvent) and their homogeneity is ensured. Next, the sample is positioned in the DLS device and the measurement parameters (laser wavelength, measurement time, temperature, etc.) are set. The device measures the fluctuations in light intensity caused by the Brownian motion of the particles and the resulting autocorrelation function is analyzed to determine the hydrodynamic diameter distribution of the particles. The resulting hydrodynamic diameter distribution reflects the actual size distribution of magnetic nanoparticles and provides information about properties such as particle size, size distribution, and agglomerate formation (Saadatkhah et al. 2020). DLS offers rapid, sensitive and reproducible measurements in the characterization of magnetic nanoparticles, and when used in conjunction with other characterization techniques, important information can be obtained for the synthesis, stabilization and applications of magnetic nanomaterials.

3.10 Thermal Analysis Techniques (DSC, TGA)

Thermal analysis techniques play an important role in the characterization of nanoparticles. These techniques include differential scanning calorimetry (DSC) and thermogravimetric analysis (TGA). DSC measures thermodynamic events such as phase change, melting, crystallization and oxidation that occur during heating or cooling of the sample, and thus properties of nanoparticles such as crystal structure, purity, crystallization behavior, and thermal stability can be determined (Majewsky et al. 2016). TGA, on the other hand, measures the mass change of the sample under temperature change, so events such as thermal stability of nanoparticles, evaporation, oxidation, and decomposition can be monitored. Additionally, the structure of coated nanoparticles and the amount of coating can be determined by TGA (Corcione and Frigione 2012). Thermal analysis techniques provide detailed information about the thermal behavior of the sample, small amounts of samples can be studied, fast and precise measurements can be made, and holistic analysis can be performed by integrating with other characterization techniques (Laborda et al. 2016). As a result, DSC and TGA are effectively used in determining important properties of nanoparticles such as crystal structure, purity, thermal stability, and phase transformations.

3.11 Properties and Applications of Nanomaterials

3.11.1 Electrical and Electronic Properties

Nanomaterials exhibit unique electrical and electronic properties that differ significantly from their bulk counterparts. These properties are affected by factors such as high surface-to-volume ratio and the presence of defects in the grain boundary region. Some important points regarding the electrical and electronic properties of nanomaterials are:

Electrical Conductivity: Nanomaterials can have improved electrical conductivity due to their nanoscale size, which can be useful for applications such as sensors and energy storage devices (Sadiq et al. 2015).

Dielectric Properties: The dielectric constant and dielectric loss of nanomaterials can vary with temperature, which is important for the design of modern electronic devices (Kumar et al. 2019).

Measurement Techniques: The quality of nanomaterials can be characterized using various measurement techniques such as broadband dielectric spectroscopy and Keithley devices to study their dielectric and electrical properties (Sadiq et al. 2015).

Device Applications: Due to their high reactivity and excellent electrical conductivity, nanomaterials are ideal for use in devices such as light-emitting diodes, photodetectors, catalysts, optoelectronic memory devices, and solar cells (Kumar et al. 2019).

3.11.2 Mechanical Properties

The mechanical properties of nanoparticles can be quite different from the properties of the materials that make them up. Nanometer-sized particles have a significant effect on mechanical properties such as friction, flexibility, hardness and yield stress due to their high surface area to volume ratio. In nanometer-scale particles, atomic-level bonding forces and surface energies play a more dominant role in determining the macroscopic-scale properties of the material. In addition, the high surface/volume ratios, interparticle interactions and aggregations of nanoparticles also affect their mechanical properties. Therefore, the mechanical behavior of nanoparticles can differ significantly from the mechanical properties of conventional materials (Guo et al. 2013).

3.11.3 Chemical Properties

The chemical properties of nanoparticles change markedly due to the decrease in the size of these materials. At the nanoscale, increasing the surface area/volume ratio of particles significantly affects their chemical properties, such as electronic, optical, catalytic, and magnetic properties (Sukhanova et al. 2018).

For example, metals traditionally considered noble, such as gold and silver, increase their chemical activity when reduced to nanoscale sizes. Nanogold and silver particles may have properties such as higher catalytic activity, enhanced antibacterial properties and enhanced optical absorbance. Similarly, nano-sized metal oxides (TiO_2, ZnO, Fe_3O_4, etc.) show higher oxidizing and adsorption capacity. Additionally, some nanoparticles (such as carbon nanotubes, graphene, quantum dots) can exhibit higher electrical conductivity, light-matter interaction, and nonlinear optical properties (Nguyen et al. 2022; Podyacheva and Ismagilov 2015).

The chemical properties of nanoparticles are affected by factors such as composition, size, surface properties, and morphology. Therefore, these parameters must be carefully controlled during the synthesis and functionalization of nanoparticles. Thus, nanoparticles can be ensured to have optimized properties for their targeted chemical, physical, and biological applications.

3.12 Application Areas of Magnetic Nanoparticles

3.12.1 Electronics and Photonics

In the field of electronics and photonics, nanomaterials are used to reduce the size of devices and improve their performance. Thanks to nanoscale structures, faster and more efficient electronic circuits can be developed. Materials such as carbon nanotubes and graphene offer promising properties for transistors and other electronic components. Additionally, nanomaterials are used in photonic devices such as light-emitting diodes (LEDs), lasers, and solar panels, helping to control light more effectively (He et al. 2012).

3.12.2 Biomedical Applications

In the biomedical field, nanomaterials are transforming diagnostic and therapeutic methods. Magnetic nanoparticles are used as contrast agents for magnetic resonance imaging (MRI). Gold nanoparticles can serve as drug carriers by targeting cancer cells. Additionally, nanoscale drug delivery systems enable the delivery of drugs directly to target tissues, reducing side effects and increasing treatment effectiveness. Nano-biosensors can detect sensitive biological signals to enable early diagnosis of diseases (Mohammed et al. 2017).

3.12.3 Environment and Energy

In the field of environment and energy, nanomaterials offer sustainable solutions. Nanoporous materials can effectively filter pollutants by using them in water purification and air purification systems. Nanotechnological materials used for solar cells contribute to renewable energy production by increasing energy conversion efficiency. Additionally, nanocatalysts can increase the efficiency of energy production processes by increasing the rate of chemical reactions and help reduce industrial waste (Kudr et al. 2017).

3.12.4 Material Science and Engineering

In the field of materials science and engineering, nanomaterials improve the mechanical, thermal, and electrical properties of materials. Carbon nanotubes and nanocomposites are used to produce lightweight and durable materials. Such materials are ideal for achieving weight reduction and fuel efficiency in the aerospace and automotive industries. Additionally, nanotechnology increases the wear and corrosion resistance of materials through coatings and surface modifications. Nanomaterials are considered to be at the forefront of innovation in these various application areas and have great potential to develop more sustainable and efficient technologies in the future (Bao et al. 2016).

3.13 Conclusion

As a result, synthesis and characterization of nanomaterials are two important issues that form the foundations of the field of nanotechnology. This book chapter has provided a comprehensive understanding of the production and analysis of nanomaterials by examining different synthesis methods and characterization techniques. The unique properties of nanomaterials make them ideal for a variety of industrial and scientific applications. Advances in the synthesis and characterization of nanomaterials will continue to shape the future of nanotechnology. Technological advances, combined with new synthesis methods and more sensitive characterization techniques, will lead to nanomaterials finding a wider range of uses in industry and science. At the same time, environmental and safety issues must also be carefully considered. The development of sustainable and safe nanotechnologies will be critical to the long-term success of advances in this field. It is aimed to keep up with nanotechnology by understanding how to synthesize a nanomaterial and how to characterize that material.

References

Alagiri M, Ponnusamy S, Muthamizhchelvan C (2012) Synthesis and characterization of NiO nanoparticles by sol–gel method. J Mater Sci Mater Electron 23:728–732
Austin LG, Klimpel RR (1964) The theory of grinding operations. Ind Eng Chem 56(11):18–29
Bao Y, Wen T, Samia ACS, Khandhar A, Krishnan KM (2016) Magnetic nanoparticles: material engineering and emerging applications in lithography and biomedicine. J Mater Sci 51:513–553
Baxter JB, Aydil ES (2005) Nanowire-based dye-sensitized solar cells. Appl Phys Lett 86(5)
Beek WJ, Wienk MM, Janssen RA (2004) Efficient hybrid solar cells from zinc oxide nanoparticles and a conjugated polymer. Adv Mater 16(12):1009–1013
Biju V (2014) Chemical modifications and bioconjugate reactions of nanomaterials for sensing, imaging, drug delivery and therapy. Chem Soc Rev 43(3):744–764
Bokov D, Jalil AT, Chupradit S, Suksatan W, Ansari MJ, Shewael IH et al (2021) Nanomaterial by sol-gel method: synthesis and application. Adv Mater Sci Eng 2021:1–21
Bracewell RN (1989) The Fourier transform. Sci Am 260(6):86–95
Bruchez M Jr, Moronne M, Gin P, Weiss S, Alivisatos AP (1998) Semiconductor nanocrystals as fluorescent biological labels. Science 281(5385):2013–2016. https://doi.org/10.1126/science.281.5385.2013
Chen DH, He XR (2001) Synthesis of nickel ferrite nanoparticles by sol-gel method. Mater Res Bull 36(7–8):1369–1377

Corcione CE, Frigione M (2012) Characterization of nanocomposites by thermal analysis. Materials 5(12):2960–2980

Davis ME, Chen Z, Shin DM (2008) Nanoparticle therapeutics: an emerging treatment modality for cancer. Nat Rev Drug Discov 7(9):771–782

El-Rafie MH, Mohamed AA, Shaheen TI, Hebeish A (2010) Antimicrobial effect of silver nanoparticles produced by fungal process on cotton fabrics. Carbohydr Polym 80(3):779–782

Epp J (2016) X-ray diffraction (XRD) techniques for materials characterization. In: Materials characterization using nondestructive evaluation (NDE) methods. Woodhead Publishing, pp 81–124

Gan YX, Jayatissa AH, Yu Z, Chen X, Li M (2020) Hydrothermal synthesis of nanomaterials. J Nanomater 2020:1–3

Guo D, Xie G, Luo J (2013) Mechanical properties of nanoparticles: basics and applications. J Phys D Appl Phys 47(1):013001

Hall JB, Dobrovolskaia MA, Patri AK, McNeil SE (2007) Characterization of nanoparticles for therapeutics

Hasan S (2015) A review on nanoparticles: their synthesis and types. Res J Recent Sci 277:2502

He L, Wang M, Ge J, Yin Y (2012) Magnetic assembly route to colloidal responsive photonic nanostructures. Acc Chem Res 45(9):1431–1440

Hoo CM, Starostin N, West P, Mecartney ML (2008) A comparison of atomic force microscopy (AFM) and dynamic light scattering (DLS) methods to characterize nanoparticle size distributions. J Nanopart Res 10:89–96

Inkson BJ (2016) Scanning electron microscopy (SEM) and transmission electron microscopy (TEM) for materials characterization. In: Materials characterization using nondestructive evaluation (NDE) methods. Woodhead Publishing, pp 17–43

Inoue M (2005) Solvothermal synthesis. Mater Eng NY 28:21

Kamins T (2012) Polycrystalline silicon for integrated circuits and displays. Springer Science & Business Media

Khan A (2008) Preparation and characterization of magnetic nanoparticles embedded in microgels. Mater Lett 62(6–7):898–902

Khan S, Hossain MK (2022) Classification and properties of nanoparticles. In: Nanoparticle-based polymer composites. Woodhead Publishing, pp 15–54

Kudelski A (2008) Analytical applications of Raman spectroscopy. Talanta 76(1):1–8

Kudr J, Haddad Y, Richtera L, Heger Z, Cernak M, Adam V, Zitka O (2017) Magnetic nanoparticles: from design and synthesis to real world applications. Nanomaterials 7(9):243

Kudyakova VS, Shishkin RA, Elagin AA, Baranov MV, Beketov AR (2017) Aluminium nitride cubic modifications synthesis methods and its features: review. J Eur Ceram Soc 37(4):1143–1156

Kumar PS, Pavithra KG, Naushad M (2019) Characterization techniques for nanomaterials. In: Nanomaterials for solar cell applications. Elsevier, pp 97–124

Kuo Y (2013) Thin film transistor technology—past, present, and future. Electrochem Soc Interface 22(1):55

Laborda F, Bolea E, Cepriá G, Gómez MT, Jiménez MS, Pérez-Arantegui J, Castillo JR (2016) Detection, characterization and quantification of inorganic engineered nanomaterials: a review of techniques and methodological approaches for the analysis of complex samples. Anal Chim Acta 904:10–32

Lai J, Niu W, Luque R, Xu G (2015) Solvothermal synthesis of metal nanocrystals and their applications. Nano Today 10(2):240–267

Law M, Greene LE, Johnson JC, Saykally R, Yang P (2005) Nanowire dye-sensitized solar cells. Nat Mater 4(6):455–459

Liu HH, Cohen Y (2014) Multimedia environmental distribution of engineered nanomaterials. Environ Sci Technol 48(6):3281–3292

Long DA (1977) Raman spectroscopy. New York, p 310

Mack C (2007) Fundamental principles of optical lithography: the science of microfabrication. John Wiley & Sons
Majewsky M, Bitter H, Eiche E, Horn H (2016) Determination of microplastic polyethylene (PE) and polypropylene (PP) in environmental samples using thermal analysis (TGA-DSC). Sci Total Environ 568:507–511
Mohammed L, Gomaa HG, Ragab D, Zhu J (2017) Magnetic nanoparticles for environmental and biomedical applications: a review. Particuology. 30:1–14
Mulvaney SP, Keating CD (2010) Raman spectroscopy. Anal Chem 72(12):145–158
Nakaso K, Han B, Ahn KH, Choi M, Okuyama K (2003) Synthesis of non-agglomerated nanoparticles by an electrospray assisted chemical vapor deposition (ES-CVD) method. J Aerosol Sci 34(7):869–881
Nakayama K, Tanabe K, Atwater HA (2008) Plasmonic nanoparticle enhanced light absorption in GaAs solar cells. Appl Phys Lett 93(12)
Nguyen THA, Nguyen VC, Phan TNH, Vasseghian Y, Trubitsyn MA, Nguyen AT et al (2022) Novel biogenic silver and gold nanoparticles for multifunctional applications: green synthesis, catalytic and antibacterial activity, and colorimetric detection of Fe (III) ions. Chemosphere 287:132271
Niederberger M (2007) Nonaqueous sol–gel routes to metal oxide nanoparticles. Acc Chem Res 40(9):793–800
Pandit C, Roy A, Ghotekar S, Khusro A, Islam MN, Emran TB et al (2022) Biological agents for synthesis of nanoparticles and their applications. J King Saud Univ Sci 34(3):101869
Pantidos N, Horsfall LE (2014) Biological synthesis of metallic nanoparticles by bacteria, fungi and plants. J Nanomed Nanotechnol. 5(5):1
Podyacheva OY, Ismagilov ZR (2015) Nitrogen-doped carbon nanomaterials: to the mechanism of growth, electrical conductivity and application in catalysis. Catal Today 249:12–22
Saadatkhah N, Carillo Garcia A, Ackermann S, Leclerc P, Latifi M, Samih S et al (2020) Experimental methods in chemical engineering: thermogravimetric analysis—TGA. Can J Chem Eng 98(1):34–43
Sadiq I, Naseem S, Ashiq MN, Khan MA, Niaz S, Rana MU (2015) Structural and dielectric properties of doped ferrite nanomaterials suitable for microwave and biomedical applications. Prog Nat Sci Mater Int 25(5):419–424
Shafey AME (2020) Green synthesis of metal and metal oxide nanoparticles from plant leaf extracts and their applications: a review. Green Process Synth. 9(1):304–339
Shinkai M, Yanase M, Suzuki M, Honda H, Wakabayashi T, Yoshida J, Kobayashi T (1999) Intracellular hyperthermia for cancer using magnetite cationic liposomes. J Magn Magn Mater 194(1–3):176–184
Shipway AN, Katz E, Willner I (2000) Nanoparticle arrays on surfaces for electronic, optical, and sensor applications. ChemPhysChem 1(1):18–52
Silva CG, Corma A, García H (2010) Metal–organic frameworks as semiconductors. J Mater Chem 20(16):3141–3156
Singh LP, Bhattacharyya SK, Kumar R, Mishra G, Sharma U, Singh G, Ahalawat S (2014) Sol-Gel processing of silica nanoparticles and their applications. Adv Colloid Interface Sci 214:17–37
Singh P, Kim YJ, Zhang D, Yang DC (2016) Biological synthesis of nanoparticles from plants and microorganisms. Trends Biotechnol 34(7):588–599
Srivastava S, Usmani Z, Atanasov AG, Singh VK, Singh NP, Abdel-Azeem AM, Bhargava A (2021) Biological nanofactories: using living forms for metal nanoparticle synthesis. Mini Rev Med Chem 21(2):245–265
Stark WJ, Stoessel PR, Wohlleben W, Hafner AJCS (2015) Industrial applications of nanoparticles. Chem Soc Rev 44(16):5793–5805
Sterling HF, Swann RCG (1965) Chemical vapour deposition promoted by rf discharge. Solid-State Electr 8(8):653–654

Suh SK, Yuet K, Hwang DK, Bong KW, Doyle PS, Hatton TA (2012) Synthesis of nonspherical superparamagnetic particles: in situ coprecipitation of magnetic nanoparticles in microgels prepared by stop-flow lithography. J Am Chem Soc 134(17):7337–7343

Sukhanova A, Bozrova S, Sokolov P, Berestovoy M, Karaulov A, Nabiev I (2018) Dependence of nanoparticle toxicity on their physical and chemical properties. Nanoscale Res Lett 13:1–21

Sundgren JE, Hentzell HG (1986) A review of the present state of art in hard coatings grown from the vapor phase. J Vac Sci Technol A 4(5):2259–2279

Tang SC, Lo IM (2013) Magnetic nanoparticles: essential factors for sustainable environmental applications. Water Res 47(8):2613–2632

Tang CY, Yang Z (2017) Transmission electron microscopy (TEM). In: Membrane characterization. Elsevier, pp 145–59

Trotta F, Mele A (2019) Nanomaterials: classification and properties. In: Nanosponges: synthesis and applications, pp 1–26

Wang Y, Cheng X, Lin Z, Zhang C, Zhang F (2003) Optimization of PECVD silicon oxynitride films for anti-reflection coating. Vacuum 72(3):345–349

Zorkipli NNM, Kaus NHM, Mohamad AA (2016) Synthesis of NiO nanoparticles through sol-gel method. Procedia Chem 19:626–631

Chapter 4
Mechanisms of Nanomaterials in Wastewater Treatment

Radha Rani Mehra, Asmita Sharma, and Devaanshi Jagwani

Abstract Water is the most important commodity for us all in different fields. Water makes our planet unique, and superior compared to other planets. There are several areas where we have resources of water, but due to recent environmental conditions water is going to become contaminated by the presence of lots of harmful chemicals. It is necessary in the present era to decrease pollutants in water, not least for the healthy life of the next generation. Nowadays nanomaterials are applied for the removal of different pollutants in wastewater. The size of nanomaterial in between 1 to 100 nm, when nanomaterials change from macro to nano size it shows drastic differences in their appearance, toughness, conductivity, and sensitivity also nanomaterials show good catalytic activities, high reactivity, and better adsorption ability, also nanoparticles that have a very small size and bigger surface area show many applications in different fields. The nano-size materials enhanced the catalytic activities and adsorption capacity of nanomaterials as compared to bulk materials, due to their small size and large surface area it has been shown that nano catalysts and adsorption power of nano-size materials can expand the oxidation ability towards pollutants from wastewater effectively. The different mechanism works for the removal of pollutants from wastewater like Oxidative neutralization, Adhesive detoxification, Adsorption to active carbon, Photocatalytic degradation, etc. In this chapter we provide an overview of the basic mechanisms of adsorption and catalytic processes of nanomaterials and gradually we discuss the role of nanomaterials in pollutant removal and degradation and finally discuss the interaction mechanism between nanomaterials and pollutant in the waste water.

Keywords Nanomaterials · Mechanisms · Industrial wastewater

R. R. Mehra · A. Sharma (✉)
Department of Chemistry, Shri Vaishnav Vidyapeeth Vishwavidyalaya, Indore, India
e-mail: asmitasharma@svvv.edu.in

R. R. Mehra
e-mail: radhamehra@svvv.edu.in

D. Jagwani
Civil Engineering Department, IPS Academy, Institute of Engineering and Science, Indore, India
e-mail: devaanshijagwani@ipsacademy.org

A. Mandpe et al. (eds.), *Nanotech Solutions for Industrial Wastewater*, Advances in Wastewater Research, https://doi.org/10.1007/978-981-96-6879-3_4

4.1 Adsorption Mechanisms

Traditional techniques for wastewater treatment are based on adsorption and chemical oxidation, but sometimes these are economically costly and laborious. Nowadays, the most popular method for wastewater treatment is nanotechnology, both in terms of quality and quantity, contributing to environmental safety and strength.

The exceptional properties, such as high affinity against pollutants, large surface area, cost-effectiveness, and reactivity of nanomaterials make them highly applicable for wastewater treatment (Kalhapure et al. 2015).

When nanoparticles are compared to microparticles, due to the big surface area and greater selectivity, nanomaterials show very good adsorbent quality. Many materials typically alter their characteristics in response to the size of the target chemical. For example 7 nm diameter, Hermite particles adsorb more Cu^{+2} ions at a lower pH than a dimeter of 25 or 88 nm, this is proof that the surface activity of ferric oxide nanoparticles altered with the diameter reducing (Madden et al. 2006).

This part of the chapter looks at the basics of the adsorption process, and the other part describes the elimination of several pollutants from wastewater and the leading classes of unique and functionalized nanomaterials that are used as adsorbents. Nano-based methodologies eliminate the most difficult nanoscale contamination.

Nanotechnologies offer benefits for the treatment of wastewater, due to their excellent physicochemical properties and dimensional characteristics. Nanoparticles exhibit shape, size, surface area, solubility, and charge variations determined by their chemical, biological, and physical attributes. Microfiltration (MF), nanofiltration (NF), ultrafiltration (UF), and reverse osmosis (RO) methods based on nanotechnology are used for the treatment of wastewater (Gupta et al. 2015).

Nanoparticles show high efficiency in adsorbing inorganic and organic toxins from contaminated water, which are challenging to eliminate. Because the nanoparticles have a significant surface area, it permits additional effective adsorbing sites and nanoparticles can be recycled which is another benefit. After wastewater treatment, some magnetic nanoparticles can be recycled, collected from the wastewater, and used again and again. Magnetic nanoparticles can eliminate heavy metals and radionuclides from sewage (Fig. 4.1).

4.1.1 Adsorption Fundamentals in Wastewater Treatment

Adsorption occurs on the surface of a solid or liquid, as it is a surface phenomenon when there is a greater concentration of a molecular species there than there is within, then adsorption happens. In the adsorption process, adsorbate is accumulated on the adsorbent surface.

Intermolecular forces of attraction between liquid–solid cause some solute molecules from the solution to be concentrated or accumulated on the adsorbent. The materials' basic atoms are fulfilled by other atoms of materials in the case

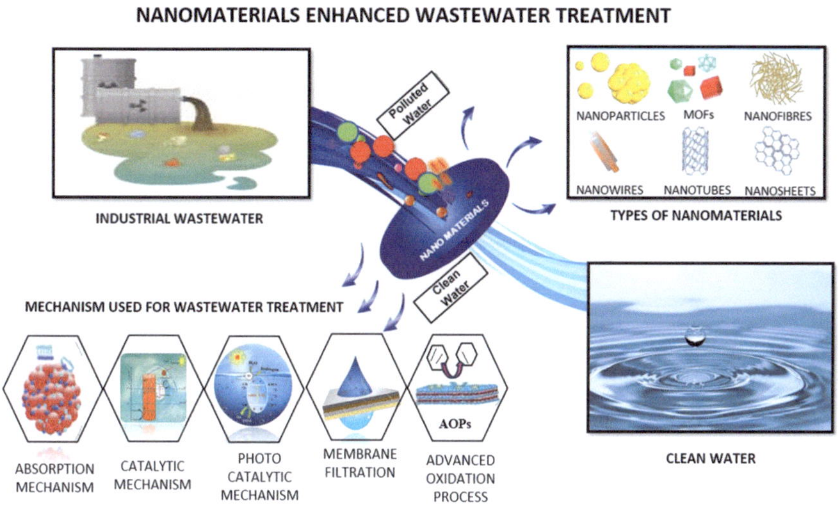

Fig. 4.1 Nanomaterials enhanced wastewater treatment

of bulk materials at all the bonding conditions (ionic, covalent, or metallic). The connecting features depend on the species involved in the process. There are two types of adsorption processes: chemisorption, also known as activated adsorption, and physisorption, also known as Van der Waals adsorption (OECD 2010).

The normal toxic materials that are present in wastewater are responsible for many problems such as heavy metals ions and dyes. Excess doses of these heavy metal ions in human beings causes health problems, for example Zn(II) ions, in small quantities this acts as a nutrient for human beings but excess doses cause health complications such as neurological signs, depression, and lethargy. Removing poisonous materials from effluent is essential for environmental protection and health. The most effective traditional approach for these purposes is the adsorption process because of its high competency and profitability, though other methods that are frequently employed include precipitation, reduction, adsorption, ion exchange, and oxidation. Wastewater is often treated with common adsorbent materials such as zeolites, activated carbon, biomaterials, and polymers. However, these materials have a relatively low adsorption efficiency. Recently, due to the enhancement of nanotechnology with nanoscience, a significant capability for the remedy of wastewater has been achieved (Chaturvedi et al. 2012).

The adsorption process is determined by the coefficient of adsorption K_d and the presentation separating of pollutants i.e. organic pollutants or heavy metals under equilibrium conditions. Activated carbon, Carbon nanotubes, Graphene, Oxides of zinc, Magnesium oxide, Silver oxides, Titanium oxide, and Ferric oxide are commonly used nanoparticles for heavy metals adsorption from wastewater. Nano-sized adsorbents have two main qualities, first intrinsic surface, and other exterior functionalization. The size, internal content, and surface structure of nano adsorbents are associated with their material, physical and chemical properties (Mirkin et al. 1996). Extraordinary adsorption ability, nontoxicity, the skill to adsorb contaminants in extremely low concentrations, multiple recycling opportunities, and ease of removal from the adsorbent surface are all necessary for the removal of heavy metals.

4.1.2 Role of Nanomaterials in the Adsorption Process

Current research shows that nanoparticles have great capability as adsorbents because of their tiny size which increases the surface area of nano adsorbents. The adsorption capacity and chemical reactivity of metals on the surface improved due to the large surface area of nanoparticles.

Based on the adsorption procedure, Nano-sized adsorbents are categorized into various groups, such as metallic nanoparticles, nano-structured oxides, and magnetic nanoparticles. Different types of silicon nanoparticles like silicon nanoparticles, silicon, and silicon nanosheets, and nanotubes are also utilized as adsorbents for wastewater treatment. Other adsorbents such as nano clays, polymer-based nanofibers, and aerogels are also useful for the removal of heavy metals in wastewater. Surface chemistry, fractal dimension, size agglomeration state, solubility, chemical composition, and crystal structure are factors that supervise nano-adsorbent properties (Khin et al. 2012).

Furthermore, recent developments in carbon nanoparticles (CNMs) contained carbon nanotubes and carbon nanosheets. Carbon Nanotubes have been considered under many types such as single-walled CNTs, composite, and multi-walled CNTs. To enhance the dispersion ability and electronic and optical properties, CNTs are modified by researchers into multi-wall carbon nanotubes (MWCNTs). Modified MWCNTs show extreme dispersion capacity and can be deleted from effluent by applying a magnet (Madrakian et al. 2011) (Fig. 4.2).

4.1.2.1 Types of Nanoparticles for Wastewater Treatment

Based on numerous studies, nanoparticles show great efficiency in the treatment of wastewater, including their recyclability, environmental friendliness, and cost-effectiveness characteristics. Sufficient knowledge of the characteristic properties and functions of different nanoparticles is required when choosing which types of nanoparticles to utilize to remove contamination.

Currently, the general studies of nanomaterials for wastewater treatment mostly incorporate zero-valent metal nanoparticles, carbon nano tubes, metal oxide nanoparticles, and nanocomposites (Table 4.1).

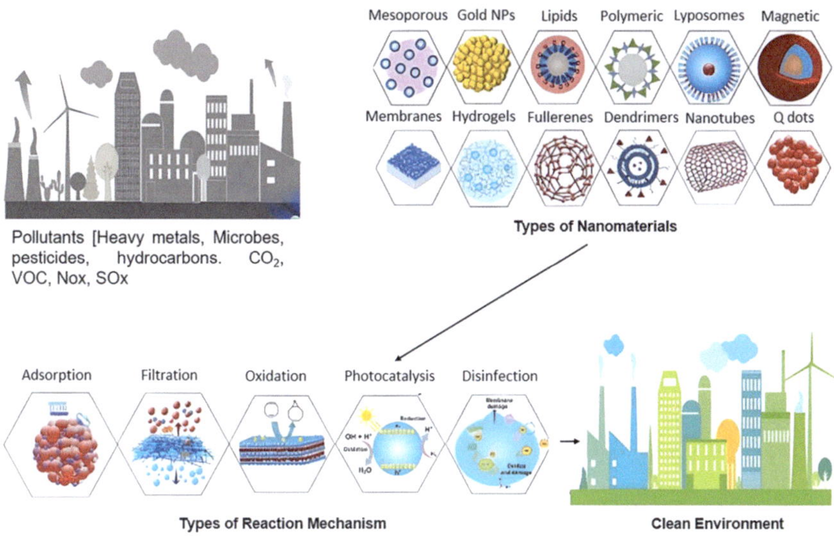

Fig. 4.2 Role of nanomaterials in the adsorption process

Table 4.1 Comparison of various nanomaterials ability to remove dye and capture metal ions based on their adsorption capability

Nanomaterials	Adsorbents	Adsorbate	Adsorption capacity (mg g^{-1})
Carbon nano tubes	AC SWCNTs MWCNTs	Reactive red M-2BE Reactive blue 29 Reactive red 18	260.7 496 355.7
Metal oxide based nanomaterials	CaO/SiO_2 nanocomposite TiO_2 Fe_3O_4 Magnetic nanoparticles Modified Fe_3O_4 microporous Functionalized Fe_3O_4 MgO	MB Cd(I) Cu(II) Mo Cu(II) Hg(II) Cr(II) Reactive blue 19	58.7 16.69 5.87 85.39 61.07 37.4 (Micromole/g) 136.98 166.7
Metal oxide hybrid and carbon nanomaterials	Graphene oxide Fe_3O_4 hybrid composite Al_2O_3/MWCNTs	MB Trichloroethylene Cd(II)	167.2 19.84 21.27

4.1.2.2 Metal Oxides and Metal

Metallic nano-sized particles have an enormous capability for different types of environmental remediation for wastewater due to their adaptability. Metal oxide and metal nanomaterials are categorized into classes of nanomaterials that consist of

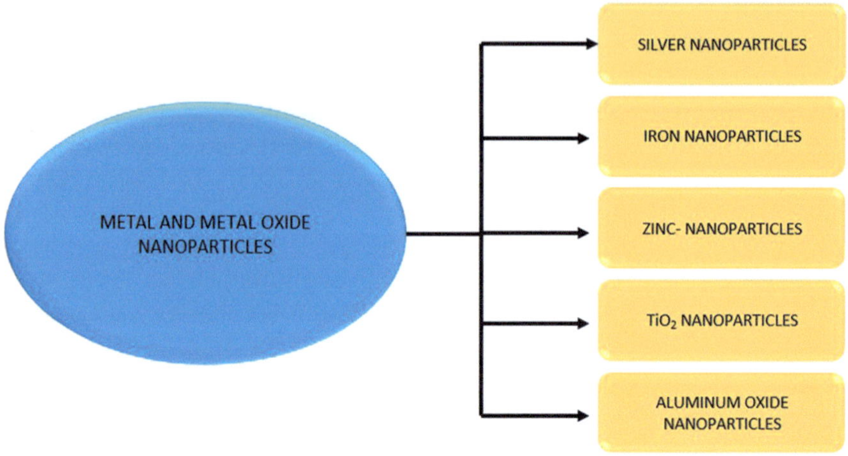

Fig. 4.3 Different metals nanoparticles

one, two, or three metals and their oxides. These nanomaterials have some characteristic qualities such as huge surface area, shape control, stability, monodisperse, etc., these beneficial characteristics can be practically used for wastewater management involving adsorption, member separation, and photo degradation chemical disinfection (Azizi-Lalabadi et al. 2019).

Nanoparticles of zinc oxide, silver oxide, and TiO_2 are the most examined metal and metal oxide nanoparticles for referring to current energy and environmental issues. They removed heavy metals and dyes from contaminated water. Metal oxide nanoparticles have minimal solubility and are not engaged in the formation of secondary contamination (Fig. 4.3).

4.1.2.3 Silver Nanoparticles

Silver nanoparticles have strong antibacterial impacts against many bacteria, fungi, and viruses. Due to its antimicrobial properties, it is used as a good antimicrobial agent for water disinfection.

As we know from several references AgNPs show strong antimicrobial activity, but the mechanism of their antimicrobial properties is not identified. AgNPs have been informed to be skilled at adhering to the cell wall of bacteria and consequently penetrating it, so structural changes in the cell wall and its permeability-growing. Further, AgNPs can produce free radicals when they interact with bacteria, these free radicals, can harm the cell membrane and are responsible for the death of cells. AgNPs can act with sulphur and phosphorus in DNA and destroy it. This is an additional description of the death of the cell caused by AgNPs. In solution AgNPs will relieve germicide Ag^+ ions, these ions relate with the thiol group of various essential enzymes, disable them, and disturb the regular function of the cell. In recent years AgNPs have been used for disinfection of wastewater and water due to their high antibacterial properties (Prabhu and Poulose 2012).

Through the reduction of silver nitrate, AgNPs have been employed on the cellulose fibers of an adsorbent blotting paper sheet. For the duration of the filtration process through the sheet, the AgNPs sheet shows antibacterial properties against *Escherichia coli* and *Enterococcus faecalis* and it disabled them. In recent ages, AgNPs have been used for the treatment of household water due to their disinfection qualities. The removal capacity of *E. coli* increases by AgNPs to ceramic filters assembled with clay and sawdust. It was noted that higher porosity filters eliminated more bacteria than lower porosity filters.

It was noted that the colloidal form of AgNPs enhanced the filter execution and the filter can eliminate *E. coli* at a rate between 97.8 and 100% (Prabhu and Poulose 2012). Further research on AgNPs will support their applications in wastewater treatment.

4.1.2.4 Iron Nanoparticles

The standard reduction potentials of zinc, aluminium, and nickel are shown in Table 4.2. In recent years, several studies have focused on metal nanoparticles, like Fe, Zn, Al, and Ni, for wastewater treatment. Fe or Zn zero-valent nanoparticles act as a reducing agent compared to many redox labile pollutants with a moderate standard reduction potential. Although it has a lower reduction power, Fe holds many major benefits over Zn for treating water contamination, involving brilliant adsorption qualities, oxidation, precipitation, and a small price. To treat wastewater, research on zero-valent iron nanoparticles has therefore proved quite interesting. nZVI has a great reduction capacity and good adsorption capabilities due to its large surface area and minuscule size.

These properties are responsible for the elimination of toxins from wastewater treatment in anaerobic conditions, Fe^{2+} and H_2 are generated by the oxidation of Fe^0 by H_2O or H^+. In the reduction–oxidation reaction connecting nZVI and waste product, Ferrous ions will be converted into ferrocenes which can change Fe $(OH)_3$ with a raised pH. Fe $(OH)_3$ helps eliminate pollutants from wastewater, as $Fe(OH)_3$is a general and good flocculant, for example, Cr (VI) (Liang et al. 2013).

Table 4.2 The standard reduction potentials—various metals

Metal	Standard reduction potential (E^0/V)
Fe	– 0.440
Zn	– 0.762
Al	– 1.677
Ni	– 0.236

Source Bratsch (1989)

nZVI can oxidize and destroy a mixture of organic compounds by ZVI transporting two electrons to O_2 to generate H_2O_2. In the presence of dissolved oxygen (DO) Eq. (4.3). By ZVI, the outcome is produced by the reduction of H_2O_2 Eq. (4.4). Additionally, Hydroxyl radicals (HO^*), show a strong oxidizing capacity for many organic compounds and it is produced by the combination of H_2O_2 and Fe^{2+} (Fenton reaction) Eq. (4.5) (Fu et al. 2014).

$$Fe^0 + 2H_2O \rightarrow H_2 + Fe^{2+} + 2OH^- \tag{4.1}$$

$$Fe^0 + 2H^+ \rightarrow H_2 + Fe^{2+} \tag{4.2}$$

$$Fe^0 + O_2 + 2H^+ \rightarrow H_2O_2 + Fe^{2+} \tag{4.3}$$

$$Fe^0 + 2H^+ + H_2O_2 \rightarrow Fe^{2+} + 2H_2O \tag{4.4}$$

$$Fe^{2+} + H_2O_2 \rightarrow Fe^{3+} + OH^- + HO^{\cdot} \tag{4.5}$$

With specific properties of adsorption, precipitation, reduction, and oxidation (in aerobic conditions), nZVI has been effectively utilized for the elimination of a variety of pollutants, including organic compounds of halogens (Liang et al. 2014), dyes (Xiong et al. 2015), nitroaromatic compounds (Yu et al. 2015), heavy metals, phenols, inorganic anions, and radio elements. In recent years, nZVI achieved uses at real water-contaminated field sites as well as experimental single scale applications. Aggregation, oxidation, and separation problems from the reduced technique are some of the drawbacks of nZVI, despite its many benefits for eliminating pollutants. Several ways of alteration are used to improve nZVI's ability to overcome these drawbacks. Emulsification, encapsulation in matrix, surface coating, and doping with additional metals are a few typical modification techniques (Stefaniuk et al. 2016).

4.1.2.5 Zinc Nanoparticles

These modified ZnO nanoadsorbents are far more effective at removing Cu (II) than commercial ZnO because of their unique nanostructure. Most of the studies on the treatment of wastewater using nZVI are alerted to the use of iron nanoparticles. However, Zn is thought to have a greater negative standard potential (see Table 4.2). In comparison to Iron, Zinc is a greater reducing agent, so the degradation rate of contaminated Zinc may be quicker than that of other nZVI. ZnO has a high surface area with a porous nanostructure for the adsorption of heavy metals. Nanoplates, nano assemblies, nanosheets, and nanorods of ZnO are commonly used as nano adsorbents to eliminate heavy metals from wastewater (Ge et al. 2011). For the removal of Cu(II) from wastewater, Wang et al. (2010) utilized ZnO nanoplates and nanosheets.

Because of their distinct nanostructure, these modified ZnO nano adsorbents exhibit a remarkable capability for removing Cu(II) compared to commercial ZnO. Zero-valent zinc has been focused on the dehalogenation reaction, which is another of it's applications. Recent research indicates that Zero-valent zinc significantly changed the reduction rate of CCl_4. Further nano assemblies of zinc were used to eliminate several heavy metals, such as Ni^{2+}, Co^{2+} Cd^{2+}, Pb^{2+}, Hg^{2+} and As^{3+}.

4.1.2.6 Metal Oxide Nanoparticles

TiO_2 nanoparticles: Photolysis of water by electrochemical process on TiO_2 semiconductor electrodes is a virtuous technology for photocatalytic degradation. In the past few years, this technique has been effectively applied for contaminant degradation for wastewater treatment. In the photolytic process, pollutants can be slowly converted into low molecular weight intermediate products by the oxidation process and finally transferred into H_2O and CO_2, as well as anions like Cl^-, $PO_4{}^{3-}$, and $NO_3{}^-$, in the presence of light and a catalyst. At present, TiO_2 is the best photocatalyst, due to its reasonable price, photostability, high photocatalytic activity, and chemical and biological stability.

Under UV radiation, TiO_2 will produce highly reactive oxygen species, which can entirely destroy pollutants in a quick time due to the large bond gap energy of TiO_2. Although TiO_2 is slightly selective, this is appropriate for the degradation of all types of impurities, like polyaromatic hydrocarbons, organic compounds (chlorinated), dyes, cyanide, phenols, arsenic, and other heavy metals. TiO_2 can also kill a variety of microorganisms, such as gram-negative and gram-positive bacteria, algae, fungi, protozoa, and viruses, due to its photocatalyst properties.

4.1.2.7 Carbon Nanotubes

In modern years, nanomaterials of carbon have been of great importance for researchers, due to their exceptional electronic characteristics and structure, so it is fascinating for basic studies and their varied applications. Especially for wastewater treatment. Carbon nanomaterials are widely used because they can eliminate other organic contaminants and heavy metals from polluted water through the adsorption process. With unique structural features and large surface area, functionalization, and chemical stability, most carbon nanomaterials, such as multiwalled carbon nanotubes (MWCNT), single-walled carbon nanotubes (SWCNT), and graphene oxide, have been used for water purification methods. SWCNT and MWCNT are the two main types of nanotubes of carbon. The cylindrical structure of these allotropes of carbon nanotubes approaches friction free transfer of water molecules to improve water penetrability during the wastewater treatment process. Additionally, through CNT, it is possible to maintain an equilibrium between the rejection of ions and permeation by revising the members to provide an appropriate pore diameter (Peng et al. 2003). Graphene sheets turned up in a cylindrical shape with a diameter of less than

1 nm are carbon nanotubes (Zhao and Stoddart 2009). Because of their large, precise surface area and rich porous structures. Carbon nanotubes have excellent adsorption capacity and great adsorption competences for several pollutants, like ethylbenzene, dichlorobenzene, Cu^{2+}, Cd^{2+}, Zn^{2+}, and different dyes (Zhao and Stoddart 2009).

Single-walled carbon nanotubes (SWCNTs) have a single layer of graphene sheet rolled into cylindrical tubes however Multiwalled carbon nanotubes (MWCNTs) contain multiple layers of circular cylinders with 0.34 nm spacing between two continuous layers. In current years, equally, Single-walled carbon nanotubes and Multiwalled carbon nanotubes have been utilized for the elimination of pollutants from wastewater (Fig. 4.4).

Carbon nanotubes combined with other metals improve their electrical, mechanical, and optical properties as well as their adsorption capacity. For example, Gupta et al. (2011) reported in their research that carbon nanotubes, with magnetic iron oxide, enhance the adsorption properties of carbon nanotubes combined with the magnetic properties of iron oxide and form a composite adsorbent that was used to eliminate chromium from water.

Despite extraordinary properties, the growth and application of carbon nanotubes are limited due to their low production and high cost.

Carbon nanotubes are widely used materials that have a great capability to eliminate heavy metals and other pollutants from wastewater via adsorption. Carbon nanotubes are, when Graphene sheets rotated up in a cylindrical shape with smaller than 1 nm a diameter are known as Carbon nanotubes. Carbon nanotubes have excellent adsorption competence and great adsorption efficiencies for many types of contaminants such as ethyl benzene, dichlorobenzene, and dyes and heavy metals like Pb(II) and Mn(II) Cu(II).

According to the literature, carbon nanotubes (CNTs) have been evaluated in aqueous conditions using single or multiple toxins, though only a limited number of

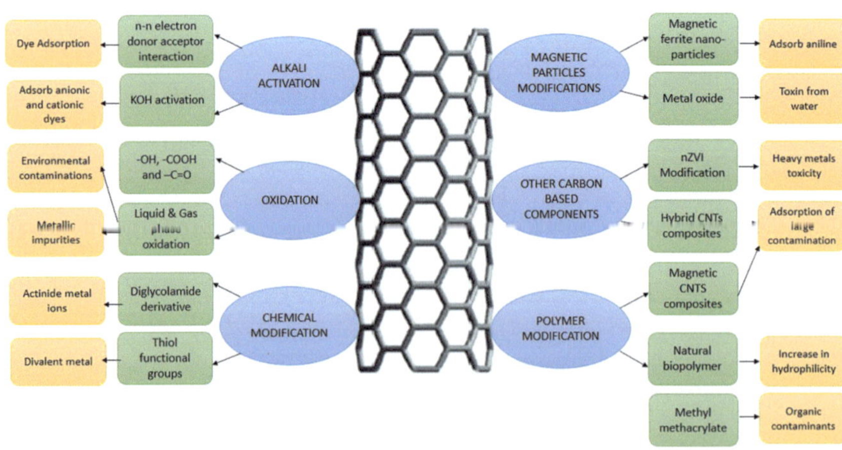

Fig. 4.4 Mechanism of CNTs for the elimination of contaminants from wastewater

studies demonstrate their use in actual wastewater treatment. In real environmental conditions, the presence of diverse pollutants and interfering chemicals, such as varying pH, ionic strength, and organic matter, can challenge CNT performance. The complex wastewater matrix contains chemicals that may react with pollutants and oxidants, necessitating the application of CNTs in practical treatment scenarios. For instance, ions like phosphates, bicarbonates, carbonates, and sulfates can compete for adsorption sites on CNTs, potentially reducing their efficiency by blocking active sites. To address this, CNT-based materials, often in composite forms, are being explored for their high adsorption capacity for organic molecules, though their widespread use in wastewater treatment is still under development.

4.1.2.8 Nanocomposites

As early studies reported, every nanomaterial has its limitations, such as nZVI having oxidation and separation difficulty from degraded systems, whereas TiO_2 and ZnO have disadvantages related to the adsorption of light due to their large band gap energies. The applicability of CNTs is limited due to their high cost. So, to overcome these limitations and attain better removal capacity of contaminants from wastewater, hybrid nanocomposites are fabricated. In modern research, the synthesis of several nanocomposites is the most appealing area in the field of nanomaterials. For example, through the chemical deposition of nZVI on CNTs, a unique nano adsorbent was made. As per the results, this adsorbent has great potential for the effective removal of nitrate in water (Azari et al. 2014).

Another example is a synthesis of polymer nanocomposites, to enhance their physicochemical, mechanical, and thermophysical qualities. For the elimination of chemical impurities from water, polymer composites are employed as adsorbents and filtering membranes.

The development of polymer nanocomposites enhanced its pollution remediation applications such as the removal of heavy metals, dyes, and hydrocarbons from contaminated water by polymer nanocomposites.

4.1.3 Factors Affecting Adsorption Efficiency

Discussed earlier in this chapter, adsorption has been utilized for the elimination of a variety of impurities from wastewater. However, the performance of adsorbents is dependent on numerous factors, such as adsorbent doses, contact time, temperature, pH, surface properties of the adsorbents used, and starting ion concentration. This segment has discussed the impact of factors affecting the adsorption efficiency of nanoscale materials (Srivastava et al. 2015).

4.1.3.1 Effect of pH of Solution

pH is one of the main factors in handling the remedy with the adsorption procedure and influences the adsorption capability. Due to environmental impacts, the typical pH level of water is varied, so, this is required to find the ideal pH value for advanced adsorption capacity. Most of the researchers reported showing a pH-dependent adsorption process for heavy metals eliminating using nanomaterials. Such as example, Zhu and Li identified the primary effect of pH on the adsorption of Cu^{2+}, Ni^{2+}, Pb^{2+}, and Cu^{2+} and analyzed it by applying nanoparticles of hydrous manganese dioxide. The researchers discovered that the adsorption process plays an important role in the adsorption process, it was dependent upon the pH and the order of adsorption capacity was $Lead^{2+}$ > $Copper^{2+}$ > $Cadmium^{2+}$ > $Nickel^{2+}$, as demonstrated (Srivastava et al. 2015).

Hayat et al. (2011) invented advanced materials for the elimination of multiple divalent cations via the fifth generation of poly (propylene imine) (PPI-G5) dendrimer connected with SiO_2 nanoparticles with the highest adsorption acceptance at neutral pH in the range of 440–500 mg/g. Another study shows that magnetic nanocomposites compiled of G3-PAMAM-graft-poly (methyl acrylate) were developed for the removal of Pb^{2+} ions, and at pH 5–6, the maximum adsorption capacity was found. It was examined in comparative research that the percentage of Cd^{2+} adsorption increased subsequently at pH 4.0–6.0 and constant (97%) at pH 9.0. The adsorption percentage decreased to 80% when the pH was raised to 11.0. Another study reported that the adsorption of Pd^{2+} and Cu^{2+} on the surface of chitosan/TiO_2 nanofibers was higher at pH 6.0. Cegłowski and co-authors reported that the two types of synthesized nano-polymeric adsorbents (HDI-IC-DETA and HDI-IC-PEHA) were used for the removal of different types of heavy metal from the wastewater (Ge et al. 2011). It was observed that the ideal pH was 3 for Cr^{6+} when applying magnetite nanoparticles. So, the above discussion concludes that the major role is played by pH in the adsorption of contaminants during the treatment of wastewater. We may suggest that it is necessary to identify the ideal pH for maximum elimination of heavy metals from wastewater. The solution pH is a very important part because it can modify the mechanism of adsorption.

4.1.3.2 Temperature

Another important factor is the temperature of the solution, which affects the adsorption process. As we all know, temperature variations change the viscosity of solutions. The primary temperature rises, decreases the viscosity of the solution, improves the adsorbate molecules' transition over the adsorbate surface, and increases the adsorption rate.

Successive temperatures alter the adsorption process differently, irrespective of whether heat is absorbed or evolved. For the endothermic process, the adsorption rises with an increase in the temperature; for the exothermic process, the adsorption rate decreases, when the temperature increases. Furthermore, valuable information such as the standard Gibbs free energy ($\Delta G°$), distribution coefficient, enthalpy ($\Delta H°$), and entropy changes ($\Delta S°$), is offered by studying the temperature of the adsorption process.

$$\mathrm{KD} = \frac{q_e}{C_e} \tag{4.6}$$

KD = $\Delta H°$ (kJ/mol) and $\Delta S°$ (kJ/mol K^{-1}) as a function of temperature.
Ce = equilibrium metal concentration in solution (mmol/L).
qe = equilibrium metal concentration on the adsorbent (mmol/L).

The following reaction can be estimated as standard free energy ($\Delta G°$) (Gupta et al. 2011)

$$\Delta G° = \Delta H° - T\Delta S \tag{4.7}$$

By the basic laws of thermodynamics, we know that the minus value of $\Delta G°$ indicates that spontaneous adsorption. The −(minus)$\Delta H°$ describes the reaction's exothermic nature. This may suggest that physical adsorption is possible when the system's temperature increases. The +(positive)$\Delta S°$ indicates an expansion in uncertainty at solution/solid with temperature rises and decreases in adsorption because of the reduction of adsorptive forces between the functioning sites of the adsorbate and adsorbent species and relating the neighboring molecules of the adsorbed.

For example, as per references, the divalent mercury ion adsorption was explained as an exothermic reaction by chitosan-alginate nanoparticles (Tahoon et al. 2020), so the adsorption process improved at rises in 10 °C, but above this temperature, the adsorption declined. The desorption of metal ions from the adsorption site is increased when increasing kinetic energy and raising temperature lower the chelated ions on chitosan–alginate nanoparticles' surfaces. In conclusion, improving the temperature improved ion transfer to more binding sites, supporting the central ions adsorption ability.

4.1.3.3 Impact of Initial Ion Concentration

Commonly, the degree of adsorption of a solute improves with increases in an adsorbent concentration because the rise in the concentration of an adsorbent explains an increase in the effective adsorption site. Nevertheless, the total solute adsorption per unit weight of an absorbent can decline because of some obstruction by the contact of the active site of the absorbent (Xie et al. 2015). Wu et al. (2013) reported that when increasing the dose of beads Fe_3O_4@Ca-Alg from 0.1 to 0.3 g/50 mL, no rises in La-adsorption were observed.

The concentration of hazardous metal ions plays a vital role in adsorption. Usually, the hazardous metal ions concentration increases the speed of the adsorption process up to a specific limit, and then the adsorption procedure reduces due to the metal ion's ideal concentration.

4.1.4 Kinetics and Equilibrium Models for Adsorption

Adsorption is a surface spectacle, in this process, the adsorbate is gathered on the adsorbent material surface. The liquid–solid intermolecular force of attraction plays a role in adsorption when the adsorbent molecules are deposited on the highly porous solid surface.

In the adsorption process, the species that are involved choose the exact nature of bonding, but basically, the adsorption process is classified into two types: physisorption and chemisorption physisorption is, when an adsorbate is bound to the surface by weak Van der Waals forces, and when an adsorbate is tied through due to electrostatic attraction or covalent bonding, it is known as chemisorption.

The following equation can calculate the equilibrium of adsorption:

$$q_e = \frac{V(C_o - C_e)}{m} \tag{4.8}$$

where, solution volume (L) = V

mass of adsorbents (g) = m.

The initial and equilibrium adsorbate concentrations = C_0 and C_e.

The equilibrium between the adsorbent and solution is reached and at equilibrium, the amount of adsorption (q_e, mmol g^{-1}) can be calculated by the above-mentioned formula (Kurniawan and Lo 2009). In addition, adsorption may be explained as a bulk amount of adsorbent transfer by a material that is transferred from the liquid/gas phase to the surface of a solid and develops bound by chemical or physical bonding (Agarwal et al. 2016). It is significant to note that adsorption capacity and surface reactivity depend upon the adsorbent surface area, the large surface area permits high adsorption ability and surface reactivity.

4.1.4.1 Models of Adsorption Isotherm

As the amount adsorbed is a function of pressure (P) and Temperature (T) because initially as discussed already that the amount adsorbed is dependent on pressure and temperature. Mathematically represent it as:

$$a = f(P, T) \tag{4.9}$$

where a is the amount adsorbed.

When the temperature is kept constant and pressure is varied, then the curve between the a and P is known as the adsorption isotherm.

Freundlich and Langmuir and, Har-kin-Jura, Temkin, and Dubinin–Radushkevich are the many isotherm models. But for the explanation of adsorption statistics, mainly Langmuir and Freundlich's models are used (Abdel Ghafar et al. 2015).

The heterogeneous surface adsorption is represented by Freundlich isotherm equation. The distribution of energy for adsorbate sites (in the Freundlich isotherm) obeys

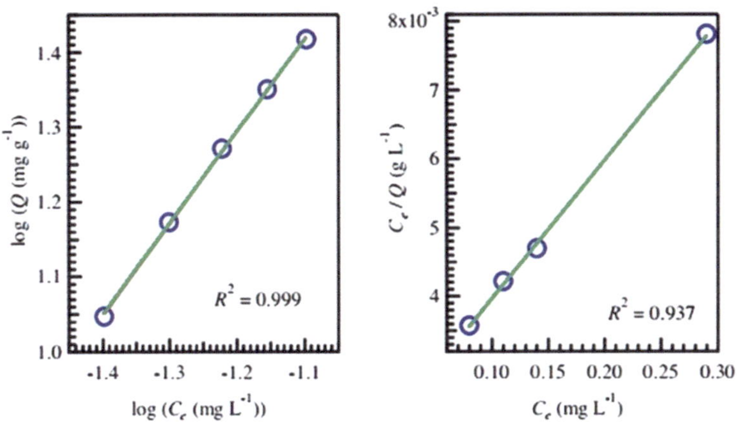

Fig. 4.5 CO_3O_4/SiO_2 nanocomposite MB adsorption is suggested to be studied using Freundlich and Langmuir isotherms. The solid lines show a linear fit. *Source* Ines et al. (2021)

a function of exponential type. The adsorption/desorption rate changes with the pressure at a constant temperature. The Freundlich equation is said as

$$\text{Log } q_e = \frac{1}{n}\log C_e + \log K_f \tag{4.10}$$

where $1/n$ are the constant characteristics of the system and k (mg g^{-1}) (Agarwal et al. 2016). This indicates that Freundlich isotherm is approximate and does not apply to the adsorption pf gases by solids at higher pressure.

The equation of Langmuir:

$$\frac{C_e}{q_e} = \left(\frac{1}{Q_{\max}}\right)K_1 + \left(\frac{1}{Q_{\max}}\right)C_e \tag{4.11}$$

q_e = amount of adsorbate adsorbed per unit mass of adsorbent (mg g^{-1}),

C_e = equilibrium concentration (mg L^{-1}),

K_1 = (affinity towards adsorbate) and $Q_{\max}$ = (monolayer adsorption capacity) are Langmuir constants (Agarwal et al. 2016). For example, the straight-line relation of Langmuir and Freundlich isotherms is indicated in Fig. 4.5 for MB adsorption on a nanocomposite of CO_3O_4/SiO_2.

4.1.4.2 Kinetic Models

A relevant kinetic model is required to investigate the mechanism and rate of adsorption processes (chemical reactions and mass transfer). Simple-first-order, pseudo-first-order, pseudo-second order, and intra-particle diffusion models, are some of the models that have been employed to express the adsorption mechanism. The simple-first-order, pseudo-first-order rate are provided with Eqs. (4.12) and (4.13) (Agarwal et al. 2016).

$$\log q_t = \log q_e + \frac{K_s}{2.303}t \tag{4.12}$$

$$\log(q_e - q_t) = \log q_e - \frac{K_1}{2.303}t \tag{4.13}$$

K_s, K_1 = rate constants $(h - 1)$

q_e, q_t = amount of adsorbate (mg g^{-1}) at time t and at equilibrium

The quasi-second-order formula is expressed as

$$\frac{t}{q_t} = \frac{1}{q_e}t + \frac{1}{K_2}q_e^2 \tag{4.14}$$

K_2 = rate constant at equilibrium (g mg^{-1} h^{-1}).

For calculating K_2 the intercept and slopes t/q versus t plot are used.

4.2 Catalytic Mechanisms

Researchers are increasingly focused on nanoscale catalysts for their wastewater treatment, mainly inorganic materials such as metal oxides and semiconductors. Photocatalysts (Liu 2019), electrocatalysts, and Fenton-based catalysts are some types of catalysts that are used for wastewater treatment.

4.2.1 Catalytic Degradation of Contaminants Using Nanomaterials

Light energy interacts with metallic nanoparticles in nano-catalysis, directing to extreme and wide activities of photocatalytic. Due to its high and wide photocatalytic activity, this treatment is gaining popularity. Hydroxyl radicals destroy bacteria and organic matter. Generally, nano-catalysts contain inorganic components like semi-conductors and metal oxides. Plenty of criteria for nano photocatalysts; include being harmless and their concentration in water and air under the limit approved level; making, precipitating, and forming ordinary particles (Fu et al. 2019).

Specific characteristics that nanomaterials have, acted as the foundation of high-performance catalytic purposes. With the large specific surface area, and a smaller size, it's advantage is the provision of a more active site for the reaction and improves their adsorption capacity (Wang et al. 2019). When the atomic ratio of the surface of particles is high the surface activity increases accordingly, and surface tension will increase at the same time (Podyacheva and Ismagilov 2015).

Eventually, the excellent implementation of the energy-gap band of the nanoparticles and the quantum size effect will develop and thus have powerful redox and other properties (Toyama et al. 2019). These minute effects afford the nanomaterials with brilliant chemical and physical properties, making them more suitable for catalytic reactions.

Whenever nanomaterials explore light energy, the electrons stimulated in the semiconductor band jump forward for the photovoltaic carrier, the valance band produces a hole (h^+) by losing electrons which makes the oxidation capacity stronger, and reduces the capacity of the conductive belt.

On the surface of nanomaterial, oxidative reduction reactions are performed on different positions, like photolysis of oxygen, water, and hydrogen and the nano catalyst will not change it's form during the reaction (Jin et al. 2020). For electrolysis, nano catalyst materials with high conductivity can be used, in between the electrolysis process, the electric field is conducive for the transfer of electrons (e.g., electrolysis in water, electrolysis in hydropower, and electrolysis in water loss), so speeds up the response rate (Ismail et al. 2012).

As per above catalytic reactions, it is realized that the choice and formation of catalysts are very significant. Nano catalysis with the absorption capacity of rapid change separation and migration are needed for attaining the target of high productivity catalyst. Commonly, the following materials are employed to complete high-competence redox strategies.

1. Graphene and other nanocarbon-related materials
2. Nano-metal elements and compounds
3. MOFs and other materials.

The primary catalyst, secondary catalyst, and carrier are the three categories into which the various materials' functions can be categorized based on the amount of nano catalyst present. When present alone, the primary catalyst component—also referred to as the activation component—can exhibit catalytic activity. The catalyst can increase the primary catalyst's activity when added, even though it frequently exists by itself without exhibiting catalytic activity.

Structured catalysts, such as iron–potassium oxide for ammonia synthesis–alum, oxide–alum, and oxide–oxide alumina catalysts, are generally employed to increase active components of specific surface are or to improve the stability of active structures. The activity of the active component can be altered by the modifier's catalyst; examples of these are oxide–alum, iron–potassium, and oxide catalysts (Wang and Zang 2017).

Tiny amounts of potassium oxide and aluminium oxide added to the medium of synthetic ammonia (iron contact) can double the onboard's catalytic activity and prolong its life by ten times (Podyacheva and Ismagilov 2015; Toyama et al. 2019). In addition, the carrier is a crucial part of the load-type catalyst; although it lacks catalytic property, it can serve as the framework for the catalyst's active ingredient (Jin et al. 2020).

Table 4.3 Standard electrode potentials (oxidants) (Akhavan 2009)

Oxidants	Oxidation potential (*V*) (hydrogen standard)
$^{\cdot}OH$	2.80
F_2	3.03
$^{\cdot}O$	2.42
$SO_4^{-\cdot}$	2.60
$S_2O_8^{2}$	2.01
O_3	2.07
Cl_2	1.36
H_2O_2	1.77
O_2	0.40

The dynamic components are evenly circulated across the catalyst's surface, creating what are known as "active points," while simultaneously acting to strengthen the catalyst's mechanical properties as nanomaterials have larger and more gaps than surface area. On the catalytic performance of the primary catalyst, various carriers, catalysts, and preparation techniques can have an effect (Kovacic 2019) (Table 4.3).

4.2.2 *Types of Catalysts*

4.2.2.1 Nanomaterials as Photocatalysts

When the light energy interacts with metallic nanoparticles, it is the base of photocatalytic nanomaterials and is of great importance due to their considerable and high photocatalytic activities towards various contaminants (Li et al. 2016). Generally, a variety of permanent organic pollutants in wastewater such as detergents, dyes, pesticides, and volatile organic compounds are destroyed by these photocatalytic properties. The simple mechanism of semiconductor-based photocatalysts is based on the photoexcitation of electrons.

In photocatalysts the excitation of electrons in the catalyst is the basic mechanism. For example, when the UV light irradiates on TiO_2, creates holes (h^+) and stimulated electrons (e^-) in the conduction band. In an aqueous medium, the hole is imprisoned by water molecules (H_2O) and forms hydroxyl radicals (OH^-). This hydroxyl radical is responsible for the oxidation of organic pollutants into water and releases gaseous degradation products.

Details of photocatalysts are described in the third section of the chapter.

4.2.2.2 Nanomaterials as Electro-catalysts

Microbial fuel cells are vital in wastewater treatment and direct electricity generation. Electrocatalysts play an important role in working microbial fuel cells. Recently, fuel cells' performance using nanomaterials as electro catalysts were improved by attaining higher surface area and a consistent supply of catalysts in reaction media. A lot of research has been shown on the progress of carbon nanomaterials as nano catalyst (Chaturvedi et al. 2012). It is reported that in glucose oxidation electrocatalysis reaction, Pt nanocatalyst supported by carbon black XC72 showed a capability up in current density to 6.2 mA cm^{-2}. Pt is the most effective co-catalyst among various types of noble metals. Even though Pt can be worked as an electro catalyst, it has various disadvantages that limit its application. For instance, Pt is a precious metal and has limited availability. The high cost narrowed the interest in applying it as a catalyst. It is described that Au electro catalyst supported on imidazolium-type ionic liquid/Polypyrrole nanotube surface was developed for the activity in fuel cell for wastewater treatment. It was noticed that because it has a smaller size and better dispersion this hybrid electrocatalyst showed high catalytic reduction in model pollutant 4-4-nitro phenol due to effective electrocatalysis of dioxygen reduction and 4-nitrophenol.

In the last decade, most researchers focused on advanced treatment methods and advanced oxidation processes for wastewater treatment. AOPs is a process that involves the production of active species such as ($^{\cdot}OH$) hydroxy radicals which are responsible for the oxidation of pollutants in wastewater. Various materials such as zeolite, carbons, metal and metal oxides, metal–organic farm work and reduced graphene oxides were denoted as a catalyst in such processes.

In current years, for the treatment of wastewater, nanomaterials have been effectively used as catalysts. It is necessary to mention that nanomaterials gain excessive density at active sites due to greater surface area. Carbonate, bicarbonate, phosphate, and sulphate, are hydroxyl radical scavengers in real wastewater that may affect the complete productivity of nano catalysts. Additionally, ammonia, heavy metals, suspended particles, and nitrogen may also impact the capability of catalysts in real wastewater. Usually, the researchers reported in the works used aqueous solutions of pollutants to observe the application of nano catalysts, but purification of real wastewater is more difficult (Ercan et al. 2015). Some other factors such as pH changes caused by a catalyst, leaching of metals, recycling of catalyst and metal oxide deposition, and the reactivity of reactive oxygen species (ROS) with organic-based nano catalysts should also be measured for real wastewater treatment.

The nanoparticles have a unique size range between 1–100 nm because they possess a high surface area and high density of active site. Various nanoparticles were used as a catalyst in the past for such applications, these include metals and metal oxides, carbon nanotubes (CNTs), zeolites, metal–organic frameworks (MOFs), etc.

4.2.2.3 Graphene-Based Materials

Graphite's allotropic form is Graphene. Graphene and its altered form used as advanced oxidation catalysts in the treatment of wastewater, it has been observed that surface hydroxyl groups, Lewis's acid sites, and π-electrons play an important role in generating hydroxyl radicals in various AOP. It has different functional groups (carbonyl, hydroxyl, epoxy) and a high surface area is responsible for adsorbing pollutants on its surface and competently destroying them through adsorbed ROS (reactive oxygen species). The nature of contaminants can affect the catalytic activity of graphene-based catalysts. For example, pollutants block the active site of the catalyst by adsorbing them on the surface of the catalyst.

4.2.2.4 Nanoscale Metal–Organic Frameworks and Other Materials

Now, the most popular semiconductor catalytic material is made of nanomaterials, like titanium dioxide and transitional metal oxide, which have been extensively researched and produced. This is because of their high stability, reasonable cost, and suitable oxidation–reduction power.

In a three-dimensional cyclic network structure, MOFs are made up of metal clusters, ions, and organic molecules. Their smaller energy gap zone improves light utilization efficiency by enabling nano catalysts to absorb more solar light. Based on the current investigation, its extreme complexity influences the light catalytic activity of pure MOF's medium luminous electron empty cave. Therefore, it's essential to insert nanoparticles into the MOF medium to remove the catalytic efficiency constraint brought on by the combined of the photovoltaic electron-empty cavity.

The advanced oxidation method can be applied in two ways: photochemical and electrochemical oxidation, based on the energy source; Fenton (Fenton-like oxidation) and ozone oxidation, based on the type of oxidant used.

Advanced oxidation technology (AOPs) is described in detail in another part of this chapter.

4.2.3 *Mechanisms Involved in Catalytic Reactions*

4.2.3.1 Oxidation (Photochemical)

Normally, activation of a photo catalyst carrier via light is a fundamental requirement for photo catalysis to occur. An optimized material should have higher porosity levels and larger surface areas coupled with other enhanced properties towards efficiently facilitating photo catalysis. Therefore, a good light catalyst carrier should display strong adsorption of the target oxides or degraded pollutants, easy solid–liquid separation, which may make it easier to go through different types of phase states, active components showing good binding strength without any reduction in catalytic activity, and long-lasting chemical stability. Photo catalysis oxidation has been categorized into three broad categories: direct photolysis oxidation, indirect photolysis catalytic oxidation, and a comprehensive photo oxidation method, including photovoltaic oxidation and photo Fenton oxidation, as the most widely known photochemical oxidation techniques used for the removal of organic contaminants.

4.2.3.2 Electrochemical Oxidation

The utilization of electric fields to expedite the rate of electron transfer in place of chemical oxidants is known as the enhanced electrochemical oxidation approach.

It directly causes the anode's reaction to oxidize or generates potent oxidizing agents in the cathode (such as $^{\cdot}OH$). It indirectly stimulates the catalytic oxidation process using a high-efficiency technique.

Among them, the cathode electrocatalytic oxidation method's synergistic usage of strong oxidative active free radicals can speed up the domestic free radical's conversion rate of oxide products like water. As an alternative, it can be applied to the treatment of organic wastewater with notable high-efficiency catalysis, cleansing, and pollution control.

4.2.3.3 Fenton Oxidation

The Fenton oxidation method, in general, describes a process that involves combining hydrogen peroxide and ferric iron in an acidic (pH 2–4) mixture, then oxidizing the organic component to a state(inorganic) utilizing a strong oxidizing hydroxyl group. In the past century, using the various types of Fenton reactions such as, optical Fenton (Wiley 2018), electric Fenton, microwave Fenton, ultrasonic Fenton, other complete technologies have been generated with the in-depth study and ongoing optimization of the Fenton reaction.

4.2.3.4 Ozone Oxidation

Oxygen molecules which carry triatomic oxygen, is another name for ozone. A normal temperature can be formed when oxygen atoms react with $^{\cdot}OH$, combining the remaining oxygen atoms to form a product that allows oxygen to reach a stable state.

Equation (4.15) indicates that there is a considerable influence on the deodorization, decolorization, and elimination of contaminants from sewage for $^{\cdot}OH$ and $^{\cdot}O$, which have a high oxidation potential. The byproduct is non-toxic and harmless to oxygen. Ozone functions as a powerful oxidant to prevent secondary contamination.

$$O_3 \rightarrow O_2 + O^{\cdot};\quad O^{\cdot} + H_2O \rightarrow OH^{\cdot} \tag{4.15}$$

There are two possible reactions to organic matter and ozone. Ozone directly reacts with organic matter present in water bodies to produce the first reaction. On organic matter containing double keys, such as aromatic hydrocarbons and unsaturated adipose hydrocarbons organic compounds, this approach is more selective and has a better oxidation-degrading impact.

The alternative method depends on alkaline circumstances for the breakdown of ozone in a water system, generating a free hydroxyl that reacts oxidatively with organic molecules Eq. (4.16).

$$H_2O + O_3 \rightarrow \left(OH^-\right)HO + \frac{1}{2}O_2 + HO_2 \tag{4.16}$$

4.2.3.5 Adsorption Catalysis

Principal heavy metal treatment techniques are chemical precipitation, biological treatment, membrane separation, adsorption, capacitive deionization (CDI), and electrodialysis. However, the lack of research they are now conducting has slowed down the development of wastewater treatment for heavy metals.

The adsorption catalysis method of nanomaterials in the treatment of heavy metal wastewater is quite broad. It has emerged as one of the most promising catalytic adsorbents for the elimination of heavy metals because of its high efficiency, low cost, simple device, and numerous other benefits.

The adsorption catalytic principle of nanomaterials primarily modifies the toxicity of active free radicals by utilizing a portion of their cationic keys, such as the surface hydroxyl group and the environment (Yu et al. 2018). This completes the primary goal of treating contaminants in water by influencing the ultimate movement and conversion of pollutants. As an illustration, mixing iron oxide magnetic material with graphite oil can improve the adsorption and elimination of heavy metal ions from water. Additionally, with UV exposure, titanium dioxide nanoparticles can decrease their oxidation and absorb organic contaminants with the interaction of surface electrostatic effects and hydrogen bonds.

Compared to a conventional catalyst, the surface area of nanomaterial adsorbents is larger. At the end, it influences the adsorption process because it frequently reunites to produce a secondary particle (Fig. 4.6).

Most nanomaterial adsorbents are dispersed in the form of powder in the water treatment environment. Even though heavy metals can be adsorbed well due to the tiny particles, separation frequently faces challenges to resolve this constraining issue. A chain of polyvinyl alcohol (PVA)/graphene oxide (GO)-sodium alginate (SA) nanocomposite hydrogel beads were made by Yu et al. (2018) using in-situ crosslinking to eliminate Pb^{2+}.

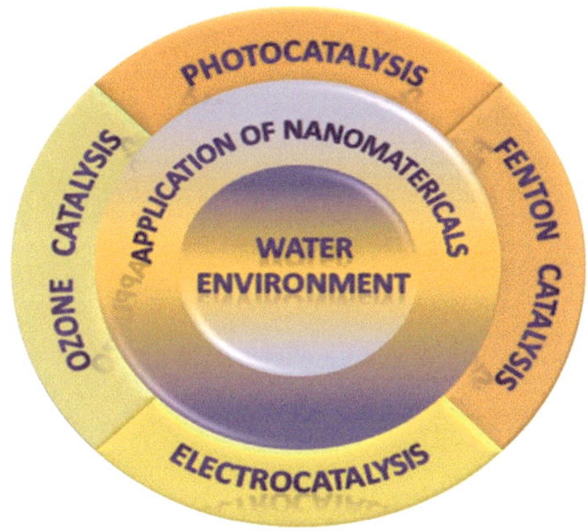

Fig. 4.6 Types of catalysts

4.2.3.6 Hydrolysis Catalysis (Hydrogen Analysis Reaction and Oxygen Analysis Reaction)

The hydrolysis reaction proceeds quite slowly. The conventional Gibbs free energy change relationship Eq. (4.17) states that $\Delta G = +237.2$ kJ mol^{-1}. As a result, more energy needs to be supplied to offset the positive change in Gibbs' freedom throughout the decomposition process. Currently, there are two primary methods of obtaining energy: (1) light catalytic hydrolysis, and (2) electrocatalytic hydrolysis.

$$H_2O \rightarrow \frac{1}{2}O_2 + H_2; \quad \Delta G = +237.2 \text{ KJ mol}^{-1} \tag{4.17}$$

An efficient plus workable approach to industrializing hydrogen energy will involve the creation of catalysts that effectively simulate the hydrolysis reaction mechanism (Chen 2020).

4.2.4 Factors Influencing Catalytic Activity

In watery conditions, it is possible that the metals, metal oxides, and their different forms will not exist independently. They may hydrate and form various complexes when water molecules are present; the process may be pH dependent. Furthermore, the pH of water can change due to the presence of pollutants and metal oxides. The pH of the water affects the mechanism and efficacy of the different AOPs because they are pH-dependent processes (Bhanvase et al. 2021). pH is a determining factor in UV-based, Fenton-like, and catalytic ozonation processes. For instance, an alkaline pH is needed for the ozonation process. On the other hand, an acidic pH is required

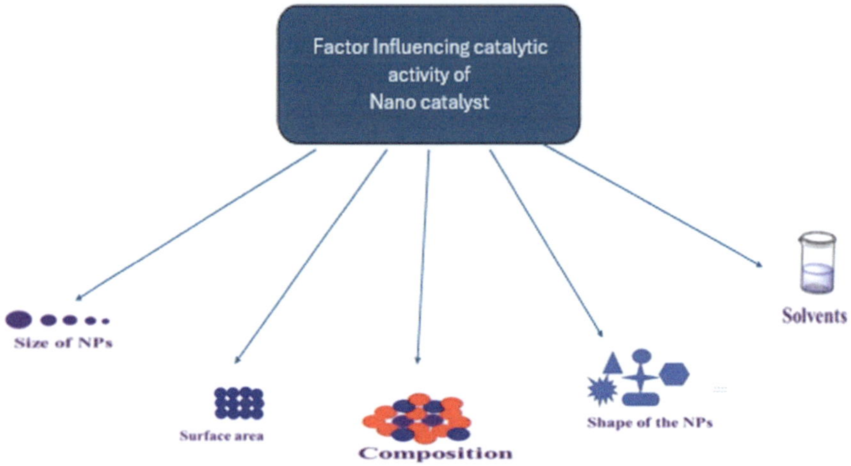

Fig. 4.7 Factors influencing catalytic activity

for the efficient generation of hydroxyl radicals, which are required for the degradation of contaminants, in the Fenton-like process of catalytic hydrogen peroxide decomposition utilizing Fe-based catalysts (Fig. 4.7).

These days, an enormous variety of catalysts (nano catalysts) are used in organic reactions. To synthesize important chemicals in industries and pharmaceutical industries some variety of metals, metal oxides, magnetic, alloyed/mixed nano catalysts are used in reactions such as reduction/hydrogenation, oxidation, condensation, C–C coupling, cyclization, with a high yield, selectivity, and reusability.

4.2.5 *Photocatalytic Mechanism*

Photocatalysis contains two words 'Photon; and 'catalysis' means when light or photon act in a reaction like a catalyst, so this type of reactions is known as Photocatalysis reactions. From the last few decades this method was widely used for wastewater treatment. This method is not only economic, easy to use, ecofriendly but also very useful for treatment of wastewater. With the use of this method we can degrade various harmful chemicals from different industries such as Pharmaceuticals, Pesticides, Textile, poultry, Paper, oil, and grease industries etc. this technique is also used in sterilizing, deodorizing, Air purifying effect, anti-fogging, self-cleaning, antifouling (Ameta et al. 2018; Zhang et al. 2018; Hussein 2012).

The idea of photocatalysis is based on natural Photosynthesis. Photocatalysis is also known as Artificial photosynthesis. This phenomenon was first suggested by Edmond Becquerel in 1839. We are blessed with an abundant Source of sunlight, this solar light we can change in heat or in the form of electricity we can use for further applications. In the recent era this is necessary and we must use natural resources as

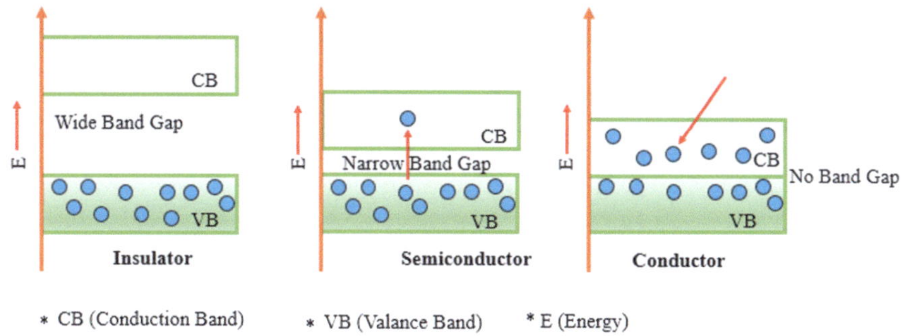

Fig. 4.8 Schematic representation of the photocatalysis process in the presence of quantum dots

energy sources and develop different techniques for reducing the pollution and for other applications (Ameta et al. 2018; Zhang et al. 2018; Hussein 2012) (Fig. 4.8).

Photocatalysis can be defined as a phenomenon where we are utilizing different types of light sources and semiconductor materials for completing the reaction. Substrate which absorbs light acts as a catalyst known as a photocatalyst in a chemical reaction. In the photocatalysis process light falls on semiconductor materials and an electron hole pair is generated (Zhang et al. 2018).

As we already know Materials are classified in three different categories like Conductor (if Value less than 1 eV), Semiconductor (if Value less than 1.5–3 eV), and insulator (if Value more than 5 eV), according to their band gap (energy difference between valance bond and conduction band) also represent as mention in Fig. 4.9 based on appearance, Photo catalysis can be classified in two broad categories (Zhang et al. 2018).

Homogenous Photocatalysis-Photocatalyst and Reactants both in same Phase (Solid, Liquid or gas). Homogenous Photocatalysis used comparatively limited heterogenous photocatalysis. The most used homogenous photocatalyst is the ozone and Photo-Fenton System (Fe^+ and Fe^+/H_2O_2) they used OH radical as Reactive Species to increase the rate of reaction.

Heterogenous Photo catalysis-Photo catalyst and Reactants both in different Phase (Solid, Liquid or gas). Heterogenous photocatalysis can be originated by water or air, may be organic, inorganic, or biological.

Because of many applications in different fields heterogenous photocatalysis has been broadly studied by many scientists. Also, lots of metals classified as semiconductors act as photocatalysts and reactants. They contain liquid easily to interact and shows better results.

Photocatalysis technique is used for self-cleaning of buildings in homes, offices cleaning purpose kitchens, washrooms, Hospitals, and many other areas.

Air Cleaning—Air Cleaning purpose in indoor and outdoor purifiers.

Water Purification—Different sources of Water cleaning.

Anti-tumor Activity—in Endoscopy type instrument for cancer Therapy.

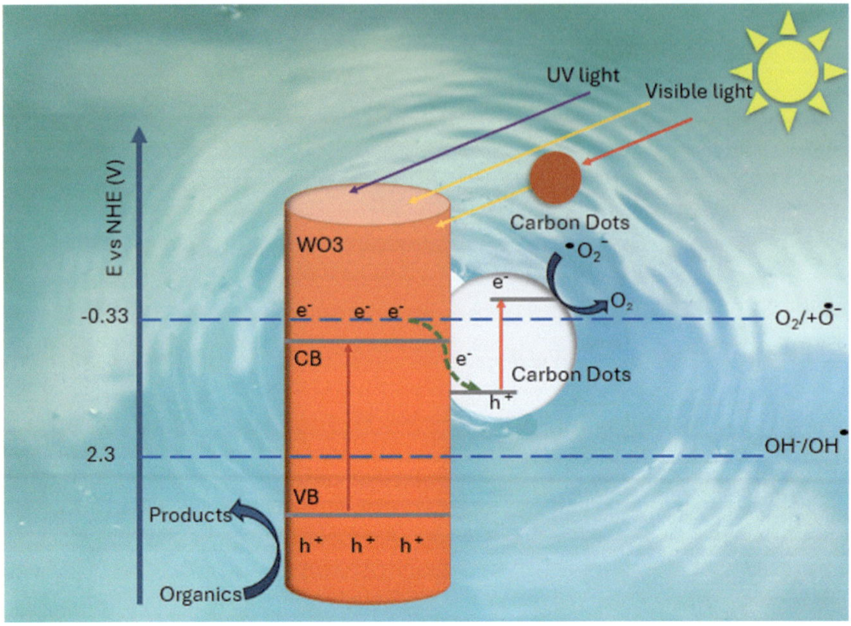

Fig. 4.9 Schematic representation of band gap differences between insulator, semiconductor and conductor

Self-Sterilizing—In Hospitals, Public rest room, Bathrooms and other, Dairy and Consumer Products, Public Transport etc.

Bio-Compatibility—For instruments contact lenses and clathrates.

These days, lots of research groups are engaged in searching for various new technologies using Photocatalysis (Zhang et al. 2018).

4.2.5.1 Photocatalysis Basics (Principles of Light Absorption, Electron–Hole Pair Generation)

The basic Principle of Photo catalysis is the same but the overall mechanism is quite complicated. Photocatalysis generally happens in semiconductors when electrons transfer from Valance band to Conduction band. When the Number of orbitals in the Valance Band known as Highest occupied molecular orbital (HOMO) and Conduction Band known as Lowest unoccupied molecular orbitals (LUMO) increases, the energy of electron shifting valance band to conduction band also decreases.

Hofmann and his research group measured that with the help of Flash photolysis at room temperature this Photo-generated electron can recombine to another species within 10–100 ns and generated heat. The recombination is the reason for the low quantum yield of Photocatalysis (Hussein 2012) (Fig. 4.10).

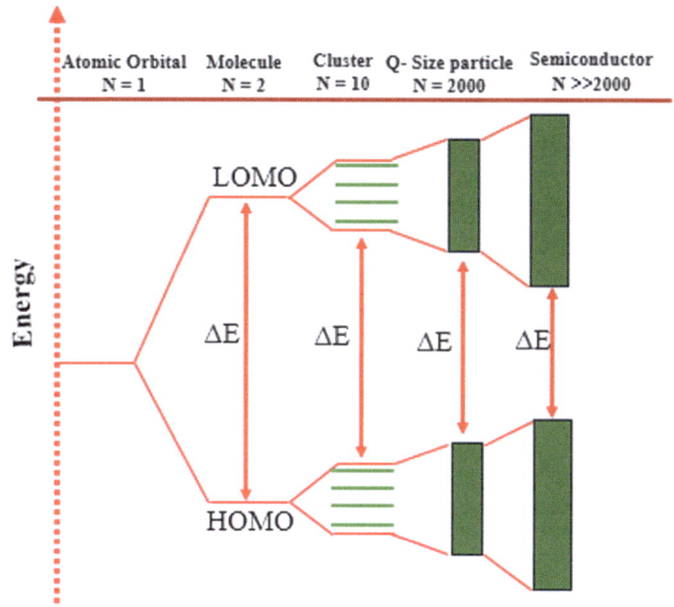

Fig. 4.10 According to change in electronic structure of compounds the number of atomic orbitals increases

When different types of light sources (Sun light, artificial light, UV, Visible, florescent lamp LED etc.) hit on the semiconductor surface (made from different metal oxides), the photon's energy might be equal or greater than the semiconductor.

In a Photo catalysis reaction when the photo catalyst strike by light with specific intensity or wavelength electrons absorbs the energy of the photon from the Valance band (VB) and after excitation electrons reach to conduction band (CB), after this incident the same number of gap or hole (h^+) will be generated to balance the band. When the hole is created by excited electron in balance band, this process induces the photo-excitation state.

Many reports suggested that in the mechanism of photocatalysis, recombination of the photo-generated electron and holes form a bulk quantity in the sample of the catalyst. If we add some appropriate trap sites the incorporation may give the best result for producing a surface of better absorption, or other sites.

$$TiO_2 + h\nu \rightarrow e^- + h^+ \quad (4.18)$$

If we provide sufficient time to the photo generated electron and holes before recombination the generated charge carriers move to catalyst surface and can go to charge transfer for start reduction oxidation reactions which can absorb different pollutants from wastewater.

Hole (h^+) of valance band having strong oxidising power and its potential range in between + 1.0 to − 3.5 V (at room temperature measured via Normal hydrogen

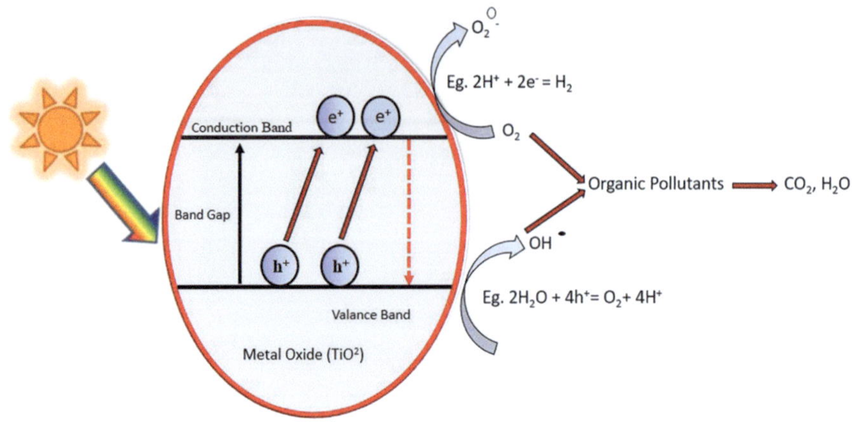

Fig. 4.11 Basic mechanism of photocatalysis

Electrode) it depends upon properties of semiconductor and pH. Therefore, hole which are present on catalyst surface plays an important role on pollutants degradation. With the help of oxidation reaction or indirect oxidation reaction, the surface attaches with hydroxyl radicals (·OH), (surface trapped h^+) before trapping from particle or from directly h^+ attached on the metal surface.

This surface generated Hydroxyl radical contains numerous efficiencies to oxidation of pollutants and so completely mineralizes them (Fig. 4.11).

Also, Photo-generated electrons react to avoid collection of excess charges on the particles of catalyst. Hence, improving the removal of electrons will improve the degradation of pollutants with the help of Photocatalytic oxidation.

Oxygen also has various applications in this reaction as we know this is very economic as well as a target to the oxidation of dyes and other organic compounds. Also, in aqueous medium it can be able to dissolve in other solutions (Bhanvase et al. 2021).

This hole is used for the oxidation of donor molecule and the excited electron is used for reducer or an acceptor, so in the photocatalyst process catalysis process reduction and oxidation both environments are going on, hence in the mechanism the photo catalyst is based on redox reactions.

With the help of titanium dioxide (TiO_2) which act as a best photocatalyst agent we will understand the mechanism of photocatalysis clearly (Zhu et al. 2017).

When titanium dioxide (TiO_2) strikes from any light source with appropriate intensity of wavelength, it emits an electron from its valance band and reaches to the conduction band, so that the electron hole is generated. Excess energy helps the conduction band of titanium dioxide by creating an electron and positive hole pair. The difference of energy between the conduction band and Valance is well known as bandgap and this stage is known as the photo excitation state.

When wastewater or contaminant react with titanium oxide so that positive hole (h^+) breaks down water molecule in hydrogen gas and OH (Hydroxyl) radical.

Hydroxyl radicals degrade the pollutant due to their strong oxidising nature. Oxygen molecules react with electrons and form superoxide anion.

This presence of light on semiconductor surfaces, produces products according to our demands.

In short, we can say the mechanism of photo catalysis is divided into five major points:

1. Charge carrier formation with the help of photon absorption.
2. Recombination of charge carrier.
3. Conduction band electron trapping from Ti(IV) site and yield Ti(III)
4. Trapping of a valance bond hole.
5. Initiation of an oxidative.

There are three different ways to interact substrate and semiconductor with each other Based on the band gap energies (Eg) arrangement of semiconductor heteroconjunction (Fig. 4.12).

1. Straddling Gap Structure—Conduction band of semiconductor 2 is more negative compared to the valance band of semiconductor 1. Based on photo catalysis electron (e^-) and (h^+) charge carrier transformation has happened in smaller energy gap (Eg) in the semiconductor hence, because of small band gap recombination of charge band gap will reduce performance of photo catalysis. In this condition Reduction of substrate takes place.
2. Staggered gap Structure—In the second type of semiconductor 2 is more negative and semiconductor 1 is more positive. Hence electron will flow semiconductor 2 to semiconductor 1 and hole will flow semiconductor 1 to semiconductor 2. Therefore, in type 2 semiconductor staggered band is better than straddling semiconductor (type 1) it separates better because of better charge carriers. Here oxidation of substrate takes place, when the redox level of substrate is higher than the valance band of the semiconductor.

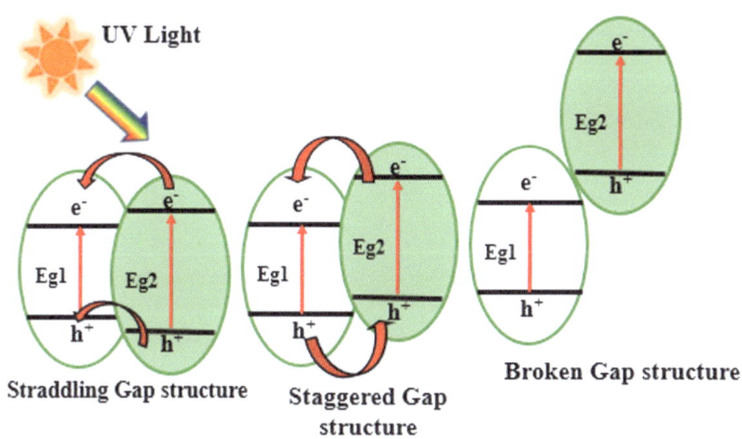

Fig. 4.12 According to band gap classification of heteroconjunction semiconductor

3. Broken gap structure—Conduction band and Valance band of semiconductor 1 is less than semiconductor 2. Hence these semiconductors are not capable of passing electron and hole interface to respective bands in the semiconductors. Oxidation and reduction are both not possible when the redox level of the substrate is higher than the conduction band and lower than the valance band of the semiconductor.

When we compare all three types of heteroconjunction structure type 2 staggered gap structure is most appropriate heteroconjunction system. It provides better separation of photogenerated electrons and in electron transfer process prevents the charge carrier recombination. It is proven in the photocatalytic reaction. As an evidence g-C_3N_4 doped copper oxide Cu_2O synthesized via solvothermal and chemisorption structure. Research reported that hybrid photolysis showed better photocatalysis in visible light after 3 h and generated hydrogen 795 micromole/g.

Different noble metals Rhodium, Titanium and silver, have some metal ions, TiO_2, Fe^{2+}, Cu^{2+}, Ir^{n+}, Ru^{3+}, Pd^{2+} and Rh^{3+} for produce Fe^{2+}/TiO_2, Cu^{2+}/TiO_2, Ir^{n+}/TiO_2, Ru^{3+}/TiO_2, Pd^{2+}/TiO_2, Rh^{3+}/TiO_2 like photo catalysis, and nitrogen, transition metals like Fe_2O_3, copper, silver and metalloids for example Platinum, Palladium, ruthenium, rhodium, metal oxides, graphene, carbon nanotubes, carbon quantum dots, clay are used for preparing Hybrid Photo catalysis (Tahoon et al. 2020).

4.2.5.2 Nanomaterials for Photocatalytic Degradation of Pollutants

Different research groups are engaged in developing advance techniques for wastewater treatment. In this category nanomaterials have related new and advance techniques and are also an interesting topic for various researchers due to the small size and large surface area, showing better efficiency in photo catalytic response. Due to the extended volume, the surface ratio enhances the number of active sites and interfacial charge-carrier transfer rates will achieve higher synergistic activities. Apart from this when nanomaterials interact with an active photo catalyst with specific physicochemical properties advantageous for photo catalysis, that will not only provide a large surface area but also improve different properties such as various morphologies, better absorption capacity and improve photocatalytic response at comparative low cost.

Carbon Nanotubes

Carbon nanotubes can be classified in single walled carbon nanotubes (SWCNTs), double walled carbon nanotubes (DWCNTs), and multi wall carbon nanotubes (MWCNTs). MWCNTs made from the rolling of single layers of graphite into concentric cylindrical. SWNTs show better synergism in constituents because there are better chances interface can contact, as compared to MWCNTs.

CNTs attract numerous research groups for developing Photocatalyst due to their special structure, and high mechanical strength makes them good candidates for advanced composite. Also, carbon nanotubes contain Superior thermal stability, outstanding transport, and electronic properties. On the other hand, CNTs having properties of semiconductor, semi-metallic, or metallic depend upon the tube diameter and helicity. They can conduct electricity due to delocalization of π-electrons.

CNTs showing good absorbent capacity due to their large and specific surface area, hollow and layered structured and presence of π-electrons on the surface. Also, there are many active sites we can create on nanotubes, making them a good candidate for promising material for environmental cleaning.

In photocatalytic oxidation, semiconductors such as TiO_2, ZnO, and WO_3 are used as photocatalysts for the degradation of various pollutants in wastewater. Upon illumination by a light source with energy equal to or greater than the band gap, photons excite electrons from the valence band to the conduction band, creating electron-hole pairs. This process involves two simultaneous reactions: oxidation by photogenerated holes and reduction by photogenerated electrons.

These holes are reacted with water molecules and formed Hydroxide Ions (OH^-) producing hydroxyl Radicals ($^{\cdot}OH$). These radicals depend upon the pH of the media. ($^{\cdot}OH$) Radicals absorbed targeted pollutants from the surface of the catalyst which will be oxidized by $^{\cdot}OH$. On the conduction band Excited electrons (e^-) generated hydroxyl ions react with oxygen which stimulated the formation of highly reactive superoxide radical ion ($O_2^{-\cdot}$) that can work as an oxidising agent for targeted pollutants.

Many research groups reported how incorporated carbon nanotubes on semiconductor. Some of them are magnetic bismuth ferrite ($BiFeO_3$) for the enhancement of Photo catalysis. Recently a research group suggested how CNT mediated semiconductor materials, with the oxidation reaction degraded and mineralized organic pollutants. Lots of research is already working on this topic and research groups have proved that when carbon nanotubes are coupled with semiconductor polymer, photocatalytic activity is promoted. They suggested SWCNTs in a different ratio on SSA (sulfosalicylic acid)—doped Polyaniline (PANI) in Photodegradation of organic dyes. SWCNT is best electron-acceptor created electron-hole pair, giving a longer time to produce more radicals. In Photocatalytic activity results showing better efficiency for the removal of Rose bengal and Methyl orange was improved from 85.2 to 98% and 57.9 to 94.35% respectively. New Graphene based Nanocomposite—Carbon quantum dots are new kinds of Nanomaterials being very nanosized at less than 10 nm. It has different properties like high quantum yield, high crystallinity, less toxic and highly stable. Carbon quantum dots used in a various field like Photocatalysis, Bioimaging, Optoelectronics, photovoltaic Devices, Biosensing and drug delivery. CQDs showing property of UCPL means wavelength emitted by CQDs upon light absorption is less than absorbed light wavelength that can be absorbed by composite and is extended to the visible region.

Apart from many properties CQDs capable to apply on photo catalysis, Murali et al. research group suggested Photogenerated holes and excited electron did this job. Properties of CQDs of UCPL enable them to use the full light spectrum of sun which may enhance light absorption and increase the photogenerated holes and electrons, hence the efficiency of photo excited electron increases also enhances the lifetime of photo-generated holes and electron.

Saleh (2013) suggested that N-doped CQDs (NCQDs) can support on TiO_2 enhance the photocatalytic performance on photoreduction of CO_2. Hence CQDs photocatalyst so that they not only provide better results, but also enhance the photocatalytic activity of other photocatalysts as a spectral converter, photosensitiser and electron mediator.

4.2.6 Graphene Based Photocatalyst

Graphene gained huge interest in the past decade due to their numerous applications in different fields. Production of graphene also increased from 14 to 1200 tons. Due to their continuous growth in commercialization they are also widely used in different scientific communities in different fields. It shows an excessively large surface area, remarkable physical properties, interesting electronic properties, advanced elasticity, is a thermal and electrical conductor and specially behaves as a charge carrier. Because of these specific properties Graphene is used for many applications such as solar cells, sensors, hydrogen production though water splitting, different catalytic activities, and energy storage devices. Graphene is a zero-gap two-dimensional semimetal with little overlap in between VB and CB and the large surface area makes them exact nano catalyst and nano absorbent.

Zhu and Wang (2017) demonstrated that the non-porous ability of N-TiO_2 and GN-TiO_2 indicated a mesoporous structure. Also, GN-TiO_2 exhibit large pore volume and like GO facilitate numerous active catalytical site for Hybrid catalysis.

4.2.7 Magnetic Nanomaterials-Based Photocatalyst

Magnetic absorbent is also one of the interesting fields of research where scientists are taking an interest in developing different magnetic absorbents with a combination of different oxides of metals like Fe, Co, Ni, Cu. These combined materials help to function more efficiently on the removal of waste-form water, when applying magnetic field nano absorbents can absorb, due to the magnetically active site of molecules. Magnetic absorbents can be used in a mineralization of contaminated water.

As we are aware about the toxic nature of different metals such as lead, chromium, mercury, arsenic, antimony, fluorides, copper, cobalt and other inorganic pollutants which not only contaminated our water resources but also generate many diseases like cancer, hormonal imbalance, skin irritation and other infections.

Some benefits for using Photocatalyst based on Magnetic nanomaterials are these Nanoparticles (NPs) can smoothly identify, separate, and recycle, have high chemical stability Nontoxic in nature, are Biocompatible, are Visible light active, do not causing Photo corrosion, and have good Magnetic properties.

As evidence Ferrite based magnetic NMs when interacted with a light source also creates hole and excited electron as like other semiconductor NMs and produces highly reactive species OH radicals and Oxygen superoxide, which can reduce the targeted pollutants more quickly.

Zhao and his research group (2009) suggested a biodegradable catalyst with the help of yeast as carbon support for TiO_2-core-shell nanoparticles. Photocatalyst was irradiated with the help of xenon lamps under 200–800 nm and using 350 W capacity of lamp it was degraded tetracycline. Its efficiency up to 86% was reported till six cycles.

Kosar Jahanara and Saeed Farhadi (2019) synthesized a magnetic $CdTiO_3$/$CuFe_2O_4$ nanocomposite using a sol-gel hydrothermal method at 180 °C, comprising spherical $CuFe_2O_4$ nanoparticles deposited on $CdTiO_3$ nanoplates. The nanocomposite achieves complete degradation of methylene blue, rhodamine B, and methyl orange under visible light in the presence of H_2O_2 within 90 minutes. Its superior photocatalytic performance, attributed to efficient electron-hole pair separation and hydroxyl radical generation, combined with magnetic recyclability, renders it highly suitable for environmental remediation applications.

Imidacloprid, the main ingredient of some insecticides generated from agricultural waste is very toxic for the environment, it can be also degrade with the help of ZnO/$CoFe_2O_3$ nanocomposite. This is suggested by Naghizadeh et al. (2019) they develop ZnO/$CoFe_2O_3$ nanocomposite with the help of the co-precipitation method. They are working on a pH level between 2–10 and the optimized pH was 10. A 50 W LED lamp was used as a source of light degradation which was very high due to the generation of OH radical, at maximum. The efficiency with the ZnO is nearly 56.24% and with $CoFeO_4$ was 58.45% and highest degradability is about 98.11% in the presence of ZnO/$CoFe_2O_3$.

Due to the low efficiency of TiO_2, due to the fast rate of electron-hole pair recombination, synthesis of Nanocomposite is one solution for improving the efficiency of TiO_2 Nanomaterials. Researcher Combination of TiO_2 and magnetic NPs is showing very good results due to lowers recombination of Electron-hole pair recombination, which enhances catalytic activities (Jahanara and Farhadi 2019; Zhao et al. 2016; Naghizadeh et al. 2019).

4.2.8 Photocatalytic Reaction Pathways

Photocatalytic reaction may divide in several steps and it depends upon various factors as to which type of product will form with the help of the photocatalysis mechanism.

Photocatalysis is a very efficient technique for wastewater purification and with the help of this technique it will not only degrade the waste as well as mineralizing the toxic pollutants. In the reaction pathway two types of process may occur, it may be degradation or mineralization.

In the degradation method organic pollutants can separate and decompose to various products and in the mineralization pathway organic pollutants after degradation, form pure water and CO_2. This process is complete in the presence of an appropriate intensity of light.

When irradiation light intensity is appropriate for penetrating the surface of the catalyst (semiconductor) then from the conduction band an electron is promoted to the valance band and a positive hole (h^+) will create on the surface of the valance band. This positive hole (h^+) will react with wastewater and the surface of the semiconductor forms OH radicals, simultaneously electrons will react with absorbed oxygen from the surface and produce perhydroxyl radicals ($HO_2^{\cdot}$) which are chemically very active species.

After the degradation process, we check the purity of water with the help of GC–MS or HPLC-MS. After that we check the toxicity of degraded products with the help of a toxicity test. These products are more or less toxic and are compared to parent compounds. As already discussed, after complete mineralization, the compound will form CO_2 and H_2O sometimes other products are also released after or during mineralization like Chloride, sulphates, ammonia, fluoride, sulphides, and nitrites. Finally with the help of oxidation and reduction, waste will be mineralized (CO_2 and H_2O) (Hussein 2012).

Second possibility which is also possible in photo catalysis reactions, is that energy may be less than the band gap of the irradiated semiconductor (Photocatalyst) in this pathway wastewater (dye) molecules decolorised process. In this process photosensitization is involved for decolorization of dye molecules for wastewater (Hussein 2012; Yaqoob et al. 2020).

4.2.9 Influence of Parameters (Light Intensity, Wavelength, Catalyst Dosage) on Photocatalytic Efficiency

Different parameters can affect efficiency of photo-catalyst. It may be pH, temperature, intensity of light, wavelength of light sources, catalyst dose and the concentration of wastewater also has an affect. With these parameters size, shape, different characters such as magnetic behaviour and luminescence properties also affect the rate of photocatalysis.

Recently various researchers suggested that the square root of light intensity depends on the photocatalyst (TiO_2) for different wavelengths of light and suggested kinetics of photocatalysis with light intensity on how they effect the degradation of dyes.

I. The rate of photocatalysis degradation at low intensity of light in between 0–20 mW/cm^2 is directly proportional with intensity of light and order of reaction being of first order.
II. The rate of photocatalysis degradation at high intensity of light in between 25 mW/cm^2 is directly proportional with square root of the intensity of light and order of reaction being of half order.
III. The rate of photocatalysis degradation at very high intensity of light is independent with intensity of light and order of reaction being of zero order.

Various research suggested that Photocatalytic properties increase with increases in the intensity of light. Rate of Photocatalytic degradation depends upon the intensity of incident light. When photocatalyst absorb quanta of light, they provide the quantum yield known as the rate of reaction of assimilation of radiations (Reza et al. 2015).

Efficiency of photocatalysis also depends on the wavelength of irradiation sources. Nguyen and his research group suggested that Electron-hole generation will enhance in the presence of a shorter wavelength of the irradiation source and will promote photo-catalysis reactions. Lower wavelength contains more energy not more photon. This was examined following photodecomposition of organic contaminants using $CaBi_2O_4$ and photo-degradation of 4 chlorophenol using TiO_2 also when Ag/TiO_2 used in photo-reduction of CO_2. Different organic compounds in the presence higher energy region of (UV light). If we irradiated at too low a wavelength we may have stopped the chance of side reactions.

Catalytic dose is one of the major parameters which will effect the rate of Photocatalysis degradation. Literature reveals that if we increase the mass of Catalyst, so the rate of reaction also increases and this increases the active sites of the surface of catalyst. As a result more contamination can be absorbed from wastewater. Also, the density of particles will improve the area of illumination.

As already discussed in the photo-catalysis process small molecules are more efficient compared to larger particles. May be the reasons behind this are:

1. More material concentration provides space for charge increase chemisorption of depletive oxygen. This enhances photohole migration and exciton dissociation.
2. In photocatalysis, holes and excited electrons are generated in bulk which makes available fewer traps and centers of recombination to be overcome before reaching the surface.

Neppolian et al. (2003) reported percentage degradation of dyes when loading TiO_2 dyes was Reactive Yellow 17 (RY17), Reactive red 2 (RR2) Reactive Blue 4 (RB4) was examined and changed the amount from 100 to 600 mg/100 ml was solution of dyes. Percent degradation was increased quickly when increasing the amount from 100 to 300 mg/ml over three days. In the case of RR2 and RB4 percentage degradation increases slightly in 300–500 mg of TiO_2. Percent concentration was constant when increase TiO_2 in the case of RY17. Represent the catalyst if on a higher level of effectiveness. Further he found if the amount of concentration increases 500–600 mg so mg in RY17, percent degradation was decreased.

This research suggested that if we increased the fixed dose, so the rate of reaction is increased but if we exceed the level of concentration, so the rate of reaction may be decreased or not affected or shows negative results. The reason behind that is when we increased the concentration of catalyst opacity of solution, we decrease the penetration of the photon flux in the reactor. Hence the rate of photocatalytic reaction decreasing. (Koe et al 2019; Reza et al. 2015).

4.3 Membrane Filtration and Separation Mechanism

Water is essential for us, with every human being dependent on water for their survival but due to different pollutants which release from various industries, agriculture waste and municipal waste quality of water is not fit for different applications. This water contains different heavy metals ions, Inorganic matter, pharmaceutical waste, organic dyes and other harmful bacteria and fungus. These chemicals cause severe health risk for living and non-living organisms and also pollute our environment drastically.

Many traditional methods such as Ion exchange, biological treatment, adsorption, chemical precipitation, and membrane filtration are used for maintaining water quality standards. Membrane filtration is used in different industries for treatment of wastewater from eighteenth century and after the second world war this filtration was used widely for purification of water.

Membrane filtration defined as a semipermeable membrane which acts as a barrier between two phases. Membrane filtrations can be classified in different categories according to their pore size and pressure. Membrane filtration is not only eco-friendly, economical, convenient, and compatible with working environment but also provides efficient results for wastewater treatment. Membrane technology can be done isothermally at low temperature and there is also no need for any additives.

This technique contains a semipermeable thin layer capable of separating contaminants as a function of their physical or chemical characteristics. Membrane filtration is used for a variety of different industrial applications. It can be broadly classified as Microfiltration, Ultrafiltration, Nanofiltration, and Reverse Osmosis. The selection of membrane is dependent upon the categorization and types of impurities present in wastewater (Fig. 4.13).

Microfiltration uses pressure driven separation techniques based on the size exclusion principle. Membrane has a symmetrical and asymmetrical porous structure, separate particles and dissolved component pore size of 0.1–1.0 micromolar. This membrane is widely used in the food industry for clarification of wastewater such as wine, juice, and beer, used in biotechnology industries for separation and harvesting of cells and is also used for the separation of recombinant proteins from cells debris and the purification of process streams for commercial and therapeutic uses. It is also used in Plasma separation from blood.

Ultrafiltration used when solutes are larger than the solvent use hydraulic pressure which allows only a pore size range of 2–50 nm. If thickness of filtration is more so filtration is less. This membrane can easily filter colloidal, particulates, fats, bacteria, and proteins and allows sugar and other low molecular weight molecules to pass through the membrane. Its pore size works between nano and micro filtration.

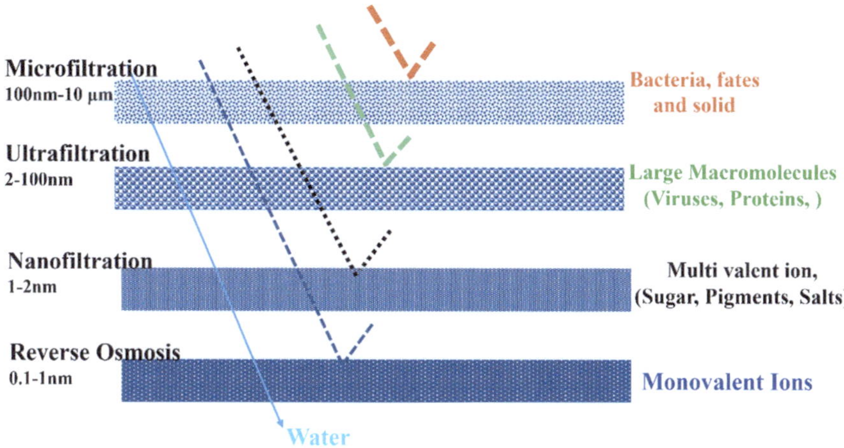

Fig. 4.13 Different membrane with pore size

It's dependent on transmembrane pressure for driving the separation and it can be operated between 50–120 PSI (3.4–8.3 bar).

Ultrafiltration can be used in Oil removal, brine clarification from wastewater (Khraisheh et al. 2021; Chadha et al. 2022).

4.3.1 Nanomaterial-Enhanced Membrane Filtration Processes

With the development in industrialization, new techniques of wastewater treatment must develop. When industrial wastewater is mixed with natural resources of water it will create numerous health issue and environmental pollution. For wastewater treatment nanotechnology also plays an important role and has significant benefits. These include enhancing the surface area, physical and chemical properties, and also has the high potential to solve many problems related to membrane technology.

Traditional methods such as Adsorption, oxidation, and extraction are effective but the problem with them is they are very costly. To overcome the cost issue we can incorporate nanomaterials in membrane filtration. Nanomaterials processes include photocatalysis, nanofiltration, adsorption electrochemical oxidation with the used of different metal oxides like TiO_2 membrane can made from ceramics, nanocomposite, and polymers.

Materials with size between 1–100 nm is consider as a NMs. They have various advantageous properties for enhancing the efficiency of membranes. Use of NMs in membrane technique is very common in these days with many scientists taking an interest for the incorporation of NMs into membrane technique for use in wastewater purification as this technique is very efficient in solving pollution issues and water scarcity. Various NMs can be used for incorporation in membrane techniques but commonly used are CNTs, Metal organic framework Graphene oxide, and other Carbon nanoparticles like fullerenes.

Surface roughness of NMs will enhance the membrane absorption capacity so that they can absorb heavy metals and organic chemicals. Also having the property of hydraulic stability make them durable in membrane filtration. In addition they are highly permeable, provide transportation of water, and some specific solute. They are also affective for fouling control, can make the membrane surface contamination free, and it will be helpful for cleaning and maintenance. NMs are highly selective so that they provide better removal of specific contaminant so that other interference of other compounds will be less. Hence, we can say that when we use NMs in membrane techniques they enhance the efficiency and effectiveness of contamination removal from wastewater. That is not only cost effective and provides better results for a healthy environment but also provides better heath of other living beings.

Nanomaterials (NMs) such as Carbon Nanomaterials are used as advanced membrane materials containing novel porous size-dependant properties which are helpful for removing inorganic as well organic contaminants from wastewater. Carbon nanomaterials are very attractive for removing impurities from water coming from various industries such as Textile industries, Hospitals, pharmaceuticals, and different laboratories, due to their excellent properties and novel features for removal of wastewater.

CNMs are classified in various types according to the arrangement of carbon atoms in that nanomaterial. Different morphology can be used in a variety of contaminants like a high concentration of heavy metals ions and different pathogens which can generate serious disease. Graphene oxide contains different functional groups in their structure, so this is compatible with more composites. Polymeric membrane are also very compatible with Graphene oxide (Alekseeva et al. 2016).

This is proven by experiments. GO contains high water flus in the fabrication of the membrane, possesses antifouling characteristics and also high resistance power in RO and NF. Carbon Dots -This is proven when hydroxyl, epoxy, and carboxyl groups are present in Carbon dots they enhance Antifouling Property, water solubility, antimicrobial properties, significant chlorine resistance. Also, the presence of hydrophilic groups makes them cost effective, reduces roughness of membrane and enhances an extraordinary separation capacity. The negative charge on the membrane surface increase anti fouling properties.

4.3.1.1 Fullerene

Fullerene is an allotrope, its Carbon formula is C_{60} also known as Buckminster fullerene. Its structure is hollow, cage-like and spherical. Fullerene is nontoxic and biocompatible in nature containing high affinity and surface volume ratio, making them perfect for separating different species in wastewater. Also considered as due to large surface, absorb industrial wastewater well. Fullerene can absorb heavy metals and pathogens can be treated with the help of Hydrophilic functionalized fullerene.

With the help of photocatalysis water soluble fullerene known as fullerols in the presence of UV or Visible light can generate Reactive Oxygen Species (ROS) this ROS can help with the removal of organic contaminants from wastewater.

When fullerene is incorporated with some Nano absorbents like zeolites and activated carbon then its absorption properties improve and showing a more porous structure will help more contaminant absorption and improve its hydrophobicity also. If fullerene is embedded with polymeric membrane for the formation of nanocomposite, it helps to enhance its absorption rate and mechanical properties (Alekseeva et al. 2016).

4.3.1.2 Inorganic NPs-Based Membranes

Nanomaterials can be used in membrane techniques in two different ways. They may be used individually as a layer or in the time of synthesis NPs we can introduce as a filler. Carbon-based NMs, metal oxides and nanosized polymers as NMs can be used for this purpose. This can easily attach to polymeric surfaces. Metal oxides such as TiO_2, ZnO_2, SiO_2 alumina can be used as a filler in membrane technique. TiO_2 can used for NaCl removal and SiO_2 can be used for aminosilanized and sodium sulphate removal from wastewater.

Silica based nanoparticle-based membrane can be used for pure gases from less contaminant content. For fuel cells proton exchange membranes are used, iron-based NMs can also be used. These NPs are useful for a variety of applications because they can catch impurities easily and they are inert. With the help of the molecular layer deposition method, scientists prepared Novel membrane, they also used TiO_2, this membrane may be used for nanofiltration membrane.

4.3.1.3 Nanofibers-Based Membranes

Cellulose is the oldest known nanofiber. Nanofibers having properties of porosity make nanofibers good candidates for selecting membrane preparation. Their diameter is less than 100 nm. Electrospinning technique is basically used for synthesis of nanofibers. These membranes are capable of removing oil from wastewater in comparatively less time, also this method is very energy efficient (Ameta et al. 2018).

4.3.2 Mechanism of Pollutants Removal Through Membrane

Membrane filtration is based on a thin semipermeable membrane which makes a barrier for solid and dissolved contaminants like colour of dye, to pass through from influent into effluent streams, because of the pressure gradient.

The membrane filtration porous layer will allow certain fluids to flow but is restricted to contaminant particles for flow. Flow and restriction are based on molecular sites, physicochemical properties, and size of particles. Previously membrane was prepared by the help of petroleum derived polymers like polypropylene (PP), Polyethylene (PE) and glass fibres.

Mechanism of filtration is based on four physical capturing processes such as sieving, diffusion, inertial impaction, and interception. Nanosized particle filtration mechanisms can be explained by Brownian motion, but this is not sufficient for the filtration mechanism. Because of Brownian motion some particles are spread to the surface of the membrane and collect both sides of the membrane, when they pass though the membrane. This happens because they pass from the distance and feel collision by capturing or inertia. This works the same as the cell walls of our bodies and other scientific membranes which are used for filtration of impurities, salts, heavy metals ion, viruses, and other particles from water. Membranes are a selective barrier in wastewater treatment. They allow the passage of pure water and stop other impurities and contaminants (Yang and Mi 2013).

However, the membrane process is further divided into integrated and non-integrated. In the integrated process for example, bioreactors work as a biological process where conventional activated sludge process and membrane is maintaining biology in the tank. Water can go down below 0–1 μ producing purified disinfected water. For example, in the Membrane Aerated Biofilm reactor (MABR) the membrane is used for aerator which effectively enhances the biological process. But there is need of further clarification of water. So that membrane bioreactors work as a diffuser for water treatment for the further process of clarification, water is transferred into a sequential polishing chamber.

Alekseeva et al. (2016) suggested that in Micellar enhanced Ultrafiltration some surfactants, when added to wastewater promote complexation between surfactants and ions or solute. When molecule of surfactants attaches with each other, they form amphiphilic aggregates that form large size micelles. Micelle structures are absorbed solutes by the hydrophobic interaction and ionic so the size of solute molecule is bigger that can be retained by the Ultrafiltration process.

4.4 Role of Nanomaterials in Enhancing Membrane Performance (Antifouling, Selectivity, Permeability)

NMs with small dimensions and being highly adjustable are excellent for antifouling efficiency, selectivity, permeability, and other properties nowadays very useful in many fields. There are a variety of Nanomaterial that we can use for incorporation in membrane for enhancing antifouling, selectivity, and permeability.

Characteristics of Membrane surface like hydrophobicity/hydrophilicity, charge and morphology, play an important role for the deposition of foulants and the formation of the fouling layer. So, modifying these characteristics through a selective

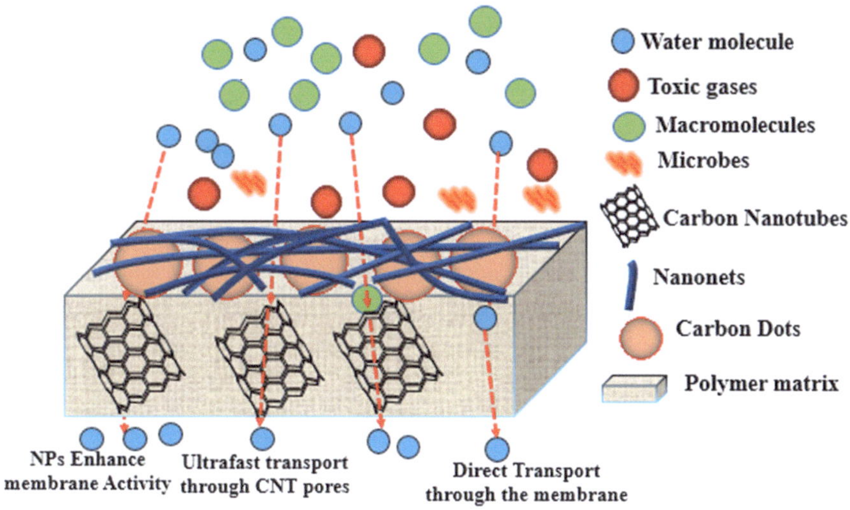

Fig. 4.14 Schematic representation of nanomaterial enhances the activity of membrane

surface modification strategy plays a vital role in improving membrane efficiency. For modification of membrane surface, we can decrease surface roughness of membrane, enhance hydrophilicity, adding some anti-adhesive monomer, polymer, and other particles on the surface of the membrane. For altering membrane surface including absorbing materials of fouling agents, chemical grafting, enhance selectivity and permeability, we can modify membrane surface with the help of different nanomaterials. We can apply two methods for the incorporation of nanomaterials in membrane surface (1) Nanostructured amphiphilic coating, which helps with the combination of hydrophobic and hydrophilic segments, to create nano-scaled chemically inert surfaces for foulants repelling. (2) Nanocomposite materials, where we can add small quantity of NMs for incorporated polymer matrix for enhance antifouling, selectivity, and permeability (Fig. 4.14).

Metal Oxide based nanomaterial are a good source for enhancing membrane properties. For example, Ag-NPs (silver nanoparticles), when incorporating silver nanoparticles into membrane, the three different processes we can choose mainly is in situ growth of Ag-NPs on membrane polymer matrix, process membrane casting solution can blend with Ag-NPs and attaching Ag-NPs on membrane surface. Swada and his research group used this approach for grafting polyacrylamide on polyether sulphone (PES) hollow fiber membrane surface and performed in situ Ag-NPs formation in the grafted polymer layer. TEM image suggested diameter 10 nm of NPs. This approach showed a nice result for the incorporation of NPs with membrane surface (Wang et al. 2024).

In membrane filtration fouling is a very common issue and must be managed. With the help of NPs, we can minimize this problem. Many researchers found Zwitterion like materials for capturing fouling of the membrane. Supramolecular chemistry, click chemistry, coupling chemistry, Bioinspired adhesion chemistry and mineralization chemistry are some fields which can track fouling issue of membrane.

When we develop new membrane, we must keep in our mind the rate of water permeation. If permeability is more, so capacity or efficiency of membrane also increased. Different types of cross-linker are used for increasing the enhancement of water permeation. PVDF coated nanocomposite or chitosan nanocomposite and the interfacial polymerization method are showing good results for increasing permeability and selectivity. GO functionalised nanosheets with O-(carboxymethyl) chitosan nanofiltration membrane are capable for permeation and salt rejection to take place. Antifouling mechanisms can be divided into three steps for better results. For the separation of fouling agents steric hindrance and hydration layer effect are working on the first step. Hydraulic pressure is also the main parameter for the efficiency of the membrane. Low surface energy because of electrostatic mechanism and hydrophobicity works on second stage for separating antifouling agent. After that in the third stage foulant will assemble and form cake (Nain et al. 2022).

Carbon-based NPs

Due to many active sites, large pore volume, low cost, excellent stability especially hydrophilicity is making them good candidates for membrane modification. Its derivative such as graphene, other derivatives of graphene, carbon nanotubes, carbon nanomaterials, nanodiamonds, cellulose nanocrystals, are widely used in various fields. Mainly surface engineering (modification of membrane surface), biomedicines, and environmental application.

Graphene

Functionalized graphene incorporated with polyether sulfone (PES) membrane checked the functional group effect and with a different concentration of graphene based nanosheets and performance of membrane, results showed that morphology of membrane was not changed but hydrophilicity, permeability, selectivity, and antifouling properties was enhanced.

In past studies GO with Polysulfone this membrane also showed good results in antifouling, permeability, and pollutant removal capacity. It also suggested that if GO was introduced with QDs in membrane that enhanced water permeability, solute selectivity antifouling and chlorine tolerance. Functional group also play important role for modifying the properties of membrane and if we introduce CNTs with carboxyl and hydroxyl group, they enhance hydrophilicity of membrane, oxygen rich group with MWCNTs improve permeability, hydrophilicity, the antifouling nature of membranes (Fig. 4.15).

Thus, we can say if we modify the membrane with the incorporated different types of NMs in Membrane according to our needs we can find better results in membrane techniques (Nain et al. 2022). Deng and his research group (2020) developed membrane with the help of MOF NPs (metal organic frameworks) for modification solubility of MOF they add polyether–sulphone membrane for uniform porous membrane. They observed permeability of this membrane and find better water flux also. Value of rejection also they observed 100% checked on bovine serum albumin.

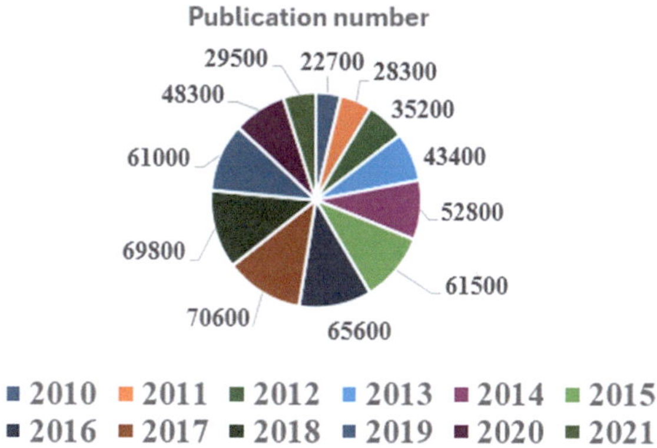

Fig. 4.15 Research paper published for membrane filtration incorporated nanomaterials in membrane technique

So, the modification with NPs improves the permeability of the membrane. With the same membrane addition of NPs also enhance hydrophilicity, porosity, flux recovery and oil water separation with the help of Phase inversion method, electrospinning method, interfacial polymerization method, self-assembly methods are used for developing membrane.

4.5 Introduction to AOPs for Wastewater Treatment

AOPs is based on the generation and use of powerful radicals mainly Hydroxyl radical (·OH). These radicals can generate with the help of different energy sources (UV, Solar energy, electrical energy, Sound energy, or may be simple chemicals like H_2O_2, Ozone) may be presence of specific catalyst or absence of catalyst. Other active oxygen containing species are hydroxyl radicals $HO_2^{\cdot}$, superoxide ion radical O_2— and its conjugated acid form also generated from AOPs, but they are comparatively less active then ·OH radical. In these days other radicals like sulphate also add with hydroxyl radical for improving the yield of AOPs. This based on normal hydrogen electrode for example ·OH radical oxidation potential is 2.8 V is stronger. Other oxidants also work very fast on organic compounds but it depends upon the oxidant and types of organic compounds.

On completion of reaction, products will be formed: water, CO_2, and salts depending upon types of pollutants. Some scavenger compounds are not degraded with the help of oxidation reaction. From this type of pollutant may be formed propionic acid, acetone etc. which further treated from other biological process. Therefore, AOPs can used as a single technique or with the addition of other treatment methods, we can use this process (Fig. 4.16).

When Hydrogen peroxide is in contact with the active surface or the surface of the catalyst is relatively unstable and liberating oxygen by continuous radicals like ·OH, ·OH_2. Then when we couple hydrogen peroxide with the appropriate catalyst it shows excellent efficiency, activity and cost effectiveness. This process is commonly used

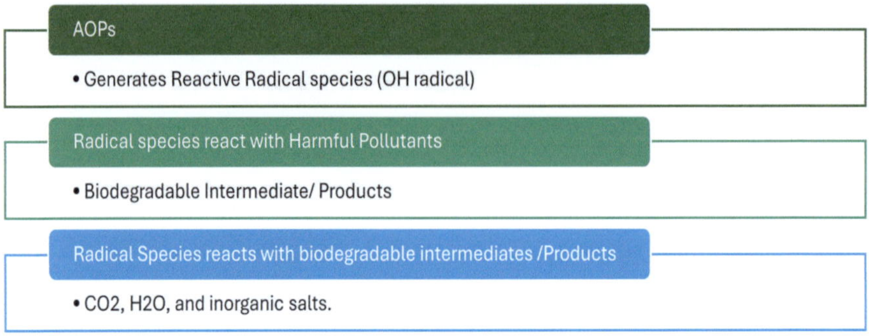

Fig. 4.16 Schematic representation steps of advance oxidation process

in contamination removing in many industries like Plastic and Petrochemical, Pharmaceuticals, Metal and metal plating, Chemicals, textile, Drying, Food processing industries, slaughterhouse waste, and removal for pathogens. (Oturan and Aaron 2014) reactivity depends upon the hydroxyl radical preparation mechanism.

In advanced oxidation processes (AOPs), both heterogeneous and homogeneous catalysts can be effectively employed. As previously reported in the literature, heterogeneous catalyst systems are generally superior to homogeneous systems due to their ease of separation and reusability, making them highly suitable for water treatment applications. It can tolerate extreme operating conditions so that secondary treatment can be avoided. Apart from this, the heterogenous catalyst system can be effective on a broader range of pH (2–9). Efficient, stable and active catalysts play an important role in the degradation of contaminants. Different catalyst may be Metal oxide, metal exchange clay, zeolites and other nanomaterials playing a significant role for AOPs (Katiyar et al. 2021; Kurian 2021).

Many Conventional methods are used for wastewater treatment which are already discussed in the previous section, among all of them Advance oxidation process is one of the most powerful and efficient methods for the purification of water. In the year 1980 AOPs was first proposed for wastewater treatment. After that this technique was frequently used for degraded organic and removal of inorganic contaminants from wastewater. AOPs mostly use Hydroxyl (OH) HO_2 and sulphate ($^{\cdot}OH$, HO_2, SO_4^{-} and other radicals) radicals as a strong oxidising agent in this process (Table 4.4).

Highly reactive radicals are responsible for the degradation of pollutants present in water. These radicals are very reactive and unstable because of their high reactivity and these radicals oxidized very easily and degraded the organic molecules present in wastewater. This method is also referred to as a versatile method due to degradation of harmful organic compounds and recalcitrant pollutants because of the production of hydroxyl radical via different systems.

This process is more efficient for producing intermediates which are showing non-toxic behaviour in reaction mixture.

When we go deep inside the AOPs Process we find that when Ozone (O_3), Hydrogen peroxide H_2O_2 or we may be in the presence of UV light. This type

Table 4.4 Values of oxidation potential of different species

Value of oxidation potential	E_0 (V, 25 °C)
Fluorine	3.03
˙OH	2.8
SO_4	2.5–3.1
Oxygen atom	2.42
Ozone	2.07
H_2O_2	1.78
Perhydroxyl radicals	1.7
Permegnate	1.68
Chlorine dioxide	1.57
Hypochlorous acid	1.49
Chlorine	1.36

of oxidation known as in situ chemical oxidation (Deng and Zhao 2015). AOPs are used and highly reactive hydroxyl radicals were produced by the In-situ method. These radicals are the strongest oxidizing agents and can oxidize any substances which are present in a mixture of water on a controlled diffusion rate. When ˙OH is formed it reacts fast with every species and at a very fast speed and changes into smaller inorganic molecules. One or more primary oxidants (e.g. Oxygen, H_2O_2, and ozone) and any energy source such as UV light or any catalyst like TiO_2 may be present when Hydroxyl radical (˙OH) are formed. If we want maximum ˙OH we must maintain all parameters which are necessary for ˙OH generation.

When we applied proper conditions for AOPs, it can have a degraded concentration of contamination from thousands to ppm. So AOPs are useful for reducing COD and TOC. AOPs also useful cleaning non-degradable and biologically toxic materials like Volatile Organic Compounds (VOC). Pesticides, Petroleum products, aromatic substances etc. this process mineralizes this substance into Water CO_2 and Salts (Katiyar et al. 2021).

4.5.1 Advantages

Advantage of this process are Easy to operate, low operational time will be required, secondary waste not generated, easily operated fluctuating flow rate and composition, eco -friendly and economical method.

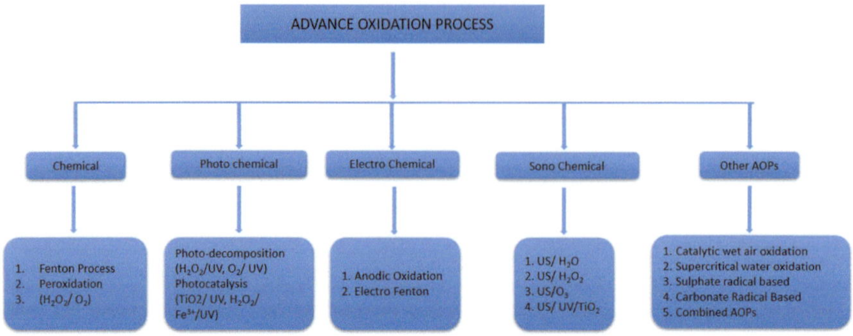

Fig. 4.17 Working mechanism of AOPs (Demir et al. 2019; Kumari and Kumar 2023)

4.5.2 Classification of AOPs

AOPs are classified depend on the type of oxidation. Oxidation may occur in four different ways (Fig. 4.17).

1. **Chemical Process**—in this process we can use different chemicals reagent and catalyst (1. Fenton's process, 2. Perozonation or ozonation (H_2O_2/O_3))
2. **Photochemical**—in the presence of solar energy or UV/Vis Source. (1. Photodecomposition H_2O_2/UV or O_3/UV, 2. Photocatalyst TiO_2/UV or H_2O_2/Fe^{2+}/UV)
3. **Electrochemical**—with the help of electric sources (1. Anodic oxidation, 2. Electro Fenton)
4. **Sono chemical**—with the help of ultrasound (US) (1. US/H_2O, 2. US/H_2O_2, 3. US/O_3, 4. US/UV/TiO_2)
5. **Other AOPs** (Catalytic wet oxidation, supercritical water oxidation, sulphate radical based, carbon radical based, Combined AOPs).

The type of activated oxygen process (AOP) to be utilized is classified based on the source of reactive oxygen species formation and depends on the type of pollutant, resources available, and process cost (Kumari and Kumar 2023).

4.5.3 Role of Nanomaterials as Catalysts in AOPs (e.g., Fenton Reaction, Ozonation)

As we already discussed earlier NMs specific surface area and surface energy, which makes them fit for high-surface activity. Nanosized particles contains high-surface area for reaction of pollutants and react actively and quickly. Nanocatalysts can operate effectively at lower temperatures, enhancing their selectivity, minimizing side reactions and byproducts, and improving efficiency in advanced oxidation processes

(AOPs). Additionally, their high recyclability reduces energy consumption. Because of these properties the researcher is also searching for various uses of NMs in green chemistry, efficient conversion of biomass, remediation of environment. Many reports reveal that Nanomaterials can enhance efficiency of AOPs.

NMs like Nanomembranes, nanotubes etc. have catalytic activity proven for degradation and detection of many biological and chemical agents such as organic substances, metals, algae, different drugs, viruses, nutrients etc. Nanomaterials enhance catalysis activity through synergistic mechanistic effect in between absorbent and catalytic NMs. This can be improved by Nanocomposite materials which contains a catalyst and are absorbent. These might be helpful for the degradation of pollutants like dye (Deng and Zhao 2015).

Some method such as Fenton process, ozonation, sonolysis, Photochemical oxidation is successfully used for degradation of harmful chemicals in wastewater (laboratory level). Different light source like UV, Visible light, or other catalyst like TiO_2 and various oxidants such as O_2, H_2O_2, O_3 (oxygen, hydrogen peroxide, and ozone) are necessary to initiate the AOPs (Cardoso et al. 2021).

Literature documented that with the help of AOPs they are capable of degradation of non-biodegradable, recalcitrant, and various organic components. Different engineered NMs owing help from electrochemical, physicochemical, and optical properties can solve many handling challenges, so incorporation of NMs in AOPs gives better results for water treatment.

NMs contains many unique features like higher electron mobility due to π-electron cloud in 2D structure of graphene. In 1D materials optical charge distribution and path length for light, higher specific surface area in 0D structures, all these properties are providing catalytic influence for the degradation of pollutants in AOPs. Also, after use NMs can change the oxidation state of degraded materials if we want to recycle. In addition, designed NMs based AOPs are being applied for the degradation of pollutants, these are (1) Peroxy-based AOPs that use Fenton, ozone, and persulphate in the presence of photocatalyst. (2) Electrochemical AOPs controlled by electricity in the presence of designed NMs. (3) Photocatalytic AOPs dependent on light and complete in the presence of light (Katiyar et al. 2021).

4.5.3.1 Fenton Process

Fenton is one of the most efficient innovative techniques for management of waste. This method is accepted as being very efficient for numerous organic pollutants. This method was firstly introduced by Fenton in 1894. In this method a Mixture of soluble iron (II) salt and H_2O_2 are known as Fenton's reagent and are used for degrading the pollutants. A very effective class of chemicals in AOPs. This method was firstly used by Fenton for oxidation degradation of tartaric acid. This method was followed by catalytic decomposition of H_2O_2 by salt of irons following a complex radical and chain mechanism (Lecompte et al. 2020).

$^{\cdot}OH$, is an oxidizing agent and it's dependent upon the Concentration of H_2O_2. If the concentration of H_2O_2 is increased till the appropriate value, so the rate of reaction

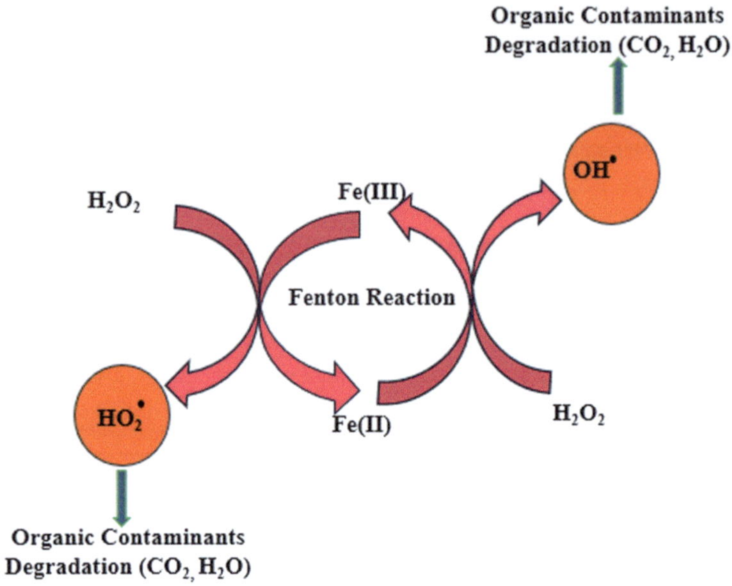

Fig. 4.18 Schematic representation of Fenton reaction

is also increased. If the radical's oxidation potentials are lower so the pollutant degradation rate is also decreased.

Recent Research suggested that some other Fenton-like process like iron NMs, Carbon nanocomposite, solid iron oxide, activated carbon and Fe_3O_4 magnetic NPs showed good results and owing to porosity, and unique surface area, they produce large number of Hydroxyl radicals and enhanced the activity of the Fenton process of wastewater (Deng and Zhao 2015) (Fig. 4.18).

Most of the Fenton reaction is done in acidic medium (pH 3) commonly. pH is very important for the Fenton reaction. ˙OH, radical reduces when $[Fe\ (II)\ (H_2O)]^{2+}$ reacts with H_2O_2 with this complex form in very low pH (2.5), on high pH more than pH 4 the rate of degradation is slow because at higher pH (pH > 4), the rate of decomposition is less due to shortage of free ion because formation of Fe(II) complexes in buffer. Ferric oxyhydroxide precipitation will initiate in range of basic pH. So, iron works as a catalyst on 3pH because precipitation is not done in this pH. We can replace iron oxide with other transition metals and they also improve efficiency for TOC removal. This is observed with a 90–100% removal result reported within 2 h. Also dose/concentration of Fe^{2+} and H_2O_2 major factor for rate of reaction also for determination cost of reaction. Rate of degradation dependant on Fe^{2+} concentration in the reaction matrix. But if we increase more concentrations beyond the limits so TDS may increase and with more H_2O_2 cleaning, ˙OH and remaining H_2O_2 increase TDS. H_2O_2 also damages the microorganisms which are responsible for the biological process in the next step.

To overcome these issues, more types of Fenton reaction were introduced by researchers for example Sono Fenton, Heterogenous Fenton reaction, heterogenous photo Fenton process, Sono-Photo Fenton process. (Katiyar et al. 2021; Kurian 2021; Pandis et al. 2022).

4.5.3.2 Heterogenous Fenton Process

Due to some drawback of the Fenton process, like regeneration of slow $Fe^{2+}$and recovery of it, dependency of pH, difficulty in handling, lacking in efficiency, make this method less attractive for waste treatment. On the other hand, in the heterogenous Fenton process, which is based on heterogenous catalyst mostly iron oxide, waste iron, iron and minerals of iron. This catalyst is appropriate for solving this issue and is applicable for the treatment of industrial waste.

Heterogenous Fenton process mechanism is not clearly resolved and one must check all intermediate actions and maintain the record of each step during the reaction. In heterogenous Fenton reactions many steps are contained. Initially, reactant absorption from the surface of catalyst, then a surfaces chemical reaction and finally degradation of the product is taking place.

In Fenton-like reactions heterogenous solid catalysts are mostly studied in a different pH range. In this reaction system Fe(III) is not mobile inside the structure catalyst pore generate hydroxyl radical from H_2O_2.This technique stops iron ions from leaching into water and causing iron hydroxide to precipitate. Other benefits of the approach includes extended activity during subsequent operations and simple recovery and separation of the spent catalysts. Because of their high surface area and porosity, which lead to a rise in the formation of ·OH radicals, a variety of nanomaterials have been tried with notable success.

Recently, nano-zero valent iron (NZVI) emerged in a family of heterogeneous Fenton reactions and oxidation of (NZVI) felicitated in situ production of Fenton Reagents. It was noticed that ultrasonic processes enhance the leaching rate of Fe^{2+}/Fe^{3+} ions from the core of the metal and diffusion through silica into bulk medium, improve degradation of sulphadiazine till multiple fold.

Photo Fenton Reactions initiate in the presence of a Source of light UV–Vis (greater than 600 nm) and Fenton reagents. Reaction benefit is preparation of more ·OH radicals can be synthesized by two different ways. First, when Fe^{3+} and Fe^{2+} are reduced in the presence of light (greater than 580 nm) and second by photolysis in the presence of UV light (greater than 310 nm).

In acidic medium (2.5–4.0 pH) Fe^{3+} firstly prepare $[Fe\ (OH)]^{2+}$. Fe^{2+} which is produced by Photolysis and Fenton reagent continuously formed ·OH radicals. They help to increase the rate of reaction and minimize the dose of iron (Katiyar et al. 2021; Deng and Zhao 2015; Govindraj et al. 2016).

4.5.3.3 Sono-Fenton Process

Sono-Fenton Process means a combination of Ultrasound with the Fenton Reagent. This method has overcome the drawback of limited pH range and low rate of Fe^{2+} diffusion. Reaction mechanism of Sono-Fenton Process is given in the scheme. In reaction H_2O_2 react with Fe^{2+} and generate $^{\cdot}OH$ radicals and Fe^{3+} ions, which stepwise react with H_2O_2 and the prepared $FeOOH^{2+}$ rate of Fe^{2+} is very slow. If we allow ultrasonication so this $FeOOH^{2+}$ divides into Fe^{2+} and $^{\cdot}OH$ radicals which increase the rate of reaction. In the presence of Ultrasonic, some part of H_2O_2 direct decompose in $^{\cdot}OH$. So, from the studies we found that when we combine ultrasound with Fenton so $^{\cdot}OH$ radicals are generated and there is a regeneration of Fe^{2+}.

Factors which effect this process are the concentration of Fenton reagent, pH, Concentration, DO level and Temperature. Akram et al. check the rhodamine B degradation and with the help of the Sono Fenton reaction they reported with or without ultra sonification Fenton process. They optimized ultrasonic alone and did not degrade Rhodamine B dye but the combination of Fenton and ultra sonification was effective for decolourization and degraded on high concentration also effectively (Katiyar et al. 2021; Pandis et al. 2022).

4.5.3.4 Sono-Photo-Fenton Process

As name suggested when Fenton reagent is combined with the Sono-photo so this combined reaction is known as the Sono-Photo-Fenton process. The basics of all these reactions we have already discussed but when we combine all these phenomenon it shows its best result for the degradation of waste. This combined process enhanced the $^{\cdot}OH$ radicals formation in aqueous solution and this radical combines with H_2O_2, also UV light dissociate H_2O_2 regenerate $^{\cdot}OH$ radicals in the reaction matrix. Ultra sonification also enhances the regeneration of $^{\cdot}OH$ radicals so the rate of degradation is increased. It is reported that Fe^{2+} also has less requirement alone in the Fenton Process. This is also experimentally proven. In a combined approach the amount of Fe^{2+} is less so the in treated water COD is also less which is important from an industrial point of view. Torres and his research group suggested degradation of bisphenol with the help of this method, and he reported 93% Dissolved organic compounds (DOC) was reduced in 4 h comparatively 5, 6 and 22% of was obtained in Photocatalysis, sonolysis and Photo-Fenton Process individually for the same treatment time. Other research groups also suggested that the combination of the Sono-Photo-Fenton-process is much more efficient compared to individual methods (Wang et al. 2010).

Many other combinations were also reported in literature where the Fenton reaction is combined with electric, known as electro-Fenton process. In this process at cathode electrochemical reduction of oxygen provide in situ production of H_2O_2 and regeneration of Fe^{2+} regularly when power supply is provided by external sources.

One more combination method in this group is Photo-electro Fenton Process when in electro photo system irradiated by UV–Vis source of light. This process can also enhance the degradation rate of contaminants (Xiong et al. 2015).

4.5.3.5 Ozonation

Ozone AOPs are a good source for the removal of major refractory chemicals from wastewater due to their strong oxidizing power and felicitate excellent disinfectant properties. For the treatment of effluent discharge generated from different industries, we can use a combination of O_3 and H_2O_2. Furthermore, we can combine this method also with a different catalyst or different energy sources (UV–Vis, Electricity, ultrasound etc.) for enhancing the activity of this method for pollutants degradation.

A low concentration of Ozone is not sufficient for the complete degradation of pollutants so the addition of hydrogen peroxide with ozone may enhance the efficiency and increase hydroxyl production and O_3 decomposition. Aqueous H_2O_2 can dissociate in hydroperoxide ion and proton. When hydrogen peroxide ions interacts with ozone, decomposition will occur. In alkaline medium (7–8) this reaction will work. Hydroxyl radicals' high concentration makes this process more efficient. Other advantages of ozonation we can mention is its convenient and bactericidal activity. This can be used for disinfection of potable water. Ozonolysis activity is limited by some factors such as ozone being less soluble in water, further depended on pH, temperature, types of micropollutants and side-reaction interference.

Several reports suggested that if we combined NMs with this process so NMs work as a catalyst on this reaction and activate the speed of degradation. Zerovalent iron and cost-effective metal slag such as Cu–S, Cd–S, Fe–S, Pb–S and Zn–S have been used with ozonation.

Other NMs like $NiFe_2O_4$, Co_3O_4, Fe_3O_4, MgO, Mn/Al_2O_3 and Fe_3O_4/carbon nanotubes also showing very efficient results.

Galey and Paslawski reported ozonation for pesticide elimination, in addition phenyl urea, Organochlorine (Lindane and Endosulfan) also on triazines (atrazine, simazine, terbutryne) from plant of wastewater efficiency or process was 80–90%. Adams and his research group checked biodegradability of 1,4 dioxane and he suggested that combination of O_3/H_2O_2 oxidant with conventional treatment of wastewater increases the reaction and also found enhancement in biodegradability (Cardoso et al. 2021; Katiyar et al. 2021; Pandis et al. 2022).

4.6 Mechanisms of Pollutant Degradation in AOPs

Advanced oxidation is a multistep process. In the first step with high reduction potential reactive oxygen species (ROS) will form in optimum quantities through oxidation reaction. Oxidant formation may be based on a chemical reaction or in the presence of catalyst may absorb energy (UV–Vis, Electrical energy, Ultrasound). After that

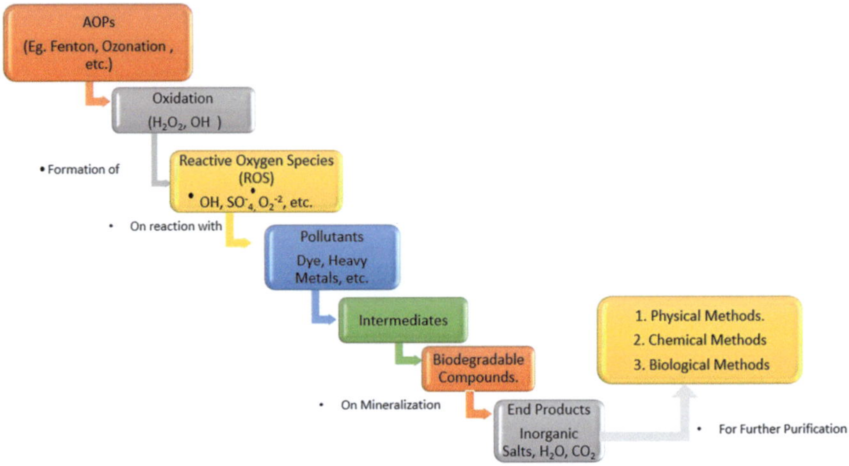

Fig. 4.19 Mechanism of pollutants degradation advance oxidation process

some oxidants will form (like $^{\cdot}OH$, $O_2^{\cdot-}$, $RO_2^{\cdot}$, $SO_4^{\cdot-}$ etc.) these oxidants will react with pollutants and form reaction intermediates (Zhou et al. 2019) (Fig. 4.19).

Each radical interacts in reaction with a different manner like $^{\cdot}OH$ radical, when it interacts with organic compounds, it may be via uptake hydrogen from C–H, N–H or OH groups or with the help of Radical. Radical interactions for example are when molecular oxygen is added they may form peroxyl radicals. With the help of direct electron transfer, an oxidized intermediate will form in the case of sulphate radical's electrophilic electron transfer will occur. Different mechanisms shows unique properties and efficiency for degradation. Oxidised Intermediates produce biodegradable compounds which pass though mineralization and form stable inorganic compounds, like water, $CO_{2,}$ and salts. These inorganic salts further react with biological (anaerobic, digestion, activated sludge, etc.) physical (filtration, sedimentation etc.) chemical (ion exchange and chemical precipitation etc.) in the figure mentioned the basic mechanism involves the degradation of Pollutants through an Advanced oxidation process (Kumari and Kumar 2023).

With the help of the Fenton reaction mechanism, we can clearly understand the process of AOPs Mechanism. Initially Fe^{2+} reacts with hydrogen peroxide in acidic medium and forms Fe^{3+} Hydroxyl anion (OH^-) and $^{\cdot}OH$ radical. Here hydrogen peroxide reacts like an oxidising agent and Ferrous ion (Fe^{2+} cation) is oxidised and ferric ion Fe^{3+} cation is prepared. Highly reactive species $^{\cdot}OH$ radical when react with H_2O_2 so they form water molecules and some other radicals also form which is a further reaction of hydrogen peroxide and forms oxygen, water and hydroxyl radical.

When Fe^{2+} ion reacts with $^{\cdot}OH$ radical they again oxidised and form Fe^{3+} and hydroxyl anion (OH^-). If $^{\cdot}OH$ radicals and hydrogen peroxide react with each other in this case, they generate cations (H^+) oxygen radical anion ($O_2^{-\cdot}$). When Fe^{3+} react with ($O_2^{-\cdot}$) oxygen radical anion so Iron reduced and form Fe^{2+} and oxygen.

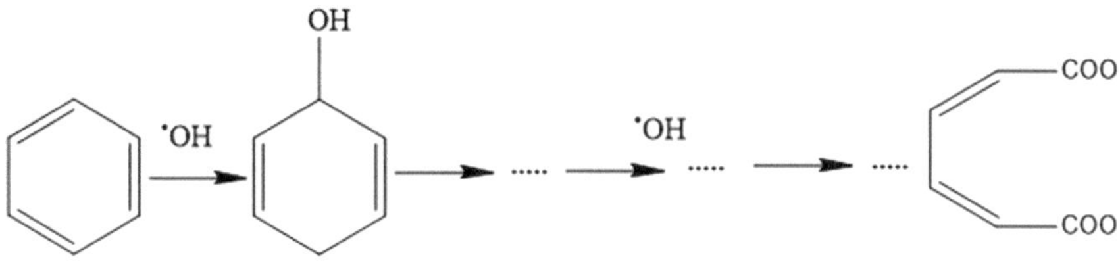

Fig. 4.20 Oxidation of an aromatic compound by ˙OH radical

Fe^{3+} combine with hydroxyl peroxide and iron again reduced in Fe^{2+} inside product hydrogen cation, HOO˙ also can prepared, when Fe^{3+} react with this HOO˙ radical they reduced and O_2, H^+ and reduced Iron will be formed.

From the reaction of ($O_2^{-\cdot}$), H_2O_2 and H^+ proton degraded water, oxygen and ˙OH radicals are formed with further reactions of ˙OH radicals with protons and reduced iron which generate degraded water and Fe^{3+} and ˙OH radicals are reacting with each other and they form hydrogen peroxide which is again used as an oxidizing agent in the reaction mixture.

The generated hydroxyl radical then interacts with contaminants, breaking them down into non-toxic and other final products. Because the hydroxyl radical has greater oxidative strength than the hydroperoxyl radical, ferrous ions are always preferred over ferric ions for Fenton's process (Kumari and Kumar 2023) (Fig. 4.20).

4.6.1 Synergistic Effects of Combining Nanomaterials with AOPs

Traditional methods are useful for treatment of water but due to some limitation's, scientists are taking an interest in developing new techniques also for the treatment of wastewater. In AOPs technique nanomaterial addition also showed a great result due to the unique surface area, and versatile nature. When AOPs hybrid techniques are used with NMs they showed high efficacy and effective treatment.

Literature proved that when we combined NMs with the Oxidation process it is showing the best result for treatment of water. Most of the research is focused on oxidation in the presence of photocatalyst, but when we add NMs in these techniques it is showing good results and can be used for large-scale treatment process.

If we want to better this result and avoid different issues such as pH, cost, temperature, pressure, kinetics, and better degradation rate, we can add NMs in AOPs this acts like a novel catalytic material.

If we focus on the other drawback of separate AOPs, its seen that if the concentration of organic pollutants is higher compared to the normal limit so activity of AOPs is reduce because of the presence of unknown oxidizing radicals, and lack of sufficient UV radiation. If we add NMs absorbing surface, they can absorb extra organic pollutant from wastewater.

When two or more methods or entity work together and their overall results are better than the individual effect, this effect is known as synergistic effect. When biological structure and medicinal drug are combined with each other, they will produce a synergistic effect. In another example which is related to the environment, when UV radiation and tobacco smoke are combined, they harm more comparatively than as individuals. One more example is when baking soda and vinegar mix with each other so CO_2 gas is released individually. They do not produce this gas otherwise, so when we add or mix something their efficiency will increase and this is known as the synergistic effect.

In the same concept when we add nanomaterials in Advanced Oxidation Process so efficiency of AOPs is increased to a very high rate. Many researchers suggested that when they combine NMs with AOPs method, so they observed better result in hybrid AOPs.

Zhao and his research group (2009) checked the synergistic effect of the Fenton reaction combined with reduced graphene oxide, co-modified titanium dioxide nanotube arrays and sun assisted electrochemical degradation of bisphenol A by CeO_2. They observed that in appropriate conditions photoelectrochemical—Fenton process degradation rate of bisphenol A was 80% increased and application of RGO-CeO_2-TNAs represent an excellent profile when working as a photoelectrochemical (PEC) catalyst. In the presence of positive charge Graphite Rod absorb RB5 dye contaminated textile water graphite rod increased the efficiency in decolorization. And TiO_2 NMs coating neglect iron reaction to 97%. This is the range of the reactor in the Fenton-based process. In this unique arrangement the researcher arranges the anode in two different sides to allow better enhancement and UV exposed side anode cover with TiO_2 NPs. Coupling in the Fenton process with other processes represents efficient water treatment because of synergistic effects within this process (Zhao and Stoddart 2009).

When we go deep inside to synergistic effect so we found that we can calculate the synergistic coefficient also with the help of the ratio of rate constant of the combined process, with the total of all individual process rate constant. If the rate constant of combined process is greater than individual process rate constant so efficiency of hybrid process is better, and shows better result in synergistic coefficients. We can understand this coefficient more clearly with the help of the son photocatalytic process.

$$\text{Synergistic coefficients} = \frac{\text{K sonophotocatalysis}}{\text{K sonolysis} + \text{K sonocatalysis} + \text{K photocatalysis}}$$

Using a visible light photocatalyst Ag_3PO_4/Bi_2S_3-HKUST-1-MOF synthesised by the ultrasound-assisted solvothermal technique, Mosleh et al. (2016) investigated the degradation of trypan blue (TB) and vesuvine (VS). The study was conducted in a continuous flow-loop photocatalytic reactor equipped with LED lights that are blue.

A maximum degradation of 98.44 and 99.36% of TB and VS, respectively, was reported for an initial dye concentration of 25 mg/L, solution flow rate of 70 mL/

min, irradiation, and sonication time of 25 min, pH of 6.0, and photocatalyst dosage of 0.25 g/L. The process parameters were optimized using central composite design by response surface methodology.

The synergetic coefficient at these ideal operating conditions was determined to be 2.53, indicating that the combined process was more efficient than the individual processes.

Chakma and Moholkar were reported Synergistic efficiency of TiO_2 and sonolysis. They checked individually Synergistic efficiency of UV and TiO_2 (photocatalyst) (UV + TiO_2) and in combined process Sonophotocatalysis (US + UV + TiO_2) they also add ZnO and Fe-doped ZnO catalyst and checked mechanistic effect of ultrasonication with catalyst and photocatalyst.

These catalysts help to generate more hydroxyl radicals which enhance the decolourization rate of wastewater from dye industries. Main reason behind that is the generation of shockwaves due to intense mixing with the help of ultrasonication it will help to increase the rate of desorption dye molecules.

They also observed negative value of Synergistic effect in sonolysis and photocatalysis individual methods when they apply on same dye molecule treatment from wastewater.

In conclusion we can say if we add more techniques with the help of NMs, so they enhance the efficiency of AOPs technique (Katiyar et al. 2021).

4.7 Conclusions

This has been important in understanding how nanomaterials function in enhancing wastewater treatment, which is essential in improving their chances of combating complex and emergent contaminants. There have been shows of efficiencies within broad scopes of mechanisms that include adsorption, catalytic degradation, disinfection, and even ion exchange. Each of these technologies exploits unique nanoscale properties, which include high surface area and associated reactivity, plus tailored surface modifications to eliminate a wide range of target pollutants, from heavy metals to persistent organic compounds and pathogens.

Many positive aspects notwithstanding, there are several challenges that have to be overcome in order for nanomaterials to see a full-scale application potential in large-scale wastewater treatments. These include the environmental and health implications of releasing nanomaterials and the problem of recovering them for recycling after use, coupled with high energy demands for some applications on nanomaterial-based treatment processes. Another issue lies in the economic viability for scaling up, especially in areas with less infrastructure or even less resources.

Further research in sustainable nanomaterials can be focused on decreased risk to the environment by introducing biodegradable or recyclable nanomaterials. Hybrid systems integrating traditional treatment methods with nanomaterials should also be explored for improved efficiency at reduced costs with minimized impacts to the environment. Improvements in synthesis techniques to optimize the stability, selectivity and regeneration capacity of nanomaterials will also enhance their application value in wastewater treatments.

Finally, further developments in nanomaterials for wastewater treatment hold hope not only for achieving good-quality water purification but also for helping to reach environmental goals in general. Thus, with innovative research, overcoming current limitations, and interdisciplinary collaboration, nanomaterials could be an important part of advanced and sustainable solutions for water treatment worldwide.

References

Abdel Ghafar HH, Ali GAM, Fouad OA, Makhlouf SA (2015) Enhancement of adsorption efficiency of methylene blue on Co_3O_4/SiO_2 nanocomposite. Desalination Water Treat 53(11):2980–2989

Agarwal S, Sadegh H, Monajjemi M, Hamdy AS, Ali GAM, Memar AO et al (2016) Efficient removal of toxic bromothymol blue and methylene blue from wastewater by polyvinyl alcohol. J Mol Liq 218:191–197

Akhavan O (2009) Lasting antibacterial activities of Ag–TiO_2/Ag/a-TiO_2 nanocomposite thin film photocatalysts under solar light irradiation. J Colloid Interface Sci 336:117–124

Alekseeva OV, Bagrovskaya NA, Noskov AV (2016) Sorption of Heavy Metal Ions by Fullerene and Polystyrene/fullerene Film Composite 52:443–447

Ameta R, Solanki MS, Benjamin S, Ameta SC (2018) Advanced oxidation process for wastewater treatment (Emerging green chemical technology), 135–175

Azari A, Babaei A-A, Kalantary RR, Esrafili A, Moazzen M, Kakavandi B (2014) Nitrate removal from aqueous solution using carbon nanotubes magnetized by nano zero-valent iron. J Mazandaran Univ Med Sci. 23(2):14–27

Bhanvase BA, Sonawane S, Pawade VB, Pandit AB (eds) (2021) Handbook of nanomaterials for wastewater treatment: fundamentals and scale-up issues. Elsevier

Bratsch G (1989) Standard electrode potentials and temperature coefficients in water at 298.15 K. J Phys Chem Ref Data 18(1):1–21

Cardoso IMF, Cardoso RMF, Esteves da Silva JCG (2021) Advance oxidation process coupled with nanomaterials for wastewater treatment. Nanomaterials Basel

Chadha U, Sevaraj SK, Thanu SV, Cholapadath V, Abraham AM, Zaiyan M, Manoharan M, Velmurugan (2022) A review of the function of using carbon nanomaterials in membrane filtration for contaminant removal from wastewater 9

Chaturvedi S, Dave PN, Shah NK (2012) Applications of nano-catalyst in new era. J Saudi Chem Soc 16:307–325

Chen H (2020) Optimization of active sites via crystal phase, composition, and morphology for efficient low-iridium oxygen evolution catalysts. Angew Chem

Demir ME, Chehade G, Dincer I, Yuzer B, Selcuk H (2019) Synergistic effects of advanced oxidation reactions in a combination of TiO2 photocatalysis for hydrogen production and wastewater treatment applications. Int J Hydrog Energy 44(43):23856–23867

Deng Y, Wu Y, Chen G, Zheng X, Dai M, Peng C (2020) Metal-organic framework membranes: recent development in the synthesis strategies and their application in oil-water separation. ChemEng J. 405:127004. https://doi.org/10.1016/j.cej.2020.127004

Deng Y, Zhao R (2015) Advanced oxidation processes (AOPs) in wastewater treatment 1:167–176

Ercan Ö, Deniz S, Yetimoğlu EK, Aydın A (2015) Degradation of reactive dyes using advanced oxidation method. Clean Soil Air Water 43:1031–1036

Fu F, Dionysiou DD, Liu H (2014) The use of zero-valent iron for groundwater remediation and wastewater treatment: a review. J Hazard Mater 267:194–205

Fu M, Liu Y, Zhang Q, Ning G, Fan X, Wang H et al (2019) Fe_2O_3 and Co bimetallic decorated nitrogen-doped graphene nanomaterial for effective electrochemical water split hydrogen evolution reaction. J Electroanal Chem 849

Ge F, Li M, Ye H, Zhao B (2011) Effective removal of heavy metal ions Cd^{2+}, Zn^{2+}, Pb^{2+}, Cu^{2+} from aqueous solution by polymer-modified magnetic nanoparticles. J Hazard Mater 211–2:366–372. https://doi.org/10.1016/j.jhazmat.2011.12.013

Govindraj D, Nambi I, Jaganathan S (2016) Nanocatalyst in Fenton based advance oxidation process for water and wastewater treatment. J Bio Nanosci

Gupta VK, Agarwal S, Saleh TA (2011) Synthesis and characterization of alumina-coated carbon nanotubes and their application for lead removal. J Hazard Mater 185:17–23

Gupta VK, Tyagi I, Sadegh H, Shahryari-Ghoshekandi R, Makhlouf ASH, Maazinejad B (2015) Nanoparticles as adsorbent; a positive approach for removal of noxious metal ions: a review. SciTechnol Dev 34:195

Hayat K, Gondal MA, Khaled MM, Ahmed S, Shemsi AM (2011) Nano ZnO synthesis by modified sol-gel method and its application in heterogeneous photocatalytic removal of phenol from water. ApplCatal a 393:122–129

Hussein FH (2012) Chapter 6 Photochemical treatments of industrial wastewater, pp 5427–5434

Industrial Water Resource Management: Challenges and Opportunities for Corporate Water Stewardship (2018) Wiley.com. https://www.wiley.com/en-us/Industrial+Water+Resource+Management%3A+Challenges+and+Opportunities+for+Corporate+Water+Stewardship-p-9781119272502

Ismail AA, Bahnemann DW, Al-Sayari SA (2012) Synthesis and photocatalytic properties of nanocrystalline Au, Pd and Pt photo-deposited onto mesoporous RuO_2-TiO_2 nanocomposites. ApplCatal A Gen. 431:62–68

Jahanara K, Farhadi S (2019) A magnetically separable plate-like cadmium titanate-copper ferrite nanocomposite with enhanced visible-light photocatalytic degradation performance for organic contaminants 9:15615–15628

Jin R, Qiu Z, Cheng W, Jin X (2020) Photocatalytic degradation of aniline by magnetic nanomaterials Fe3O4@SiO2@BiO1.8·0.04H2O/Ag3PO4. Chem Phys Lett 137747

Kalhapure RS, Sonawane SJ, Sikwal DR et al (2015) Solid lipid nanoparticles of clotrimazole silver complex: an efficient advances in materials science and engineering 7 nano antibacterial against *Staphylococcus aureus* and MRSA. Colloids Surf B Biointerfaces 136:651–658

Katiyar J, Bargole S, George S, Bhoi R, Saharan VK (2021) Advanced technologies for wastewater treatment: new trends. In: Handbook of nanomaterials for wastewater treatment, pp 85–133

Khan MR, Kurny ASW, Gulshan F (2015) Parameters affecting the photocatalytic degradation of dyes using TiO_2 7:1569–1578

Khin MM, Nair AS, Babu VJ, Murugan R, Ramakrishna S (2012) A review on nanomaterials for environmental remediation. Energy Environ Sci 5(8):8075–8109

Khraisheh M, Elhenawy S, AlMomani F, Al-Ghouti M, Hassan MK, Hameed BH (2021) Recent progress on nanomaterial-based membranes for water treatment

Koe WS, Lee JW, Chong WC, Pang YL, Sim LC (2019) An overview of photocatalytic degradation: photocatalysts, mechanism, and development of photocatalytic membrane 27:2522–2565

Kovacic M (2019) Advanced oxidation processes. In: Encyclopedia of water. Wiley, Hoboken, NJ, USA, pp 1–15

Kumari P, Kumar A (2023) Advance oxidation process: a remediation technique for organic and non-biodegradable pollutants 11 (Surface and Interface)

Kurian M (2021) Advance oxidation process and nanomaterials. Elsevier, Cleaner Eng Technol

Kurniawan TA, Lo W-H (2009) Removal of refractory compounds from stabilized landfill leachate using an integrated H_2O_2 oxidation and granular activated carbon (GAC) adsorption treatment. Water Res 43(16):4079–4091

Lecompte C, Bustillo-Lecompte C, Ghime D, Ghosh P, Zhang T, Zhang Z et al (2020) Advanced oxidation processes—applications, trends, and prospects. In: IntechOpen eBooks. https://doi.org/10.5772/intechopen.85681

Li XY, Ai LH, Jiang J (2016) Nanoscale zerovalent iron decorated on graphene nanosheets for Cr(VI) removal from aqueous solution: surface corrosion retard induced the enhanced performance. ChemEng J 288:789–797

Liang DW, Yang YH, Xu WW, Peng SK, Lu SF, Xiang Y (2014) Nonionic surfactant greatly enhances the reductive debromination of polybrominated diphenyl ethers by nanoscale zero-valent iron: mechanism and kinetics. J Hazard Mater 278:592–596

Liang H, Xu B, Wang Z (2013) Self-assembled 3D flower-like α-Fe_2O_3 microstructures and their superior capability for heavy metal ion removal. Mater Chem Phys 141(2–3):727–734

Liu J (2019) Recent progress in graphene-based noble-metal nanocomposites for electrocatalytic applications. Adv Mater 31:e1800696

Madden AS, Hochella MF Jr, Luxton TP (2006) Insights for size-dependent reactivity of hematite nanomineral surfaces through Cu^{2+} sorption. GeochimCosmochimActa 70(16):4095–4104

Madrakian T, Afkhami A, Ahmadi M, Bagheri H (2011) Removal of some cationic dyes from aqueous solutions using magnetic-modified multi-walled carbon nanotubes. J Hazard Mater 196:109–114

Mautner A (2020) Nanocellulose water treatment membranes and filters: a review. Polym Int 69(9):741–751

Mirkin CA, Letsinger RL, Mucic RC, Storhoff JJ (1996) A DNA-based method for rationally assembling nanoparticles into macroscopic materials. Nature 382(6592):607–609

Mosleh S, RahimiGhaedi M, Dashtian K (2016) Sonophotocatalytic degradation of trypan blue and vesuvine dyes in the presence of blue light active photocatalyst of Ag_3PO_4/Bi_2S_3-HKUST-1-MOF: central composite optimization and synergistic effect study. UltrasonSonochem 32:387–397

Naghizadeh M, Taher MA, Tamaddon AM (2019) Facile synthesis and characterization of magnetic nanocomposite ZnO/$CoFe_2O_4$ hetero-structure for rapid photocatalytic degradation of imidacloprid. Heliyon 5(11)

Nain A, Sangili A, Hu SR, Chen CH, Chen YL, Chang HT (2022) Recent progress in nanomaterial-functionalized membranes for removal of pollutants. iScience 25(7):104616

Neppolian B, Kanel SR, Choi HC, Shankar MV, Arabindoo B, Murugesan V (2003) Photocatalytic degradation of reactive yellow 17 dye in aqueous solution in the presence of TiO_2 with cement binder. Int J Photoenergy 5(2):45–49

OECD (Organisation for Economic Co-operation and Development) (2010) List of manufactured nanomaterials and list of endpoints for phase one of the sponsorship programme for the testing of manufactured nanomaterials: revision; series on the safety of manufactured nanomaterials 27. Organisation for Economic Co-operation and Development, Paris

Oturan MA, Aaron J-J (2014) Advanced oxidation process in water/wastewater treatment: principles and applications. Taylor and Francis, pp 2577–2641

Pandis PK, Kalogirou C, Kanellou E, Vaitsis C, Savvidou MG, Sourkouni G, Zorpas AA, Argirusis C (2022) Key points of advanced oxidation processes (AOPs) for wastewater, organic pollutants, and pharmaceutical waste treatment: a mini review. ChemEng 6(1):8

Peng X, Li Y, Luan Z, Di Z, Wang H, Tian B, Jia Z (2003) Adsorption of 1,2-dichlorobenzene from water to carbon nanotubes. Chem Phys Lett 376(1–2):154–158

Podyacheva OY, Ismagilov ZR (2015) Nitrogen-doped carbon nanomaterials: to the mechanism of growth, electrical conductivity, and application in catalysis. Catal Today 249:12–22

Prabhu S, Poulose EK (2012) Silver nanoparticles: mechanism of antimicrobial action, synthesis, medical applications, and toxicity effects. Int Nano Lett 2(1). https://doi.org/10.1186/2228-5326-2-32

Saleh TA (2013) Synthesis and application of carbon nanotubes and their applications. The role of carbon nanotubes in enhancement of photocatalysis

Srivastava RR, Kim MS, Lee JC, Ilyas S (2015) Liquid–liquid extraction of rhenium (VII) from an acidic chloride solution using Cyanex 923. Hydrometallurgy 157:33–38

Stefaniuk M, Oleszczuk P, Ok YS (2016) Review on nanozerovalent iron (nZVI): From synthesis to environmental applications. ChemEng J. 287:618–632

Tahoon MA, Siddeeg SM, Alsaiari NS, Mnif W, Rebah FB (2020) Effective heavy metals removal from water using nanomaterials: a review. Processes 8(6):645
Toyama H, Bessho K, Huang L, Hirota SK, Kano Y, Mase K et al (2019) The effects of water pollution on the phylogenetic community structure of aquatic plants in the East Tiaoxi River, China. Freshw Biol 65:632–645
Wang C, Kim J, Malgras V, Na J, Lin J, You J, Zhang M, Li J, Yamauchi Y (2019) Metal-organic frameworks and their derived materials: Emerging catalysts for a sulfate radicals-based advanced oxidation process in water purification. Small 15:e1900744
Wang N, Zhang R, Liu K, Zhang Y, Shi X, Sand W, Hou B (2024) Application of nanomaterials in antifouling: a review
Wang T, Zhang S (2017) Binary NiCu layered double hydroxide nanosheets for enhanced energy storage performance as supercapacitor electrode. Sci China Mater 61:296–302
Wang X, Cai W, Lin Y, Wang G, Liang C (2010) Mass production of micro/nanostructured porous ZnO plates and their strong structurally enhanced and selective adsorption performance for environmental remediation. J Mater Chem 20:8582–8590
Xie J, Lin Y, Li C, Wu D, Kong H (2015) Removal and recovery of phosphate from water by activated aluminum oxide and lanthanum oxide. Powder Technol 269:351–357
Xiong Z, Lai B, Yang P, Zhou Y, Wang J, Fang S (2015) Comparative study on the reactivity of Fe/Cu bimetallic particles and zero valent iron (ZVI) under different conditions of N2, air or without aeration. J Hazard Mater 297:261–268
Yang Q, Mi B (2013) Nanomaterials for membrane fouling control: accomplishments and challenges 20:536–555
Yaqoob AA, Parveen T, Umar K, Mohamad Ibrahim MN (2020) Role of nanomaterials in the treatment of wastewater: a review. Water 12(2):495
Yu D, Bai J, Liang H, Wang J, Li C (2015) A new fabrication of AgX (X = Br, I)–TiO_2 nanoparticles immobilized on polyacrylonitrile (PAN) nanofibers with high photocatalytic activity and renewable property. RSC Adv 5(111):91457–91465. https://doi.org/10.1039/c5ra19235h
Yu Y, Zhang G, Ye L (2018) Preparation and adsorption mechanism of polyvinyl alcohol/graphene oxide-sodium alginate nanocomposite hydrogel with high Pb(II) adsorption capacity. J ApplPolym Sci 136
Zhang J, Tian B, Wang L, Xing M, Lei J (2018) Phase control of TiO_2 photocatalyst. In: Photocatalysis: fundamentals, materials, and applications, pp 133–172
Zhao YL, Stoddart JF (2009) Noncovalent functionalization of single-walled carbon nanotubes. AccChem Res 42(8):1161–1171
Zhao X, Lu Z, Kan J, Yi C, Gao N, Zhu Z, Song M, Yan Y, Wang Y, Zhang T (2016) Synthesis of stable core-shell structured TiO_2@Fe_2O_3 based on carbon derived from yeast with enhanced photocatalytic ability. 6:46889–46899
Zhu S, Wang D (2017) Photocatalysis: basic principle, diverse forms of implementations and emerging scientific opportunities. Adv Energy Mater

Chapter 5
Application of Nanomaterials for Remediation of Emerging Pollutants in Industrial Wastewater

Mudita Nagpal, Nidhi Sharma, and Tanuja Nautiyal

Abstract Urbanization, population expansion, changes in consumption patterns and advancements in industrial processes have resulted in a drastic increase in the amount of industrial effluent being generated. The composition of effluents changes significantly based on the specific industrial processes, raw materials employed, and the nature of pollutants generated. Emerging pollutants in wastewater refer to contaminants that have recently been found in wastewater effluents and have potential health and environmental impacts. Therefore, there is a growing need for advanced technologies for the treatment of these emerging pollutants and protect human health and ecosystems. Nanomaterials offer various applications in wastewater treatment. Large surface area, reactivity, and selectivity of nanomaterials make them outstanding candidates for wastewater treatment. Further, the cost-effectiveness of the nanomaterials makes them an economically viable solution. In this article, we have discussed the recent reports in emerging pollutants removal with the help of nanomaterials. The studies focus on the remediation of five classes of emerging pollutants: industrial chemicals, pesticides, pharmaceutical compounds (drugs and antibiotics), endocrine disruptors and hormones and steroids. Further, the advantages offered by the advanced nanomaterials over traditional methods have also been discussed. The challenges and the future prospectives in this direction have also been highlighted.

Keywords Nanomaterials · Remediation · Industrial wastewater

M. Nagpal (✉) · N. Sharma
Department of Applied Sciences, Vivekananda Institute of Professional Studies, Delhi, India
e-mail: mudita.nagpal@vips.edu

T. Nautiyal
Dr. Akhilesh Das Gupta Institute of Technology and Management, Delhi, India

A. Mandpe et al. (eds.), *Nanotech Solutions for Industrial Wastewater*, Advances in Wastewater Research, https://doi.org/10.1007/978-981-96-6879-3_5

5.1 Introduction

In recent years, the substantial growth in population, industrial activity, and economic development has raised significant concerns about the pollutants discharge in environment (Hettige et al. 2019). The release of effluents from both industrial and residential sources which leads to the accumulation of harmful substances in water, which is severely impacts the aquatic ecosystems and health of humans (Chowdhary et al. 2022).

According to a UN annual report, an average of 1500 cubic kilometers of wastewater is produced annually, which is approximately six times the volume of all existing river waters worldwide (Lima 2018). Around 90% water in urban areas is polluted, with 70% attributed to sewage from houses and 33% to untreated wastewater from industries discharged into rivers and lakes (Geissen et al. 2015).

The term "emerging pollutants" pertains to newly identified pollutants found in aquatic environments and organisms, or to novel attributes and effects of compounds already existing in the environment. Emerging pollutants (EPs) have also been defined as substances detected in the environment that are not currently part of regular environmental monitoring programs. These substances might be considered for future regulation due to their harmful effects and/or persistence (Teodosiu et al. 2018) (Fig. 5.1). The European Environmental Agency emphasizes the importance of closely monitoring emerging pollutants (EPs), also known as "hazardous substances and chemicals," due to their rising presence and impacts in water bodies throughout the European Union (European Environment Agency, 2012).

It has been reported that the EPs present in water accumulate in the organisms, resulting in risks to the health of humans (Khan et al. 2022a, b). Various pharmaceutical compounds such as ibuprofen, diclofenac, galaxolide, chloroform, bromoform, caffeine, naproxen, methyl dihydro jasmonate, and butylated hydroxyanisole can disrupt aquatic ecosystems, affecting the health and behavior of aquatic organisms. This disruption results in negative effects in the entire food web, thus disturbing the ecosystem and biodiversity (Gopal and Mukherjee 2023). Hormones act as endocrine disruptors, potentially affecting the survival and reproductive health of fish species and indigenous fish populations (Marlatt et al. 2022). Pesticides applied to agricultural fields are released into water bodies through runoff and erosion, as well as leach into groundwater, contaminating drinking water supplies. Certain pesticides can induce hormonal imbalances in both humans and animals, potentially resulting in severe consequences, including fatalities (Sabarwal et al. 2018). Each year, around 4.6 million tons of various types of pesticides, including herbicides, rodenticides, insecticides, molluscicides, fungicides, nematocides, are applied, with a substantial portion ending up in water bodies (Ahmad et al. 2019).

The treatment of EPs in water is becoming increasingly urgent due to the absence of effective treatment facilities. Research on EPs is expanding, driven by the need to develop appropriate methods for wastewater treatment (Naidu et al. 2016). Current treatment methods for EPs include various approaches such as activated sludge-based systems, membrane filtration, adsorption on activated carbon, and treatments

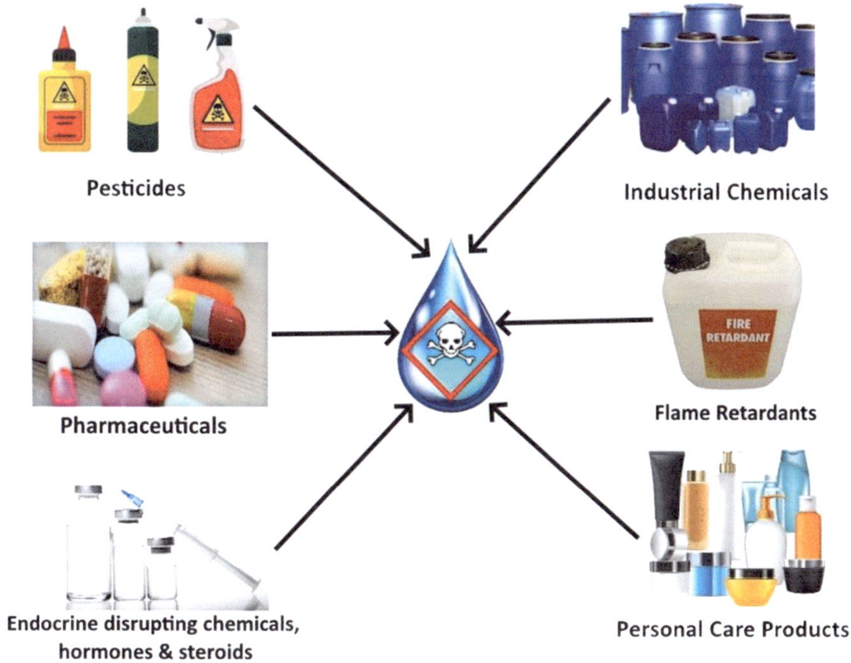

Fig. 5.1 Emerging pollutants (EPs) in wastewater

like permanganate oxidation, ozonation, chlorination, and UV irradiation (Mashile et al. 2022). However, these conventional methods sometimes fall short in adequately reducing EPs to levels necessary for reuse of wastewater. Consequently, the need of the hour is novel treatment technologies to ensure the provision of water of good quality fit for humans and use in environment.

Advancements in nanotechnology have created significant methods for next-generation process for treatment of water. Compared to traditional bulk materials, certain nanomaterials exhibit enhanced potential for treating EPs in studies at the laboratory level due to their unique properties which are dependent on size (Mauter et al. 2018). Nanomaterials possess attributes such as large area, elevated free energy, abundant sites for reaction and rapid dissolution (Zhang et al. 2016). Numerous nanotechnology reports focus on various aspects of water and wastewater treatment, including filtration using membranes, photocatalysis, adsorption, disinfection, microbial control (Bhat et al. 2022; Jain et al. 2021). Particularly, photocatalysis and adsorption utilizing nanomaterials, are largely employed for water treatment. Adsorption processes offer efficient pollutants removal from water (Sahoo and Prelot 2020), while photocatalysis not only eliminates stable and non-biodegradable pollutants but also effectively degrades organic contaminants in trace quantities with minor conditions, resulting in minimal byproducts (Zare et al. 2021).

This review aims to gather and analyze recent research findings related to the use of various nanostructures, including metal organic frameworks, metal-based nanoparticles, inorganic-biopolymer hybrid nanomaterials and materials composed of carbon like graphene, carbon nanotubes, and graphene oxide (GO), in the treatment of water with EPs. Each type of nanostructure possesses unique properties that differentiate them from one another to a significant extent. Various characteristics which influence the interactions between pollutants and nanostructures, will be discussed. In addition, mechanism of removal and advantages of using these nanomaterials in treatment processes, this article will also highlight the challenges in current research and suggest potential avenues for further exploration (Fig. 5.2).

5.2 Nanomaterials Used for the Remediation of Emerging Pollutant

5.2.1 Carbon-Based Materials

5.2.1.1 Carbon Nanotubes

Carbon nanotubes (CNTs) are cylinder-shaped and composed of hexagonally arranged atoms of carbon. CNTs can be made of a single wall (SWCNTs), or multi-walled, i.e., multiple layers of carbon atoms (MWCNTs) (Rahman et al. 2019). They exhibit extraordinary properties, which makes them a potential candidate for large number of applications. CNTs possess remarkable mechanical strength, with tensile strength surpassing that of any known material. They exhibit excellent thermal conductivity, electrical conductivity, chemical stability, biocompatibility and light absorption and emission in the near-infrared and visible regions (Ding et al. 2020; Amiot et al. 2008; Liang and Cheng 2010). They are commonly used in large number of fields, including medicine, materials science, nanotechnology, and electronics (Jain et al. 2022). Moreover, CNTs are effective at removing various emerging pollutants in wastewater (Kurwadkar et al. 2019), due to their ability to be transformed in numerous ways to modify their properties. Additionally, inert carbon nanomaterials' catalytic activity can be boosted through methods like introducing defects, incorporating heteroatoms, and adsorbing organic molecules (Yan et al. 2015). CNTs have exhibited a strong attraction to polar organic compounds due to different mechanisms, including the, π–π interactions, hydrogen bonding, Lewis acid–base interactions, and hydrophobic effect (Yu et al. 2014a, b).

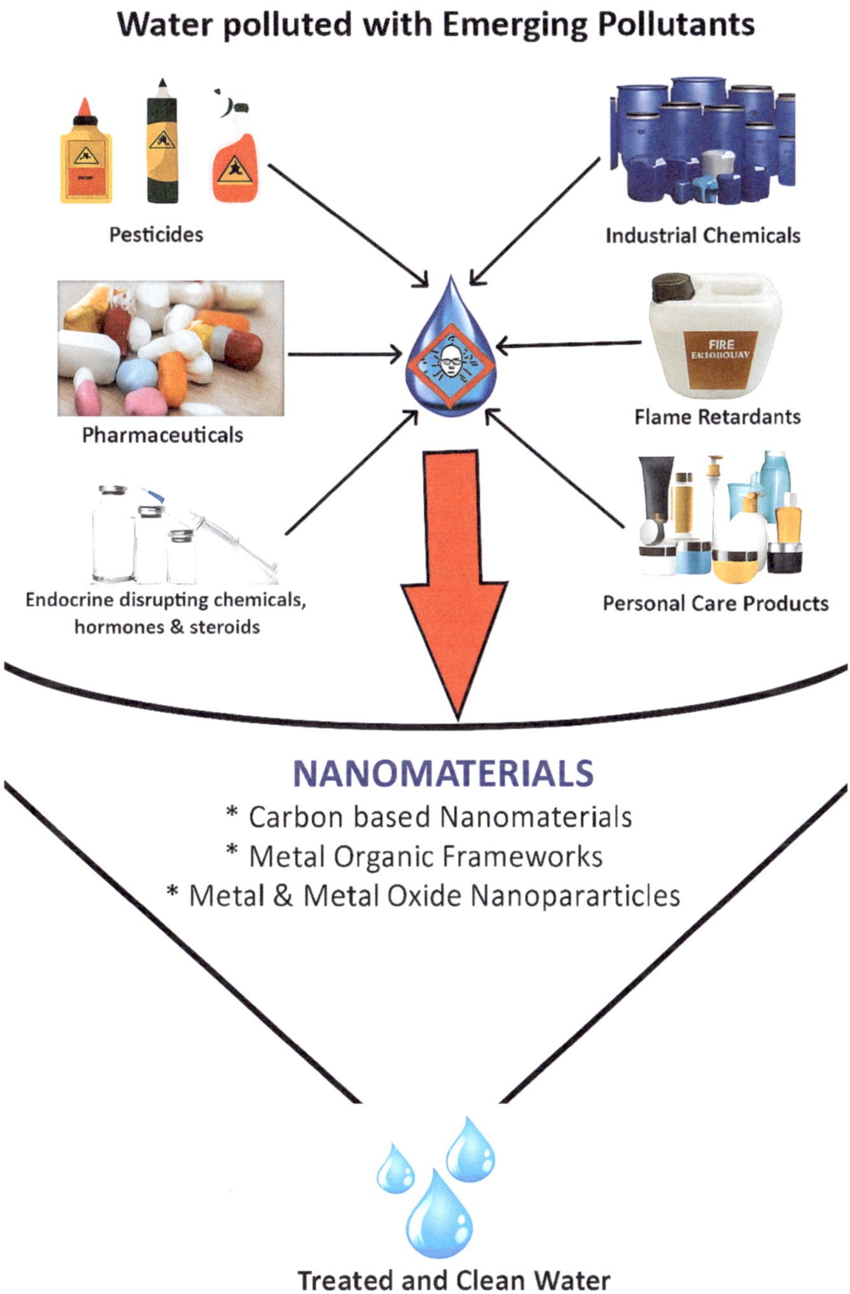

Fig. 5.2 Remediation of wastewater

5.2.2 Graphene and Graphene Oxide-Based Composites

Graphene is a carbon allotrope which is 2D in structure. It is composed of sp^2-hybridized atoms of carbon, forming a single layer, resulting in a honeycomb like structure (Xu et al. 2013). Graphene exhibits remarkable properties, including excellent thermal and electrical conductivity, exceptional strength, flexibility, chemical stability, efficient adsorption, and biocompatibility (Gurunathan and Kim 2016). Graphene has been combined with metal oxides and been widely utilized in the breakdown of environmental pollutants (Thakur and Kandasubramanian 2019).

GO is a graphene derivative, consisting of sheets of graphene with functional groups containing oxygen, such as epoxide and hydroxyl groups, dispersed throughout its structure. It is typically produced by graphite oxidation and exfoliation, resulting in single-layer or few-layer sheets (Aliyev et al. 2019). The functional groups containing oxygen impart hydrophilicity to graphene oxide and also make it highly reactive, allowing for various chemical modifications and functionalizations (Guo et al. 2022). GO readily combines with a variety of materials, resulting in nanocomposites that exhibit improved properties. It is utilized for adsorption and catalysis during the removal of harmful pollutants from air and water, due to its large surface area, reactivity, and contaminants adsorption ability (Ali et al. 2019a; b). The highly delocalized π-electrons in graphene/graphene oxide play a key role in their potential for degrading pollutants. These materials exhibit a significant adsorption rate due to π–π interactions, strong hydrogen bonding, electrostatic interactions between adsorbed pollutants and the graphene/GO sheets (Ersan et al. 2017).

5.2.3 Metal Organic Frameworks (MOFs)

MOFs are crystalline and porous in nature and are made up of metal clusters or ions forming coordination bonds with organic ligands (Qin and Zheng 2017). These materials have been used widely due to excellent characteristics, like large area, tunable porosity, and large number of functionalities (Zhang et al. 2022). The structure of MOFs can be precisely controlled, allowing for the design of materials with pore shapes and sizes which can tailored (Chen et al. 2017). This tunability enables selective adsorptive gas separation, liquids, and ions, making MOFs promising for applications in purification, gas storage, drug delivery and sensing and detection (Kumar et al. 2017).

For treating emerging organic pollutants, MOFs are utilized through adsorption and advanced oxidation processes. Adsorption primarily relies on electrostatic, hydrophobic, acid–base interactions, π–π interactions, and hydrogen bonding (Zango et al. 2020). In advanced oxidation processes catalyzed by MOFs, key photocatalytic mechanisms involve charge transfer from ligand to the metal and metal-oxo clusters excitation caused by photons, particularly in the presence of visible and UV light (Younis et al. 2020).

5.2.4 Metal and Metal Oxide Nanoparticles

Metal and metal oxide nanoparticles (NPs) are nanoscale particles composed of metallic elements or their oxides, respectively. NPs based on metals are very effective as they offer a large surface area and a high reactivity which is dependent on size of particles, dispersion regimes and forms (Jain et al. 2020). Nanometals possess a wealth of reactive sites for interacting with various chemicals owing to their large specific area. Furthermore, as size decreases, surface energy increases significantly (Zhang et al. 2020). These characteristics make nanometallic adsorbents highly promising for water treatment processes (Pandey et al. 2022). Many metal and metal oxide NPs have been employed for the treatment of organic pollutants. Common removal mechanisms include adsorption, photocatalysis, and advanced oxidation processes (Rani and Shanker 2020). They offer various advantages like low cost, effectiveness for the remediation of pollutants, high selectivity (Nagpal and Kakkar 2019) and antifouling and antibacterial properties (Agrawal et al. 2021). These properties enable them to remove very low concentrations of organic pollutants (Soni et al. 2020). Further, metal-based nanomaterials can be used to treat pollutants in various molecular sizes and hydrophobic rate (Khan et al. 2022a, b).

5.2.5 Biopolymer-Inorganic Hybrid Nanomaterials

Biopolymer-inorganic hybrid nanomaterials refer to composite materials composed of both biopolymers, such as proteins, polysaccharides, or nucleic acids, and inorganic components, like metal oxide or metal nanoparticles or nanostructures (Zhao et al. 2018). These hybrid materials allow a unique combination of properties of both components, offering synergistic effects and expanded functionalities. Biopolymers provide flexibility, biocompatibility, and biodegradability (Brigham 2018), while inorganic components contribute to mechanical strength, stability, and durability (Liu et al. 2023). The combination of these properties results in hybrid materials with enhanced mechanical performance, which makes them useful in materials science, tissue engineering, sensing, biomedical devices and energy conversion and storage (Park et al. 2020; Vijayakanth et al. 2022). Biopolymer-inorganic hybrids are also explored for applications in remediation, including purification of water and pollutant removal (Kalita and Barman 2019). The combination of biopolymer matrices with inorganic nanoparticles enhances adsorption capacity, photocatalytic activity, and stability, facilitating the contaminant removal from air and water (Das et al. 2018).

5.3 Application of Nanomaterials for Removal of Toxic Emerging Pollutants

5.3.1 *Pesticides*

Pesticides have had detrimental impacts on humans and the overall ecosystem. While they effectively eliminate pests and boost production of crops, they also contribute to environmental pollution due to their composition. These harmful chemicals pervade the water, plants, air, food, soil, and animal feed, and detrimental to animal and human health (Chaud et al. 2021). They are carcinogenic and through means including agricultural products, air and water, as well as the surrounding environment, they can pervade the human body. Once inside the body, pesticides accumulate in tissues and may be eliminated through various processes, but their toxic effects persist. Organs are all profoundly affected by long-term exposure, leading to conditions like cancer, tumors, mutations, and potentially death (Shukla et al. 2022).

Advanced nanotechnology has brought about efficient methods to combat the detrimental and damaging impacts of pesticides infiltrating the environment (Mishra and Khare 2021). Multi-walled CNT's have been reported to exhibit 100% removal of organophosphorus pesticides, Diazinon (initial concentration = 0.3 mg/L) and Malathion (initial concentration = 6 mg/L), with a contact time of only 15 min and 30 min, respectively (Dehghani et al. 2017, 2019). The selectivity and adsorption ability of CNT's can be enhanced through functionalization of surface using carboxyl groups, hydroxyl groups and amines. MWCNT's functionalized with hydroxyl and carboxyl groups and uniformly impregnated with Fe_3O_4 nanoparticles, have been used successfully for atrazine adsorption. 2,4-D. MWCNT-OH-Mag performed as a better adsorbent than MWCNT-COOH-Mag and exhibited an adsorption efficiency of 47.7 and 51.4 mg g^{-1} for atrazine and 2,4-D, respectively. Electrostatic interactions and complexation were observed to be the two possible interactions between the functionalized MWCNT's and the pesticides (Pereira et al. 2023).

Ali and group reported the fenuron pesticide adsorption on CNTs. Fenuron adsorption was observed to be affected by various parameters including pH, contact time, concentration of fenuron and CNTs and temperature. At the optimized conditions of pH 7, time of contact of 60 min, 2 g L^{-1} CNT dose and temperature 250 °C, 90% removal of the pesticide was observed. The major hydrophobic interactions between the pesticide and CNTs were observed to be π–π T shaped, π–σ, π-alkyl, and π–π stacked (Ali et al. 2019a; b). SWCNT's and MWCNT's have also be reported for the remediation of other pesticides including dichlobenil, diuron (Chen et al. 2011), and isoproturon (Sotelo et al. 2012).

Zhu and group utilized a Zr-based MOF known as UiO-67 to remove glyphosate and glufosinate from water solutions. They found that the adsorbent they synthesized had impressive adsorption capacities, with 537 mg/g and 360 mg/g for glyphosate and glufosinate, respectively. These remarkable capacities were likely due to the abundant hydroxyl groups in the MOF structure. Furthermore, UiO-67 demonstrated consistent performance at a wide pH range and in the presence of high electrolyte concentrations (Zhu et al. 2015). Zr clusters within MOFs have been reported to function as Lewis acid sites, capable of interacting with phosphate functional group found in glyphosate, acting as Lewis basic sites. According to subsequent findings, Zr-based MOF, NU-1000 exhibited a greater maximum adsorption capacity (8.97 mmol/g) compared to UiO-67 calculated using Langmuir isotherm. This difference was due to the large energy of adsorption and larger diameters of pores present in NU-1000 (Pankajakshan et al. 2018). Abdelhameed and group developed MOFs by bonding cotton with cellulose functional groups and copper (Cu) resulting in Cu-BTC@Cotton MOF. This was used to adsorb 14C-ethion organophosphate insecticides. They reported that the MOF bonded with ethion via sulfur, achieving a high adsorption capacity of 182 mg/g and a removal exceeding 97% for ethion. The Cu-BTC@Cotton MOF exhibited desirable characteristics such as high reusability and stability when utilizing acetonitrile for the solvent system (Abdelhameed et al. 2021).

To mitigate the secondary risks associated with the MOFs nanoparticles use as adsorbents, a new approach was developed recently, and involved loading two MOFs (UiO66-NH_2 or ZIF-8) onto carbon nanotube aerogels (MPCA) through the growth of MOF nanoparticles on CNTs. In the investigation of herbicides adsorption (alachlor and chipton), it was found that the MOF@MPCA adsorption capacity was greater than individual MOF NPs, suggesting a synergism between MPCA and MOFs. Specifically, the chipton adsorption capacity exhibited by UiO66-NH_2@MPCA was measured at 227.3 mg/g, and the material could be conveniently used for 5 cycles with a minimum decrease in adsorption capacity (Liang et al. 2021).

In water purification, graphene, a carbon-based nanoparticle, has been employed for pesticide removal, exhibiting an adsorption capacity in the range of 200 to 600 mg/g (Paramasivan et al. 2019). Additionally, research by Liu and group demonstrated enhanced organophosphorus pesticides adsorption in water by using graphene modified with silica (Liu et al. 2013). A study by Maliyekkal and group highlighted an exceptional adsorption of endosulfan (ES), chlorpyrifos (CP), and malathion (ML) on rGO and GO from water. Remarkably, the adsorption capacities for ES, ML, and CP reached approximately 1100, 800, and 1200 mg/g, respectively. It was noted that adsorption remains unaffected by pH levels or ions in the background. Moreover, the adsorbent proves that it can be regenerated and can be deployed in practical applications with appropriate changes. Through a first-principle density functional analysis, it was revealed that graphene-water-pesticide interactions are facilitated by water, while interactions between the pesticides and graphene are comparatively weak or improbable (Maliyekkal et al. 2013). In a separate investigation conducted by Zhang and group, a cellulose/graphene composite was prepared and studied for its efficacy in removing six triazine pesticides (ametryn, cyprazine, simeton, atrazine, prometryn, and simazine). The adsorption process, using a 3 g/L dosage, obeyed the Langmuir model and was identified as endothermic, physical, favorable, and spontaneous. Furthermore, it was observed that the electron-donating groups such as S, Cl, and O, along with π-bonding, facilitated the adsorption process (Zhang et al. 2015).

Metal nanoparticles, including silver and gold shown effectiveness in removing toxic pesticides from drinking water (Pradeep 2009). The adsorption of various pesticides on noble metal nanoparticles (MNPs) tends to vary, with some exhibiting higher affinity compared to others. Several factors, including composition, size, structure, and stability, influence the adsorption of pesticides onto nanoparticles (Intisar et al. 2022). Saifuddin and group developed a chitosan-silver nanoparticle composite, which is both non-toxic and biodegradable, for the purpose of removing atrazine from wastewater. In a solution with 1 ppm pesticide concentration, 98% of the pesticide was removed when 2 g of crosslinked silver nanoparticle microbeads were added to 1 L of the solution. Batch studies further revealed that the adsorbent capacity exhibited by cross-linked silver nanoparticle microbeads for atrazine was observed to be 0.5 mg/g within 65 min (Saifuddin et al. 2011).

Dehaghi and group synthesized composite beads of chitosan-ZnO nanoparticles for removing permethrin, achieving a 99% removal rate under optimized conditions. Complete adsorbent recovery was achieved by treating it with 0.1 M NaOH (Dehaghi et al. 2014). Chitosan impregnated with Ag NPs have been reported for removing permethrin neurotoxin insecticide, resulting in 49% higher adsorption capacity than pure chitosan (Rahmanifar and Moradi 2014). Graca and group evaluated zerovalent monometallic zinc (ZVZ), iron (ZVI), and copper (ZVC), along with copper-coated bimetallic ZVZ (ZVZ/Cu) and ZVI (ZVI/Cu) for their efficacy in degrading the pesticide, chlorpyrifos. All nanometals facilitated the chlorpyrifos dechlorination, with reactivity following the order: ZVC > ZVZ > ZVI for monometals, and ZVZ/Cu > ZVI/Cu for bimetals. Chlorpyrifos degradation was analyzed using LC/ESI-TOF, which revealed dechlorinated products and a hydrolyzed product which was found to be 3,5,6-trichloro-2-pyridinol (TCP). Detection and quantification of these products were estimated at 0.49 mg/L and 1.46 mg/L, respectively (Graça et al. 2020) (Fig. 5.3).

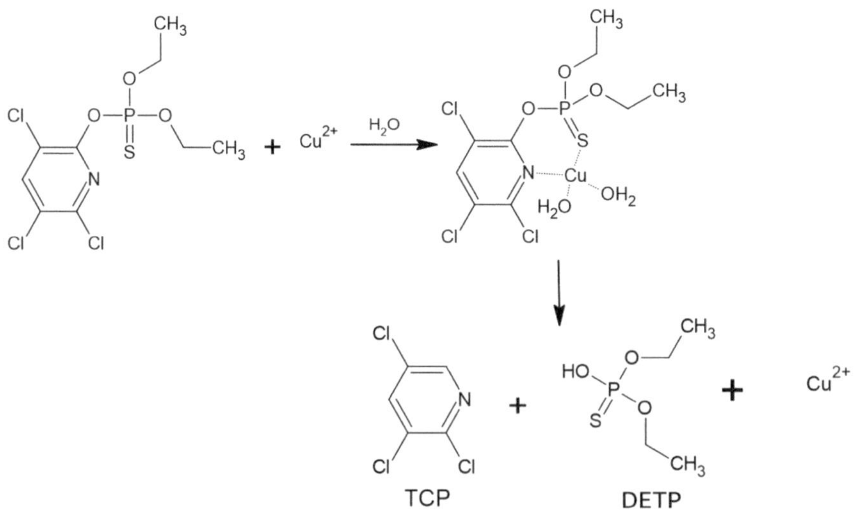

Fig. 5.3 Hydrolysis of chlorpyrifos catalyzed by Cu^{2+}

5.3.2 Pharmaceutical Compounds (Drugs and Antibiotics)

Among the major classes of newly discovered environmental micropollutants are pharmaceutical products (Ahmad et al. 2019; Homem and Santos 2011; Zaib et al. 2013). Drugs are produced, consumed, and released into the environment on a yearly basis in tons. The management of drugs is crucial to both human and animal health. These pharmaceuticals are made with precise effects on the host organisms in mind. These chemicals bioaccumulate in tissues and animals and are pervasive, mobile, persistent, and cause other problems that raise concerns about public safety. Their presence in water bodies, is well-documented globally and is a big problem because it negatively impacts the reproduction, growth, and non-target organisms' behavior, particularly aquatic ones. For pharmaceutical micropollutants, a variety of ecotoxicological effects have been observed (Kayode-Afolayan et al. 2022; Sorensen et al. 2015; Van Doorslaer et al. 2014; Raja et al. 2018). The pharmaceutical-based contaminants are beta-blockers (propranolol), anticonvulsants (phenytoin), hormones (estradiol), antibiotics (tetracycline), anti-inflammatories (ibuprofen), lipid regulators (fenofibric acid), and antidepressants (Carbamazepine) (Rasheed et al. 2020). These dangerous pollutants cause a serious risk to humans as well as the health of entire living ecosystems because they alter the chemical and physical characteristics of water (Santhosh et al. 2016; Joel and Fansen 2013). This is partly due to the fact that the majority of them are thought to be pollutants in the environment that are either persistent or developing and are hard to fully degrade (Wang et al. 2013)) (Fig. 5.4).

Pharmaceutical removal from wastewater is crucial owing to the significant risk they provide to ecosystems and human health. Furthermore, they are resistant to standard chemical and biological cleanup approaches. Because of their huge surface areas, electrical conductivities, and partial optical transparency, nanomaterials represent a viable method to address this difficulty (Luo et al. 2014).

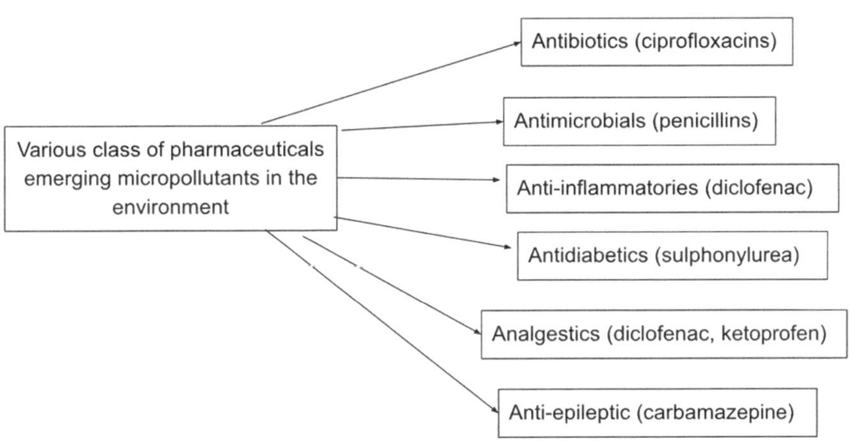

Fig. 5.4 Various classes of pharmaceutical compounds

CNTs have been found to be highly efficient for removing drugs and antibiotics owing to their considerable surface area, small dimensions, and substantial porosity. In their study, Yu and group employed oxidized MWCNTs with varying contents of oxygen to adsorb tetracycline (TC) from aqueous solutions. Using Langmuir isotherm model, the adsorption capacities (Q_{max}) were determined to be 217.8, 269.25, 217.56, and 210.43 mg g^{-1} for CNTs containing 2.0%, 3.2%, 4.7%, and 5.9% oxygen, respectively (Yu et al. 2014a, b). In another report, MWCNTs were used for TC removal from water. These CNTs underwent oxidation using a mixture of 70:30 (v/v) H_2SO_4/H_2O_2 under reflux conditions at 120 °C for 2 h. This procedure resulted in C–O and COOH groups on the surface of CNTs. According to the Liu isotherm model, the Q_{max} was calculated as 253.4 mg g^{-1} (Babaei et al. 2016). CNTs have also been used for the treatment of diclofenac (Gil et al. 2018), sulfapyridine (Tian et al. 2013), tylosin (Xia et al. 2019), sulfadimethoxine (Rahmani Sani et al. 2018), oxytetracycline (Yu et al. 2021), and carbamezapine (Deng et al. 2019).

Metal and their oxide nanoparticles are used in the remediation of EPs owing to their cost-effectiveness, high surface areas, and efficient and selective removal of pollutants. Magnetite nanoparticles (Fe_3O_4) have been identified as highly efficient adsorbents for removing various antibiotics such as rifampicin, trimethoprim, tetracycline, tazobactam, erythromycin, ampicillin, piperacillin, and sulfamethoxazole. They demonstrate a removal efficiency exceeding 90% (Wanakai et al. 2023). In a recent study, de Escobar and group found that prepared photocatalysts loaded with molecularly imprinted TiO_2 exhibited significantly higher selectivity and degradation of diclofenac compared to fluoxetine and paracetamol. The researchers attributed the higher diclofenac selectivity and the improved degradation to the presence of specific sites for binding within the prepared nanocomposites (de Escobar et al. 2018).

In a study by El-Maraghy and group, levofloxacin was effectively treated using a composite made of GO and ZnO NPs, achieving removal efficiencies of 99.6% and 99.2% for 400 μg/mL and 50 μg/mL concentrations, respectively (El-Maraghy et al. 2020). Similarly, removal efficiencies of over 85% were observed for β-lactam antibiotics, namely ceftriaxone and cefadroxil, by zerovalent copper nanoparticles (Oliveira et al. 2018). Zhu and group examined the pharmaceutical metformin adsorption on GO. Their findings demonstrated that GO could reduce the metformin concentration very rapidly. The uptake of metformin by GO was affected by various indicators such as pH, ionic strength, composition of electrolyte and temperature. Adsorption kinetics revealed that 80% metformin was removed within 20 min at pH 6.0 and a temperature of 15 °C, possibly attributable to hydrogen bonding and π–π interactions. Thermodynamic analysis suggested that the adsorption was exothermic and instantaneous (Zhu et al. 2017).

Lin and group investigated the removal of ketoprofen, furosemide, indomethacin, ibuprofen, naproxen, acetophenone, and salicylic acid from aqueous solutions using three Zr(IV)-based MOFs: MOF-808, MOF-802, and UiO-66. They found that UiO-66 showed the best removal for naproxen, ketoprofen, indomethacin furosemide, ibuprofen, while MOF-808 showed a preference for naproxen, furosemide, ketoprofen, ibuprofen, salicylic acid indomethacin, and acetophenone. The researchers proposed that π–π interactions between the MOF ligands and benzene rings present in the pharmaceutical compounds are important for the high adsorption efficiency exhibited by the synthesized MOFs. Additionally, they attributed the lower adsorption capacity of MOF-802 to its smaller pores and thinner scale of inner pores in its structure (Lin et al. 2018). In a separate investigation, Jin and group synthesized Cr based MOF decorated with CuCo bimetallic NPs, referred to as CuCo/MIL-101, with the aim of removing tetracycline (TC) from water. The study revealed that MIL-101 after modification exhibited a significantly higher adsorption capacity, recording a 140% increase compared to neat MIL-101. This enhanced TC removal was due to the synergism between the Co and Cu NPs in CuCo/MIL-101 and TC (Jin et al. 2019).

5.3.3 Industrial Chemicals

Industrial chemicals are substances produced or used in various industrial processes for manufacturing goods, processing materials, or performing other functions. These chemicals encompass a broad range of substances with diverse applications across industries such as manufacturing, agriculture, pharmaceuticals, electronics, and construction. Common classes of industrial chemicals include polychlorinated alkanes, flame retardants, plasticizers, polycyclic aromatic hydrocarbons, gasoline additives, and so on (Table 5.1).

Various nanomaterials including magnetic nanoparticles, GO, CNTs, nanometals and metallic oxides, hybrid nanostructures, nano-polymers, silica, and other materials are widely employed to treat chemicals in industrial wastewater. According to a study by Xu and group, graphene exhibited a notable adsorption capacity of 182 mg g^{-1} for bisphenol A (BPA), at 302 K. This capacity is higher than other carbon-based materials explored in previous research. The enhanced BPA adsorption on graphene is likely due to factors such as hydrogen bonding, π–π interactions, and its unique single-layer sp^2-hybridized structure (Xu et al. 2012). Jin and coworkers conducted a study on the adsorption behavior of BPA and 4-n-nonylphenol (4-n-NP) on magnetic rGOs. They investigated the impact of contact time, humic acid, pH, and ionic strength on the process of adsorption using a batch adsorption system. The results revealed that at pH 6.5 and 293 K, the Q_{max} of magnetic rGOs were 48.74 mg g^{-1} and 63.96 mg g^{-1} for BPA and 4-n-NP, respectively. The higher removal of 4-n-NP on rGOs can be attributed to flexible long alkyl chain present in 4-n-NP and $\pi-\pi$ stacking. Conversely, BPA adsorption on rGOs can be due to a lying-down configuration due to dispersion forces and $\pi-\pi$ stacking, as evidenced by FTIR analysis (Jin et al. 2015). Reduced GO/FeO·Fe_2O_3 composites have been reported to be effective adsorption of naphthalene, 1-naphthylamine, and 1-naphthol, from water. Through their study, they determined the Q_{max} based on the Langmuir adsorption isotherm, finding that naphthalene exhibited the highest value at 2.85 mmol g^{-1}, followed by 1-naphthol at 2.70 mmol g^{-1}, and then 1-naphthylamine at 2.63 mmol g^{-1}. They proposed that electron-donor–acceptor (EDA) interactions as the mechanism for adsorbing aromatic compounds, with the adsorption increasing with the adsorbate's dipole moment (Yang et al. 2013).

Table 5.1 Common classes of industrial chemicals

S. No.	Industrial chemicals class	Compounds
1	Plasticizers	Bisphenol A, phthalates
2	Corrosion inhibitors	1,2,3-benzotriazole, benzothiazole-2-sulfonic acid
3	Flame retardants	PBDE, C_{10}–C_{13} polychlorinated alkanes, 3,3′,5,5′-tetrabromobisphenol-A, tris(2-chloroethyl phosphate)
4	Polycyclic aromatic hydrocarbons (PAHs)	Benzo[a]pyrene, naphthalene
5	Perfluoroalkyl compounds	Perfluorooctane sulfonic acid and derivatives (PFOS)
6	Volatile organic compounds	Toluene, $CHCl_3$, benzene, styrene, 1,2-dichloroethane, tetrachloroethene (PCE), xylenes
7	Antifouling compounds	Irgarol, dibutyl tin ion
8	Gasoline additives	Methyl tert-butyl ether (MTBE), dialkyl ethers
9	Alkylphenol ethoxylate surfactants	Octyl phenols, nonyl phenols
10	Antioxidants	2,6-Di-tert-butylphenol
11	Others	2,4-Dinitrophenol, PCBs, pentachlorophenol, hexabromocyclodecane, chloropicrin

The creation of three-dimensional macrostructures (3D GTs) by combining 2D GO nanosheets with 1D CNTs resulted in exceptional adsorption capacities for diethyl phthalate (680 mg/g). The adsorption capacity observed using 3D GTs surpasses that of previously reported adsorbents based on CNTs and graphene/GO. This heightened performance is credited to the synergism between CNTs and GO within the nano-substrate, active sites, and micro-environment (Shen et al. 2017). In a study by Wei and coworkers, composite adsorbents comprising MWCNTs-SiO_2, denoted as MWCNTs-SiO_2-6, MWCNTs-SiO_2-4, MWCNTs-SiO_2-2 (CS6, CS4, CS2), were prepared by sol–gel procedure, with MWCNTs molar percentages ranging from 38 to 66%. The findings revealed that the adsorption capacity for toluene at temperatures between 30 °C and 60 °C followed the order of activated carbon (AC) < CS2 < CS4 < CS6, with CS6 exhibiting an adsorption capacity reaching 50.28 mg/g. Additionally, with a rise in temperature by 10 °C, the time for penetration lowered by 10 – 20 min, and with every 1% rise in concentration of water vapor, the amount of adsorption lowered by 3.5% (Wei et al. 2021).

Zheng and coworkers explored the removal and detection of perfluorinated alkyl compounds using Fe_2O_3 nanoparticles in conjunction with guanidinocalix [5]arene. The study specifically focused on perfluorooctanoic acid (PFOA) and perfluorooctane sulfonate (PFOS). Fluorescence detection facilitated the identification of the fluorinated components, while magnetic removal and filtration were employed for their removal. The efficacy of the removal process was boosted through solid-phase extraction of the perfluorinated substances (Zheng et al. 2019). Qin and group developed Fe_3O_4/mSiO_2-Me-PTSA by coating magnetic nanoparticles with methyl dimethoxy modified silica, where p-toluenesulfonic acid (PTSA) acting as a catalyst to accelerate the rates of reaction. This prepared material was utilized for adsorption of polychlorinated biphenyls from water, leveraging its large area and strong magnetic field. Within just 10 min, the adsorbent showed an impressive adsorption capacity of 46.3 mg/g, with a detection limit in the range of 0.16 to 0.91 ng/L (Qin et al. 2018). Zirconia nanoparticles modified with hydroxyapatite has been successfully reported to remove 95, 98, 97% of 4-chlorophenol, phenol, and 4-nitrophenol, respectively, in the presence of UV light (Ekka et al. 2018).

Sarro and group developed more intricate hybrid materials by combining soybean peroxidase (SBP) and zinc oxide nanoparticles with poly(styrene-co-maleic anhydride) nanofibers. Experimental results with this hybrid material in aqueous solution demonstrated the complete bisphenol A and 2,4-dichlorophenol degradation in 24 h (Sarro et al. 2018). A novel yolk–shell Co_3O_4@MOFs nanoreactor was developed using a simple procedure to incorporate sulfate radical-based advanced oxidation processes (SR-AOPs) within its inner cavity. The nanoreactor was used to carry out the catalysis of 4-chlorophenol (4-CP) degradation using peroxymonosulfate (PMS). The effectiveness of SR-AOPs within nanoreactor was evidenced by achieving nearly 100% degradation of 4-CP in 60 min by Co_3O_4@MOFs catalysts, while only 59.6% degradation was exhibited using bare Co_3O_4 nanoparticles, in presence of similar conditions (Zeng et al. 2015).

5.3.4 Disrupting Chemicals, Hormones, and Steroids

Endocrine disruptors (EDCs) as new age pollutants have been identified as EPs due to their utility in everyday life, adverse health effects and their environmentally persistent nature in waste water (Kalofiri et al. 2023). EDCs are the types of chemical substance of natural or synthetic origin which disrupts the enzymatic pathways and metabolic activities and interferes with hormonal secretions either by mimicking the activities of endogenous hormones or interferes with the biochemical reaction of hormones. The most significant endocrine disruptors which we found use in day-to-day life are as under:

- Personal care products like cosmetic, shampoos, perfumes (Dodds and King 2001)
- Food products/dairy products
- Steroids hormones.

EDCs enters into human body through food and water intake, dust inhalation, suspensions or particulate matter inhalation or dermal contact in the form of skin creams, lotion, soap, and other personal care products. Thus, these chemicals if end up in receiving water bodies are public health hazards (Gonsioroski et al. 2020). It finally finds entry in the food chain through food crops and bioaccumulate inside the

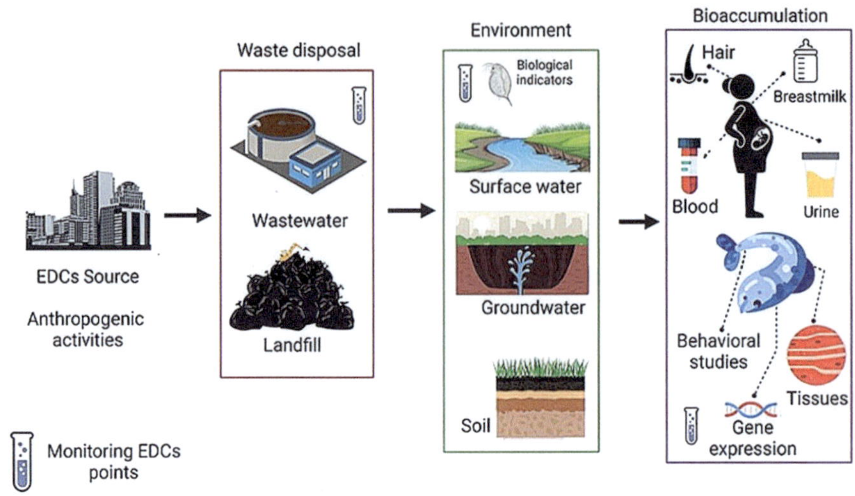

Fig. 5.5 Endocrine disruptors migration pathways from sampling point source to the final destination and its adverse impact on target tissues

fatty tissues of the human body. EDCs in contaminated water find its traces in dairy products and aquaculture products through food chain. In addition, inhalation of aerosol from perfumes, cosmetic fragrances, deodorant and hair spray and ingestion through lip color and lipstick also results in exposure to EDCs (Archer 2010) as depicted in Fig. 5.5.

Various studies are being conducted to ascertain the environmental risk but being the part of personal hygiene, various lotions, sunscreen, and soaps become the part of water pollutants in fresh water or marine ecosystem through divers, navigators, and swimmers. Some of these parabens and cosmetic items get accumulated in landfills and become the primary reason of soil contamination (Loretz et al. 2005; Feng et al. 2015; Ismail et al. 2019; Lincho et al. 2021). Being bioaccumulative in nature, its concentration keeps on increasing in the body of tertiary consumers. Moreover, they are never completely excreted out of the body. These EDCs post several debilitating health effects ranging from altered metabolism to immunosuppression, increasing risk of cancer and genetic mutation also. (La Merrill et al. 2020). As per the reports of statutory global health and environmental agencies like UNEP and WHO in 2013, EDCs are responsible for inducing disorders related to endocrine gland in human beings and animals, various cardiovascular and neurodegenerative diseases (Darbre 2019; Charman and Williams 2003).

Thus, strains of EDCs in water bodies are detrimental to the environment and society at large. The categories of endocrine disruptors comprise of bisphenols, estradiols, parabens, phthalates and nonylphenols. Various methods of water treatment are being explored to counter this emerging new age pollutant. Steroids are a natural micropollutant which found their way into receiving waterbodies and act as environment hazard as well as threat to human health also. They find active use in hormonal imbalance and further treatments related to growth and development and problems pertaining to fertility (Zhang et al. 2021). The emission rate can be quantified on the basis of rate of excretion of human population and also animal sources (Yu et al. 2019). The contamination due to steroid hormone pose serious threat due to the fact that at a concentration level as low as part per trillion can adversely affect aquatic life, affecting their reproductive organs.

Three main category of steroid hormones that are vital for human beings are glucorticoids, mineralocorticoids, and sex hormones (Fig. 5.6). The sex hormones may further be divided into estrogens, progesterone, and androgens. Nowadays, the usage of progesterone in medical field has been increased due to potent hormonal activity as compared to androgen and estrogen specifically in the field of fertility treatment and hormone replacement. The concentration of sex hormones in food chain increases due to their bioaccumulating properties and thus pose as health hazard to humans and aquatic environment (Yu et al. 2019; Paulos et al. 2010). Reports have suggested that norethindrone and levonorgestrel at the level of even 1 ng/L can adversely impact the production of egg in fathered minnows (Zhao et al. 2019). Similarly, the presence of steroids dydrogesterone and medroxyprogesterone acetate can cause alteration in ovaries of zebra fish even at a very low concentration of 4.5 ng/L (Liao et al. 2022). Thus, it has become imperative to search for novel methods for removal of steroid based pollutants. Carbon-based nanomaterials prove to be a promising alternative. For sustainable and long-term removal of EDC from waste water, absorption methods utilizing nanocomposites, carbon-based nanomaterials viz. nanofiber, CNTs, magnetic carbon nanomaterial, and carbon dots are being explored (Arbabi 2020).

Engineered nanomaterials (ENMs) are considered as a novel alternative to counter the environmental challenges posed by endocrine disrupting chemicals. CNTs have shown itself to be a promising method due to its porosity and strong absorptive properties against organic persistent chemicals like bisphenol A (Iavicoli et al. 2013). Though a few studies pertaining to risk assessment of engineered nanomaterials have pointed toward endocrine disrupting properties of CNT as they behave like

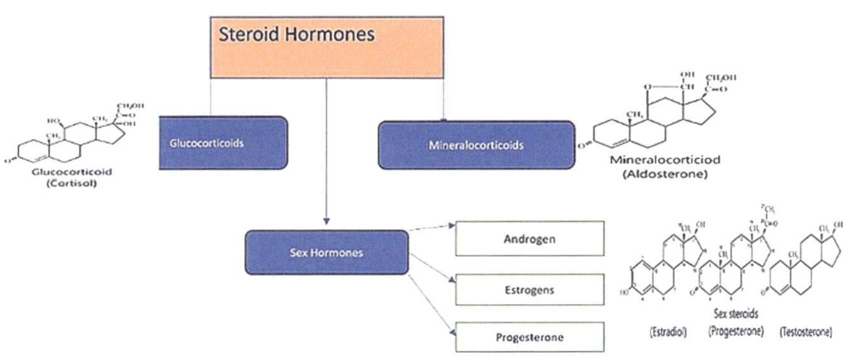

Fig. 5.6 Type of steroid hormones

body hormones by interacting with hormone receptors in animals and humans both (Ghasemi et al. 2019). MWCNTs have reportedly found widespread use in destroying EDCs from effluent water after being chemically reacted with polydopamine, molecular imprinting polymers, and magnetic particles (Wang and Wang 2016). Some of the factors which enhances the adsorptive capacity and efficiency of CNT are physical and chemical properties of adsorbates viz. functional groups, pore size, pKa value, and hydrophobic nature of adsorbates (Jung et al. 2015). Additionally, well-defined adsorption sites lead to enhanced adsorption capacity and thus facilitates effective EDCs removal during treatment of wastewater (Azizi et al. 2022). Due to heterogeneity of micropore in nanofiltration membranes, steroid removal gets challenging. By designing the membrane with balancing of hydrodynamic drag on vertically-aligned carbon nanotube membranes (VaCNT) and frictional forces on steroid hormones in waste water, rapid adsorption and removal can be made possible. The interplay of repulsive and adhesive forces between inner carbon nanotube and steroid hormone plays a significant role in designing VaCNT membrane for steroid removal (Nguyen et al. 2024).

Graphene and GO-based composites provide an edge over CNTs due to its greater adsorption surface area and low cost. Chemical properties of graphene and GO-based composites as well as its efficacy in providing chemical modification and fabrication of composites makes it the desirable and efficient alternative for removal of EDCs in waterbodies (Bielawski et al. 2010). GO sheet contains numerous hydrophilic groups containing oxygen on the surface viz. hydroxyls and epoxides (Sweetman et al. 2017) which increases the water permeability of GO membrane (Jayakaran et al. 2019). Functional groups containing oxygen on GO surface facilitate the EDCs adsorption due to existence of hydrogen bonding between GO and EDCs. GO composites have also been used for steroids contaminants adsorption. Organic pollutants which are polar in nature readily adsorb on graphene while ionic pollutants adsorb better on the charged GO surface. Studies have shown that the number and special arrangement of aromatic rings in the molecule positively impacts the adsorption of organic pollutant pollutants (Baig et al. 2019).

In recent times, the studies were carried out on nano-zero-valent iron (nZVI) for treatment of various PAC present in water. The mechanism of its removal is based on primarily adsorption and thereafter, the degradation of pharmaceuticals with nZVI occurred by simultaneous adsorption and reduction process, but no significant development took place in this process and till now it is limited to pilot scale investigation only (Hanay and Türk 2019).

New range of nanostructured polymeric hybrid materials known as MOF and covalent organic framework (COF) have both adsorptive and oxidative properties. MOF have tunable pore size and extremely large surface area while COF is crystallized organic polymer with adjustable porosity and designable topologies. It has been employed for PAC removal from aqueous solutions (Suyambulingam et al. 2023). Utilization of polysaccharide-based biopolymers in removal of steroids and EDCs has gained importance due to its nontoxic nature and complete biodegradability. These materials undergo both physical and chemical interaction with different molecules for binding with them. For removal of metal ions, water-soluble pollutants as well as organic pollutants, chitosan, and chitin are the most preferred polysaccharides by the researchers. One of the steroids namely estradiol, could be removed by employing the chitosan/$Al_2O = -HA$ (hydroxyapatite) composite beads with Q_{max} of about 39.78 mg/g (Al Sharabati et al. 2021). Further use of biodegradable polymer and their nanocomposites are being explored for EDCs removal from water. This novel technology is still in nascent stage and still being explored for treatment of composite waste water by employing both natural and synthetic biodegradable polymer (Werkneh et al. 2022). As majority of novel research is conducted in simulated medium and not on real EDC containing water, more extensive studies are required in the area of nanotechnology along with other adsorption methods using combination of waste water treatment technologies.

5.4 Conclusions and Future Perspectives

The global challenge of water contamination is becoming more pressing due to the widespread and intensive use of pharmaceuticals, personal care products, pesticides, industrial chemicals, hormones, and steroids. These pollutants are recognized as emerging pollutants (EPs). Addressing this issue requires efficient, cost-effective, and user-friendly methods for mitigating EPs contamination in water. This review explores a range of nanomaterial-based approaches for degradation, adsorption, and filtration of EPs from water sources. Both adsorption and photocatalysis methods demonstrate significant promise for removing EPs from water. In the adsorption approach, there has been considerable interest in adsorbents based on metal oxides, CNTs, and GO, with key mechanisms including the π–π interactions, hydrophobic effect, electrostatic interactions, hydrogen bonding, and covalent bonding. Recently, MOF nanomaterials have also shown remarkable capabilities for adsorption-based removal of these pollutants. Nanomaterial-based methods offer highly efficient, rapid, environmentally friendly, and energy-saving approaches for wastewater treatment. Despite their effectiveness and cost efficiency, these techniques have not yet been widely adopted for large-scale industrial wastewater purification due to some challenges.

Testing the efficiency of these new technologies on a large scale, particularly in real wastewater and in the presence of other coexisting contaminants is crucial. Regeneration procedures for the nanomaterials have been insufficiently described in existing studies, warranting further study in future research. Future studies should also aim to understand the fate of EPs in ecosystems and their potential transformation into more hazardous compounds post-treatment. In addition, instead of relying solely on a specific single treatment strategy, multiple treatment strategies must be employed in combination to effectively treat EPs.

When utilizing nanomaterials for mitigating hazardous EPs from the environment, it is essential to consider their potential adverse effects. The toxicity of nanoparticles is affected by parameters such as size (particles larger than 20 nm are hazardous), concentration, and composition (ideally less than 100 ppm). Therefore, the potential harmful impact of nanomaterials on the environment and these aforementioned factors should be carefully considered when employing them for pollution control. Consequently, there is a critical requirement to categorize and synthesize non-toxic nanomaterials for EPs removal from water.

References

Abdelhameed RM, Abdel-Gawad H, Emam HE (2021) Macroporous Cu-MOF@ cellulose acetate membrane serviceable in selective removal of dimethoate pesticide from wastewater. J Environ Chem Eng 9(2):105121

Agrawal A, Sharma A, Awasthi KK, Awasthi A (2021) Metal oxides nanocomposite membrane for biofouling mitigation in wastewater treatment. Mater Today Chem 21:100532

Ahmad J, Naeem S, Ahmad M, Usman AR, Al-Wabel MI (2019) A critical review on organic micropollutants contamination in wastewater and removal through carbon nanotubes. J Environ Manag 246:214–228

Al Sharabati M, Abokwiek R, Al-Othman A, Tawalbeh M, Karaman C, Orooji Y et al (2021) Biodegradable polymers and their nano-composites for the removal of endocrine-disrupting chemicals (EDCs) from wastewater: a review. Environ Res 202:111694

Ali I, Alharbi OM, ALOthman ZA, Al-Mohaimeed AM, Alwarthan A (2019a) Modeling of fenuron pesticide adsorption on CNTs for mechanistic insight and removal in water. Environ Res 170:389–397

Ali I, Mbianda XY, Burakov A, Galunin E, Burakova I, Mkrtchyan E et al (2019b) Graphene based adsorbents for remediation of noxious pollutants from wastewater. Environ Int 127:160–180

Aliyev E, Filiz V, Khan MM, Lee YJ, Abetz C, Abetz V (2019) Structural characterization of graphene oxide: surface functional groups and fractionated oxidative debris. Nanomaterials 9(8):1180

Amiot CL, Xu S, Liang S, Pan L, Zhao JX (2008) Near-infrared fluorescent materials for sensing of biological targets. Sensors 8(5):3082–3105

Arbabi BS (2020) Endocrine disrupting effects of carbon nanotubes: a systematic review on next generation nanotechnology based agrochemicals. J Water Environ Nanotechnol 5(2):102–113

Archer CB (2010) Functions of the skin. In: Burns T, Breathnach S, Cox N, Griffiths C (eds) Rook's textbook of dermatology. Blackwell Publishing Ltd., Singapore, pp 1–11

Azizi D, Arif A, Blair D, Dionne J, Filion Y, Ouarda Y et al (2022) A comprehensive review on current technologies for removal of endocrine disrupting chemicals from wastewaters. Environ Res 207:112196

Babaei AA, Lima EC, Takdastan A, Alavi N, Goudarzi G, Vosoughi M et al (2016) Removal of tetracycline antibiotic from contaminated water media by multi-walled carbon nanotubes: operational variables, kinetics, and equilibrium studies. Water Sci Technol 74(5):1202–1216

Baig N, Sajid M, Saleh TA (2019) Graphene-based adsorbents for the removal of toxic organic pollutants: a review. J Environ Manag 244:370–382

Bhat SA, Sher F, Hameed M, Bashir O, Kumar R, Vo DVN et al (2022) Sustainable nanotechnology based wastewater treatment strategies: achievements, challenges and future perspectives. Chemosphere 288:132606

Bielawski CW, Dreyer DR, Park S, Ruoff RS (2010) The chemistry of grapheme oxide. Chem Soc Rev 39(1):228–240

Brigham C (2018) Biopolymers: Biodegradable alternatives to traditional plastics. In: Green chemistry. Elsevier, pp 753–770

Charman C, Williams H (2003) The use of corticosteroids and corticosteroid phobia in atopic dermatitis. Clin Dermatol 21(3):193–200

Chaud M, Souto EB, Zielinska A, Severino P, Batain F, Oliveira-Junior J et al (2021) Nanopesticides in agriculture: Benefits and challenge in agricultural productivity, toxicological risks to human health and environment. Toxics 9(6):131

Chen GC, Shan XQ, Pei ZG, Wang H, Zheng LR, Zhang J, Xie YN (2011) Adsorption of diuron and dichlobenil on multiwalled carbon nanotubes as affected by lead. J Hazard Mater 188(1–3):156–163

Chen L, Luque R, Li Y (2017) Controllable design of tunable nanostructures inside metal–organic frameworks. Chem Soc Rev 46(15):4614–4630

Chowdhary P, Bharagava RN, Mishra S, Khan N (2020) Role of industries in water scarcity and its adverse effects on environment and human health. In: Environmental concerns and sustainable development: volume 1: air, water and energy resources, pp 235–256

Darbre PD (2019) The history of endocrine-disrupting chemicals. Curr Opin Endocr Metab Res 7:26–33

Das S, Chakraborty J, Chatterjee S, Kumar H (2018) Prospects of biosynthesized nanomaterials for the remediation of organic and inorganic environmental contaminants. Environ Sci Nano 5(12):2784–2808

de Escobar CC, Ruiz YP, dos Santos JHZ, Ye L (2018) Molecularly imprinted TiO_2 photocatalysts for degradation of diclofenac in water. Colloids Surf A Physicochem Eng Asp 538:729–735

Dehaghi SM, Rahmanifar B, Moradi AM, Azar PA (2014) Removal of permethrin pesticide from water by chitosan–zinc oxide nanoparticles composite as an adsorbent. J Saudi Chem Soc 18(4):348–356

Dehghani MH, Niasar ZS, Mehrnia MR, Shayeghi M, Al-Ghouti MA, Heibati B et al (2017) Optimizing the removal of organophosphorus pesticide malathion from water using multi-walled carbon nanotubes. Chem Eng J 310:22–32

Dehghani MH, Kamalian S, Shayeghi M, Yousefi M, Heidarinejad Z, Agarwal S, Gupta VK (2019) High-performance removal of diazinon pesticide from water using multi-walled carbon nanotubes. Microchem J 145:486–491

Deng Y, Ok YS, Mohan D, Pittman CU Jr, Dou X (2019) Carbamazepine removal from water by carbon dot-modified magnetic carbon nanotubes. Environ Res 169:434–444

Ding H, Zhou XX, Wei JS, Li XB, Qin BT, Chen XB, Xiong HM (2020) Carbon dots with red/near-infrared emissions and their intrinsic merits for biomedical applications. Carbon 167:322–344

Dodds L, King WD (2001) Relation between trihalomethane compounds and birth defects. Occup Environ Med 58(7):443–446

EEA (2012) Water resources. Eur. Context vulnerability

Ekka B, Nayak SR, Achary LSK, Kumar A, Mawatwal S, Dhiman R et al (2018) Synthesis of hydroxyapatite-zirconia nanocomposite through sonochemical route: a potential catalyst for degradation of phenolic compounds. J Environ Chem Eng 6(5):6504–6515

El-Maraghy CM, El-Borady OM, El-Naem OA (2020) Effective removal of levofloxacin from pharmaceutical wastewater using synthesized zinc oxide, graphene oxide nanoparticles compared with their combination. Sci Rep 10(1):5914

Ersan G, Apul OG, Perreault F, Karanfil T (2017) Adsorption of organic contaminants by graphene nanosheets: a review. Water Res 126:385–398

European Environment Agency. Water assessment. EEA Rep No 11/2012. Available from: http://www.eea.europa.eu/publications/waterresources-and-vulnerability/at_download/file

Feng M, He Q, Meng L, Zhang X, Sun P, Wang Z (2015) Evaluation of single and joint toxicity of perfluorooctane sulfonate, perfluorooctanoic acid, and copper to Carassius auratus using oxidative stress biomarkers. Aquat Toxicol 161:108–116

Geissen V, Mol H, Klumpp E, Umlauf G, Nadal M, van der Ploeg M et al (2015) Emerging pollutants in the environment: a challenge for water resource management. Int Soil Water Conserv Res 3(1):57–65

Ghasemi SS, Hadavifar M, Maleki B, Mohammadnia E (2019) Adsorption of mercury ions from synthetic aqueous solution using polydopamine decorated SWCNTs. J Water Process Eng 32:100965

Gil A, Santamaría L, Korili SA (2018) Removal of caffeine and diclofenac from aqueous solution by adsorption on multiwalled carbon nanotubes. Colloid Interface Sci Commun 22:25–28

Gonsioroski A, Mourikes VE, Flaws JA (2020) Endocrine disruptors in water and their effects on the reproductive system. Int J Mol Sci 21(6):1929

Gopal G, Mukherjee A (2023) Removal of emerging pollutants by clay and clay-nZVI nanocomposites-a review. Environ Technol Rev 12(1):476–492

Graça CA, Mendes MA, Teixeira ACS, de Velosa AC (2020) Anoxic degradation of chlorpyrifos by zerovalent monometallic and bimetallic particles in solution. Chemosphere 244:125461

Guo S, Garaj S, Bianco A, Ménard-Moyon C (2022) Controlling covalent chemistry on graphene oxide. Nat Rev Phys 4(4):247–262

Gurunathan S, Kim JH (2016) Synthesis, toxicity, biocompatibility, and biomedical applications of graphene and graphene-related materials. Int J Nanomedicine 1927–1945

Hanay Ö, Türk H (2019) An overview on usage of nanoscale zero valent iron for pharmaceuticals elimination. Eskişehir Teknik Üniversitesi Bilim Ve Teknoloji Dergisi B-Teorik Bilimler. 7(2):222–239

Hettige H, Mani M, Wheeler D (2019) Industrial pollution in economic development: the environmental Kuznets curve revisited. In: The economics of water quality. Routledge, pp 27–58

Homem V, Santos L (2011) Degradation and removal methods of antibiotics from aqueous matrices—a review. J Environ Manag 92:2304–2347. https://doi.org/10.1016/j.jenvman.2011.05.023

Iavicoli I, Fontana L, Leso V, Bergamaschi A (2013) The effects of nanomaterials as endocrine disruptors. Int J Mol Sci 14(8):16732–16801

Intisar A, Ramzan A, Sawaira T, Kareem AT, Hussain N, Din MI et al (2022) Occurrence, toxic effects, and mitigation of pesticides as emerging environmental pollutants using robust nanomaterials–a review. Chemosphere 293:133538

Ismail NAH, Wee SY, Kamarulzaman NH, Aris AZ (2019) Quantification of multi-classes of endocrine-disrupting compounds in estuarine water. Environ Pollut 249:1019–1028

Jain M, Mudhoo A, Ramasamy DL, Najafi M, Usman M, Zhu R et al (2020) Adsorption, degradation, and mineralization of emerging pollutants (pharmaceuticals and agrochemicals) by nanostructures: a comprehensive review. Environ Sci Pollut Res 27:34862–34905

Jain K, Patel AS, Pardhi VP, Flora SJS (2021) Nanotechnology in wastewater management: a new paradigm towards wastewater treatment. Molecules 26(6):1797

Jain N, Gupta E, Kanu NJ (2022) Plethora of Carbon nanotubes applications in various fields–a state-of-the-art-review. Smart Sci 10(1):1–24

Jayakaran P, Nirmala GS, Govindarajan L (2019) Qualitative and quantitative analysis of graphene-based adsorbents in wastewater treatment. Int J Chem Eng

Jin Z, Wang X, Sun Y, Ai Y, Wang X (2015) Adsorption of 4-n-nonylphenol and bisphenol-A on magnetic reduced graphene oxides: a combined experimental and theoretical studies. Environ Sci Technol 49(15):9168–9175

Jin J, Yang Z, Xiong W, Zhou Y, Xu R, Zhang Y et al (2019) Cu and Co nanoparticles co-doped MIL-101 as a novel adsorbent for efficient removal of tetracycline from aqueous solutions. Sci Total Environ 650:408–418

Joel AB, Fansen T (2013) Pattern and disposal methods of municipal waste generation in Kaduna metropolis of Kaduna state, Nigeria. Int J Educ Res 1(12)

Jung C, Son A, Her N, Zoh KD, Cho J, Yoon Y (2015) Removal of endocrine disrupting compounds, pharmaceuticals, and personal care products in water using carbon nanotubes: a review. J Ind Eng Chem 27:1–11

Kalita A, Barman P (2019) Natural polymer-based bionanocomposites as smart adsorbents for removal of metal contaminants from water. In: Hybrid nanocomposites. Jenny Stanford Publishing, pp 343–366

Kalofiri P, Biskanaki F, Kefala V, Tertipi N, Sfyri E, Rallis E (2023) Endocrine disruptors in cosmetic products and the regulatory framework: public health implications. Cosmetics 10(6):160

Kayode-Afolayan SD, Ahuekwe EF, Nwinyi OC (2022) Impacts of pharmaceutical effluents on aquatic ecosystems. Sci Afr 17
Khan A, Malik S, Shah S, Ali N, Ali F, Ghotekar S et al (2022a) Nanoadsorbents for environmental remediation. In: Nanotechnology for environmental remediation, pp 103–120
Khan S, Naushad M, Govarthanan M, Iqbal J, Alfadul SM (2022b) Emerging contaminants of high concern for the environment: current trends and future research. Environ Res 207:112609
Kumar P, Vellingiri K, Kim KH, Brown RJ, Manos MJ (2017) Modern progress in metal-organic frameworks and their composites for diverse applications. Microporous Mesoporous Mater 253:251–265
Kurwadkar S, Hoang TV, Malwade K, Kanel SR, Harper WF, Struckhoff G (2019) Application of carbon nanotubes for removal of emerging contaminants of concern in engineered water and wastewater treatment systems. Nanotechnol Environ Eng 4:1–16
La Merrill MA, Vandenberg LN, Smith MT, Goodson W, Browne P, Patisaul HB et al (2020) Consensus on the key characteristics of endocrine-disrupting chemicals as a basis for hazard identification. Nat Rev Endocrinol 16(1):45–57
Liang F, Chen B (2010) A review on biomedical applications of single-walled carbon nanotubes. Curr Med Chem 17(1):10–24
Liang W, Wang B, Cheng J, Xiao D, Xie Z, Zhao J (2021) 3D, eco-friendly metal-organic frameworks@ carbon nanotube aerogels composite materials for removal of pesticides in water. J Hazard Mater 401:123718
Liao Z, Zi Y, Zhou C, Zeng W, Luo W, Zeng H et al (2022) Recent advances in the synthesis, characterization, and application of carbon nanomaterials for the removal of endocrine-disrupting chemicals: a review. Int J Mol Sci 23(21):13148
Lima EC (2018) Removal of emerging contaminants from the environment by adsorption. Ecotoxicol Environ Saf 150:1–17
Lin S, Zhao Y, Yun YS (2018) Highly effective removal of nonsteroidal anti-inflammatory pharmaceuticals from water by Zr (IV)-based metal–organic framework: adsorption performance and mechanisms. ACS Appl Mater Interfaces 10(33):28076–28085
Lincho J, Martins RC, Gomes J (2021) Paraben compounds—part i: an overview of their characteristics, detection, and impacts. Appl Sci 11(5):2307. https://doi.org/10.3390/app11052307
Liu X, Zhang H, Ma Y, Wu X, Meng L, Guo Y et al (2013) Graphene-coated silica as a highly efficient sorbent for residual organophosphorus pesticides in water. J Mater Chem A 1(5):1875–1884
Liu P, Miao L, Sun Z, Chen X, Si Y, Wang Q et al (2023) Inorganic–organic hybrid multifunctional solid electrolyte interphase layers for dendrite-free sodium metal anodes. Angew Chem 135(47)
Loretz LJ, Api AM, Barraj LM, Burdick J, Dressler WE, Gettings SD et al (2005) Exposure data for cosmetic products: lipstick, body lotion, and face cream. Food Chem Toxicol 43(2):279–291
Luo Y, Guo W, Ngo HH, Nghiem LD, Hai FI, Zhang J et al (2014) A review on the occurrence of micropollutants in the aquatic environment and their fate and removal during wastewater treatment. Sci Total Environ 473–474:619–641
Maliyekkal SM, Sreeprasad TS, Krishnan D, Kouser S, Mishra AK, Waghmare UV et al (2013) Graphene: a reusable substrate for unprecedented adsorption of pesticides. Small 9(2):273–283
Marlatt VL, Bayen S, Castaneda-Cortès D, Delbès G, Grigorova P, Langlois VS et al (2022) Impacts of endocrine disrupting chemicals on reproduction in wildlife and humans. Environ Res 208:112584
Mashile GP, Selahle SK, Mpupa A, Nqombolo A, Nomngongo PN (2022) Remediation of emerging pollutants through various wastewater treatment processes. In: Emerging freshwater pollutants, pp 137–150
Mauter MS, Zucker I, Perreault F, Werber JR, Kim JH, Elimelech M (2018) The role of nanotechnology in tackling global water challenges. Nat Sustain 1(4):166–175
Mishra D, Khare P (2021) Emerging nano-agrochemicals for sustainable agriculture: benefits, challenges and risk mitigation. In: Sustainable agriculture reviews 50: emerging contaminants in agriculture, pp 235–257

Nagpal M, Kakkar R (2019) Use of metal oxides for the adsorptive removal of toxic organic pollutants. Sep Purif Technol 211:522–539

Naidu R, Espana VAA, Liu Y, Jit J (2016) Emerging contaminants in the environment: risk-based analysis for better management. Chemosphere 154:350–357

Nguyen MN, Jue ML, Buchsbaum SF, Park SJ, Vollnhals F, Christiansen S et al (2024) Interplay of the forces governing steroid hormone micropollutant adsorption in vertically-aligned carbon nanotube membrane nanopores. Nat Commun 15(1):1114

Oliveira LM, Nascimento MA, Guimarães YM, Oliveira AF, Silva AA, Lopes RP (2018) Removal of beta-lactams antibiotics through zero-valent copper nanoparticles. J Braz Chem Soc 29:1630–1637

Pandey P, Khan F, Agarwal S, Singh M (2022) Nano adsorbents in wastewater treatment: a new paradigm in wastewater management. Lett Appl Nanobiosci 12:125

Pankajakshan A, Sinha M, Ojha AA, Mandal S (2018) Water-stable nanoscale zirconium-based metal–organic frameworks for the effective removal of glyphosate from aqueous media. ACS Omega 3(7):7832–7839

Paramasivan T, Sivarajasekar N, Muthusaravanan S, Subashini R, Prakashmaran J, Sivamani S et al (2019) Graphene family materials for the removal of pesticides from water. In: A new generation material graphene: applications in water technology, pp 309–327

Park W, Shin H, Choi B, Rhim WK, Na K, Han DK (2020) Advanced hybrid nanomaterials for biomedical applications. Prog Mater Sci 114:100686

Paulos P, Runnalls TJ, Nallani G, La Point T, Scott AP, Sumpter JP, Huggett DB (2010) Reproductive responses in fathead minnow and Japanese medaka following exposure to a synthetic progestin. Norethindrone Aquat Toxicol 99(2):256–262

Pereira HA, da Boit MK, Vieira Y, Diel JC, Netto MS, Reske GD et al (2023) Adsorptive behavior of multi-walled carbon nanotubes immobilized magnetic nanoparticles for removing selected pesticides from aqueous matrices. Chemosphere 325:138384

Pradeep T (2009) Noble metal nanoparticles for water purification: a critical review. Thin Solid Films 517(24):6441–6478

Qin L, Zheng HG (2017) Structures and applications of metal–organic frameworks featuring metal clusters. CrystEngComm 19(5):745–757

Qin SB, Fan YH, Li XS, Zhang Y, Qi SH (2018) Rapid preparation of methyltrimethoxy-modified magnetic mesoporous silica as an effective solid-phase extraction adsorbent. J Sep Sci 41(3):669–677

Rahman G, Najaf Z, Mehmood A, Bilal S, Shah AUHA, Mian SA, Ali G (2019) An overview of the recent progress in the synthesis and applications of carbon nanotubes. C 5(1):3

Rahmani Sani A, Hosseini Bandehgharaei A, Naeemi M, Navidzadeh A, Agheli E (2018) Removal of sulfadimethoxine antibiotic from aqueous solutions using carbon nanotubes. J Res Environ Health. 4(1):11–22

Rahmanifar B, Moradi DS (2014) Removal of organochlorine pesticides by chitosan loaded with silver oxide nanoparticles from water. Clean Technol Environ Policy 16:1781–1786

Raja S, Mohapatra S, Kalaiselvi A, Rani RJ (2018) Awareness and disposal practices of unused and expired medication among health care professionals and students in a tertiary care teaching hospital. Biomed Pharmacol J 11(4):2073–2078

Rani M, Shanker U (2020) Remediation of organic pollutants by potential functionalized nanomaterials. In: Handbook of functionalized nanomaterials for industrial applications, pp 327–398

Rasheed T, Bilal M, Hassan AA, Nabeel F, Bharagava RN, Ferreira LFR et al (2020) Environmental threatening concern and efficient removal of pharmaceutically active compounds using metal-organic frameworks as adsorbents. Environ Res 185:109436

Sabarwal A, Kumar K, Singh RP (2018) Hazardous effects of chemical pesticides on human health–Cancer and other associated disorders. Environ Toxicol Pharmacol 63:103–114

Sahoo TR, Prelot B (2020) Adsorption processes for the removal of contaminants from wastewater: the perspective role of nanomaterials and nanotechnology. In: Nanomaterials for the detection and removal of wastewater pollutants, pp 161–222

Saifuddin N, Nian CY, Zhan LW, Ning KX (2011) Chitosan-silver nanoparticles composite as point-of-use drinking water filtration system for household to remove pesticides in water. Asian J Biochem 6(2):142–159

Santhosh C, Velmurugan V, Jacob G, Jeong SK, Grace AN, Bhatnagar A (2016) Role of nanomaterials in water treatment applications: a review. Chem Eng J 306:1116–1137

Sarro M, Gule NP, Laurenti E, Gamberini R, Paganini MC, Mallon PE et al (2018) ZnO-based materials and enzymes hybrid systems as highly efficient catalysts for recalcitrant pollutants abatement. Chem Eng J 334:2530–2538

Shen Y, Zhu X, Zhu L, Chen B (2017) Synergistic effects of 2D graphene oxide nanosheets and 1D carbon nanotubes in the constructed 3D carbon aerogel for high performance pollutant removal. Chem Eng J 314:336–346

Shukla A, Malhotra S, Kumar M, Singla N (2022) Pesticides and human health: the noxious impact on maternal system and fetal development. In: Pesticides in the natural environment, p 209–226

Soni R, Pal AK, Tripathi P, Lal JA, Kesari K, Tripathi V (2020) An overview of nanoscale materials on the removal of wastewater contaminants. Appl Water Sci 10(8):189

Sorensen JPR, Lapworth DJ, Nkhuwa DCW, Stuart ME, Gooddy DC, Bell RA et al (2015) Emerging contaminants in urban groundwater sources in Africa. Water Res 72:51–63

Sotelo JL, Rodríguez AR, Mateos MM, Hernández SD, Torrellas SA, Rodríguez JG (2012) Adsorption of pharmaceutical compounds and an endocrine disruptor from aqueous solutions by carbon materials. J Environ Sci Health B 47(7):640–652

Suyambulingam I, Gangadhar L, Sana SS, Divakaran D, Siengchin S, Kurup LA et al (2023) Chitosan biopolymer and its nanocomposites: emerging material as adsorbent in wastewater treatment. Adv Mater Sci Eng 2023:1–20

Sweetman MJ, May S, Mebberson N, Pendleton P, Vasilev K, Plush SE, Hayball JD (2017) Activated carbon, carbon nanotubes and graphene: materials and composites for advanced water purification. C 3(2):18

Teodosiu C, Gilca AF, Barjoveanu G, Fiore S (2018) Emerging pollutants removal through advanced drinking water treatment: a review on processes and environmental performances assessment. J Clean Prod 197:1210–1221

Thakur K, Kandasubramanian B (2019) Graphene and graphene oxide-based composites for removal of organic pollutants: a review. J Chem Eng Data 64(3):833–867

Tian Y, Gao B, Morales VL, Chen H, Wang Y, Li H (2013) Removal of sulfamethoxazole and sulfapyridine by carbon nanotubes in fixed-bed columns. Chemosphere 90(10):2597–2605

Van Doorslaer X, Dewulf J, Van Langenhove H, Demeestere K (2014) Fluoroquinolone antibiotics: an emerging class of environmental micropollutants. Sci Total Environ 500:250–269

Vijayakanth T, Liptrot DJ, Gazit E, Boomishankar R, Bowen CR (2022) Recent advances in organic and organic–inorganic hybrid materials for piezoelectric mechanical energy harvesting. Adv Funct Mater 32(17):2109492

Wanakai IS, Kareru GP, Sujee MD, Madivoli SE, Gachui ME, Kairigo KP (2023) Kinetics of rifampicin antibiotic degradation using green synthesized iron oxide nanoparticles. Chem Africa 6(2):967–981

Wang J, Wang S (2016) Removal of pharmaceuticals and personal care products (PPCPs) from wastewater: a review. J Environ Manage 182:620–640

Wang S, Sun H, Ang HM, Tadé MO (2013) Adsorptive remediation of environmental pollutants using novel graphene-based nanomaterials. Chem Eng J 226:336–347

Wei L, Zhuang L, Zhang S, Jian WW (2021) Adsorption performance of multi-walled carbon nanotube-SiO_2 adsorbent for toluene. J Fuel Chem Technol 49(6):861–872

Werkneh AA, Gebru SB, Redae GH, Tsige AG (2022) Removal of endocrine disrupters from the contaminated environment: public health concerns, treatment strategies and future perspectives—a review. Heliyon 8(4)

Xia T, Yan N, Li S, Lin Y, Su T (2019) Adsorption of tylosin and sulfamethazine by carbon nanotubes and titanium dioxide nanoparticles: pH-dependent mechanisms. Colloids Surf A Physicochem Eng Asp 581:123851
Xu M, Liang T, Shi M, Chen H (2013) Graphene-like two-dimensional materials. Chem Rev 113(5):3766–3798
Xu J, Wang L, Zhu Y (2012) Decontamination of bisphenol A from aqueous solution by graphene adsorption. Langmuir 28(22):8418–8425
Yan Y, Miao J, Yang Z, Xiao FX, Yang HB, Liu B, Yang Y (2015) Carbon nanotube catalysts: recent advances in synthesis, characterization and applications. Chem Soc Rev 44(10):3295–3346
Yang X, Li J, Wen T, Ren X, Huang Y, Wang X (2013) Adsorption of naphthalene and its derivatives on magnetic graphene composites and the mechanism investigation. Colloids Surf A Physicochem Eng Asp 422:118–125
Younis SA, Kwon EE, Qasim M, Kim KH, Kim T, Kukkar D et al (2020) Metal-organic framework as a photocatalyst: progress in modulation strategies and environmental/energy applications. Prog Energy Combust Sci 81:100870
Yu F, Ma J, Han S (2014a) Adsorption of tetracycline from aqueous solutions onto multi-walled carbon nanotubes with different oxygen contents. Sci Rep 4(1):5326
Yu JG, Zhao XH, Yang H, Chen XH, Yang Q, Yu LY et al (2014b) Aqueous adsorption and removal of organic contaminants by carbon nanotubes. Sci Total Environ 482:241–251
Yu Q, Geng J, Zong X, Zhang Y, Xu K, Hu H et al (2019) Occurrence and removal of progestagens in municipal wastewater treatment plants from different regions in China. Sci Total Environ 668:1191–1199
Yu R, Yu X, Xue B, Liao J, Zhu W, Fu J (2021) Adsorption of oxytetracycline from aquaculture wastewater by modified carbon nanotubes: kinetics, isotherms and thermodynamics. Fullerenes Nanotubes Carbon Nanostruct 29(1):28–38
Zaib Q, Mansoor B, Ahmad F (2013) Photo-regenerable multi-walled carbon nanotube membranes for the removal of pharmaceutical micropollutants from water. Environ Sci Process Impacts 15:1582–1589
Zango ZU, Jumbri K, Sambudi NS, Ramli A, Abu Bakar NH, Saad B et al (2020) A critical review on metal-organic frameworks and their composites as advanced materials for adsorption and photocatalytic degradation of emerging organic pollutants from wastewater. Polymers 12(11):2648
Zare EN, Iftekhar S, Park Y, Joseph J, Srivastava V, Khan MA et al (2021) An overview on non-spherical semiconductors for heterogeneous photocatalytic degradation of organic water contaminants. Chemosphere 280:130907
Zeng T, Zhang X, Wang S, Niu H, Cai Y (2015) Spatial confinement of a Co_3O_4 catalyst in hollow metal–organic frameworks as a nanoreactor for improved degradation of organic pollutants. Environ Sci Technol 49(4):2350–2357
Zhang C, Zhang RZ, Ma YQ, Guan WB, Wu XL, Liu X et al (2015) Preparation of cellulose/graphene composite and its applications for triazine pesticides adsorption from water. ACS Sustain Chem Eng 3(3):396–405
Zhang Y, Wu B, Xu H, Liu H, Wang M, He Y, Pan B (2016) Nanomaterials-enabled water and wastewater treatment. Nanoimpact 3:22–39
Zhang LH, Shi Y, Wang Y, Shiju NR (2020) Nanocarbon catalysts: recent understanding regarding the active sites. Adv Sci 7(5):1902126
Zhang QQ, Xing C, Cai YY, Yan XT, Ying GG (2021) How much do human and livestock actually contribute to steroids emission and surface water pollution from past to the future: a global research. Sci Total Environ 772:145558
Zhang W, Taheri-Ledari R, Saeidirad M, Qazi FS, Kashtiaray A, Ganjali F et al (2022) Regulation of porosity in MOFs: a review on tunable scaffolds and related effects and advances in different applications. J Environ Chem Eng 10(6):108836
Zhao S, Malfait WJ, Guerrero-Alburquerque N, Koebel MM, Nyström G (2018) Biopolymer aerogels and foams: chemistry, properties, and applications. Angew Chem Int Ed 57(26):7580–7608

Zhao X, Grimes KL, Colosi LM, Lung WS (2019) Attenuation, transport, and management of estrogens: a review. Chemosphere 230:462–478

Zheng Z, Yu H, Geng WC, Hu XY, Wang YY, Li Z et al (2019) Guanidinocalix [5] arene for sensitive fluorescence detection and magnetic removal of perfluorinated pollutants. Nat Commun 10(1):5762

Zhu X, Li B, Yang J, Li Y, Zhao W, Shi J et al (2015) Effective adsorption and enhanced removal of organophosphorus pesticides from aqueous solution by Zr-based MOFs of UiO-67. ACS Appl Mater Interfaces 7(1):223–231

Zhu S, Liu YG, Liu SB, Zeng GM, Jiang LH, Tan XF et al (2017) Adsorption of emerging contaminant metformin using graphene oxide. Chemosphere 179:20–28

Chapter 6
Nanomaterials in the Advanced Treatment Processes

Doyel Bhattacharya, Shubhajit Halder, and Rohit Maind

Abstract Access to safe water is a fundamental human need, but with such skyrocketing contamination levels of this lifeblood resource, its implications threaten serious danger to global health, and by extension, that calls for the urgent application of emerging technologies. Water sources are contaminated by a variety of organic, inorganic, and biological agents. The main organic pollutant from among these contaminants is oily sludge from petroleum refineries, which contributes significantly to hazardous waste, thereby necessitating proper disposal. While several advanced wastewater treatment methods have been developed to enable reuse of water, most of these are limited by the continued presence of pollutants. Innovation in technology coupled with monitoring has been perceived as critical to effective water management. Optimization of the use of nanomaterials for its efficiency in treating different concentrations of pollutants is key in obtaining environmental sustainability. The occurrence of nanomembranes has generated new opportunities for softening water, along with the removal of a great range of contaminants, in wastewater treatment. It is proper to appraise these technologies regarding their sustainability so that undesirable consequences are not encouraged. The chapter discusses an overview of different nanomaterials and how to present the effectiveness of such nanomaterials for advanced wastewater treatment concerning environmental, economic, and social factors.

Keywords Advanced processes · Wastewater treatment · Nanomaterials · Sustainability · Nanomembranes

---

D. Bhattacharya (✉)
Department of Chemistry, Dada Ramchand Bakhru Sindhu Mahavidyalaya, Nagpur, Maharashtra, India
e-mail: doyelnagpur@gmail.com

S. Halder · R. Maind
Department of Chemistry, Hislop College, Nagpur, Maharashtra, India

A. Mandpe et al. (eds.), *Nanotech Solutions for Industrial Wastewater*, Advances in Wastewater Research, https://doi.org/10.1007/978-981-96-6879-3_6

6.1 Introduction

One of the major developments of sustainable growth goals of the United Nations is the easy and cheap access to clean water that stands as a major global concern in the twenty-first century. Reusing wastewater is frequently recommended as a solution to enhance water supply and fulfil the increasing demand, especially to deal with potable water scarcity (Angelakis et al. 2018). Technologically enhanced treatment plants for wastewater can recover energy, chemicals, fertilizers, and other resources in addition to enabling the reuse of potable water (Diaz-Elsayed et al. 2018; Mo and Zhang 2013). Already, drier and more populous regions across the globe engage in drinkable reuse of wastewaters in hopes of somehow reducing water shortages. Examples include Big Spring, Texas; Beaufort West, South Africa; and Windhoek, Namibia (Gerrity et al. 2013; Lahnsteiner et al. 2018). Throughout the years, newly established treatment processes like MF, UF, MBR, RO, GAC, and UV disinfection have been developed and tested (Roccaro 2018; Voulvoulis 2018). For example, Gómez-López et al. (2009) have provided comparative account of six different methods of disinfection methods by multicriteria decision analysis for the optimization of wastewater treatment for its reuse in Cartagena, Spain. The authors have considered important thirteen criteria wherein the water quality, economic, social, and environmental aspects are assessed. The outcome favored chlorination as the most suitable option for the urban water reuse by virtue of the economic and environmental aspect in comparison with the other methods. As of now, the concept of greenness in infrastructure has gained momentum because of their environmentally friendly processes. Green infrastructure is exploiting the already existing human resource in the discovery and exploitation of the natural elements that adds to the natural resources as part of sustainable growth and development.

Other examples of this type include all types of rooftop gardening, intensive afforestation, green plains to prevent soil erosion, stormwater detention ponds, and even the development of wetlands as a way to mitigate some of the effects of urbanization. There are over 75,000 lakes and stormwater ponds in Florida alone. Overall, 2.7% of water bodies occupy land in the total urban area of the state, (Sinclair et al. 2020). However, they are not just the problem causers for water quality problems. Total 20–40% of total nitrogen and 60–70% of total phosphorus is removed by using stormwater system in Florida (Harper and Baker 2007). Three urban stormwater detention ponds located in Minnesota holds 48% and 58% P and N_2, respectively. Directly related to the hydrology and the storage capability of the ponds, the release difference occurred. In coastal urban sites, stormwater uses the nitrification and denitrification ponds (Rivers et al. 2018; Yazdi et al. 2021), uptake by the macrophytes (Schwammberger et al. 2020) and sedimentation in the ponds (Lusk and Chapman 2021; Schroer et al. 2018) for nutrient removal. Though natural stormwater extends easily to natural wastewater treatment, most studies about large urban systems are scant. The other ancient method that is gaining popularity is the reuse of treated water (Angelakis et al. 2018). This is because there is a growing need to conserve water (Maniam et al. 2022; Partyka and Bond 2022; Almuktar et al. 2019; Rao et al. 2022).

Treated effluent from the wastewater facility is what is referred to as "reclaimed water." This water is recycled over time and supplemented to groundwater withdrawals to supplement landscape irrigation demands. The landscape and green infrastructure systems are maintained in place as lake ecosystems in which the treated wastewater is retained for later use (Lee et al. 2003). These topographical elements, such as mechanical filtering, aeration, chlorination-based disinfection, and a subterranean transport pipe system, are meant to impart value on traditional grey infrastructure. Other researchers have also reported effects of high salt concentration on agricultural productivity (Liu et al. 2023). Another, less well-known perspective, like the control of water quality and influence of the ecosystems where water is reused of reused storage lakes' ecosystems, can have equally important management implications. Nutrient retention mechanisms for reclaimed water lakes might be different because of differences in environmental conditions and systems. These may include phytoplankton, macrophytes, wetlands, biogeochemical processes and microbial communities. As the report points out, the waters or any other processes that comprise reclaimed water is passed through green infrastructure so that its quality is improved regardless of mediating mechanisms. This has huge management implications throughout the world. There are other benefits entailed with integrating green infrastructure into recovered water recycling systems.

According to Amann et al. (2022); Zhang et al. (2021), grey infrastructure-based advanced treatment methods are often associated with great energy requirements, GHG emissions, environmental pollution, including heavy metals, and low cost-effectiveness. All these factors can be improved by executing green infrastructure as shown by Jayasooriya et al. (2016), Sturiale and Scuderi (2019), Wang et al. (2020). Gray infrastructure typically consumes more labor, electricity, and chemicals when natural ecological processes are allowed to advance naturally. Water need for irrigation and fertilizer needs of landscaping and agriculture could be supplied using reclaimed water hence reducing the additional exploitation of resources. An optimal loading rate for nutrients thus prevents downstream eutrophication. If reuse of recovered water shuns detrimental externality, such as eutrophication downlines due to acceptable stacking rates of supplement, then the use of reclaimed water can meet the demands of supplement and water in the framework and supplement supplies for arranging or agrarian purposes and, thus, can lower the demand for the application of supplements (Liu et al. 2023; Narain-Ford et al. 2021; Partyka and Bond 2022; Rao et al. 2022; Zhu and Dou 2018; Zurita and White 2014). Such programs of the usage of recovered or reclaimed water by avoiding negative externalities lead to feasible and circular economy as in the following ways (Estevez et al. 2021). On the other hand, during the last 100 years, the rate of increase has been exponential because of population growth, changes in consumption patterns and also due to economic developments. Personal care products (POPs) along with products of pharmaceutical importance (PPs) have highly polluted the freshwater resources across the globe and became the largest group of pollutants. Nowadays, their persistence, bioaccumulation, and toxicity thresholds. Although they have properties like POPs, they are not POPs.

Prescription drugs and veterinary pharmaceuticals cures or prevents diseases in humans and animals whereas personal care products (PPs) are primarily meant to enhance quality of life. The hazardous pharmaceutical contaminants disturb aquatic ecosystem by contaminating the water bodies and thus it is a potential threat to several aquatic creatures. This is a severe problem because products of pharmaceutical significance, PPs, are becoming part of ever-increasing human and animal medical devices thus releasing these chemicals into the environment on a regular basis. PPCP may exist in water at concentrations of mg/L to μg/L. because of this, conventional water treatment methods yield poor outcomes since their treatment parts are not designed to extract impurities at a constantly low quantities. The overwhelming increasing rate of antibiotic resistance genes, pesticides, PPs, steroids, emerging organic contaminants, and nitrogen pose a challenge to the scientific community to develop suitable and feasible alternative methods (Asif et al. 2018; Chen et al. 2020; García et al. 2020; Sun et al. 2019). For the last ten years, the latest developments in nanotechnology offer the impetus for leaping in constructing water supply networks. The paper is a review of current advancements in water and wastewater treatment based on the principles of nanotechnologies. The unit activity functions of large nanomaterial technologies serve as the foundation for an objective study of the technology. The following sections of this review paper cover the different types of water treatment techniques. A possible hazard and current issues outline concerning the application of nanotechnology have been provided.

6.2 Role of Nanomaterials in Wastewater Treatment

Nano-adsorbents are tiny particles made either from organic or inorganic materials with a high ability to attract and hold substances. This is evident since nano-adsorbents have a great pore volume, minute dimensions, and an active surface that enables them to remove contaminants with different molecular sizes and degrees of hydrophobicity, hence improving the effectiveness of manufacturing processes by minimizing harmful toxic emissions (Pacheco et al. 2006). They act quickly and can accumulate enormous levels of pollutants, and even after being saturated, they can be chemically regenerated (Yang et al. 2008). For this reason, nanotechnology has gained interest rapidly across the world among scholars. Materials at the nanoscale have exceptional properties and because of their tiny dimensions, they offer an enormous surface area and "surface area to volume" ratio (Hristovski et al. 2007). These properties enable nanoparticles to improve their adsorption capacity. Particles exist with some unique features; these include high reactivity and catalytic property besides their high surface area, which, in turn, makes them significantly more efficient as adsorbents than materials commonly used. High surface area nanoparticles provide large numbers of active sites for their strong interaction with various

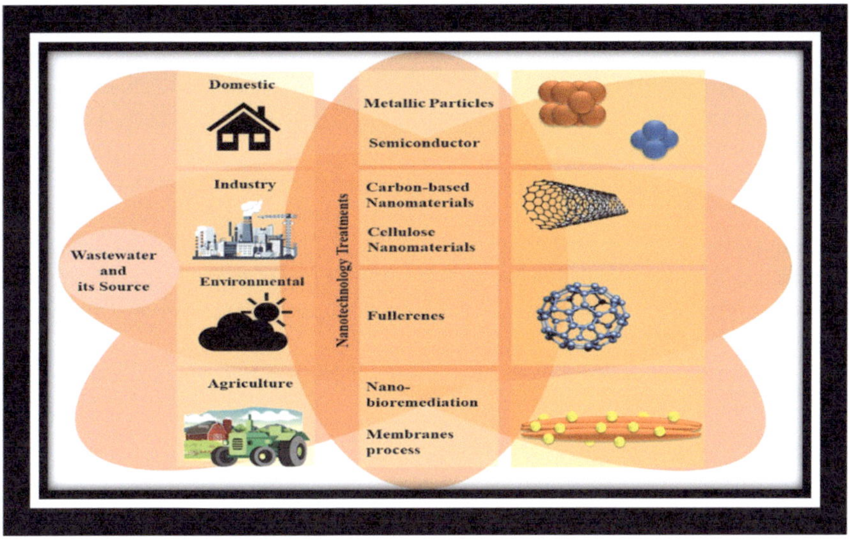

Fig. 6.1 Nanomaterials in wastewater treatment

chemical species (Khaleel et al. 1999; Li et al. 2014). Nanomaterials is a new field of materials generally smaller than 100 nm. With their dimensional properties being dimensionally dependent, nanomaterials have large surface areas and a high reactivity profile. With immense prospects in wastewater and water treatment, lots of work has been carried out on it. Other characteristic striking feature features exhibited by nanomaterials include super para-magnetism and quantum containment. The most widely used nanomaterials can be broadly categorized into four major categories, namely carbon-dependent nano-adsorbents, metal-based nano-adsorbents, polymer-based nano-adsorbents, and zeolites and their composites. All of these categories have shown promising output in the treatment of wastewater that will be briefly discussed in the subsequent section (Fig. 6.1).

6.2.1 Carbon-Based Materials

The properties of nanoparticles like the wide surface area, pore volume, and appropriate functionalities decide the efficiency of any successful adsorption process. Porous materials eliminate harmful contaminants from soil, water, and the air. Different types of porous materials having differing degrees of success include polymers, activated carbon, zeolites, pillared clays, mesoporous oxides, and metal–organic frameworks (Wang et al. 2020; Wan Ngah et al. 2011; Sen Gupta and Bhattacharyya 2012; Adeyemo et al. 2012; Kim and Yi 2004). Carbon nanomaterials in the form of activated carbon, carbon nanotubes, fullerenes, and graphene are carbon-based adsorbents that are often known for their high adsorption capacity and thermal stability (Ren et al. 2011; Rao et al. 2022; Wang et al. 2020). Among the nanomaterial-based adsorbent, superior adsorption properties of carbon-based materials for both organic and inorganic contaminants removal have been reported. However, because of the discovery of fullerene and carbon nanotubes (CNTs), these nanomaterials were widely exploited as promising adsorbents. However, due to some economic constraints, their large-scale applicability becomes limited, and thus it is very hard to construct the adsorbents at a lower cost. Studies have established that multiwall carbon nanotubes doped with magnetic nanomaterials provide very efficient removal of inorganic metal ions (Yang et al. 2008).

6.2.2 Carbon Nanotubes (CNTs)

Among a number of nanomaterial-based adsorbents, the outstanding adsorbent characteristic of carbon-based materials for the removal of organic and inorganic pollutants is investigated. Carbon nanotubes (CNTs) are one of the allotropes of carbon. These allotropes of carbon have a cylindrical tube-like appearance. CNTs have two types which are differentiated on the basis of their walls. If the CNTs are made of a single roll-up graphene sheet then they are referred to as single-walled (SWCNTs) and multiwalled CNTs (MWCNTs) comprises of numerous roll-up graphene sheets. SWCNTs and MWCNTs had high surface-active site to volume ratio with controlled pore size distribution which led to high exceptional sorption capability and high sorption efficiency compared to conventional granular and powder activated carbon with a certain limitation of surface-active sites and the activation energy of sorption. Surface oxidized carbon nanotubes by $KMnO_4$, H_2O_2, and HNO_3 were used for removal of Cd^{2+} ion from aquatic solutions (Slobodian et al. 2013).

The rate of kinetics for the metal ions could be faster with increased adsorption capacities after oxidation process on CNTs. Functional groups such as hydroxyl, carboxylic acids, and carbonyl are present on the surfaces of oxidized carbon nanotubes. In such cases, the adsorption capacity of heavy metal ions is greater at higher values of pH for oxidized carbon nanotubes than the isoelectric threshold. It is found that the heavy metal ions, Pb^{2+}, Cu^{2+}, Zn^{2+}, and Cd^{2+}, are adsorbed extremely strongly by the CNTs (Abukhadra et al. 2019; Hayati et al. 2017; Xu et al. 2018). An injectable carbon nanotube sponge coated with boron has good ability for both water and oil adsorption. These sponges are promising to be employed in order to put an end to the leakage of crude oil for its recovery and then they can be reused once the crude oil drains out from it. Several studies indicate that the properties of sorbate and surface functional groups of CNTs influence each other's ability to adsorb. For example, the level of carboxylic, lactonic, and phenolic groups favors the adsorption of polar substances. Conversely, it is evident that the unfunctionalized surface of CNT has a better adsorption ability of non-polar substances like polycyclic aromatic hydrocarbons (Stafiej and Pyrzynska 2008).

For polar chemicals, CNT sorption behavior typically involves chemical contact; for nonpolar compounds, they include physical contact. CNTs have an effective pH range for sorption very wide, such that the highest performance can be achieved at the pH range of 7 to 10, although ionization and competition between species ionic outside this pH range will occur. The sorption and desorption cycles of CNTs are more efficient than that of the traditional activated carbon, but it is costlier. Among the various nanomaterial-based adsorbents, carbon-based materials have been researched as excellent adsorbents for the removal of organic and inorganic pollutants. Carbon nanotubes are one of the allotropes of carbon. They are distinct structural forms of the element carbon. Carbon nanotubes are essentially tubes made of rolled-up cylindrical forms. There are two types of carbon nanotubes, including both single-walled (SWCNTs) and multiwalled (MWCNTs). Single-walled CNTs are made from a single roll-up graphene sheet, whereas multiwalled CNTs are made from various roll-up graphene sheets. Exceptional sorption capability and high sorption efficiency with regard to conventional granular and powder activated carbon have SWCNTs and MWCNTs despite having some intrinsic limitations such as surface-active sites and activation energy of sorption. Surface-oxidized carbon nanotubes with H_2O_2, $KMnO_4$, and HNO_3 are used for removal of Cd^{2+} from aqueous solutions. The oxidation of carbon nanotubes leads to the increased adsorption capacities toward the metal ions with faster kinetics. Oxidized carbon nanotubes' surfaces carry functional groups in the form of carbonyl, hydroxyl, and carboxylic acids. It has high potential to adsorb heavy metal ions provided that the pH of oxidized carbon nanotubes is greater than the isoelectric threshold. Many studies have shown that carbon nanotubes had excellent heavy metal ion adsorption with very strong nanoparticles, such as Pb^{2+}, Cu^{2+}, Zn^{2+}, and Cd^{2+}; (Abukhadra et al. 2019; Hayati et al. 2017; Xu et al. 2018).

The absorption of water oil by carbon nanotube sponge with sprinkle boron was excellent. These sponges promise to cease oil leakages for oil recovery and can be used after the oil has been drained. Many studies have come up with the findings that the characteristics of the sorbate influence the other adsorption capacity of the surface functional groups of CNTs. For instance, the adsorption ability is influenced by the amount of surface acidity, including the carboxylic, lactonic, and phenolic groups. Yu et al. addressed the four possible adsorption sites on CNT—interstitial channels, outer surface, grooves, and inner cavities. Outer surfaces of CNTs were found to be more attractive to nonpolar materials including polycyclic aromatic hydrocarbons because of their fundamental structure that might be a limiting factor for the application of such materials in adsorption (Iqbal and Ashiq 2007; Stafiej and Pyrzynska 2008). The polar chemical adsorption for using CNT is provided with the surface functionalization, which makes the functional groups such as phenolic, carboxylic, and lactonic groups available to use. There are various techniques through which the functionalizing of the CNTs is done, such as physical changes like impregnating, coating, or chemical treatment like oxidization and chemical grafting.

Li et al. reported graphitized, carboxylized, and hydroxylized CNT for ciprofloxacin adsorption in 2014. It is observed that the surface area increased by a greater degree with carboxylized CNT in comparison to graphitized CNTs. However, adsorption capacity was much higher in the graphitized compared to the carboxylized. This could be because the hydrophobicity of the CNTs is considerably decreased by the two OH and COOH. When CNTs are replaced with OH, the benzene ring is considered as an electron donor, hence forming π–π bonds which facilitate the rate of adsorption. It has been researched on how 8-hydroxyquinoline impacts the functionalization of CNT for regulated removal of heavy metals (Kosa et al. 2012) characteristics of concentration.

Graphene is usually oxidized chemically and then graphite of high purity is oxidized in order to form graphene oxide (GO). The oxygenated groups, similar to those found within the GO sheets, bond with the species adsorbed through the carboxylic, epoxide, and hydroxyl groups (Zhao et al. 2012). It has been claimed that these functional groups are responsible for the hydrophilic character of the GO sheets along with their ease of modification in surface properties. The basic surface chemistry of GO shows two disparate adsorption properties: an aromatic matrix and functional groups containing oxygen. The functional groups are the entities that attract hydrophilic organisms, due to hydrogen bonds or electrostatic interactions. The hydrophobic organic species are attracted by the formation of π–π bonds or hydrophobic interaction. The equilibrium time was below five minutes for methylene blue adsorption onto pure GO (Zhang et al. 2011). It was also observed that as the pH of the solution raised from 1.5 to 11, the adsorption ability of GO enhanced. This is because at higher pH values, the GO surface will be negative; thus it has greater electrostatic attraction of the adsorbate.

There has been a much reported recent work on efforts to enhance GO's adsorption capacity. In one of the methylene blue adsorption studies, Wu et al. (2014) functionalized the surface of GO with a biosurfactant rhamnolipid and brought about a change in zeta potential of the functionalized GO from 0 to 2 mV against an equivalent rise in pH from 3 to 11. A 500 mg/g improvement in the adsorption capacity. Adhesion was caused by the hydrogen bond, π–π contact, and electrostatic attraction between GO and adsorbate. One study found that GO is utilized for the absorption of tetracycline medicines like doxycycline, oxytetracycline, and tetracycline. In order to functionalize GO for its corresponding adsorption of lead ions, a silanization reaction of hydroxyl groups on the surface of GO with N-(trimethoxysilylpropyl) ethylenediamine triacetate, EDTA-silane, was followed (Madadrang et al. 2014). This is due to the chelating property of ethylene diamine triacetic acid. EDTA with Si_2O_2C bonds with Si_2OH and C_2OH can easily reach the graphene surface. Multiwalled carbon nanotubes (MWCNTs) purified by mixed HNO_3/H_2SO_4 solution were used as adsorbents. Equal percentages of $CHCl_2Br$, $CHCl_3$, $CHBr_3$, and $CHClBr_2$, were used in the THM solution.

Studies on kinetics show that adsorption of THM onto CNTs is governed by diffusion mechanism. $CHCl_3$ is favored over other THMs such as $CHCl_2Br$, $CHClBr_2$, and $CHBr_3$ since due to its small size it can penetrate the internal cavity through pores. At 5 °C, maximum amount adsorbed for $CHCl_3$ was 3.158 mg/g whereas for $CHBrCl_2$, $CHBr_2Cl$, and $CHBr_3$ the amount adsorbed was 2.016 mg/g, 2.008 mg/g, and 1.976 mg/g, respectively. Another sharp decrease of the adsorption capacity was observed with a rise in temperature from 5 °C to 35 °C. The maximum adsorption capacity calculated by the model simulated Langmuir of $CHCl_3$ reduce from 3.158 to 2.826 mg/g when the temperature rise from 5 to 35 °C. This result demonstrates how highly effective CNTs can serve as adsorbents. These outcomes indicate the effectiveness of CNTs as adsorbents, since they are two to three times greater than commercial PAC, which evaluates at 1.32 mg/g under 25 °C. They also indicate that the kinetic process of adsorption was controlled by the first-rate law. In addition, it was found that there was a significant reduction in the adsorption of $CHCl_3$ on the carbon spheres from the mixed solution. This accounts for about 40% of the adsorption capacity compared to pure $CHCl_3$ solution. Besides this, there are quite a few other papers (Chen et al. 2007; Lu et al. 2006) discussing the use of carbon nanoparticles for the removal of aquatic pollutants. Carbon nanoparticles, surely must be one of the best adsorbents for the removal of divalent metal ions, dyes, non-organic metals (NOMs), and trihalomethanes (THMs) from aqueous solutions. However, their actual usage is limited due to their relatively high unit price. Moreover, since raw CNTs contain metal catalysts they could be little toxic but chemically functionalized CNTs have yet to show any toxicity till now (Chen and Wang 2006). Thus, for the practical application of CNTs as a sorbent in water and wastewater treatment, low-cost means of producing carbon nanotubes (CNTs) and nontoxicity of CNTs and CNT-based materials are needed. Below here, a short overview of the diverse organic micropollutants for which CNTs have been used for removal from water and wastewater aside from metals is presented. The majority of organic dyes have significant toxicity at lower concentrations in water, which may limit their potential carcinogenic effects on living beings. For these reasons, most of the researchers are interested in studying the application of adsorbents with different characteristics for the treatment of organic dyes before they are discharged into the environment.

Various adsorbents, including zeolites, clays, polymers, and activated carbon, have been used to remove organic dyes from wastewaters. The dyes, which are a water pollutant, are typically encountered in the effluents of food processing, paper, leather, textile, dyeing, and cosmetics sectors. Colored dye effluents cause harm to the aquatic life forms which further agitates the natural equilibrium and the symbiotic process. CNTs have large surface area that aids in the adsorption of organic dyes from the effluents. Due to their unique properties, CNTs are one of the most promising adsorbents that can be used to purify organic dyes (Yang and Xing 2007; Yao et al. 2011). Other than dyes, water supplies in particular have been widely reported as threatened by phenols, pesticides, medicines, aromatic amines, as well as many other hazardous organic pollutants. Olaquindox (OLA) is one of the well-known food ingredients but is extremely phototoxic, mutagenic, genotoxic, and carcinogenic. Tetracycline (TC) is one of the most used antibiotics in the world but has major adverse effects on human health and may harm the environment if it builds up in water systems. Thus, it is very important to remove organic pollutants from contaminated water as effectively as possible.

Creating new cost-effective and adsorbent materials with highly efficient adsorption and removal properties has now become a top priority for the efficient management of these environmental pollutants. Even though carbon nanotubes have many important advantages, the cost of producing them is so high that their use in practical applications at the industrial level is not expected to happen soon (Volder et al. 2013). Point-of-use applications that consume lesser amounts of CNTs are more economical; such uses include the remediation of easily biodegradable contaminants like much of drugs and antibiotics (Rassoulinejad-Mousavi et al. 2020; Rehman et al. 2020). Therefore, more research is needed to find practicable and cost-effective methods for water treatment using carbon nanotubes.

6.2.3 Metal Oxide-Based Nanomaterials

6.2.3.1 Nano-metal Oxides

In the last two decades, several wastewater treatment technologies were researched and developed that put emphasis on advanced oxidation processes (AOPs). These include ozonation, Fenton, cavitation, and photocatalytic oxidation; they have been applied effectively in the in vitro degradation of refractory organic contaminants. Advanced oxidation processes, whether homogeneous or heterogeneous, have been well explored in the field of wastewater treatment. The activation processes of AOPs generally need UV or visible light, different oxidants (O_2, H_2O_2, O_3), and catalysts (TiO_2). Hydroxyl radicals ($^{\cdot}OH$) are generated during the activation of AOPs, and if dissolved oxygen exists under aqueous conditions, these radicals may interact with organic molecules. Since they use an enormous amount of organic colors in their manufacturing procedures, the printing, paper, and dyeing industries have been highlighted as the most environmentally harmful sectors in the industrial chain of pollution. The natural colorants present in the water bodies retard penetration and reflection of sunlight, thereby raising the concentration of toxins and creates more problems for the aquatic life. Organic dyes can be completely mineralized with the utilization of effective AOPs. Many studies have shown the ways by which various AOPs function on the effective degradation of dye pollutants in aqueous conditions. The complicated chain reaction initiated by photocatalytic advanced oxidation process has the potential to yield colorless organic intermediates. At times, the intermediates might be even more dangerous than the original substance. Complete mineralization of the contaminant is the primary goal of the sophisticated consequent process of oxidation. While many studies took advantage of micro-sized catalysts in photocatalytic advanced oxidation processes over the past decades to investigate the degradation of organic molecules, very few studies utilized nanomaterials as nano-photocatalysts in AOPs for wastewater treatment (Geng et al. 2009; Chang et al. 2009).

Nanosized metal oxides, which include nanosized ferric oxides, manganese oxides, aluminium oxides, titanium oxides, magnesium oxides, and cerium oxides, are some of the most promising adsorbents to be employed in eliminating contaminants from aqueous systems. It is partly because of the vast surface areas and high activity of the size quantization effect. Some recent studies show that some

NMOs may exhibit favorable sorption to heavy metals in terms of high capacity and selectivity. Thus, their application could result in higher removal of harmful metals to comply with laws that are becoming increasingly stringent. Unfortunately, the increase in surface energy leads to the lower stability of metal oxides, particularly when their sizes approach the nanoscale from the micrometer scale. NMOs are therefore prone to agglomeration due to forces of van der Waals or other interactions (Pradeep and Anshup 2009), and the outstanding selectivity and capacity of NMOs would be drastically diminished or perhaps lost altogether. Also, because of their weak mechanical strength and large pressure drops (which make them difficult to remove from aqueous streams), NMOs are not feasible for fixed beds or other flow-through technologies. They were also impregnated into massive porous supports to create composite adsorbents for actual wastewater treatment (Pan et al. 2009).

Porous supports that are often used include synthetic polymeric hosts, natural materials, and activated carbon. Besides conventional NMOs, magnetic NMOs also gain ever-increasing attention because of the ease of separation from water in a magnetic field (Mahdavian and Mirrahimil 2010). It was also simple to separate magnetic NMO-based composite adsorbents for recycling or regeneration from aqueous solutions (Zhao et al. 2011). It is this kind of straightforward separation that can allow reduced cost and enhanced process efficiency in the water and wastewater treatment process. There are so many of them; therefore, comparison of the capacities of different NMO types applied for the removal of heavy metals is important. TiO_2 and ZnO are the two major utilized nano-photocatalysts due to their higher photolytic activities. The only limitation to apply these photocatalysts is due to their application in the presence of UV rays. This could be as a result of their large band gap, such as that of TiO_2 at 3.2 eV. These photocatalysts must be functionalized or modified in order to lower their band-gap energy and become useable in sunlight to increase activity. Some of the methods that modify photocatalysts include doping metal impurities, dye sensitization, and composites including semiconductors with small band-gap anions.

Chiou and Juang et al. reported degradation of phenol from aqueous solution is achievable using Pr-doped TiO_2 nanoparticles. Doping TiO_2 (1 g/L) with Pr (III) (0.072 mol%) shifts the absorption band into the visible band region due to the crystal lattice. Studies of degrading showed that Pr(III) interacts with superoxide radicals and halts the holes electrons (h^+/e^-) recombination which leads to the breakdown of phenol. A different report uses new metals for photodegradation, such as Au, Pd, and alloys of Au Pd. The nanoparticles of TiO_2 are used in the decomposition of phenol. The comparative degradation of pristine and metal doped TiO_2 samples showed the increase in degradation by about 40%. The primary cause of the increase in conversion was the reversal of the process of benzoquinone to hydroquinone conversion by the photocatalyst. The study revealed that Pd and Au–Pd-decorated TiO_2 could oxidize 90% of phenol to CO_2.

An existing study conducted using lanthanum-doped ZnO nanoparticles for acetaminophen (paracetamol) reported 99% efficacy (Thi and Lee 2017). Preparation of the La-doped ZnO nanophotocatalyst involved precipitation technique. The outcomes endowed La-doped ZnO with the ability to restrain electron–hole pairs from their recombination. A 1% (by weight) doping was found to offer maximum efficiency for the degradation process. Two degradation processes were likely in consideration for the intermediate or by-product for mating. Degradation, as seen from the proposed pathways, occurs after adsorption.

Photocatalysts of 25 nm diameter were used in the studies. Additionally, it was determined that ampicillin's adsorptive clearance was two to four times higher in this case than nanoparticles of the size 50 nm. The most recent studies used historical data design with response surface modeling to model the response of ZnO nanoparticles for the genetic algorithm (GA) and face methodology (RSM) degradation of basic dyes (Mahmoodi et al. 2017). Basic Blue 41 and Basic Red 46 dyes were used and their corresponding removal efficiency percentages are 72.56% and 67.89%. Experiments by RSM and GA optimization studies revealed very close results.

6.2.3.2 Nano-metal Oxides Support

Metal oxides have received a lot of attention as highly effective adsorbents for the removal of contaminants from wastewater and water. Some of their benefits includes fast kinetics, enhanced adsorption property, and selective contaminants sorption from water and wastewater. However, there are still some technological obstacles that need to be overcome in order to boost the metal oxide-based application such as the reduction of various pollutants. For example, metal oxides put in an aqueous solution have a tendency to aggregate into large-size particles and it appears inevitable that their capacity will be lost. Furthermore, it is still an intriguing but difficult issue to effectively isolate water and wastewater by separating metal oxides. The extensively utilized supports come in the form of synthetic polymer hosts like cross-linked ion-exchange resins (Li et al. 2014; Aguila et al. 2017; Mon et al. 2017), natural hosts like bentonite (Xu et al. 2015; Cheng et al. 2016), sand (Zou et al. 2016; Pouran et al. 2015), and metallic oxide materials like Al_2O_3 membrane (Lv et al. 2019), porous manganese oxide complex (Li et al. 2014).

Creating novel composite adsorbents based on metal oxides appears to be a successful strategy for addressing all of the aforementioned technical issues. Complete elimination and 8.58, 702 and 704 μg/g. For the development of simpler process of composite adsorbents, there is a need to highlight on some issues such as the interaction between supported metal oxides and the hosts, prolong activity of the composite adsorbents, and the remediation of heavy metal-contaminated water (Karn et al. 2009). Because magnetic adsorbents are simple to separate and collect with a magnet, there has been a lot of interest in employing them for water treatment recently. An efficient way to get around the separation issues with graphene is to combine magnetic particles with GO or GNs. Simultaneous addition of the magnetite nanoparticles can prevent or lessen restacking of the graphene sheets and the significant agglomeration which in turns increase the adsorption capacity and give a larger accessible surface area (Lakshmipathiraj et al. 2006).

A magnetite Fe_3O_4/GO composite (M/GO) for the removal of Co^{2+} from aqueous solutions has been described by Iram et al. (2010). For solutions and looked into equilibrium, thermodynamics, and adsorption kinetics. Spontaneous process of magnetic separation followed by adsorption could separate and recover M/GO as suggested by the reports. The sorption of Pb and Cd ions onto surface of $MnFe_2O_4$-G was explain using pseudo second order kinetic model. It was examined that the single adsorption capacities of Cd and Pb on the surface of $MnFe_2O_4$-G at pH 5 and 7 were calculated to be 100 and 76.90 mg/g, respectively. Antibacterial effect of $MnFe_2O_4$-G's against the bacterial strain of *E. coli* was also examined. In two hours, 82% of the bacteria's capability was lost with the treatment of 100 mg/L of $MnFe_2O_4$-G against the given bacterial strain. Heavy metal adsorption ability of $CoFe_2O_4$-G and $NiFe_2O_4$-G to adsorb Pb and Cd ions was successfully tested. The maximum adsorption of Pb ions for $CoFe_2O_4$-G and $NiFe_2O_4$-G at pH 5 and 310 K was determined to be 142.8 and 111.1 mg/g, respectively. At pH 7 and 310 K, the values for Cd ions were discovered to be 105.26 and 74.62 mg/g, respectively. The adsorption reaction was exothermic and spontaneous, as indicated by thermodynamic characteristics (Mak and Chen 2004).

It has also been possible to synthesize magnetite nanoparticles (MNPs) at a minimal cost (Cumbal and Sengupta 2005). During the investigation, the main source of metal nanoparticles, i.e., iron ore tailings (IOT) was selected. The dyes Congo red (CR) and methylene blue (MB) were used as the model pollutants. HCl was used to recover all of the iron from IOT. Ultimately, after being treated with concentrated ammonia, ferric (hydro) oxide (FHO), which is mainly composed of goethite, was produced. Sodium dodecyl sulfonate (SDS) was used to stabilize the adsorbent pH after it was elevated with NaOH. The overall iron content of the synthesized MNPs, according to the chemical analysis, was nearly at the theoretical value of 72.36%. It was discovered that MNPs have a pHpzc of 7.1. The adsorption process using MB and CR dyes claims the kinetics of the process to be fast which suggest that adsorption occurred exclusively on the surface of MNPs. Additionally, it was shown that MNPs' adsorption capacities vary with pH and are pHpzc dependant (pHpzc: ~ 7.1). Higher cationic MB adsorption was observed at pHpzc while higher anionic CR adsorption was observed in the lower pH range. It was found that the optimal pH for the adsorption of MB and CR dyes was roughly 9.2 and 6.2, respectively. It was observed that MNPs aggregated as their dose increased, resulting in a non-linear dye adsorption process.

Dyes absorption by MNPs increased as dye concentrations increased, while the proportion of adsorption fell. The results showed that MNPs had monolayer adsorption capabilities of 70.4 and 172.4 mg/g for MB and CR, respectively. In additional desorption investigations, it was discovered that 95% of the CR dye desorbs at pH values greater than 8.0. However, MB desorption was more successful (85% MB desorption) when 4% (v/v) acetic acid was added to a methanolic solution. The eluent utilized was acetic acid in methanol. Poly(γ-glutamic acid) (PGA) coated iron oxide nanoparticles were created by Inbaraj and Chen (2011) in an attempt to alter the nanoparticles so that they would be more successful for dye removal. Various aquatic matrices such as tap water, acidic and basic solutions, river water, and deionized water are used in leaching test which revealed that PGA coating increased the stability of the MNPs. The adhesion kinetics of MB onto synthesized nanomaterials were quick, taking only five minutes to reach equilibrium. Coexisting ions reduced the amount of MB adsorption. Because of the valence effect, there was a falling trend in the presence of $CaCl_2$ that was more pronounced than in the presence of NaCl. An acidic pH ranging between 1.0 and 3.0 was not optimal for the adsorption of MB dye onto PGA-MNPs, the adsorbent's surface because of the opposition between dye cations and hydrogen ions in the solution for the α-carboxylate anions on the adsorbent surface.

6.2.4 Nanomembranes

Elimination of undesirable constituents from water is essentially the fundamental objective of water treatment. The thickness of membranes apart, these offer constituents a physical barrier toward fresh water sources. It is generally capable of offering a level of automation that calls for decreased land area where the primary component of water treatment and reuse is scalable design, which is enabled through modular design and chemical intake. The risk is still inherent between permeability and selectivity of the membranes. Techniques employing pressurized membranes consume much energy and remain the major barrier for widespread applications of membranes. Fouling of membranes makes the design, operation, and energy consumption of the process much more complicated. Hence, membranes and their modules have a shorter lifespan. Only the membrane material contributes significantly to the performance of the systems.

These nanomaterials, as membrane applications, offer great scope for an increase in membrane permeability, mechanical stability, thermal stability, and fouling resistance (Lu and Astruc 2020). Membranes are divided into several types, including nano-filtration membranes, nanocomposite membranes, and self-assembly membranes; aquaporin-based membranes, biological membranes, thin film composites, and reverse osmosis membranes. There are many types of membranes. The following section details the production and application of several types of membranes in wastewater treatment. In pressurized nanofiltration, contaminants larger than the pores are forced out of the water, using a thin polymer with pores 1 to 10 nm in size under pressure. Among the membranes that use nanofiltration to dissolve organic compounds, dissolved solids, and water-soluble cations, are aquaporin-based membranes (Balkenov et al. 2020), biological membranes (Saleh et al. 2020), thin film composites and opposite osmosis membranes (Vinardell et al. 2019). Nanofiltration is useful in dissolving organic waste, dissolved particles, and water-soluble cations. Also, with the Mg and Ca ions the monovalent ions present in a liquid state in water can get diluted in the water and this specific property of nanofiltration is helpful in water inhibition. Nanofiltration membranes are used by chemical and pharmaceutical companies for the recovery of the solvents. They are also used in the petrochemical processes for purifying gas condensate sectors. Filtration membrane can be fabricated and their surface properties can be modified through various methods (Werner et al. 2018).

Yang and coworkers used a polydopamine and polyethyleneimine interlayer deposition to add a thin, composite nano-filtered membrane film on top of an ultrafiltration substrict (Yang et al. 2017). These membranes removed Na_2SO_4 by 97% and that $MgSO_4$ and $MgCl_2$.NaCl reject the Na_2SO_4 salt. Many ions, such as Na, K, and RO chloride, can be extracted with great utility. Hence, it is presumed to have many applications in marine desalination. Serious attempts have been made during the last few decades for developing inexpensive, highly efficient membrane filtration systems for desalination. The majority of industrial RO membranes employ thin film composite architectures in which the polyamide layer of microporous support membrane is protected. The polyamide does an effective job of regulating the selectivity of penetration of polyethyleneiminic deposition of polydopamine.

These membranes show the 97% removal of Na_2SO_4 and the orderly rejection of Na_2SO_4 salt by $MgSO_4$, $MgCl_2$, and NaCl. Good utility exists for extracting ions such as Na, K, and RO chloride. Therefore, several applications are envisioned in marine desalination. In the past couple of decades, a lot of research has been carried out in formulating inexpensive yet highly efficient membrane filtration techniques for desalination. Furthermore, nano-composite materials have also been reported to form the basis of novel membrane formulations (Al-Najar et al. 2020). It has been demonstrated that GO works as a double-layer protective membrane, leading to enhanced resistance against chlorine and anti-fouling membrane without the loss of separation efficiency. Functioned carbon nanotubes have been widely used in composite RO membranes over the past years. An interesting work shows that by using polyamide membrane as zwitterion-friendly nanotubes, water permeability, and salt rejection could be dramatically improved together.

Composite membranes with functional carbon nanotubes into polyamide-RO membranes are used for greater water flows. Moreover, such membranes with infused carbon nanotubes exhibited more strength and chemical resistance compared to the blank counterparts. The same strength of interactions between carbon nanotubes with the active levels of polyamide could be the reasons for the benefits of composite membranes. This was achieved through the uniformity of their membrane as well as the carbon nanotubes. There is a gradual entry of biochar adsorption technology into wastewater treatment. Standard animals' wastewater contaminants such as nitrogen, phosphate, heavy metals, and organic pollutants are known to be adsorbed by biochar (Zubair et al. 2020). However, biochar-based nanocomposites have been found to be useful in remediation of wastewater and thus the development of these nanocomposites needs further research in this aspect for its practical application.

Mixed-matrix reverse osmosis membranes are currently commercially available for applications in which such super-hydrophilic nanoparticles are impregnated into polyamide films in order to reject microbes, solutes, and contaminants. Titanium dioxide and nano-silver-coated composite RO membranes are usually provided to avoid the fouling of organic impurities by oxidation degradation. Lately, membrane modifications by biomolecules, such as proteins, were prepared to enhance selectivity. Self-assembling copolymer block membranes have also been attempted for ultrafiltration applications and have been reported to provide higher permeability and selectivity. Biological membranes are highly permeable and selective. Aquaporins' protein channels are responsible for controlling the water flow across cell membranes, and these are also required for the removal of contaminants. High an exciting approach to increase the efficiency of membranes in polymer membranes combines selectivity with permeability. When the amphiphilic triblock water permeability polymer vesicles are mixed with the original vesicles, Escherichia coli aquaporin Z rejects glycerol, urea glucose and salt to the smallest degree feasible, at least in orders of magnitude.

One of the greatest future prospects is the development of aquaporin-coated and encapsulated lipid bilayers on industrial nanofiltration membranes. There is a single-file water molecules through highly ordered water that is wrapped in the carbon nanotubes with permeation rates that are really so fast that they have been aligned in their rapid chemical atomic sweetening of the nano-size conductance channel well beyond what the Hagen-Poiseuille equation has ever predicted. A membrane whose effective symmetric surface area comprised only 0.03% of the interconnected CNT area was predicted to surpass fluxes reported for commercially available RO seawater membranes. However, due to a lack of carbon nanotubes with a uniformly sub-nanometer width, the aligned carbon nanotube membranes waste resources inefficiently reject salt or smaller chemicals. The selectivity of the carbon nanotube membranes aligning can be enhanced by functional group gating taking into account the opening nanotube. Donnan exclusion could be used, and it would allow for the rejection of $[Fe(CN)_6]_3$–98.8% at low ionic strength.

6.3 Advanced Oxidation Processes (AOPs)

Advanced oxidation processes are chemical treatments that help in the removal of organic and inorganic contaminants in water and wastewater through oxidation, primarily through the highly reactive hydroxyl radicals (·OH).

- Reactive Species: Hydroxyl radicals (·OH) and other oxidative species
- Primary Oxidants: Ozone (O_3), hydrogen peroxide (H_2O_2), and ultraviolet (UV) light
- Applications: Good for treatment of biologically toxic or non-degradable pollutants in wastewater.
- Contaminant Reduction: Can remove contaminant concentrations from hundreds of parts-per-million to less than 5 parts-per-billion
- Mineralization Goal: To convert the pollutants into stable inorganic compounds like water (H_2O) and carbon dioxide (CO_2).

AOPs are particularly efficient in accomplishing an all-around complete degradation of persistent contaminants so as to have an environment with purer and safer water. Advanced oxidation processes (AOPs) refer to a treatment technique whereby organic and inorganic contaminants are to be removed from water and wastewater through oxidation–reduction reactions mainly involving hydroxyl radicals (·OH). AOPs have gained importance because they can easily break down complex structures of pollutants that are very reluctant to be decomposed through standard treatment techniques, thus being very effective in water treatment processes (Advanced Oxidation Processes—Definition I AWC 2020).

AOPs are especially beneficial in the treatment of different types of waste waters, particularly those containing the following substances:

- **Biologically Toxic or Non-degradable Materials**: AOPs may therefore target pesticides and other compounds of abiotic origin, such as VOCs, as well as petroleum-based materials.
- **Refractory Pollutants**: Petrochemical, pharmaceutical, and textile industries also release the most recalcitrant contaminants in their wastewater effluents, which AOPs can readily oxidize.

6.3.1 Advantages of AOPs Over Traditional Methods (Kamaraj et al. 2022)

(a) High Efficiency: AOPs are capable of attaining extremely low contaminant concentrations-most typically less than 5 ppb-and are effective against a broad spectrum of contaminants.
(b) Minimal Secondary Waste: Unlike biotic treatments, which generate sludge, AOPs generate very minimal amounts of secondary waste, thus reducing the disposal requirements.
(c) Faster Reaction Kinetics: AOPs occur much faster than biotic processes, thus providing quicker, better treatment.

Despite all these benefits, AOPs have several drawbacks:

6.3.2 Drawbacks of AOPs (Deng and Zhao 2015)

(a) High Running Costs: The need for regular chemical charges also increases the overall running costs of AOP systems, particularly in large-scale applications.
(b) Pre-treatment Needed: Some wastewaters need a pre-treatment for maximum efficiency with the AOP, and this enhances the complexity of the treatment process.
(c) Not much Commercial Application: High cost and technical limitations have restrained the usage of AOPs on a large commercial scale—especially in developing countries.

6.4 Photocatalytic Treatment with TiO_2

TiO_2 nanoparticles have been favored in photocatalytic treatment because they are very strong oxidizing agents, highly chemically stable, and can degrade organic contaminants in a slow but effective manner. The fundamental mechanism of the operation of TiO_2-based photocatalysis involves the ultraviolet light-availability mechanism for creating ROS, namely·OH radicals and superoxide ions, through decomposition of the widest variety of contaminants into nontoxic harmless by-products, including water and carbon dioxide (Rayaroth et al. 2022).

6.4.1 *Mechanism of TiO_2 Photocatalysis*

Nanoparticles of TiO_2 absorb photons from UV light, thereby forming an electron–hole pair (e^-/h^+). The e^-/h^+ react with the surrounding H_2O and O_2 molecules to give radicals that are extremely reactive and destroy and break down pollutants.

The conduction band electron can reduce molecular oxygen (O_2) to produce superoxide radicals ($O_2^-\cdot$), and the hole in the valence band can oxidize water (H_2O) or hydroxide ions (OH^-) with the formation of hydroxyl radicals ($\cdot OH$), which are the major active oxidizing species.

6.4.2 *Wastewater Treatment Processes*

- Degradation of Organic Pollutants: TiO_2 photocatalysis is highly effective for the degradation of organic pollutants, including organic dyes, pesticides, and residues from pharmaceuticals. It is particularly suited to the industrial treatment of wastewaters from industry, pharmaceuticals, and agriculture, which normally contains harmful organic contaminants difficult to degrade with conventional methods.
- TiO_2 can also be used for the removal of heavy metals in the presence of UV light. Heavy metals do not break down; it only transforms metal ions into harmless or even less soluble forms, making the removal from water easier.
- Example—Treatment of Textile Industry Effluent: TiO_2 photocatalysis was explored for the treatment of effluent of a textile industry known for emission that consists of high percentages of dyes as well as heavy metals. Under UV irradiation, remarkable removal efficiencies were shown by TiO_2 nanoparticles:
- Dye Degradation: TiO_2 catalyzed the degradation of a set of synthetic dyes including methylene blue, rhodamine B, and azo dyes, and these dyes degraded into harmless byproducts. Dyes are stable and do not easily degrade through biological processes; thus, it is challenging to treat them traditionally. TiO_2 exhibited rapid degradation of color and toxicity.

- Heavy Metal Detoxification: In the same research study, it was proved that TiO_2 could be used for detoxifying heavy metals like lead (Pb) and chromium (Cr) in wastewater also. Surface adsorption coupled with photocatalytic reactions degraded the heavy metal ions into much less toxic species, which further supplemented the advantages of TiO_2 photodriven catalysis.

6.4.3 *Advantages of TiO_2 Photocatalysis*

- Photocatalytic Process: TiO_2 is nontoxic and chemically stable, and all the photocatalytic processes occur without any dangerous byproducts. Hence, the process is quite eco-friendly.
- Wide Range of Application: It can be applied for a wide range of substances ranging from organic compound inorganic ions, to microbial pathogens and hence applicable for all types of wastewaters.
- Self-cleaning Functionality: The photocatalytic surfaces of TiO_2 self-regenerate; thus, they keep on acting without repeating any kind of cleaning or replacement over time.

6.4.4 *Challenges and Improvements*

Although TiO_2 photocatalysis is excellent in effectiveness, there still prevail a number of challenges that need improvements:

UV Light Requirement: The main disadvantage is that it requires UV light to activate it, which limits the efficiency under visible light or sunlight. Researchers are working on the doping of TiO_2 with other elements, like nitrogen or metals that can extend its activity into the visible light range thus making it more practical for use in real-world applications.

Recombination of Electron–Hole Pairs: The recombination of electron–hole pairs can be quite rapid and is a frequent limiting factor in photocatalytic reactions. Attempts to counter such limitations involved the introduction of co-catalysts, the formation of heterojunctions, or the use of nanocomposites for enhancing the charge separation and, in consequence, photocatalytic efficiency.

6.4.5 *Future Prospects*

Future of TiO_2 application in wastewater treatment application of TiO_2 in water treatment is highly promising while research expands on improving efficiency and versatility. Large-scale application of TiO_2-based systems may become widespread for industrial uses, leading to more sustainable and cost-efficient water purification methods when solar-driven photocatalysis becomes applied. Improved material science will mean that TiO_2 and photonic material like it is part of the solution set for the global challenge of water pollution.

6.5 Fenton's Reaction Enhanced by Fe_3O_4 Nanoparticles

Fenton's reaction is one of the most widely practiced AOPs, where hydroxyl radicals (·OH) are generated through the interaction of hydrogen peroxide (H_2O_2) and iron ions (Fe^{2+}) for the degradation of different organic pollutants. This activity is very effective in the degradation of a variety of pollutants and can be enhanced further with iron-based nanomaterials, like Fe_3O_4 nanoparticles. These nanoparticles improve the efficiency of a catalyzed reaction with added advantages like increased reusability, easy magnetic recovery, and an increased surface area in reaction-based applications (Cardoso et al. 2021).

6.5.1 *Mechanism of Fe_3O_4-Enhanced Fenton's Reaction*

Fe_3O_4 nanoparticles critically aid the speeding of the Fenton process by acting as a catalyst. In an ordinary Fenton reaction, Fe^{2+} reacts with H_2O_2 to form hydroxyl radicals (OH), which degrade the pollutants in water. Some way through which incorporation of Fe_3O_4 nanoparticles enhanced the process significantly:

High Surface Area: The surface area of Fe_3O_4 nanoparticles is much greater compared to the bulk iron. As a result, the number of available reactive sites for the catalytic decomposition of H_2O_2 also increase.

Fe_3O_4 nanoparticles can catalyze the Fe^{2+} as well as the Fe^{3+} cycles in the Fenton reaction. In the latter, the reaction responsible for the production of hydroxyl radicals is between Fe^{2+} and H_2O_2, and the Fe_3O_4 nanoparticles can reduce Fe^{3+} back to Fe^{2+} which keeps cycling the species without terminations in order to maintain constant pollutant degradation.

Magnetic Properties: One of the major advantages of Fe_3O_4 nanoparticles is the magnetic nature, which makes it easy to separate and recover the particles from treated water, thus enhancing efficiency and saving money through reuse of the nanoparticles.

For instance, degradation of phenolic compounds and COD reduction were seen in a recent study in which Fe_3O_4 nanoparticles were used in the presence of hydrogen peroxide to treat phenolic compound-containing wastewater. Phenolic compounds are common organic pollutants seen in industrial effluents coming from petrochemical, pharmaceutical, and dye manufacturing sectors. These compounds are notoriously toxic and resistant to conventional treatments.

6.5.2 Benefits

The introduction of Fe_3O_4 nanoparticles into Fenton's reagent resulted in several remarkable effects:

Improved Degradation: The Fe_3O_4 nanoparticle-H_2O_2 system gave way to improved phenolic compound degradation by the cleavage of these compounds into other, even smaller degrading species. This phenomenon has been attributed to higher catalytic activity of the nanoparticles in comparison to the conventional Fenton reaction with solubilized iron salts, giving rise to more hydroxyl radicals.
COD Reduction: COD is one of the most important parameters that are used to measure the amount of organic pollutants present in wastewater. The COD levels dramatically decreased due to the degradation process, and high levels of non-toxic, easily biodegradable byproducts produced as a result of phenolic compound degradation directly contributed to lower COD levels, resulting in drastically improved water quality after treatment.

Fe_3O_4 Nanoparticles-Based Advantages Used in Fenton's Reaction

Improved Efficiency: The Fe_3O_4 nanoparticles improve the generation of hydroxyl radicals, leading to faster degradation combined with higher efficiencies in pollutant removal.
Reusability: Magnetic properties of Fe_3O_4 mean it is easy to recover using an external magnet, and the nanoparticles can be reused many times, saving from the treatment.
Green Environment: Nanoparticles can work under relatively mild conditions, close to neutral pH, rather than highly acidic conditions characterizing the traditional Fenton reactions. Therefore, pH adjustment can be minimized, which makes the process more environmentally friendly.

6.5.3 Challenges and Concerns

Though the results seem promising, the Fe_3O_4-Fenton process does carry some challenges:

Agglomeration in Nanoparticles: The Fe_3O_4 nanoparticles tend to agglomerate, which lowers the surface area and efficiency as a catalyst. This can be avoided by using surface modification or dispersing agents for keeping the nanoparticles stabilized in solution.

Iron Leaching: Although nanoparticle suspensions of Fe_3O_4 happen to be relatively stable to the soluble iron salts, this still leads to a secondary contamination in the treated water due to the leaching of iron. This must be prevented by careful operational parameters control and appropriate recovery technique.

6.5.4 *Future Perspectives*

Fe_3O_4 nanoparticles as catalysts for Fenton's reaction: There is thus a bright future in wastewater treatment, particularly in high organic pollutant containing industries, since Fenton's reaction can be developed enhanced by Fe_3O_4 nanoparticles. The current research work primarily involves optimization of size, surface properties, and stability of the Fe_3O_4 nanocatalyst in order to achieve the highest catalytic performance with reduced environmental risks. Fenton's process enhanced using Fe_3O_4 could be hybridized with other remediation methods such as photocatalysis or adsorption to further enhance the treatment efficiency and broaden its practical applicability.

This, with the shown efficacy of Fe_3O_4 in pollutant removal, puts an Fe_3O_4-enhanced Fenton's reaction in the category as a promising technology for advanced wastewater treatment, holding both environmental and economic benefits.

6.6 Use of ZnO Nanoparticles in AOPs

These ZnO nanoparticles have found wide-ranging applications in advanced oxidation processes for their high photocatalytic activities, chemical stability, and comparative cost-effectiveness. Being a semiconductor of wide band gap, ZnO has been established to be highly efficient in degrading a variety of micropollutants, such as organic dyes, pharmaceutical residues, pesticides, and other persistent contaminants found in wastewater (Taie et al. 2021).

6.6.1 *Mechanism of ZnO Nanoparticles in Photocatalytic AOPs*

When ZnO nanoparticles are under irradiation by UV light, they absorb photons and give rise to electron–hole pairs. This is the precursor for a set of redox reactions in which holes photogenerated in the valence band oxidize water molecules or hydroxide ions to produce very reactive hydroxyl radicals (·OH), whereas the electrons, which are present in the conduction band, reduce molecular oxygen (O_2) into superoxide radicals ($O_2^{-}\cdot$). Hydroxyl and superoxide radicals play an important role in oxidative degradation of organic pollutants (Kurian 2021).

The key steps in ZnO photocatalysis are as follows (Ong et al. 2018):

- Absorption of UV Light: The ZnO nanoparticles absorb the UV light, which leads to the generation of electron–hole pairs.
- Formation of Radicals: After the formation of these electron–hole pairs, redox reactions take place, which result in the generation of ROS, such as·OH and O_2^-. These are highly contributing species for pollutant degradation.
- Degradation of Pollutants: ROS break the chemical bonds of organic pollutants and convert them into harmless products, like water (H_2O), and carbon dioxide (CO_2).

6.6.2 Applications of ZnO Nanoparticles in Wastewater Treatment

ZnO nanoparticles hold excellent potential in removing a wide range of micropollutants present within wastewater, particularly in UV light. The combination of ZnO nanoparticles with UV light provides an efficient photocatalytic system capable of degrading even the recalcitrant pollutants that are resistant to common treatment procedures. Notable applications include (Mustapha et al. 2020):

- Organic Pollutant Degradation: ZnO nanoparticles have been widely used for degradation of many organic pollutants, including dyes, pharmaceuticals, and pesticides. In recent study, it was identified that UV-irradiated ZnO nanoparticles have high efficiency in degrading organic pollutants; examples include industrial dyes like methylene blue and rhodamine B, which are usually released into the environment through textile wastewater. These dyes, known for their stability and resistance to biodegradation, were decomposed into nontoxic by-products with ZnO photocatalysis—using which was established that nanoparticle-based technologies are capable of degrading such challenging wastewater streams.
- Removal of Pharmaceutical Residues: Persistent and toxic effects of pharmaceutical contaminants, which include antibiotics, hormones, and analgesics, bring a serious threat to water bodies. ZnO nanoparticles activated with UV light have been found to be effective in degrading pharmaceutical residues. Major oxidative power of hydroxyl radicals formed from ZnO makes complex pharmaceutical molecules degrade into much simpler, non-toxic forms of compounds for cleaner water resources.
- The studies also show the application of ZnO nanoparticles in the treatment of pesticide-contaminated wastewater where harmful chemicals applied in agricultural activities were degraded. Less harmful substances were produced since atrazine is not easy to break and toxic when compared to other pollutants. Benefits of ZnO Nanoparticles in AOPs
- High Photocatalytic Activity: ZnO nanoparticles possess excellent photocatalysis activity, thus the efficient decomposition of a large variety of organic pollutants by UV light.

- Cost Effectiveness: It is less costly, thus an inexpensive material that is economically sound to be used for mass-scale applications in wastewater treatments.
- Chemical Stability: ZnO nanoparticles are chemically stable and contain low toxicity, thus contributing to their use in environmental remediation in a safe manner.
- Wastewater containing dyes, pharmaceuticals, pesticides, and other micropollutants: ZnO NPs have wide applicability to wastewater containing dyes, pharmaceuticals, pesticides, and other micropollutants.

6.6.3 Difficulties and Limitations

Although ZnO NPs have these mentioned advantages; however, there are several difficulties in practical applications (Singh et al. 2019):

- UV Light Requirement: In the case of ZnO nanoparticles, like TiO_2 nanoparticles, activation requires UV light, reducing their efficiency in visible light. This is a problem for large-scale applications with solar-driven systems. But researchers have continued to investigate this by doping ZnO with nitrogen or metals for better photocatalytic activity under visible light.
- Agglomeration of Nanoparticles: ZnO nanoparticles have the tendency of agglomeration in water-based suspensions. This drastically reduces the surface area and is directly proportional to less photocatalytic efficiency. This may be overcome through surface modification or added dispersants, thus stabilizing the nanoparticles more easily in suspension.
- Recycling and Reusability: Although zinc oxide nanoparticles are highly efficient, recycling them after one round of treatment is not that easy. Many researchers now try to modify the magnetic or structural properties of ZnO for easier recovery and reuse.
- Case Study: ZnO Photocatalysis in Treatment of Textile Wastewater. In one publication reporting the result for textile industry wastewater, ZnO nanoparticles were applied in the presence of UV irradiation for the degradation of organic dyes and other contaminants. Quite interestingly, high removal efficiencies for the ZnO photocatalytic system were observed.
- Dye Degradation: ZnO nanoparticles have been able to decompose synthetic dyes, including methylene blue and rhodamine B. This caused a decrease in both the color as well as the toxicity of the wastewater. Zinc oxide proved to be able to handle the complex mixture found in textile effluent.
- COD and BOD reduction: The levels of both COD and BOD were greatly reduced, showing complete breakdown of contaminants into less harmful byproducts, signifying that the degree of organic pollutants in water was very low.

6.6.4 Future Directions

With continued scientific growth, ZnO NPs are expected to gain a more significant role in the field of wastewater treatment technologies in the near future with special importance in such areas as visible-light-driven photocatalysis and development of composite nanomaterials. Hybrid systems based on ZnO NPs in combination with other AOPs or adsorption techniques are expected to optimize pretreatment of highly polluted streams of wastewater not only toward environmental sustainability but also as an economical measure. ZnO nanoparticles are an effective solution that could overcome the difficulties associated with micro-pollutant removal toward cleaner water resources and more sustainable industrial practices.

6.7 Ozonation Coupled with Nanomaterials

Ozonation is an advanced oxidation process in which ozone (O_3) present within water degrades organic and inorganic pollutants. Once dissolved in water, ozone produces highly reactive oxygen species, notably hydroxyl radicals, along with ozone itself, which is capable of breaking down different contaminants; however, in many conventional ozonation processes, inefficiencies may arise owing to rapid decomposition of ozone or due to the fact that little contact time occurs between ozone and pollutants (Cardoso et al. 2024).

These limitations stimulate researchers to use the combination of ozonation and nanomaterials in order to enhance the oxidation process of pollutants in water. The nanomaterial features—enlarged surface area, high activity, and catalytic properties—will be taken into account. The combination of ozonation with nanomaterials may ensure a significant improvement of oxidation processes efficiency: faster pollutant degradation can be facilitated.

6.7.1 Mechanism of Ozonation

When nanomaterials are combined with ozone, they can function as a catalyst or adsorption agent to enhance oxidation reactions. The following mechanisms apply (Guo et al. 2012):

- Larger Surface Area: Nanomaterials, for example carbon nanotubes (CNTs), and metal oxides, generally provide larger surface area adsorption for pollutants, ensuring that the pollutants receive high exposure to ozone molecules and reactive species. As a result, the pollutant degrades rapidly.
- Catalytic Activation: Some nanomaterials, such as nanoparticles of iron, copper, or silver, act as catalysts, causing ozone to decompose to hydroxyl radicals. These

radicals are orders of magnitude more reactive than ozone and readily oxidize even the most resistant pollutants, thereby enhancing overall treatment efficiency.

- Enhanced Radical Generation of Hydroxyl Radicals: The nanostructures could also enhance more the generation of hydroxyl radicals. Such hydroxyl radicals are the key species in breaking up complex organic molecules into much simpler and nontoxic compounds. With this synergistic effect, oxidation reactions could be enhanced beyond what is achievable through ozonation alone.

6.7.2 Application

- One of the interesting examples of nanomaterial-based integrated ozonation is activated carbon nanotubes (CNTs). CNTs have been extensively investigated as adsorbents and for their diversified physicochemical properties, and are good candidates for ozone dosage in treatment of water applications. In the study, activated carbon nanotubes were integrated with ozone for aqueous solutions containing organic dyes, which are typical products of wastes of textile industries (Fan et al. 2023).
- The carbon nanotubes provided a large surface area for the organic dye molecules to absorb onto the surface of the CNTs. This mechanism increased the residence time of the pollutants near the ozone molecules, thus providing a better opportunity for the ozone molecules to react with the contaminant.
- Accelerated Ozone Decomposition: The CNTs also acted as a catalyst for the acceleration of ozone decomposition into hydroxyl radicals. That would mean their rapid formation was the basis for more comprehensive breakup of such complex structures within the molecules of the dye into nontoxic smaller byproducts.
- Increased Degradation Rates: Ozonation plus CNTs improved dramatically the degradation of organic dyes as compared to classical ozonation. Within this context, dyes like methylene blue, that were previously considered insensitive to most of the traditional wastewater treatment techniques, were degraded and significantly reduced the color and COD of the treated water.

6.7.3 Advantages of Nanomaterials in Ozonation Process

A. Enhanced Degradation Efficiencies: The nanomaterials-combined ozonation enhances the overall efficiency by which pollutants degrade because organic molecules are degraded much faster and systematically.

B. Increased Elimination of Pollutants: Nanomaterials, especially the activated carbon nanotubes, have been characterized by very high surface areas, and this will increase the adsorption of pollutants, and thus, a larger percentage of the contaminants will now interact with reactive oxygen species such as hydroxyl radicals.

C. Synergistic Effects: In some nanomaterials, especially metal-based nanoparticles, the generation of hydroxyl radicals was able to boost oxidation and mineralization rates of pollutants.
D. Ozone Consumption Cut: Kinetics enhancement for the reaction results in ozone and subsequently consumption cutback.
E. Minimal Secondary Pollution: Ozonation has minimal generation of secondary waste, and the use of nanomaterials does not contribute to unnecessary materials, and the cleanliness of the process of treatment is still intact.

6.7.4 Challenges and Opportunities

Despite this good outlook for integration of ozonation with nanomaterials, the following challenges need to be addressed for wider commercial applications (Fan et al. 2023).

(1) Although highly efficient, these nanomaterials like carbon nanotubes and metal nanoparticles are pretty pricey at the cost, which serves as a barrier for their application on a massive scale. The scientists continue to work on much cheaper and scalable synthesis routes to continue working on these materials.
(2) Post-treatment Recovery and Reuse of Nanomaterials: The recovery and reuse of nanomaterials after the treatment process can be problematic. These problems can be alleviated for example, by designing nanomaterials that offer a possibility of easy separation from treated water. Magnetic nanoparticles, for example, could be so designed to be recoverable with an external magnetic field.
(3) Possible Environmental Impact: The potential risk associated with its use is the egression of nanomaterials into the environment. The effect of these materials on ecosystems in the long term is not well known, and hence for sustainable implementation, it is very important that the nanomaterials remain within the treatment system and do not leach into the environment.

6.7.5 Future Prospects

The future prospect of ozonation integrated with nanomaterials is an evolving technology that will promote further efficient, cost-effective, and environmentally friendly treatment methodology for wastewater. The development of new approaches to the synthesis of nanomaterials, for example, visible-light-responsive catalysts or hybrid nanomaterials within multiple functionalities, would certainly enhance the degradation efficiency of pollutants and expand the applicability of this technique.

Hybrid systems that contain the strong oxidative power of ozone together with catalytic and adsorption properties of nanomaterials present highly attractive hybrids.

One of the most challenging wastewater streams both from textile, pharmaceutical, and petrochemical sectors can be addressed by this hybrid system. While the costs of nanomaterials are going down and the recovery methods are well developed, ozonation coupled with nanomaterials could be an important constituent of advanced water treatment systems throughout the world.

6.8 Hybrid Processes Comprising Activated Carbon and Advanced Oxidation Processes (AOPs)

Activated carbon has been widely used in water and wastewater treatment processes. Activated carbon boasts an impressive ability to adsorb a large range of organic and inorganic pollutants. It entails the vast surface area coupled with the presence of functional groups on its surface, hence ideal for pollutant capture. In recent years, not only have modified activated carbon improved the performance of wastewater treatment through nanomaterial modification but have also been implemented in hybrid processes, combining modified activated carbon with advanced oxidation processes (AOPs). These processes have contributed very considerably to higher removal efficiencies of pollutants, especially for the most persistent and complex kinds of contaminants (Atalay and Ersöz 2021).

6.8.1 Mechanism of Hybrid AOPs with Activated Carbon and Nanomaterials

In hybrid systems, activated carbon is also modified by the addition of nanomaterials, usually metal nanoparticles or metal oxides that increase the adsorption capacities in addition to being catalysts to improve the oxidative degradation of pollutants during an AOP. The most common mechanism typically involves two steps (Yan et al. 2023):

1. Adsorption: Activated carbon adsorbs contaminants from water and concentrates them on the surface of activated carbon. This step in general increases the contact time between the contaminants and reactive species produced during the AOP.
2. Oxidation: These sorbed pollutants are further degraded by extremely reactive oxidative species such as hydroxyl radicals (·OH), produced by AOPs (such as UV/H_2O_2, ozonation, Fenton reactions). The modification of activated carbon with nanomaterial enhances the effectiveness of this step through catalytic mediation in the generation of reactive oxygen species and, thus, accelerates the degradation process.

6.8.2 Applications

(a) Silver Nanoparticles Modified Activated Carbon (Ag-NPs/AC): The activation of the modified carbon matrix by the addition of silver nanoparticles to it is considered quite promising. Silver nanoparticles were known for antibacterial action and catalytic activity in reactions. Therefore, they are very effective for use in water treatment. So, both the possibility of using activated carbon to adsorb huge amounts of the pollutants and ability of Ag-NPs to catalyze reactions can be more effectively enlarged by combining both with AOPs (Mokhtar et al. 2021).

(b) Removal of Heavy Metals and Organic Pollutants: Hybrid processes based on activated carbon modified with silver nanoparticles have achieved remarkable efficiency in the treatment of complex wastewater streams. This hybrid approach has successfully been applied to remove heavy metals, including lead, cadmium, and mercury as well as organic pollutants, such as dyes, pharmaceuticals, and residues of petrochemicals (Palani et al. 2023).

(c) Selective Heavy Metal Removal: Heavy metals are the most common elements that cannot be removed from water with conventional treating processes. In a model wastewater system, the prepared activated carbon doped with silver nanoparticles demonstrated heavy metal removal efficiencies greater than 90% (Pandey et al. 2019). The process operates on the principle whereby metal ions are brought on to the surface of the activated carbon, and the oxidation processes thereafter catalyzed by the silver nanoparticles make them amenable to removal from the water. The strong affinity of silver nanoparticles for heavy metals contributes highly toward their removal efficiency.

(d) Degradation of Organic Pollutants: Apart from heavy metals, this hybrid system was very successful in degrading organic pollutants. These included persistent organic dyes from textile industries effluent and residues of pharmaceuticals (Anandan et al. 2020), which are normally resistant to biological treatment, and were demonstrated to degrade effectively when exposed to hybrid treatment of the modified activated carbon with AOPs. Here, silver nanoparticles catalyzed the production of hydroxyl radicals ·OH from hydrogen peroxide H_2O_2 or ozone O_3, thereby seriously accelerating the degradation of these organic contaminants.

6.8.3 Advantages

- High Removal Efficiency: Hybrid systems can achieve a removal efficiency of more than 90% for several types of contaminants; in particular, high concentrations of heavy metals and organic pollutants, important microbial contaminants, such systems are highly useful for treatment of complex wastewater streams coming from a broad spectrum of such industries as textiles, pharmaceuticals, and petrochemicals.

- Activation for Faster Degradation of Contaminants: The hybrid system was enhanced through the inclusion of nanomaterials such as silver nanoparticle. Nanomaterials lead to a more potent formation of reactive species, which catalytically degrade and remove pollutants at a faster rate than the typical adsorption or AOP methods.
- Synergetic Effect: Adsorption in combination with oxidation has a synergetic effect because the pollutants are adsorbed on activated carbon to concentrate them in close proximity to the oxidative species formed during the AOP process. This increases the probability of pollutant degradation and prevents the lateral escape of remaining residuals, especially in such processes.
- Low Secondary Waste: Secondary waste is the most significant problem related to conventional treatment processes: in sludge, which has to be further disposed of. Hybrid processes normally produce a low secondary waste because the nanomaterials applied as well as the activated carbon can be recovered and reused easily. Hence, this will be a more environmentally friendly and cheaper process in the long run.
- Antibacterial Activity: Silver nanoparticles also have an extremely significant antibacterial activity in the process of catalyzing oxidation reactions. Thus, hybrid systems made from silver-modified activated carbon are capable of removing chemical pollutants and bacterial and microbial contaminants from water, offering both removal and decontamination functions.

6.8.4 Challenges

Despite the many advantages of hybrid AOPs, which involves activated carbon combined with nanomaterials, there are still several challenges and problems in pursuing their wider application toward their goal:

(a) Nanomaterials Cost: Adding silver nanoparticles and other nanomaterials can increase the overall cost of the treatment process in its entirety. Currently, work is in progress to find cheaper nanomaterials or minimize the number of nanomaterials used without sacrificing the efficiency of the system.
(b) Recovery and Reusability of Nanomaterial: Recovery and reusability of the nanomaterials with stability throughout the process without losing them is essential for economic and environmental sustainability. Techniques that are developed to recover and reuse nanomaterials, such as magnetic separation in the case of magnetic nanoparticles, have to be developed to their greatest extent for their practical applications at scale.
(c) Nanomaterials released into the environment pose a problem since long-term ecologic impacts are not well understood. Nanomaterials should be restricted in a treatment system and should not leak into any water bodies.

6.8.5 *Future Prospects*

There is good promise for future applications of hybrid processes combining activated carbon and nanomaterials in AOPs. With further research, the aim includes efficiency enhancement, decreasing costs, and minimal impact on the environment, making hybrid systems more feasible for scaling up to real wastewater treatment applications. Advances in the synthesis of such nanomaterials—for example, highly durable and regenerable nanomaterials—will be important in the practical application of hybrid systems. Hybrid AOPs, especially combining activated carbon with nanomaterials, will likely be at the core of the next generation of water treatment technologies as more industries seek sustainable and effective methods for treating complex wastewater. Their wide-spectrum treatment ability with high efficiency and minimal formation of secondary wastes makes them a promising solution to the ever-mounting challenges of water pollution and scarcity.

6.9 Conclusion

From the above sections on industrial pollutants and nanomaterials, it is evident that pollution abatement is effective. There are numerous studies that speak to the specific advantages of nanomaterials and nanotechnology regarding water treatment via photocatalysis, nanomembranes, and adsorption, among many others. Several engineered nanomaterials have been researched for the removal of pharmaceuticals, heavy metals, dyes, and phenolic compounds. Design nanomaterials in such a manner that removes the pollutants optimally while achieving economic efficiency even at lower concentration levels of pollutants and preventing byproducts or secondary pollutants.

Promising solutions in nanotechnology such as metal nanoparticles, carbon-based nanomaterials, zeolites, and dendrimers have been proven to effectively extract toxic metal ions and remove organic, inorganic, and microbial contaminants from water with advanced treatment techniques. Literature is very comprehensive in proving the advantages of nanomaterial-based adsorption, which include an improved efficiency and consistency of the method, cost-effectiveness, and the absence of sludge production. Carbon nanomaterials are distinctly superior to the conventional adsorbents with a better capacity. However, the continuous occurrence in the water bodies like carbon nanotubes (CNTs) would raise a warning for their entry into the food chain.

Doping semiconductor materials changes their band gap for photocatalytic degradation tests relying on photoluminescence but increases surface vacancies upon light exposure, producing greater degradation efficiencies. However, there are no ecological risks associated with recovered pollutants and depleted nanomaterials, although the toxicological risk needs to be resolved through ecological waste management.

Nanoparticle-based water treatment, though great promise is observed in the field, will need research collaboration among academia, industrial development, and innovation initiatives before it can develop economical, efficient, and scalable solutions for water treatment. This review highlights the importance of nanomaterials in water contaminant removal particularly through adsorption, photocatalytic, and antimicrobial properties. Some nanomaterials show higher selectivity for contaminants and higher removal efficiency compared to traditional approaches, making it an excellent alternative. However, there are technical challenges related to lack of scalability, environmental impact, and cost-effectiveness in large amounts; most nanoscale products are not yet ready to be commercialized, and most applications are not yet ready to be widely applied on the market.

References

Abukhadra MR, Bakry BM, Adlii A, Yakout SM, El-Zaidy ME (2019) Facile conversion of kaolinite into clay nanotubes (KNTs) of enhanced adsorption properties for toxic heavy metals (Zn^{2+}, Cd^{2+}, Pb^{2+}, and Cr^{6+}) from water. J Hazard Mater 374:296–308

Adeyemo AA, Adeoye IO, Bello OS (2012) Metal organic frameworks as adsorbents for dye adsorption: overview, prospects and future challenges. Toxicol Environ Chem 94:1846–1863

Advanced Oxidation Processes—Definition | AWC (2020, September 14) American Water Chemicals, Inc. https://www.membranechemicals.com/water-treatment/advanced-oxidation-plants/

Aguila B, Sun Q, Perman JA, Earl LD, Abney CW et al (2017) Efficient mercury capture using functionalized porous organic polymer. Adv Mater 1700665:1–6. https://doi.org/10.1002/adma.201700665

Almuktar SAAA, Abed SN, Scholz M (2019) Wetlands for wastewater treatment and subsequent recycling of treated effluent: a review. Environ Sci Pollut Res 25:23595–23623. https://doi.org/10.1007/s11356-018-2629-3

Al-Najar B, Peters CD, Albuflasa H, Hankins NP (2020) Pressure and osmotically driven membrane processes: a review of the benefits and production of nano-enhanced membranes for desalination. Desalination 479:114323. https://doi.org/10.1016/j.desal.2020.114323

Amann A, Weber N, Krampe J, Rechberger H, Peer S, Zessner M et al (2022) Systematic data-driven exploration of Austrian wastewater and sludge treatment-implications for phosphorus governance, costs and environment. Sci Total Environ 846. https://doi.org/10.1016/j.scitotenv.2022.157401

Anandan S, Ponnusamy VK, Ashokkumar M (2020) A review on hybrid techniques for the degradation of organic pollutants in aqueous environment. Ultrason Sonochem 67:105130

Angelakis AN, Asano T, Bahri A, Jimenez BE, Tchobanoglous G (2018) Water reuse: from ancient to modern times and the future. Front Environ Sci 6:1–17. https://doi.org/10.3389/fenvs.2018.00026

Asif MB, Hai FI, Kang J, van de Merwe JP, Leusch FD et al (2018) Biocatalytic degradation of pharmaceuticals, personal care products, industrial chemicals, steroid hormones and pesticides in a membrane distillation-enzymatic bioreactor. Bioresour Technol 247:528–536. https://doi.org/10.1016/j.biortech.2017.09.129

Atalay S, Ersöz G (2021) Hybrid application of advanced oxidation processes to dyes' removal. In: Green chemistry and water remediation: research and applications. Elsevier, pp 209–238

Balkenov A, Anuarbek A, Satayeva A, Kim J, Inglezakis V, Arkhangelsky E (2020) Complex organic fouling and effect of silver nanoparticles on aquaporin forward osmosis membranes. J Water Process Eng 34:101177. https://doi.org/10.1016/j.jwpe.2020.101177

Cardoso IMF, Cardoso RMF, Da Silva JCG (2021) Advanced oxidation processes coupled with nanomaterials for water treatment. Nanomaterials 11(8):2045. https://doi.org/10.3390/nano11082045
Cardoso RM, Esteves da Silva JC, Pinto da Silva L (2024) Application of engineered nanomaterials as nanocatalysts in catalytic ozonation: a review. Materials 17(13):3185
Chang JH, Yang TJ, Tung CH (2009) Treatment of real printing wastewater with an electrocatalytic process. J Hazard Mater 163:152–157. https://doi.org/10.1016/j.seppur.2013.07.028
Chen C, Wang X (2006) Adsorption of Ni(II) from aqueous solution using oxidized multiwall carbon nanotubes. Ind Eng Chem Res 45:9144–9149. https://doi.org/10.1021/ie060791z
Chen W, Duan L, Zhu D (2007) Adsorption of polar and nonpolar organic chemicals to carbon nanotubes. Environ Sci Technol 41:8295–8300. https://doi.org/10.1021/es071230h
Chen C, Chen L, Chen S, Yu Y, Weng D, Mahmood A, Wang G, Wang J (2020) Preparation of underwater superoleophobic membranes via TiO_2 electrostatic self-assembly for separation of stratified oil/water mixtures and emulsions. J Memb Sci 602. https://doi.org/10.1016/j.memsci.2020.117976
Cheng W, Ding C, Wang X, Wu Z, Sun Y et al (2016) Competitive sorption of As(V) and Cr(VI) on carbonaceous nanofibers. Chem Eng J 293:311–318. https://doi.org/10.1016/j.cej.2016.02.073
Cumbal L, Sengupta AK (2005) Arsenic removal using polymer-supported hydrated iron(III) oxide nanoparticles: role of Donnan membrane effect. Environ Sci Technol 39(17):6508–6515. https://doi.org/10.1021/es050175e
De Volder MFL, Tawfick SH, Baughman RH, Hart AJ (2013) Carbon nanotubes: present and future commercial applications. Science 339:535–539. https://doi.org/10.1126/science.1222453
Deng Y, Zhao R (2015) Advanced oxidation processes (AOPs) in wastewater treatment. Curr Pollut Rep. https://doi.org/10.1007/s40726-015-0015-z
Diaz-Elsayed N, Rezaei N, Guo T, Mohebbi S, Zhang Q (2018) Wastewater-based resource recovery technologies across scale: a review. Resour Conserv Recycl 145:94–112. https://doi.org/10.1016/j.resconrec.2018.12.035
Estevez S, Gonzalez-Garcia S, Feijoo G, Moreira M (2021) How decentralized treatment can contribute to the symbiosis between environmental protection and resource recovery. Sci Total Environ 812. https://doi.org/10.1016/j.scitotenv.2021.151485
Fan X, Wei G, Quan X (2023) Carbon nanomaterial-based membranes for water and wastewater treatment under electrochemical assistance. Environ Sci Nano 10(1):11–40
García J, García-Galán MJ, Day JW, Boopathy R, White JR, Wallace S, Hunter RG (2020) A review of emerging organic contaminants (EOCs), antibiotic resistant bacteria (ARB), and antibiotic resistance genes (ARGs) in the environment: Increasing removal with wetlands and reducing environmental impacts. Bioresour Technol. https://doi.org/10.1016/j.biortech.2020.123228
Geng B, Jin Z, Li T, Qi X (2009) Kinetics of hexavalent chromium removal from water by chitosan-Fe0 nanoparticles. Chemosphere 75:825–830. https://doi.org/10.1016/j.chemosphere.2009.01.009
Gerrity D, Pecson B, Shane Trussell R, Rhodes TR (2013) Potable reuse treatment trains throughout the world. J Water Supply Res Technol 62:321–338. https://doi.org/10.2166/aqua.2013.041
Gómez-López MD, Bayo J, García-Cascales MS, Angosto JM (2009) Decision support in disinfection technologies for treated wastewater reuse. J Clean Prod 17:1504–1511. https://doi.org/10.1016/j.jclepro.2009.06.008
Guo Y, Yang L, Cheng X, Wang X (2012) The application and reaction mechanism of catalytic ozonation in water treatment. J Environ Anal Toxicol 2(7):2161–2525
Harper HH, Baker DM (2007) Evaluation of current stormwater design criteria within the state of Florida
Hayati B, Maleki A, Najafi F, Daraei H, Gharibi F, McKay G (2017) Super high removal capacities of heavy metals (Pb^{2+} and Cu^{2+}) using CNT dendrimer. J Hazard Mater 336:146–157
Hristovski K, Baumgardner A, Westerhoff P (2007) Selecting metal oxide nanomaterials for arsenic removal in fixed bed columns: from nanopowders to aggregated nanoparticle media. J Hazard Mater 147:265–274. https://doi.org/10.1016/j.jhazmat.2007.01.017

Inbaraj BS, Chen BH (2011) Dye adsorption characteristics of magnetite nanoparticles coated with a biopolymer poly(γ-glutamic acid). Bioresour Technol 102:8868–8876. https://doi.org/10.1016/j.biortech.2011.06.079

Iqbal MJ, Ashiq MN (2007) Adsorption of dyes from aqueous solutions on activated charcoal. J Hazard Mater 139(1):57–66

Iram M, Guo C, Guan Y, Ishfaq A, Liu H (2010) Adsorption and magnetic removal of neutral red dye from aqueous solution using Fe_3O_4 hollow nanospheres. J Hazard Mater 181:1039–1050. https://doi.org/10.1016/j.jhazmat.2010.05.119

Jayasooriya VM, Ng AWM, Muthukumaran S, Perera BJC (2016) Green infrastructure practices for improvement of urban air quality. Urban for Urban Green 21:4–47. https://doi.org/10.1016/j.ufug.2016.11.007

Kamaraj M, Aravind J, Sudhakar S (2022) Endocrine-disrupting pollutants in domestic and industrial wastewater: occurrence and removal by advanced treatment system for wastewater reuse. In: Elsevier eBooks, pp 177–192. https://doi.org/10.1016/b978-0-323-88507-2.00006-3

Karn B, Kuiken T, Otto M (2009) Nanotechnology and in situ remediation: a review of the benefits and potential risks. Environ Health Perspect 117(112):1823–1831. https://doi.org/10.1289/ehp.0900793

Khaleel A, Kapoor PN, Klabunde KJ (1999) Nanocrystalline metal oxides as new adsorbents for air purification. Nanostruct Mater 11:459–468. https://doi.org/10.1016/S0965-9773(99)00329-3

Kim Y, Yi J (2004) Advances in environmental technologies via the application of mesoporous materials. J Ind Eng Chem 10:41–51

Kosa SA, Al-Zhrani G, Salam MA (2012) Removal of heavy metals from aqueous solutions by multi-walled carbon nanotubes modified with 8-hydroxyquinoline. Chem Eng J 181–182:159–168

Kurian M (2021) Advanced oxidation processes and nanomaterials—a review. Clean Eng Technol. https://doi.org/10.1016/j.clet.2021.100090

Lahnsteiner J, van Rensburg P, Esterhuizen J (2018) Direct potable reuse–a feasible water management option. J Water Reuse Desalin 8:14–28. https://doi.org/10.2166/wrd.2017.172

Lakshmipathiraj P, Narasimhan BRV, Prabhakar S, Raju GB (2006) Adsorption of arsenate on synthetic goethite from aqueous solutions. J Hazard Mater 136:281–287. https://doi.org/10.1016/j.jhazmat.2005.12.015

Lee P, Smyth C, Boutin S (2003) Quantitative review of riparian buffer width guidelines from Canada and the United States. J Environ Manag 70:165–180. https://doi.org/10.1016/j.jenvman.2003.11.009

Li H, Zhang D, Han X, Xing B (2014) Adsorption of antibiotic ciprofloxacin on carbon nanotubes: pH dependence and thermodynamics. Chemosphere 95:150–155. https://doi.org/10.1016/j.chemosphere.2013.08.053

Liu X, Shang L, Yang F, Zhou W (2023) Determining reclaimed water quality thresholds and farming practices to improve food crop yield: a meta-analysis combined with random forest model. Sci Total Environ 862:160774. https://doi.org/10.1016/j.scitotenv.2022.160774

Lu F, Astruc D (2020) Nanocatalysts and other nanomaterials for water remediation from organic pollutants. Coord Chem Rev 408:213180. https://doi.org/10.1016/j.ccr.2020.213180

Lu C, Chung YL, Chang KF (2006) Adsorption thermodynamic and kinetic studies of trihalomethanes on multiwalled carbon nanotubes. J Hazard Mater 138:304–310. https://doi.org/10.1016/j.jhazmat.2006.05.076

Lusk MG, Chapman K (2021) Chemical fractionation of sediment phosphorus in residential urban stormwater ponds in Florida, USA. Urban Sci 5. https://doi.org/10.3390/urbansci5040081

Lv Z, Fan Q, Xie Y, Chen Z, Alsaedi A et al (2019) MOFs-derived magnetic chestnut shell-like hollow sphere NiO/Ni @ C composites and their removal performance for arsenic (V). Chem Eng J. https://doi.org/10.1016/j.cej.2019.01.046

Madadrang CJ, Kim HY, Gao G, Wang N, Zhu J, Feng H et al (2014) Adsorption behaviour of EDTA-graphene oxide for Pb(II) removal. J Appl Mater Interface 4:186–1193. https://doi.org/10.1021/am201645g

Mahdavian AR, Mirrahimil MAS (2010) Efficient separation of heavy metal cations by anchoring polyacrylic acid on superparamagnetic magnetite nanoparticles through surface modification. Chem Eng J 159:264–271. https://doi.org/10.1016/j.cej.2010.02.041

Mahmoodi NM, Keshavarzi S, Ghezelbash M (2017) Synthesis of nanoparticle and modelling of its photocatalytic dye degradation ability from coloured wastewater. J Environ Chem Eng 5(4):3684–3689. https://doi.org/10.1016/j.jece.2017.07.010

Mak SY, Chen DH (2004) Fast adsorption of methylene blue on polyacrylic acid-bound iron oxide magnetic nanoparticles. Dyes Pigments 61:93–98. https://doi.org/10.1016/j.dyepig.2003.10.008

Maniam G, Zakaria NA, Leo CP, Vassilev V, Blay KB, Behzadian K et al (2022) An assessment of technological development and applications of decentralized water reuse: a critical review and conceptual framework. Wiley Interdiscip Rev Water 9. https://doi.org/10.1002/wat2.1588

Mo W, Zhang Q (2013) Energy-nutrients-water nexus: integrated resource recovery in municipal wastewater treatment plants. J Environ Manag 12:255–267. https://doi.org/10.1016/j.jenvman.2013.05.007

Mokhtar MA, Ali RR, Isa ED, Lei Z, Prasetyoko D (2021) Silver nanoparticles loaded activated carbon synthesis using clitorea ternatea extract for crystal violet dye removal. J Res Nanosci Nanotechnol 3(1):26–36

Mon M, Qu X, Ferrando-Soria J, Pellicer-Carreño I, Sepúlveda-Escribano A et al (2017) Fine-tuning of the confined space in microporous metal–organic frameworks for efficient mercury removal. J Mater Chem A. https://doi.org/10.1039/C7TA06199D

Mustapha S, Ndamitso MM, Abdulkareem AS, Tijani JO, Shuaib DT, Ajala AO et al (2020) Application of TiO_2 and ZnO nanoparticles immobilized on clay in wastewater treatment: a review. Appl Water Sci 10:1–36

Narain-Ford DM, Bartholomeus RP, Raterman BW, Van Zaanen I, Ter Laak TT, van Wezel AP et al (2021) Shifting the imbalance: intentional reuse of Dutch sewage effluent in sub-surface irrigation. Sci Total Environ 752. https://doi.org/10.1016/j.scitotenv.2020.142214

Ong CB, Ng LY, Mohammad AW (2018) A review of ZnO nanoparticles as solar photocatalysts: synthesis, mechanisms and applications. Renew Sustain Energy Rev 81:536–551

Pacheco S, Medina M, Valencia F, Tapia J (2006) Removal of inorganic mercury from polluted water using structured nanoparticles. J Environ Eng 132:342–349. https://doi.org/10.1061/(ASCE)0733-9372(2006)132:3(342)

Palani G, Trilaksana H, Sujatha RM, Kannan K, Rajendran S, Korniejenko K et al (2023) Silver nanoparticles for waste water management. Molecules 28(8):3520

Pan B, Zhang W, Lv L, Zhang Q, Zheng S (2009) Development of polymeric and polymer-based hybrid adsorbents for pollutants removal from waters. Chem Eng J 151:19–29. https://doi.org/10.1016/j.cej.2009.02.036

Pandey P, Karki B, Lekhak B, Koirala AR, Sharma RK, Pant HR (2019) Comparative antibacterial study of silver nanoparticles doped activated carbon prepared by different methods. J Inst Eng 15(1):187–194

Partyka ML, Bond RF (2022) Wastewater reuse for irrigation of produce: a review of research, regulations, and risks. Sci Total Environ 828. https://doi.org/10.1016/j.scitotenv.2022.154385

Pouran SR, Abdul Aziz AR, Wan Daud WMA (2015) Review on the main advances in photo-Fenton oxidation system for recalcitrant wastewaters. J Ind Eng Chem 21:53–69. https://doi.org/10.1016/j.jiec.2014.05.005

Pradeep T, Anshup (2009) Noble metal nanoparticles for water purification: a critical review. Thin Solid Films 517:6441–6478. https://doi.org/10.1016/j.tsf.2009.03.195

Rao AP, Patel J, Pradhan AK (2022) Application of alternative sources of water in agricultural food production-current trends and future prospects. https://doi.org/10.1016/j.cofs.2022.100877

Rassoulinejad-Mousavi SM, Azamat J, Khataee A, Zhang Y (2020) Molecular dynamics simulation of water purification using zeolite MFI nanosheets. Sep Purif Technol 234:116080. https://doi.org/10.1016/j.seppur.2019.116080

Rayaroth MP, Aravindakumar CT, Shah NS, Boczkaj G (2022) Advanced oxidation processes (AOPs) based wastewater treatment—unexpected nitration side reactions—a serious environmental issue: a review. Chem Eng J. https://doi.org/10.1016/j.cej.2021.133002

Rehman F, Thebo KH, Aamir M, Akhtar J (2020) Chapter 8—Nanomembranes for water treatment. In: Amrane A, Rajendran S, Nguyen TA, Assadi AA, Sharoba AM (eds) Micro and nano technologies. Elsevier, pp 207–240

Ren X, Chen C, Nagatsu M, Wang X (2011) Carbon nanotubes as adsorbents in environmental pollution management: a review. Chem Eng J 170:395–410

Rivers EN, McMillan SK, Bell CD, Clinton SM (2018) Effects of urban stormwater control measures on denitrification in receiving streams. Water 10. https://doi.org/10.3390/w10111582

Roccaro P (2018) Treatment processes for municipal wastewater reclamation: the challenges of emerging contaminants and direct potable reuse. Curr Opin Environ Sci Heal 2:46–54. https://doi.org/10.1016/j.coesh.2018.02.003

Saleh IA, Zouari N, Al-Ghouti MA (2020) Removal of pesticides from water and wastewater: chemical, physical and biological treatment approaches. Environ Technol Innov 19:101026. https://doi.org/10.1016/j.eti.2020.101026

Schroer WF, Benitez-Nelson CR, Smith EM, Ziolkowski LA (2018) Drivers of sediment accumulation and nutrient burial in coastal stormwater detention ponds, South Carolina, USA. Ecosystems 21:1118–1138. https://doi.org/10.1007/s10021-017-0207-z

Schwammberger PF, Yule CM, Tindale NW (2020) Rapid plant responses following relocation of a constructed floating wetland from a construction site into an urban stormwater retention pond. Sci Total Environ 699. https://doi.org/10.1016/j.scitotenv.2019.134372

Sen Gupta S, Bhattacharyya KG (2012) Adsorption of heavy metals on kaolinite and montmorillonite: a review. Phys Chem Chem Phys 14:6698–6723

Sinclair JS, Reisinger AJ, Bean E, Adams CR, Reisinger LS, Iannone BV (2020) Stormwater ponds: an overlooked but plentiful urban designer ecosystem provides invasive plant habitat in a subtropical region (Florida, USA). Sci Total Environ 711. https://doi.org/10.1016/j.scitotenv.2019.135133

Singh S, Kumar V, Romero R, Sharma K, Singh J (2019) Applications of nanoparticles in wastewater treatment. In: Nanobiotechnology in bioformulations, pp 395–418

Slobodian P, Riha P, Olejnik R, Cvelbar U, Saha P (2013) Enhancing effect of $KMnO_4$ oxidation of carbon nanotubes network embedded in elastic polyurethane on overall electro-mechanical properties of composite. Compos Sci Technol 81:54–60

Stafiej A, Pyrzynska K (2008) Solid phase extraction of metal ions using carbon nanotubes. Microchem J 89:29–33

Sturiale L, Scuderi A (2019) The role of green infrastructures in urban planning for climate change adaptation. Climate 7. https://doi.org/10.3390/cli7100119

Sun H, Wang T, Yang Z, Yu C, Wu W (2019) Simultaneous removal of nitrogen and pharmaceutical and personal care products from the effluent of waste water treatment plants using aerated solid-phase denitrification system. Bioresour Technol 287. https://doi.org/10.1016/j.biortech.2019.121389

Taie M, Fadaei A, Sadoghi M, Hemati S, Mardani G (2021) Comparison of the efficiency of ultraviolet/zinc oxide (UV/ZnO) and ozone/zinc oxide (O_3/ZnO) techniques as advanced oxidation processes in the removal of trimethoprim from aqueous solutions. Int J Chem Eng 2021(1):9640918

Thi VHT, Lee BK (2017) Effective photocatalytic degradation of paracetamol using La-doped ZnO photocatalyst under visible light irradiation. Mater Res Bull. https://doi.org/10.1016/j.materresbull.2017.04.028

Vinardell S, Astals S, Mata-Alvarez J, Dosta J (2019) Techno-economic analysis of combining forward osmosis-reverse osmosis and anaerobic membrane bioreactor technologies for municipal wastewater treatment and water production. Bioresour Technol 297:122395. https://doi.org/10.1016/j.biortech.2019.122395

Voulvoulis N (2018) Water reuse from a circular economy perspective and potential risks from an unregulated approach. Curr Opin Environ Sci Heal 2:32–45. https://doi.org/10.1016/j.coesh.2018.01.005

Wan Ngah WS, Teong LC, Hanafiah MAK (2011) Adsorption of dyes and heavy metal ions by chitosan composites: a review. Carbohydr Polym 83:1446–1456

Wang J, Liu JH, Wang H, Mei C (2020) Approaches to multi-objective optimization and assessment of green infrastructure and their multi-functional effectiveness: a review. Water 12. https://doi.org/10.3390/w12102714

Werner S, Kätzl K, Wichern M, Buerkert A, Steiner C, Marschner B (2018) Agronomic benefits of biochar as a soil amendment after its use as wastewater filtration medium. Environ Pollut 233:561–568. https://doi.org/10.1016/j.envpol.2017.10.048

Wu Z, Zhong H, Yuan X, Wang H, Wang L, Chen X et al (2014) Adsorptive removal of methylene blue by rhamnolipid-functionalized graphene oxide from wastewater. Water Res 67:330–344

Xu D, Zhu K, Zheng X, Xiao R (2015) Poly(ethylene-co-vinyl alcohol) functional nano fiber membranes for the removal of Cr(VI) from water. Ind Eng Chem Res. https://doi.org/10.1021/acs.iecr.5b00995

Xu J, Cao Z, Zhang Y, Yuan Z, Lou Z, Xu X et al (2018) A review of functionalized carbon nanotubes and graphene for heavy metal adsorption from water: preparation, application, and mechanism. Chemosphere 195:351–364

Yan H, Lai C, Liu S, Wang D, Zhou X, Zhang M et al (2023) Metal-carbon hybrid materials induced persulfate activation: application, mechanism, and tunable reaction pathways. Water Res 234:119808

Yang K, Xing B (2007) Desorption of polycyclic aromatic hydrocarbons from carbon nanomaterials in water. Environ Pollut 145:529–537

Yang K, Wu W, Jing Q, Zhu L (2008) Aqueous adsorption of aniline, phenol, and their substitutes by multi-walled carbon nanotubes. Environ Sci Technol 42:7931–7936. https://doi.org/10.1021/es801463v

Yang Z, Wu Y, Guo H, Ma XH, Lin CE et al (2017) A novel thin-film nano-templated composite membrane with in situ silver nanoparticles loading: Separation performance enhancement and implications. J Memb Sci 544:351–358. https://doi.org/10.5004/dwt.2023.29180

Yao Y, Bing H, Feifei X, Xiaofeng C (2011) Equilibrium and kinetic studies of methyl orange adsorption on multiwalled carbon nanotubes. Chem Eng J 170:82–89. https://doi.org/10.1016/j.cej.2011.03.031

Yazdi MN, Scott D, Sample DJ, Wang XX (2021) Efficacy of a retention pond in treating stormwater nutrients and sediment. J Clean Prod 290. https://doi.org/10.1016/j.jclepro.2021.125787

Zhang W, Zhou C, Zhou W, Lei A, Zhang Q, Wan Q et al (2011) Fast and considerable adsorption of methylene blue dye onto graphene oxide. Bull Environ Contam Toxicol 87:86–90

Zhang Q, Smith K, Zhao X, Jin XK, Wang SY, Shen JJ et al (2021) Greenhouse gas emissions associated with urban water infrastructure: what we have learnt from China's practice. Wiley Interdiscip Rev Water 8. https://doi.org/10.1002/wat2.1529

Zhao X, Lv L, Pan B, Zhang W, Zhang S, Zhang Q (2011) Polymer-supported nanocomposites for environmental application: a review. Chem Eng J 170:381–394. https://doi.org/10.1016/j.cej.2011.02.071

Zhao G, Wen T, Yang X, Yang S, Liao J, Hu J et al (2012) Preconcentration of U(VI) ions on few-layered graphene oxide nanosheets from aqueous solutions. Dalton Trans 41:6182–6188

Zhu ZF, Dou J (2018) Current status of reclaimed water in China: an overview. J Water Reuse Desalin. 8:293–307

Zou Y, Wang X, Khan A, Wang P, Liu YH et al (2016) Environmental remediation and application of nanoscale zero-valent iron and its composites for the removal of heavy metal ions: a review. Environ Sci Technol. https://doi.org/10.1021/acs.est.6b01897

Zubair M, Ihsanullah I, Aziz HA, Ahmad MA, Al-Harthi MA (2020) Sustainable wastewater treatment by biochar/layered double hydroxide composites: progress, challenges, and outlook. Bioresour Technol 319:124128. https://doi.org/10.1016/j.biortech.2020.124128

Zurita F, White JR (2014) Comparative study of three two-stage hybrid ecological wastewater treatment systems for producing high nutrient, reclaimed water for irrigation reuse in developing countries. Water 6:213–228. https://doi.org/10.3390/w6020213

Chapter 7
Environmental Implications and Safety Considerations of Nanotechnology in Wastewater Treatment

Subramaniam Karuppannan and Lalitesh Sinha

Abstract Nanotechnology greatly enhances wastewater treatment by employing nanomaterials like nanoparticles, carbon nanotubes, and graphene derivatives to improve pollutant removal through better adsorption, degradation, and filtration processes. However, these materials also pose environmental and safety concerns due to their minuscule size, high reactivity, and potential for persistence in aquatic environments. Issues include toxicity to aquatic life, risks of bioaccumulation, and health hazards from exposure. Addressing these challenges requires effective risk management strategies, including robust regulatory frameworks and safety practices. Nanomaterials, such as metal nanoparticles, carbon-based materials, polymer-based nanomaterials, composites, and quantum dots, offer various mechanisms for pollutant removal like adsorption, photocatalysis, and magnetic separation. Nevertheless, their environmental and health impacts necessitate comprehensive risk assessment and responsible use. A thorough understanding of the lifecycle of nanomaterials, from synthesis to disposal, is essential for mitigating these risks. The discussion underscores the benefits, risks, and necessary management strategies for nanomaterials in wastewater treatment, highlighting the need for improved safety protocols and regulatory measures.

Keywords Nanotechnology · Environmental implications · Safety considerations

---

S. Karuppannan
Environmental Management and Research Association (ENSEARCH), Petaling Jaya, Malaysia

L. Sinha (✉)
IPS Academy Institute of Engineering and Science, Indore, Madhya Pradesh, India
e-mail: laliteshsinha@ipsacademy.org

A. Mandpe et al. (eds.), *Nanotech Solutions for Industrial Wastewater*, Advances in Wastewater Research, https://doi.org/10.1007/978-981-96-6879-3_7

7.1 Introduction

Nanotechnology has surfaced as a compelling response for treating wastewater, presenting advanced materials and approaches that effectively eliminate pollutants more accurately than traditional practices. Employing nanomaterials like nanoparticles, carbon nanotubes, and graphene derivatives can significantly improve the adsorption, degradation, and filtration of contaminants, resulting in purer waste effluents. Nevertheless, the impact on the environment and the safety aspects related to these technologies raise considerable worries. Because of their tiny size and high reactivity, nanomaterials can threaten ecosystems if they are not managed suitably. The possible toxicity effects on marine organisms, their accumulation in food chains, and the uncertainties surrounding their long-term impacts bring up concerns regarding their safe disposal and how they interact with the environment. Consequently, although nanotechnology holds significant potential, it is essential to thoroughly evaluate its environmental impacts, implement safe handling measures, and establish safe handling procedures to ensure its responsible use in wastewater management.

7.2 Environmental Implications

In treatment of wastewater the use of nanotechnology substantially increases the contaminant removal efficiency, which can result to improved water quality and resource recovery. By using processes including adsorption, catalysis, and improved filtration, the use of nanomaterials improves the efficiency of pollutant removal for complex pollutants, i.e., heavy metals and pathogens. Two ideas on the potential effects of nanotechnology on the environment were explored by Theodore et al. (2005): One is positive and the other is negative. Despite there are several factors on which environment implications of nanotechnology effect treatment of waste water as follows.

7.2.1 Persistence and Mobility

(a) Nanoparticles

The persistence of nanoparticles in aquatic environments is influenced by various environmental factors, such as pH, ionic strength, chemical composition, agglomeration behavior, and interactions with organic matter. The physicochemical properties of nanoparticles employed in wastewater treatment contribute to their environmental persistence (Mohajerani et al. 2019; Dhull et al. 2024). Besides size and stability, other factors that impact nanoparticles behavior during wastewater treatment include aggregation state, shape, fractal dimension, and interactions with pollutants. For instance, aggregated nanoparticles may have reduced mobility, making them easier to remove (Walser et al. 2012).

(b) Nanocomposites

It is well known that nanocomposites have increased stability, which adds to their durability in a variety of applications. Nevertheless, in spite of their stability, these materials may eventually degrade and release nanoparticles into the environment (Duncan 2015).

(c) Nanotubes

Long-term environmental hazards can arise from carbon nanotubes (CNTs) because of their stability and resistance to degradation (Mordkovich et al. 2024). CNTs have shown great removal efficiency and reusability in their role as excellent adsorbents for eliminating a variety of contaminants from water. Furthermore, CNTs demonstrate antibacterial characteristics by engaging with microbial cells to disrupt membranes and impede development; these attributes can be amplified by additions such as nanometals (Jamil et al. 2024).

7.2.2 Ecotoxicity

(a) Nanoparticles

The degree of ecotoxicity exhibited by nanomaterials is largely determined by their structural composition. In both aquatic and terrestrial ecosystems, various compositions can affect the level of toxicity, bioaccumulation, and interactions with biological systems. This characteristic has the potential to cause a number of toxicological consequences, such as oxidative stress and cellular process disruption in aquatic species (Hyseni 2016).

(b) Nanocomposites

In aquatic environments, nanocomposites have been found to demonstrate ecotoxicity, which has significant adverse impacts on a range of organisms including algae, plants, bacteria, invertebrates, and fish. Ecological health may suffer if these materials are released into water bodies and accumulate there15. Studies have indicated that the existence of nanocomposites in the environment may have dangers, such as acute toxicity, impaired growth, and even organ failure in aquatic species (Guerranti and Renzi 2015; Hyseni 2016).

(c) Nanotubes

In aquatic environments, carbon nanotubes (CNTs) exhibit ecotoxicological consequences, with freshwater invertebrates being more affected than vertebrates. (Freixa et al. 2018a, b)

7.2.3 Bioaccumulation

(a) Nanoparticles

Nanoparticles Bioaccumulation in aquatic environments, the bioaccumulation of nanoparticles can have important ecological ramifications. Nanoparticles can be absorbed by organisms such as algae and zooplankton. These particles then go up the food chain to higher trophic levels, where fish and other larger predatory species can find them. This mechanism may lead to biomagnifications, a process whereby nanoparticles concentrations rise at every trophic level. This increases the risk of toxicity for higher organisms and may have an impact on biodiversity (Ajith and Rajamani 2021).

(b) Nanocomposites and Nanotubes

The enhanced adsorption capabilities and structural features of nanocomposites are major factors in improving wastewater treatment. Because of their large surface area and strong affinity for contaminants, they are very good in eliminating pollutants from wastewater, including harmful metal ions and dyes (Ansari et al. 2023). The complicated problem of CNT bioaccumulation in aquatic species affects wastewater treatment procedures. While CNTs can improve treatment systems by serving as effective adsorbents for a range of pollutants, their presence may have a negative impact on biological processes that are necessary for the removal of nutrients, such as the biological removal of phosphorus and nitrogen (Jackson et al. 2013).

7.2.4 Transformation and Byproducts

(a) Nanoparticles

During wastewater treatment, nanoparticles undergo various transformation processes, including aggregation, interactions with solids, and alterations to their physicochemical properties. These changes can significantly influence the effectiveness of nanoparticles in pollutant removal. Byproducts such as metal sulfides, which may be less toxic and more stable than the original nanoparticles, can form as a result. For example, transformation processes involving titanium dioxide and silver nanoparticles during treatment can lead to the creation of new compounds (Wang et al. 2017).

(b) Nanocomposites and Nanotubes

Systems for treating wastewater incorporate nanocomposites to improve the elimination of organic and heavy metal pollutants. The processes of adsorption, filtration, and photocatalytic degradation are enhanced by their incorporation (Mittal et al. 2024). This transition addresses issues with water quality and scarcity by enabling wastewater treatment technology to operate more effectively and sustainably (Tripathi et al. 2024)

7.3 Safety Considerations

The small size and reactivity of nanomaterials, which can be hazardous through ingestion, skin contact, or inhalation, make nanotechnology unique in terms of safety considerations. Although their long-term impacts are still being investigated, nanoparticles may cause environmental contamination, toxicity, or respiratory problems. These risks can be mitigated by the introduction of engineering controls, personal protective equipment, and real-time monitoring of workplace conditions. Safe design and appropriate waste disposal are also crucial in minimizing environmental and health effects. Guidance from regulatory agencies is still in development, and standards remain in flux as new research emerges.

7.3.1 Human Health Risks

(a) Nanoparticles

Exposure to nanoparticles is building due to their rising use in a diversity of applications, which could result in health problems due to their peculiar properties. Studies reveal that nanoparticles may have detrimental biological interactions with the body that cause illnesses, inflammation, and impairment to important organs. Because of their unique characteristics, which set them apart from larger particles, such as their high capacity and reactivity to enter biological structures, nanoparticles are responsible for these interactions (Aziz et al. 2021).

(b) Nanocomposites and Nanotubes

Exposure to toxic compounds, especially heavy metals like arsenic, mercury, and chromium, which are known to cause serious health problems, can result from using nanocomposites. Long-term exposure to these compounds has been connected to neurological illness and cancer among other health issues. Besides, non-target creatures are exposed to the toxicity of nanocomposites, which could have an effect on human health through environmental pathways (Mishra 2014).

7.3.2 Occupational Safety

Occupational safety hazards must be carefully assessed and managed because using nanoparticles, nanocomposites, and nanotubes in wastewater treatment may expose workers to harmful materials. Using personal protection equipment (PPE) to reduce exposure and receiving proper training on handling nanoparticles are essential for workers participating in the wastewater treatment process (Oghyanous 2021).

7.3.3 Regulatory and Ethical Considerations

(a) Regulation

The current laws governing the application of nanotechnology to wastewater treatment put a lot of emphasis on developing moral and legal frameworks on a global scale. These frameworks manage the possible environmental effects of using nanotechnology in water treatment procedures while guaranteeing its efficient deployment (Bhat et al. 2022).

(b) Ethics

Nanomachines used in environmental settings are required by ethical norms to be specialized, non-self-replicating, and tagged for tracking. These recommendations seek to evaluate how nanotechnology affects the environment and address any possible threats to ecosystems and public health (Gardner and Dhai 2014).

7.3.4 Risk Assessment and Management

Although there are advantages to using nanotechnology in wastewater treatment, there are also threats to treatment plants' ability to purify wastewater. Some nanoparticles, including zinc oxide and silver, can suppress bacterial growth in the biological treatment phase, which could lower the efficiency of the purification procedure as a whole (Olawade et al. 2024). The disposal of waste containing nanomaterials after wastewater treatment presents a significant challenge. Nanoparticles can accumulate in sewage sludge, which may be used as fertilizer, thus leading to potential soil contamination and environmental accumulation (Rath et al. 2024).

Risk assessment and management are essential components of nanotoxicology that help to properly understand and mitigate the risks associated with nanoparticles. This highlights the need for careful investigation and practical solutions to protect ecological balance and human health (Isibor et al. 2024).

7.4 Types of Nanomaterials with Their Application and Mechanism Used in Treatment of Water and Wastewater

Owing to their special qualities, which include high surface area, reactivity, and the capacity to target certain pollutants, a variety of nanomaterials are employed in treatment. A few common types of nanomaterials used in the industry are listed below.

7.4.1 Metal and Metal Oxide Nanoparticles

Metal and metal oxide nanoparticles are becoming more and more known for their special qualities and important uses in a wide range of industries, especially in the environmental, industrial, and biological domains. These nanoparticles are effective in various fields, such as imaging, bio sensing, and drug delivery, because to their superior stability, large surface area, and nanoscale size (Negrescu et al. 2022).

7.4.1.1 Silver Nanoparticles (AgNPs)

Applications

I. Antimicrobial

In addition to their well-known broad-spectrum antibacterial qualities, silver nanoparticles are suitable in a diversity of bacterial control applications. They are often used for dental work, surgical procedures, and burn and wound care. The main process is the continuous production of highly poisonous free silver ions from the nanoparticles, which are harmful to microbes (Iravani et al. 2014).

II. Biomedical

Silver nanoparticles are being investigated in the realm of biomedicine for improved uses, including medication delivery systems, imaging agents, and biosensors (Burdușel et al. 2018).

III. Optical Probes

Silver nanoparticles are effective probes for Surface-enhanced Raman Scattering (SERS) and metal-enhanced fluorescence (MEF). Their enhanced properties over other noble metal nanoparticles—such as sharper extinction bands and greater extinction coefficients—help them function better as optical probes (Keat et al. 2015).

IV. Properties of Catalysis

Additionally, these nanoparticles show remarkable catalytic redox capabilities, which make them helpful in chemical reactions involving chemical and biological substances like benzene and dyes (Iravani et al. 2014).

Mechanism

They release silver ions, which are toxic to microorganisms. When silver nanoparticles (AgNPs) come into touch with moisture or are in an aqueous environment, they continuously release silver ions (Ag^+), which is the primary way that they demonstrate their toxicity. These released ions penetrate bacterial cell membranes, causing damage and ultimately cell death (More et al. 2023).

7.4.1.2 Titanium Dioxide (TiO_2) Nanoparticles

Applications

TiO_2 nanoparticles use their photocatalytic characteristics to effectively break down organic contaminants, making them crucial for the removal of pollutants and heavy metals from water in wastewater treatment (Waghmode et al. 2019).

Mechanism

With regard to its varied characteristics and interactions in a variety of applications, titanium dioxide (TiO_2) nanoparticles display distinct mechanisms. Their main mode of action is their photocatalytic properties, wherein (TiO_2) functions as a catalyst in the presence of light to produce reactive species capable of breaking down organic contaminants (Bejaoui et al. 2023).

7.4.1.3 Zinc Oxide (ZnO) Nanoparticles

Owing to its special qualities and efficiency in eliminating different contaminants, zinc oxide (ZnO) nanoparticles are becoming more and more popular in the wastewater treatment industry (Spoiala et al. 2021). Zinc oxide (ZnO) nanoparticles, especially when carbon-modified (ZnO-C), are highly effective in wastewater treatment for the removal of heavy metal ions like Zn(II), Cd(II), Co(II), and Mn(II). Their large surface area, nanoscale morphology, and abundance of functional groups (e.g., hydroxyl groups) make them well-suited for adsorption-driven environmental remediation (Alanazi et al. 2024).

Applications

Copper and cadmium are two heavy metals that can be removed from contaminated water using ZnO nanoparticles as adsorbents. ZnO nanoparticles' capacity to degrade organic substances, such as colors and medications, confirms its usefulness as a flexible agent in treatment procedures wastewater (Spoiala et al. 2021; Haidri et al. 2023).

Mechanism

ZnO nanoparticles mainly work by adsorbing and photo catalyzing substances, which breaks down organic pollutants and heavy metals in wastewater (Haidri et al. 2023; Gangwar et al. 2023).

7.4.1.4 Iron Oxide Nanoparticles (Fe_3O_4, Fe_2O_3)

The practice of iron oxide nanoparticles in wastewater treatment is a rapidly growing field that has the possible to approximately boost the effectiveness and efficiency of contaminant removal techniques.

Applications

The distinctive qualities of iron oxide nanoparticles, namely their high surface-area-to-volume ratio and good magnetic properties, are making them more recognized for their efficacy in wastewater treatment for the treatment of wastewater on an industrial scale, iron oxide nanoparticles (IONMs) are particularly appealing because of their low cost, potent adsorption properties, and simplicity of separation (Navratil 1999; Xu et al. 2012). By employing these nanoparticles as catalysts and nanoadsorbents, heavy metals and organic contaminants may be effectively eliminated from wastewater (Xu et al. 2012; Guo et al. 2013).

Mechanism

Iron oxide nanoparticles (IONPs) are used in wastewater treatment as adsorbents and coagulants to remove impurities (Santhosh et al. 2019). Through processes including reduction, oxidation, and complication, they make different pollutants easier to adsorb, which increases their efficacy in cleaning procedures. (Biftu et al. 2020). These nanoparticles also help remove certain ions, such as fluoride, through their effectiveness in ionic exchange and physisorption processes (Aragaw et al. 2021).

7.4.2 Carbon-Based Nanomaterials

7.4.2.1 Carbon Nanotubes (CNTs)

Applications

The following are some of the wastewater treatment procedures that use carbon nanotubes:

I. Heavy Metal Adsorption: CNTs often outperform conventional adsorbents in the removal of heavy metal ions from aqueous solutions, such as lead, cadmium, and nickel (Mishra and Sundaram 2023).
II. Nanofiltration: Membrane filtering methods that provide microbial inactivation and improve water quality use CNT-based membranes (Elsehly and Chechenin 2024).
III. Nanocomposites: They are added to composites to enhance their overall functional qualities and treatment efficiency, which qualifies them for a range of environmental applications (Mishra et al. 2023).

Mechanism

Carbon nanotubes (CNTs) play a crucial role in wastewater treatment through various techniques, including adsorption, catalysis, separation, and disinfection. Their functionalization and unique structural properties significantly enhance their capacity to interact with and eliminate contaminants from water (Mishra and Sundaram 2023; Mishra et al. 2023).

7.4.2.2 Graphene Oxide (GO)

Graphene Oxide (GO) has demonstrated remarkable efficacy in wastewater treatment because of its special qualities, including its hydrophilicity, mechanical strength, and material compatibility. It can be utilized in membrane technologies as well as serving as an adsorbent for pollutants. GO is known for its functional surface properties and scalability, is extensively utilized in epoxy resins, primarily as epoxy/GO nanocomposites or coatings. Depending on the application, GO improves mechanical strength, thermal conductivity, and corrosion resistance. Its performance is strongly influenced by surface modification and dispersion techniques. Nonetheless, further research is needed to address existing challenges and optimize its application (Hou et al. 2020).

Applications

I. Potential Candidate for treatment of wastewater: The special features of graphene have made it a fair contender for applications involving the remediation of greasy wastewater. High surface area, mechanical strength, and the capacity to create stable composites with other materials are some of these characteristics (Avornyo and Chrysikopoulos 2024).
II. Effective Adsorbent Characteristics: The effectiveness of graphene oxide-based composite materials as adsorbents in desalination, purification, and wastewater treatment procedures has been highlighted (Asghar et al. 2022).
III. Membrane Technology: One major technological advance for the treatment of industrial wastewater is the development of economically priced graphene oxide membranes (An et al. 2023).

Mechanism

The primary mechanism by which GO adsorbs contaminants from wastewater is through its several oxygen-containing functional groups, including hydroxyl, carboxyl, and epoxy groups. These functional groups make GO more hydrophilic and permit powerful electrostatic interactions with different contaminants, which promote efficient adsorption (Asghar et al. 2022; Avornyo and Chrysikopoulos 2024).

7.4.2.3 Fullerenes (C_{60})

Fullerenes, in particular C_{60}, have remarkable qualities that make them advantageous for a range of wastewater treatment uses in the field of nanotechnology (Baby et al. 2019; Kausar 2024; Chae et al. 2024).

Application

Fullerenes are being used more frequently in wastewater treatment due to their capacity to improve the effectiveness of contaminant removal when included into nanocomposite membranes.

I. Membranes of Fullerene Nanocomposite: Wastewater treatment requires the use of nanocomposite membranes, which are made possible by the use of fullerenes (Kausar 2024). The enhanced properties of these membranes, including porosity, selectivity, and penetration, increase the effectiveness of water purifying procedures.
II. Efficiency of Contaminant Removal: Fullerenes have demonstrated an amazing ability to eliminate a wide range of dangerous contaminants from wastewater, especially heavy metals like lead, mercury, and arsenic.

Mechanism

Adsorption processes are the main way that fullerenes interact with contaminants, which enables them to bind to organic pollutants efficiently. Additionally, they exhibit photocatalytic properties that allow these pollutants to oxidize and increase the effectiveness of their removal (Freixa et al. 2018a, b; Kausar 2024).

7.4.3 Polymer-Based Nanomaterials

Since polymer-based nanoparticles have special qualities and mechanisms that facilitate the removal of different contaminants and increase treatment efficiency, they are being used more and more in nanotechnology for wastewater treatment.

7.4.3.1 Dendrimers

The substantial uses of dendrimers in the arena of nanotechnology, especially in the treatment of wastewater, are becoming more widely acknowledged. Their special qualities make them useful agents for enhancing the elimination of pollutants by facilitating a variety of methods (Perez-Ferreiro et al. 2023).

Application

I. Dendrimers function as efficient flocculants, helping to clump together contaminants like metal ions and dyes to make them easier to remove from wastewater. When treating industrial effluents, this application stands out because it removes impurities quickly and effectively.
II. When it comes to the adsorption of heavy metals from contaminated water, dendrimers have demonstrated remarkable powers.
III. Dendrimers can be specifically engineered for certain purposes because of their well-defined architecture and functional groups, which enables the targeted adsorption of different organic pollutants and toxins in wastewater treatment procedures.

Mechanism

In wastewater treatment, dendrimers interact with contaminants through several mechanisms. The main procedures consist of (Abbasi et al. 2014) as follows.

I. Dendrimers have the ability to attach to pollutants, greatly increasing the effectiveness of their removal. Their several terminal functional groups provide a range of interactions that can be tailored to target particular contaminants.
II. Heavy metals can be extracted from wastewater by using dendrimers' ability to bind metal ions. This is the process by which the metal ions and the functional groups on the dendrimers create stable complexes.
III. Certain dendrimers are amphiphilic, which means that they can solubilize organic compounds and make them easier to remove from aqueous environments. This characteristic helps with hydrophobic contaminants and allows for a more thorough approach to wastewater treatment.

7.4.3.2 Nanosponges

Polymer-based nanomaterials, especially nanosponges, are crucial for wastewater treatment.

Application

Nano sponges are helpful in various wastewater treatment applications because of their exceptional adsorption and filtration capabilities. They are capable to efficiently eliminating, organic pollutants, heavy metals, and particular contaminants like pesticides and dyes from wastewater their three-dimensional cross-linked structure, which offers a huge surface area for pollutant collection, is credited with this efficacy (Iravani and Varma 2022).

Mechanism

Adsorption and filtration processes are the main ways that nanosponges remediate wastewater. Pollutants are drawn to and retained on the surface of the nanosponges through adsorption, and they are physically separated from the water by filtering, which takes advantage of their porous nature.

7.4.4 Composite Nanomaterials

An important development in nanotechnology applications for wastewater treatment is represented by composite nanomaterials. Their special composition and characteristics enable them to remove pollutants more effectively, which makes them appropriate for dealing with a variety of pollutants in water systems.

7.4.4.1 Metal–Organic Frameworks (MOFs)

Metal–organic frameworks (MOFs), recently developed in nanotechnology, hold great promise for wastewater treatment due to their high porosity, unique structural characteristics, and tenable properties.

Application

MOFs are used to remove a variety of contaminants from wastewater, including organic coloring, heavy metal ions, and antibiotics. Their use involves serving as membranes, adsorbents, and catalytic materials, all of which greatly increase the effectiveness of several water purification procedures (Yusuf et al. 2022).

Mechanism

Adsorption, mainly through electrostatic interactions with pollutants, is the main method by which MOFs remove toxins from wastewater (Liu et al. 2023). MOFs are efficient at eliminating a range of organic and inorganic pollutants from water due to their vast surface area and potent adsorption properties (Perera et al. 2023). To effectively remove harmful compounds, other mechanisms like coordination interactions and hydrogen bonding also play a role (Xie et al. 2024).

7.4.4.2 Magnetic Nanocomposites

Due to their special characteristics, magnetic nanocomposites have shown to be a significant breakthrough in nanotechnology for the treatment of wastewater, exhibiting extraordinary efficacy in the removal of a variety of contaminants.

Application

Magnetic nanocomposites have been demonstrated through research to be an excellent tool for the elimination of heavy metals, pharmaceuticals, and dyes, among other contaminants. The elimination of heavy metals from water (Zhao et al. 2023) and the treatment of wastewater containing dyes (Nikic et al. 2024) are two notable applications that highlight their versatility, particularly when used in adsorption procedures.

Specifically, the application in adsorption processes for the elimination of heavy metals from water and the treatment of dye-containing wastewater is another prominent application, showcasing their versatility.

Mechanism

Magnetic nanocomposites function primarily by means of adsorption onto their surface, catalytic degradation of contaminants, and a strong magnetic response that makes recovery from use simple (Niculescu et al. 2024).

7.4.5 Nanofibers and Membranes

The utilization of nanofibers and membranes has the likely to improve water quality and plays a critical role in wastewater treatment procedures by providing efficient solutions for the removal of contaminants.

7.4.5.1 Electrospun Nanofibers

Application

I. The main applications for electrospun nanofibers are as filtration membranes and efficient adsorbents. Heavy metals, colors, and organic contaminants can be adsorbed more effectively due to their large surface area and high porosity. This feature makes it possible to significantly lower the quantities of pollutants in wastewater, improving treatment efficiency (Rocha et al. 2024; Radoor et al. 2024).
II. Oil–water mixture separation is one particular use for electrospun nanofibers. These nanofibers' special structure makes it easier to extract oil from water effectively, which makes them useful for cleaning up oil spills and industrial effluents (Li et al. 2023)
III. The flexibility in design and fabrication of electro spun nanofibers leads to customizable membranes that can be modified for specific treatment applications. These membranes can be engineered for enhanced affinity toward certain pollutants or to incorporate additional functionalities, such as antimicrobial properties (Nayl et al. 2022; Xu et al. 2023).

Mechanism

Electrospinning techniques are among the mechanisms used in their manufacturing and application; these procedures result in the formation of fibers with distinctive qualities including large surface area and porosity, which boost their effectiveness in removing pollutants. A polymer solution is first visible to a high voltage during the Electrospinning process, creating an electrically charged jet (Lopez-Covarrubias et al. 2019).As the jet travels in the direction of the collector, the surface tension of the material is released, causing it to elongate into fibers that finally solidify. Nanofibers with widths ranging from nanometers to micrometers can be produced in a controlled manner using this technique (Nayl et al. 2022).

7.4.5.2 Nanocomposite Membranes

In order to develop their functional capabilities for a variety of applications, particularly in nanotechnology, innovative filtration materials known as 'nanocomposite membranes' incorporate nanofillers into a polymeric or inorganic oxide matrix.

Application

Water treatment benefits significantly from the excellent selectivity and antifouling properties of nanocomposite membranes. These membranes' resistance to fouling is facilitated by the antibacterial qualities of the nanoparticles incorporated in them, which raises the general effectiveness of the water purification operations (Wen et al. 2019).

Mechanism

The efficiency of nanocomposite membranes in separation processes is driven by several mechanisms, including the adsorption of different components from the feed, diffusion of components along activity gradients within the membrane matrix, and desorption into the permeate gas phase. These mechanisms enhance membrane performance in applications such as gas separation and wastewater treatment (Agboola et al. 2021).

7.4.6 Quantum Dots (QDs)

Since quantum dots (QDs) have special qualities like low toxicity, high solubility, and biocompatibility, they have become important instruments in nanotechnology for wastewater treatment. The primary focus of their application in this industry is the identification and removal of pollutants, suggesting a potential future for sustainable water management.

7.4.6.1 Cadmium-Based Quantum Dots (CQDs)

Application

Cadmium-based quantum dots have gathered consideration in the field of nanotechnology, particularly for their uses in wastewater treatment. These semiconductor nanoparticles possess unique optical and electronic properties due to their nanoscale size, which can be tuned for specific applications (Abd Rani et al. 2020; Mohamed et al. 2021).

Mechanism

When exposed to light, cadmium-based quantum dots work as photocatalysts, producing reactive species that aid in the breakdown of organic pollutants in wastewater (Samadi-Maybodi and Sadeghi-Maleki 2016; Singh et al. 2023). When electrons from the valence band move into the conduction band due to light excitation, QDs produce pairs of electrons and holes. Reactive oxygen species (ROS) is produced when these pairs interact with water and oxygen molecules, and they play a critical role in oxidizing organic contaminants and causing their destruction.

7.4.6.2 Graphene Quantum Dots (GQDs)

In nanotechnology, GQDs are becoming key components, especially for applications involving wastewater treatment. Their special qualities make it possible for different pollutants to be removed and degraded efficiently, underscoring their potential to improve water purification techniques.

Application

Many people use Graphene and its compounds, containing Graphene oxide, to eliminate both organic and inorganic pollutants from wastewater. Over 90% separation efficiency has been achieved by Graphene-based materials, especially Graphene oxide, in oil–water separation, demonstrating amazing efficiency (Avornyo and Chrysikopoulos 2024).

Mechanism

The main way that Graphene quantum dots (GQDs) work in the water treatment industry is that they may absorb pollutants and promote chemical processes that break down pollutants, such as heavy metals and organic dyes (Tshangana et al. 2021).

Nanomaterials used in water and wastewater treatment exhibit distinct properties that make them effective for removing a variety of contaminants. Each type of nanomaterials has specific mechanisms, such as adsorption, photocatalysis, and magnetic separation, which enable efficient treatment of water and wastewater as illustrated in Table 7.1.

7.5 Environmental and Safety Issues with Nanomaterials Used in Water/Wastewater Treatment

The success of nanomaterials in eliminating pollutants has led to their increasing utilization in water/wastewater treatment. But they are dangerous for the environment and public health. Aquatic life toxicity, persistence in ecosystems, and possible bioaccumulation are among the environmental issues. Exposure by ingestion, inhalation, and the possibility of cytotoxicity and genotoxicity pose serious risks to human health. When it comes to water treatment, there is a chance that inappropriate disposal or leaching of nanomaterials into treated water could contaminate the ecosystem. Regulatory frameworks for controlling these risks are inadequate, underlining the need for greater study and proper safety practices to reduce potential hazards. Table 7.2 provides an illustration of the issues numerous nanomaterials pose to safety and the environment.

Table 7.1 Comprehensive overview of various nanomaterials application in water and wastewater treatment with their characteristics

Nanomaterials	Types	Applications in water/ Wastewater treatment	Key characteristics	References
Activated carbon nanotubes (CNTs)	Carbon-based nanotubes	Adsorption of organic pollutants, heavy metals	High adsorption capacity, high surface area	Mishra et al. (2023)
Graphene oxide (GO)	Carbon-based sheets	Filtration membranes, adsorption of pollutants	High surface area, high mechanical strength	Liu (2017)
Silver nanoparticles (AgNPs)	Metal nanoparticles	Disinfection, antimicrobial treatment	Strong antimicrobial properties, high reactivity	Yu et al. (2022)
Titanium dioxide (TiO_2) nanoparticles	Metal oxides	The breakdown of organic pollutants using photocatalysis	Photocatalytic activity under UV light	Levchuk and Sillanpää (2020)
Zinc oxide (ZnO) nanoparticles	Metal oxides	Photocatalysis, disinfection, removal of organic pollutants	High photocatalytic activity, broad UV absorption	Spoiala et al. (2021)
Magnetic nanoparticles (Fe_3O_4)	Magnetic nanoparticles	Removal of contaminants via magnetic separation	Magnetic properties, high surface area	Farahbakhsh et al. (2021)
Alumina nanoparticles (Al_2O_3)	Metal oxides	Filtration, adsorption of pollutants, catalyst support	High surface area, high chemical stability	Sharma et al. (2008)
Silica nanoparticles (SiO_2)	Metal oxides	Adsorption, filtration, and as carrier for catalysts	High surface area, tunable pore size	Yehia and Said (2023)
Copper nanoparticles (CuNPs)	Metal nanoparticles	Disinfection, removal of heavy metals	Antimicrobial properties, good reactivity	Cheng et al. (2019)
Fullerenes (C_{60})	Carbon nanostructures	Adsorption, removal of organic pollutants	High surface area, electron-rich structure	Chae et al. (2014)

Table 7.2 Overview of various nanomaterials and their environmental and safety issues

Nanomaterials	Types	Environmental issues	References	Safety issues	References
Metal and metal oxides nanomaterials	Silver nanoparticles (AgNPs)	*Ecotoxicity:* • Effect on higher trophic levels in microorganisms in aquatic environments • Bioavailability and interactions with biological systems are determined by various parameters *Persistence:* • Stability and resistance to degradation	Mat Lazim et al. (2023) Wang et al. (2012), Forstner et al. (2020) Cervantes-Aviles et al. (2019), Ratpukdi et al. (2014)	*Human Health:* • Cause cytotoxicity, genotoxicity, and potentially lead to antimicrobial resistance *Occupational Hazards:* • Skin or respiratory problems through inhalation or dermal exposure • Dermatological problems	Calderon-Jimenez et al. (2017) Noga et al. (2023) Alatawi et al. (2024)
	Titanium dioxide (TiO_2) nanoparticles	*Photoxicity:* Cause oxidative stress in living things, Impact on cellular function, generating reactive oxygen species (ROS) under UV light *Accumulation:* Detrimental impacts on the benthic creatures	Shah et al. (2017) Cruz-rosa and Perez-reyes (2023), Baranowska-Wójcik et al. (2020)	*Human Health:* Oxidative stress and inflammation *Occupational Hazard:* Decline in health, functional ability, dermal contact	Skocaj et al. (2011), Baranowska-Wójcik et al. (2020)

(continued)

Table 7.2 (continued)

Nanomaterials	Types	Environmental issues	References	Safety issues	References
	Zinc oxide (ZnO) nanoparticles	*Ecotoxicity:* Negative effect on aquatic life and present serious environmental problems *Bioavailability:* Major impact on the bioavailability in wastewater treatment	Franklin et al. (2007) Ma et al. (2013)	*Human Health:* Skin irritation, respiratory issues, testicular tissues, potential systemic effects, drop in sperm count, and testosterone levels *Occupational Hazards:* Cause metal fume fever and inflammation of the airways	Vandebriel and Jong (2012) Monse et al. (2021)
	Iron oxide nanoparticles (Fe_3O_4, Fe_2O_3)	*Sedimentation:* Impacts sediment-dwelling organisms *Transformation:* Redox reactions influencing local geochemistry	Phenrat et al. (2007) Liu et al. (2012)	*Human Health:* Low toxicity but respiratory irritation from fine particles *Occupational Hazards:* Risks during manufacturing and application	Lewinsk et al. (2013). Kornberg et al. (2017) Leonel et al. (2022)
Carbon-based nanomaterials	Carbon nanotubes (CNTs)	*Persistence*: Environmental impact and long-term viability *Ecotoxicity*: Physical damage and oxidative stress to organisms	Petersen et al. (2008) Kolosnjaj et al. (2007)	*Human Health*: Lung inflammation, fibrosis, cancer *Occupational Hazards:* High risk during production; stringent safety required	Lam et al. (2004) CDC (2013)

(continued)

Table 7.2 (continued)

Nanomaterials	Types	Environmental issues	References	Safety issues	References
	Graphene oxide (GO)	*Ecotoxicity*: Causes physical and chemical stress to cell membranes *Bioaccumulation*: Affect food chains and disturb aquatic habitat	Ou et al. (2016) Seabra et al. (2014)	*Human Health:* Respiratory problems and DNA cleavage *Occupational Hazards*: Cause fibrosis and lung inflammation	Fadeel et al. (2018)
Polymer-based nanomaterials	Fullerenes (C_{50})	*Solubility*: Forms colloidal suspensions, potentially toxic *Photoreactivity*: Generates ROS under light exposure	Oberdörster (2004) Zhao et al. (2005)	*Human Health*: Oxidative stress, cytotoxicity *Occupational Hazards:* Safety protocols necessary for exposure management	Sayes et al (2004) Mori et al. (2006)
	Dendrimers	*Persistence*: Stable structure leading to environmental persistence *Toxicity*: Potential toxicity depending on functional groups	Roberts et al. (1996)	*Human Health*: Cytotoxicity and potential organ toxicity *Occupational Hazards*: Requires protective measures during handling	Malik et al. (1999) Duncan and Izzo (2005)
	Nanosponge	*Degradation*: Risk of toxic byproducts release *Bioaccumulation*: Potential long-term ecological effects	Swaminathan et al. (2010) Ansari et al. (2011)	*Human Health*: Health issues from ingestion or inhalation *Occupational Hazards*: Safety protocols for handling and disposal	Sharma and Pathak (2011) Sherje et al. (2017)

(continued)

Table 7.2 (continued)

Nanomaterials	Types	Environmental issues	References	Safety issues	References
Composite nanomaterials	Metal–organic frameworks (MOFs)	*Stability*: Potential release of organic linkers metal *Ecotoxicity*: Toxicity of released components	Kadhom and Deng (2018) Canivet et al. (2014)	*Human Health*: Respiratory and systemic toxicity *Occupational Hazards*: Protective measures needed during handling	Howarth et al. (2016) Liu et al. (2023)
Nanofibers and membranes	Magnetic nanocomposites	*Sedimentation*: Accumulation in sediments affecting benthic organisms *Magnetic Interference*: Potential impact on aquatic life	Laurent et al. (2008)	*Human Health*: Health issues from inhalation or ingestion *Occupational Hazards:* Safe handling procedures essential	Gupta and Gupta (2005) Sun et al. (2008)
	Electro spun nanofibers	*Degradation*: Release of nanoparticles into the environment *Ecotoxicity*: Impact of released nanoparticles on aquatic organisms	Malara (2024) Krysanov et al. (2010)	*Human Health*: Respiratory issues from inhalation of fibers *Occupational Hazards:* Proper handling and disposal required	Senthil et al. (2022) Balusamy et al. (2017)
	Nanocomposite membranes	*Leaching*: Potential leaching of nanoparticles into treated water *Ecotoxicity*: Impact of leached nanoparticles on aquatic organisms	Sahu et al. (2023) Dube and Okuthe (2023)	*Human Health*: Risks from exposure to leached nanoparticles *Occupational Hazards*: Safety measures needed during installation and maintenance	Drabik et al. (2021) Wen et al. (2019)

(continued)

Table 7.2 (continued)

Nanomaterials	Types	Environmental issues	References	Safety issues	References
Quantum dots (QDs)	Cadmium-based quantum dots (CQDs)	*Toxicity*: Release of cadmium ions, highly toxic to aquatic life *Persistence:* High stability and environmental persistence	Hu et al. (2021) Bechu et al. (2021)	*Human Health*: Severe health issues including carcinogenic *Occupational Hazards:* Stringent safety protocols required	Singh et al. (2023) Wang et al. (2022)
	Graphene quantum dots (GQDs)	*Bioavailability*: Potential to accumulate in organisms *Ecotoxicity*: Impact on aquatic life due to small size and reactivity	Arvidsson et al. (2013)	*Human Health*: Oxidative stress and inflammation from exposure *Occupational Hazards:* Essential safety measures for exposure management	Arvidsson et al. (2013)

7.5.1 Issues and Details

The use of nanomaterials in water treatment raises issues related to the environment and human health, including possible risks to aquatic life, bioaccumulation, and problems with the skin, eyes, and respiratory systems. PPE, appropriate confinement, labeling, and storage are necessary for safe handling. Disposal needs to adhere to rules and use decontamination techniques. Quick clean-up, exposure protection, and spill control are essential for emergency safety. The issues and details of nanomaterials with regard to distinct nanomaterials categories are presented in Table 7.3.

7.5.2 Applications of Nanomaterials and Their Properties

Nanomaterials like silver nanoparticles, titanium dioxide, zinc oxide, carbon nanotubes, iron oxide, graphene oxide, and fullerenes are used in water treatment due to their unique properties. They offer benefits like antimicrobial action, UV protection, adsorption, filtration, and magnetic separation, making them effective for various applications in environmental and purification technologies as illustrated in Table 7.4.

7.5.3 Environmental Impacts of Nanomaterials

Nanomaterials can harm the environment because of their toxicity to aquatic life, capacity to build up in the food chain, and persistence in ecosystems. In addition, they may have an unfavorable effect on the quality of soil and sediment and, when exposed to light, form hazardous reactive oxygen species, which emphasizes the need for cautious control and regulation. Table 7.5 illustrates the impact of nanomaterials across ranges.

7.5.4 Risks, Hazards, and Handling Recommendations of Nanomaterials

To improve the safety of individuals working with or exposed to nanomaterials, it is important to thoroughly study and address health risks and occupational hazards through proper handling practices, as outlined in Table 7.6.

Table 7.3 Categories of nanomaterials with issues and details

Category	Issues	Details
Environmental impacts	Toxicity to aquatic life	Possible damage to fish and other aquatic life
	Bioaccumulation	Accumulation of nanomaterials in the food chain
	Persistence in the environment	Longevity of nanomaterials in natural systems
	Soil and sediment impact	Effects on soil health and sediment quality
Health and safety concerns	Inhalation risks	Potential respiratory hazards from inhaling nanomaterials
	Skin contact	Risks of skin irritation or absorption
	Eye exposure	Risks of eye irritation or damage
	Long-term health effects	Potential chronic health issues from prolonged exposure
Handling and storage issues	Containment	Ensuring proper containment to prevent leaks and spills
	Personal protective equipment (PPE)	Use of appropriate PPE to protect workers
	Safe storage practices	Proper storage conditions to prevent accidents
	Labeling and documentation	Clear labeling and documentation for safe handling
Disposal and decontamination	Disposal regulations:	Compliance with regulatory requirements for disposal
	Specialized Disposal services:	Use of services specialized in handling nanomaterials
	Decontamination methods	Techniques to decontaminate areas and equipment
Emergency response	Spill containment	Immediate actions to contain spills
	Clean-up procedures:	Methods for safely cleaning up nanomaterials spills
	Exposure treatment:	First aid and treatment for exposure incidents
	Emergency contacts:	First aid and treatment for exposure incidents

Table 7.4 Summary of nanomaterials applications and their properties

Nanomaterials	Applications	Properties	References
Silver nanoparticles	Antimicrobial agents, water treatment	High surface area, reactivity	Sondi and Salopek-Sondi (2004)
Titanium dioxide	Photocatalysis, UV protection	Photo stability, non-toxicity	Haider et al. (2019)
Zinc oxide	UV filters, antibacterial agents	Wide band gap, high exaction energy	Moezzi et al. (2012)
Carbon nanotubes	Adsorption, filtration	High strength, electrical conductivity	Iijima (1991)
Iron oxide	Magnetic separation, catalysis	Super paramagnetic properties	Gupta and Gupta (2005)
Graphene oxide	Adsorption, membrane technologies	Large surface area, mechanical strength	Novoselov et al. (2004)
Fullerenes	Water purification, drug delivery	High electron affinity	Kroto et al. (1985)

Table 7.5 Environmental impacts of nanomaterials

Nanomaterials	Ecotoxicity	Bioaccumulation	Phototoxicity	References
Silver nanoparticles	High toxicity to aquatic organisms	Moderate	Low	Yan and Chen (2019)
Titanium dioxide	Moderate (depending on size/form)	Low	High	Sharma (2009)
Zinc oxide	High (releases Zn^{2+} ions)	Moderate	Moderate	Kaya et al. (2015), Beegam et al. (2016)
Carbon nanotubes	Low to moderate	High	Low	Flahaut (2011)
Iron oxide	Low	Low	Low	Campos et al. (2015)
Graphene oxide	Moderate	Moderate	Low	Seabra et al. (2014)
Fullerenes	High (under certain conditions)	Moderate	High (produces ROS)	Lahir (2017)

Table 7.6 Risks, hazard, and handling recommendations of nanomaterials

Nanomaterials	Human health risks	Occupational hazards	Handling recommendations	References
Silver nanoparticles	Cytotoxicity, genotoxicity	Inhalation/dermal exposure risks	Use PPE, proper ventilation	Samberg et al. (2012)
Titanium dioxide	Respiratory issues, potential carcinogenicity	Inhalation/skin contact risks	Use masks, gloves	Papillomaviruses (2011)
Zinc oxide	Skin irritation, respiratory issues	Handling risks	Proper protective equipment	Droepenu et al. (2022)
Carbon nanotubes	Lung inflammation, fibrosis	Production/ handling risks	Use of fume hoods, protective gear	Donaldson et al. (2006)
Iron oxide	Respiratory irritation	Exposure risks	Standard laboratory safety measures	Marval and Tronville (2022)
Graphene oxide	Inflammation, oxidative stress	Protective equipment needed	Use gloves, masks, and goggles	Sharma and Mondal (2020)
Fullerenes	Oxidative stress, cytotoxicity	Exposure risks	Safe storage, minimize dust generation	Aschberger et al. (2010)

7.5.5 Regulatory Standards and Guidelines for Nanomaterials

Regulating rules for nanomaterials is crucial to controlling dangers to human health and the environment. By handling, labeling, and disposing of materials properly, they guarantee safety, stop contamination, and encourage standard procedures. These rules, which clearly define safety and regulatory frameworks, also foster innovation. The safety guidelines and standards available for nanomaterials are illustrated in Table 7.7.

7.6 Nanomaterials Applications in Water Treatment

The applications highlight the diverse and valuable roles of nanomaterials in enhancing the processes of adsorption, filtration, and disinfection in water treatment as shown in Fig. 7.1.

Table 7.7 Regulatory standards and guidelines

Nanomaterials	Regulatory body	Standard/Guideline	Permissible limits	References
Silver nanoparticles	EPA, WHO	Drinking water standards, workplace exposure	0.1 mg/L (drinking water), 0.01 mg/m^3 (air)	Health Canada (2005)
Titanium dioxide	OSHA, NIOSH	Limits on occupational exposure	15 mg/m^3 (total dust)	Michaels and Wagner (2020)
Zinc oxide	EPA, OSHA	Water quality criteria, workplace exposure	5 mg/L (water), 5 mg/m^3 (air)	USEPA (2004)
Carbon nanotubes	NIOSH, EU	Workplace exposure limits	1 μg/m^3 (air)	NIOSH (2011)
Iron oxide	OSHA	Occupational exposure limits	10 mg/m^3 (fumes)	Desa et al. (2013)
Graphene oxide	NIOSH	Recommendations for workplace exposure	To be determined	NIOSH (2011)
Fullerenes	EU, EPA	Environmental safety standards	To be determined	Henshaw and O'Carroll (2009)

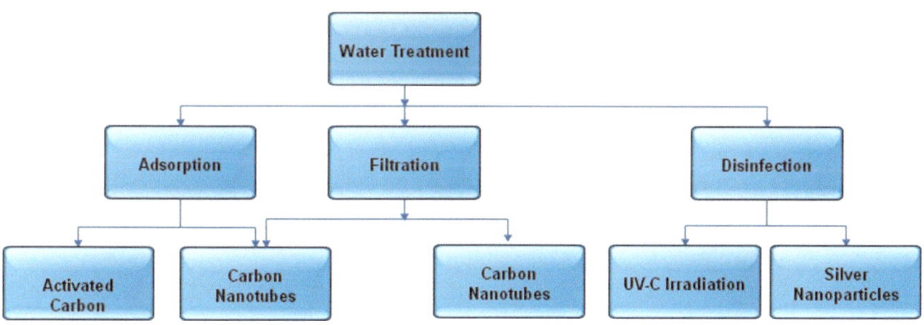

Fig. 7.1 Nanomaterials applications in water treatment

7.6.1 Overview of Components

1. **Adsorption**
   (a) Activated Carbon: Used for removing organic contaminants, chlorine, and some heavy metals through adsorption.
   (b) Carbon Nanotubes: Because of their high adsorption capacity and vast surface area, they are used for high-capacity adsorption of organic contaminants and heavy metals.

2. **Filtration**
   (a) Membrane Nanofibers: Employed in advanced filtration systems such as nanofiltration and ultra filtration membranes. They help in removing particles, bacteria, and viruses from water.
   (b) Graphene Oxide Membranes: These membranes are used for selective filtration and desalination due to their high permeability and selectivity.
3. **Disinfection**
   (a) UV-C Irradiation: Utilizes UV-C light to kill microorganisms and pathogens in water. Often combined with other nanomaterials for enhanced efficacy.
   (b) Silver Nanoparticles: Known for their antimicrobial properties, silver nanoparticles are used to disinfect water by targeting bacteria and viruses.

7.6.2 Lifecycle of Nanomaterials

The lifecycle of nanomaterials as shown in Fig. 7.2 is crucial for effectively managing risks, safeguarding the environment, ensuring health safety, and promoting sustainable practices.

7.6.2.1 Lifecycle Stages of Nanomaterials

1. Raw Material Synthesis: The initial stage where raw materials are sourced and processed to create nanomaterials. This may involve chemical reactions, physical processes, or biological methods.
2. Manufacturing and Processing: Involves the fabrication of nanomaterials into the desired form and structure. This can include nanoparticles synthesis, nanocomposite preparation, and quality control.
3. Use/Application: The stage where nanomaterials are incorporated into products or technologies. For instance, nanomaterials may be used in water treatment systems, electronics, or medicine.
4. Storage/Handling: Refers to the conditions under which nanomaterials are stored and handled during their use. Proper storage and handling are crucial to prevent accidental release or exposure.
5. End-of-Life (Disposal, Recycling, and Treatment): The final stage where nanomaterials are disposed of, recycled, or treated. This includes waste management practices and considerations for minimizing environmental impact.
6. Environmental Release/Impact: Illustrates how nanomaterials might be released into the environment during any stage of their lifecycle, potentially affecting soil, water, and air.
7. Potential Health Effects: Potential risks to human health due to exposure to nanomaterials. This includes inhalation, ingestion, or dermal contact.

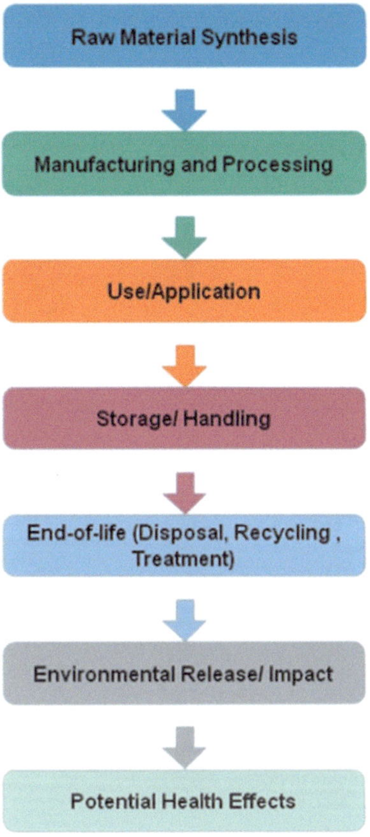

Fig. 7.2 Lifecycle diagram of nanomaterials

7.6.2.2 Toxicity Pathways and Mechanisms

To evaluate the health concerns associated with nanomaterials, it is crucial to comprehend their toxicity pathways and processes. It facilitates the understanding of the ways in which nanomaterials interact with tissues and cells to cause genotoxicity, oxidative stress, inflammation, and organ-specific toxicity. This information is essential for creating safety protocols, risk management plans, and legal requirements that safeguard public health and the environment from potential harm. The schematic diagram of nanomaterials interacts with biological systems as shown in Fig. 7.3.

7.6.2.3 Elements of Toxicity Pathways and Mechanisms

1. Nanomaterials Entry: The initial exposure pathways for nanomaterials, which can occur through inhalation (airborne particles), ingestion (contaminated food or water), or dermal contact (skin exposure).

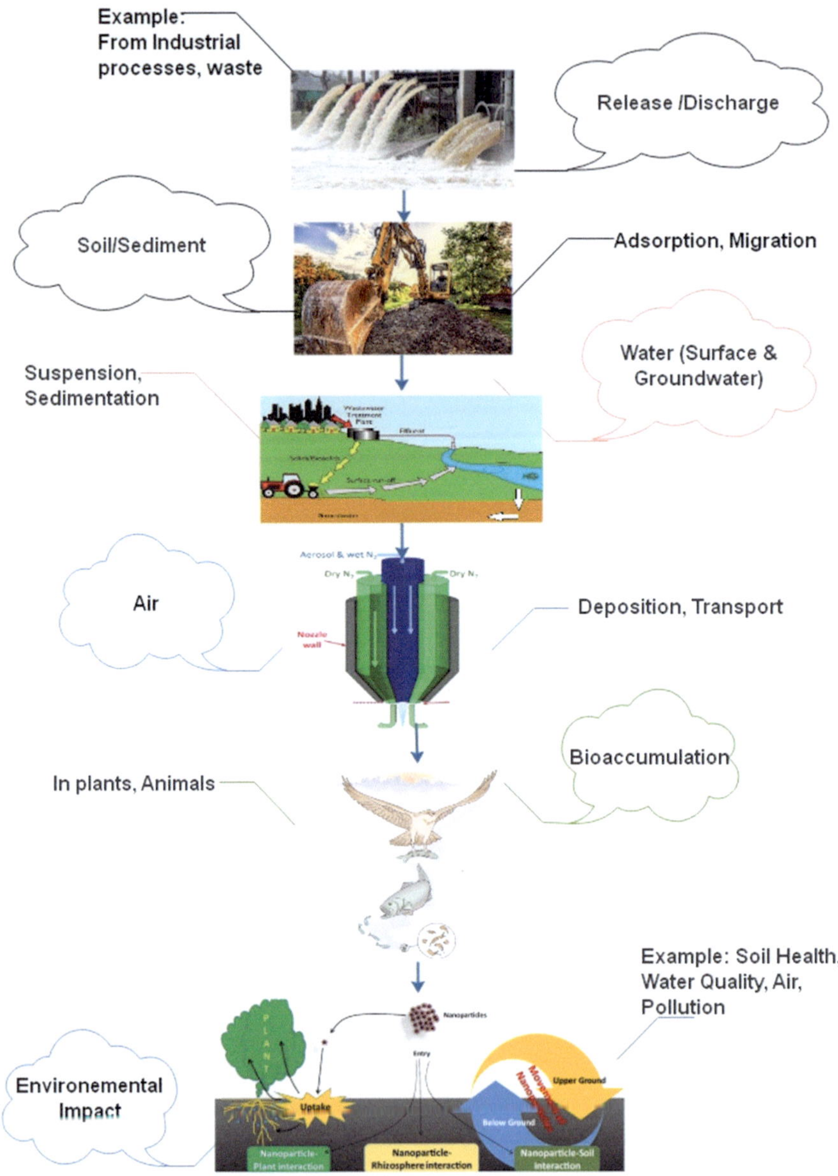

Fig. 7.3 Interaction of nanomaterials with biological systems at cellular and molecular levels

2. Cellular Uptake: The process by which nanomaterials enter cells. This can involve endocytosis (internalization by cells), membrane permeation, or direct interaction with cellular components.
3. Cellular Stress: Once inside cells, nanomaterials can induce various types of stress, including oxidative stress and general cellular stress.
4. Systemic Effects: The impact of nanomaterials on organ systems and overall health. This includes potential harm to organs such as the liver, kidneys, or cardiovascular system, and systemic inflammation or immune responses.
5. Genetic Effects: The likely for nanomaterials to cause genetic damage, such as mutagenesis (genetic mutations) or chromosomal aberrations, which can have long-term consequences.
6. Long-term Health Effects: Chronic exposure to nanomaterials may lead to extended health effects, including chronic diseases or cancer.

7.6.2.4 Environmental Fate and Transport Pathways

Visualizes how nanomaterials move through and impact different environmental compartments (air, water, soil). These routes aid in managing risks related to nanoparticles and evaluating their influence on the environment. The visualization of nanomaterials pathways as shown in Fig. 7.4.

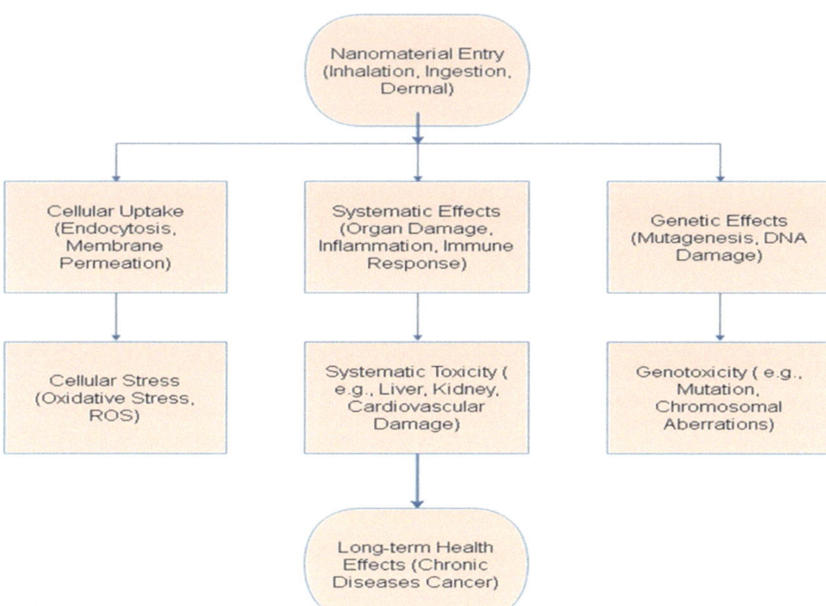

Fig. 7.4 Schematic diagram of environmental fate and transport pathways of nanomaterials

7.6.2.5 Phases of Environmental Fate and Transport Pathways

1. **Release/Discharge**: The initial point where nanomaterials are introduced into the environment. This can occur through industrial methods, wastewater discharge, or accidental spills.
2. **Soil/Sediment**: Nanomaterials may adsorb to soil or sediment particles. They can migrate within the soil or sediment matrix, affecting soil health and potentially entering groundwater.
3. **Water (Surface and Groundwater)**: Nanomaterials can be suspended in surface water (e.g., rivers, lakes) or infiltrate into groundwater. They may also settle out of suspension (sedimentation) or remain mobile in aquatic environments.
4. **Air**: Nanomaterials can become airborne through processes like aerosolization. Once in the air, they can be transported over long distances before settling (deposition) or affecting air quality.
5. **Bioaccumulation**: Nanomaterials can accumulate in living organisms, including plants and animals, through various pathways. In the food chain, this may result in bioaccumulation and even biomagnifications.
6. **Environmental Impact**: The final stage where the effects of nanomaterials on the environment are observed. This includes impacts on soil health, water quality, and air pollution.

7.7 Risk Assessment Framework

Outlines the steps involved in assessing the risks associated with the use of nanomaterials in water treatment.

1. Identify the Nanomaterials and Its Applications

   (a) Characterization: Find out the type, size, shape, and chemical composition of the nanomaterials.
   (b) Applications: Identify how the nanomaterials are used in water treatment processes (e.g., filtration, adsorption, disinfection).

2. Conduct Hazard Identification

   (a) Toxicological Assessment: Review existing data on the toxicological profiles of the nanomaterials, including potential effects on human health and the environment.
   (b) Chemical Properties: Analyze the chemical reactivity and potential for generating harmful byproducts.

3. Evaluate Exposure Scenarios

   (a) Exposure Pathways: Identify how and where exposure might occur (e.g., during production, application, or disposal).

   (b) Exposure Levels: Assess the concentration of nanomaterials in different phases (e.g., in water, air, or soil) and the potential for human and ecological exposure.

4. Assess Environmental Fate and Transport

   (a) Behavior in Water: Examine how nanomaterials behave in aquatic environments, including their stability, aggregation, and sedimentation.
   (b) Transport Pathways: Evaluate how nanomaterials may move through the environment (e.g., from water to soil or air).

5. Evaluate Potential Risks

   (a) Risk Assessment Models: Utilize models to forecast possible hazards by taking into account toxicity and exposure levels.
   (b) Ecological Impact: Assess potential impacts on aquatic organisms, soil microorganisms, and ecosystems.
   (c) Human Health Risks: Evaluate risks related to ingestion, inhalation, or dermal exposure to nanomaterials.

6. Implement Risk Management Strategies

   (a) Regulatory Compliance: Ensure that the use of nanomaterials complies with existing environmental and health regulations.
   (b) Mitigation Measures: Develop strategies to minimize risks, such as using lower concentrations, improving containment, or adopting safer handling practices.
   (c) Monitoring and Surveillance: Establish protocols for ongoing monitoring of nanomaterials levels and their effects on water quality and ecosystems.

7. Review and Update Risk Assessments

   (a) Continuous Review: Frequently review and update risk assessments based on new data, technological advancements, or changes in regulations.
   (b) Stakeholder Input: Engage with stakeholders (e.g., industry experts, regulatory bodies, and the public) to address concerns and incorporate feedback.

8. Document and Communicate Findings

   (a) Reporting: Document the risk assessment process, findings, and management strategies in detailed reports.
   (b) Communication: Communicate risks and mitigation measures to relevant stakeholders, including regulators, water treatment operators, and the public.

Comparative Impact Assessment

Comparing and contrasting the possible effects of various nanomaterials on the environment, human health, and safety is known as a comparative impact assessment. Understanding their relative risks and benefits, assisting in decision-making, and

Table 7.8 Assessing the environmental and health effects of nanomaterials

Nanomaterials	Applications in water treatment	Environmental impacts	Health impacts	References
Activated carbon nanotubes (CNTs)	Adsorption of organic pollutants, heavy metals	Potential for accumulation in soil and water; long-term effects on ecosystems unknown	Limited data; potential for respiratory irritation if inhaled	Lei et al. (2014)
Graphene oxide (GO)	Filtration membranes, adsorption	May impact aquatic life if released; potential for bioaccumulation	Potential for skin irritation and respiratory effects	Ghulam et al. (2022)
Silver nanoparticles (AgNPs)	Disinfection, antimicrobial treatment	Toxic to aquatic organisms; risk of accumulation and toxicity in the environment	Potential for antimicrobial resistance, skin irritation, and systemic toxicity	Yan and Chen (2019)
Titanium dioxide (TiO_2) nanoparticles	Photocatalysis for degradation of organic contaminants	May cause harm to aquatic life if released; potential for soil contamination	Potential respiratory and skin effects; low systemic toxicity	Fiordaliso et al. (2022), Rashid et al. (2021)
Zinc oxide (ZnO) nanoparticles	Photocatalysis, disinfection	Toxic to aquatic organisms; risk of soil contamination	Potential respiratory issues, skin irritation, and systemic effects	Beegam et al. (2016)
Magnetic nanoparticles (Fe_3O_4)	Removal of contaminants via magnetic separation	May impact soil and water systems if not properly managed	Potential for iron overload; low systemic toxicity	Tate et al. (2009), Zhang et al. (2023)

guaranteeing safe and sustainable use all depend on this assessment as given in Table 7.8.

7.8 Safe Handling and Disposal Practices

Nanomaterials must be handled and disposed of safely, which requires the use of engineering controls, appropriate work practices, and personal protective equipment (PPE). Following legal requirements, separating waste, and employing specialist services are all necessary for proper disposal. To control spills and exposure, decontamination and emergency response protocols are crucial. The steps guarantee the preservation of the ecosystem and public health as shown in Fig. 7.5.

Fig. 7.5 Steps of safe handling and disposal practices

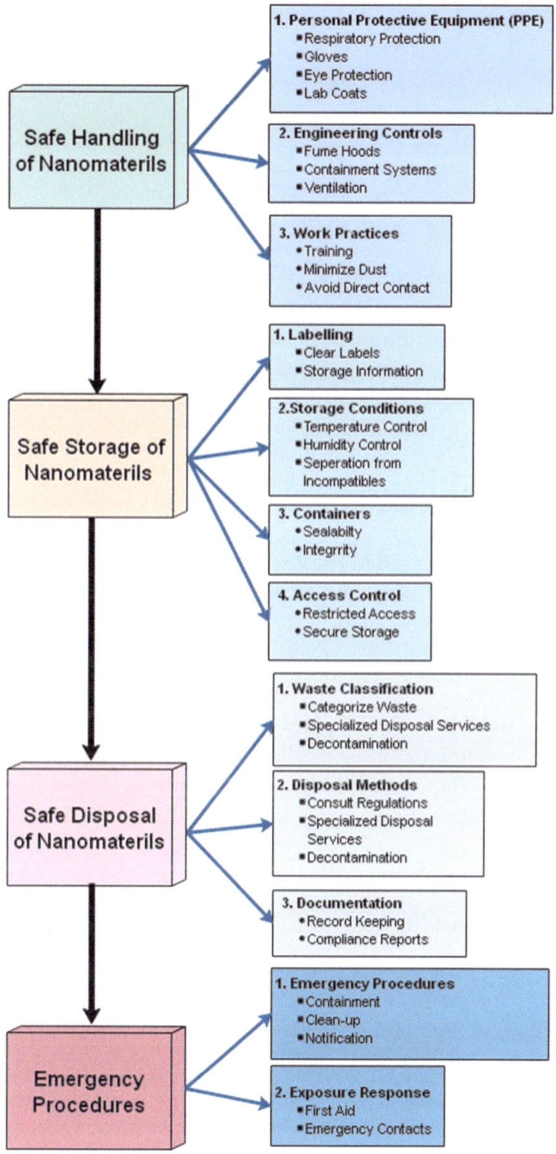

7.8.1 Overview of Essential Procedures for Secure Handling and Disposal

1. Safe Handling of Nanomaterials

(a) Personal Protective Equipment (PPE): Emphasizes the need for respiratory protection, gloves, eye protection, and lab coats.
(b) Engineering Controls: Highlights the importance of fume hoods, containment systems, and proper ventilation.
(c) Work Practices: Includes training, dust minimization, and avoiding direct contact.

2. Safe Storage of Nanomaterials

(a) Labeling: Ensures clear labeling and includes storage information.
(b) Storage Conditions: Focuses on temperature control, humidity control, and separation from incompatible substances.
(c) Containers: Addresses the need for salability and container integrity.
(d) Access Control: Stresses restricted access and secure storage.

3. Safe Disposal of Nanomaterials

(a) Waste Classification: Involves categorizing waste and labeling waste containers.
(b) Disposal Methods: Covers consultation with regulations, using specialized disposal services, and decontamination.
(c) Documentation: Ensures record-keeping and compliance reporting.

4. Emergency Procedures

(a) Spill Response: Includes containment, clean-up, and notification procedures.
(b) Exposure Response: Focuses on first aid and having emergency contacts readily available.

7.9 Conclusion

In conclusion, nanotechnology marks a major leap forward in wastewater treatment through the use of nanomaterials such as nanoparticles, carbon nanotubes, and graphene derivatives. These materials significantly improve pollutant removal by enhancing adsorption, degradation, and filtration processes, leading to more purified effluents than traditional methods. Nevertheless, their small size and high reactivity raise important environmental and safety concerns. Key risks include their persistence and mobility in aquatic environments, potential ecotoxicity, and the possibility of bioaccumulation in food chains, as well as health hazards from exposure. To tackle these issues, it is crucial to implement effective management strategies that include strong regulatory frameworks, ethical considerations, and stringent safety practices. A thorough understanding of the nanomaterials' lifecycle—from production to disposal—is vital for mitigating risks and ensuring sustainability. This involves addressing environmental impacts, health risks, and ensuring safe handling and disposal. Developing comprehensive risk assessment and management strategies is essential to balance the advantages of nanotechnology with its potential drawbacks. Improved safety protocols and regulatory measures are necessary to ensure the responsible application of nanomaterials in wastewater treatment, thereby protecting both the environment and public health.

References

Abbasi E, Aval SF, Akbarzadeh A, Milani M, Nasrabadi HT, Joo SW, Hanifehpour Y, Nejati-Koshki K, Pashaei-Asl R (2014) Dendrimers: synthesis, applications, and properties. Nanoscale Res Lett 9:1

Abd Rani U, Ng LY, Ng CY, Mahmoudi E (2020) A review of carbon quantum dots and their applications in wastewater treatment. Adv Coll Interface Sci 1(278):102124

Agboola O, Fayomi OS, Ayodeji A, Ayeni AO, Alagbe EE, Sanni SE, Okoro EE, Moropeng L, Sadiku R, Kupolati KW, Oni BA (2021) A review on polymer nanocomposites and their effective applications in membranes and adsorbents for water treatment and gas separation. Membranes 11(2):139

Ajith MP, Rajamani P (2021) Nanotechnology for water purification–current trends and challenges. J Nanotechnol Nanomater 2(2):88–91

Alanazi AG, Habila MA, Alothman ZA, Badjah-Hadj-Ahmed AY (2024) Synthesis and characterization of zinc oxide nanoparticle anchored carbon as hybrid adsorbent materials for effective heavy metals uptake from wastewater. Crystals 14(5):447

ALAtawi MK, Alasmari AA, AlAliany AD, Almajed MM, Sakran MI (2024) Silver nanoparticles forensic uses and toxicity on vital organs and different body systems. Adv Toxicol Toxic Effects 8(1):15–29

An YC, Gao XX, Jiang WL, Han JL, Ye Y, Chen TM, Ren RY, Zhang JH, Liang B, Li ZL, Wang AJ (2023) A critical review on graphene oxide membrane for industrial wastewater treatment. Environ Res 15(223):115409

Ansari KA, Vavia PR, Trotta F, Cavalli R (2011) Cyclodextrin-based nano sponges for delivery of resveratrol: in vitro characterization, stability, cytotoxicity and permeation study. AAPS PharmSciTech 12:279–286

Ansari MA, Khan ME, Mohammad A, Baig MT, Chaudary A, Tauqeer M (2023) Application of nanocomposites in wastewater treatment. In: Nanocomposites-advanced materials for energy and environmental aspects, pp 297–319. Woodhead Publishing

Aragaw TA, Bogale FM, Aragaw BA (2021) Iron-based nanoparticles in wastewater treatment: a review on synthesis methods, applications, and removal mechanisms. J Saudi Chem Soc 25(8):101280

Arvidsson R, Molander S, Sandén BA (2013) Review of potential environmental and health risks of the nanomaterial graphene. Hum Ecol Risk Assess Int J 19(4):873–887

Aschberger K, Johnston HJ, Stone V, Aitken RJ, Tran CL, Hankin SM, Peters SA, Christensen FM (2010) Review of fullerene toxicity and exposure–appraisal of a human health risk assessment, based on open literature. Regul Toxicol Pharmacol 58(3):455–473

Asghar F, Shakoor B, Fatima S, Munir S, Razzaq H, Naheed S, Butler IS (2022) Fabrication and prospective applications of graphene oxide-modified nanocomposites for wastewater remediation. RSC Adv 12(19):11750–11768

Avornyo A, Chrysikopoulos CV (2024) Applications of graphene oxide (GO) in oily wastewater treatment: recent developments, challenges, and opportunities. J Environ Manage 353:120178

Aziz T, Azad S, Nair SP, Verma JS, Unnarkat AP, Pan S, Namdeo A (2021) Potential risk and safety concern of nanomaterials used for wastewater treatment. In: Handbook of nanomaterials for wastewater treatment, pp 59–83. Elsevier

Baby R, Saifullah B, Hussein MZ (2019) Carbon nanomaterials for the treatment of heavy metal-contaminated water and environmental remediation. Nanoscale Res Lett 14(1):341

Balusamy B, Senthamizhan A, Uyar T (2017) In vivo safety evaluations of electro spun nanofibers for biomedical applications. In: Electrospun materials for tissue engineering and biomedical applications, pp 101–113. Woodhead Publishing

Baranowska-Wójcik E, Szwajgier D, Oleszczuk P, Winiarska-Mieczan A (2020) Effects of titanium dioxide nanoparticles exposure on human health—a review. Biol Trace Elem Res 193:118–129

Bechu A, Liao J, Huang C, Ahn C, McKeague M, Ghoshal S, Moores A (2021) Cadmium-containing quantum dots used in electronic displays: implications for toxicity and environmental transformations. ACS Appl Nano Mater 4(8):8417–8428

Beegam A, Prasad P, Jose J, Oliveira M, Costa FG, Soares AM, Gonçalves PP, Trindade T, Kalarikkal N, Thomas S, Pereira MD (2016) Environmental fate of zinc oxide nanoparticles: risks and benefits. In: Toxicology-new aspects to this scientific conundrum, p 33

Bejaoui B, Bouchmila I, Nefzi K, Slimen IB, Koumbad S, Martin P, Joly N, M'Hamdi N (2023) Nanostructured titanium dioxide (NS-TiO_2)

Bhat SA, Sher F, Hameed M, Bashir O, Kumar R, Vo DV, Ahmad P, Lima EC (2022) Sustainable nanotechnology-based wastewater treatment strategies: achievements, challenges and future perspectives. Chemosphere 288:132606

Biftu WK, Ravindhranath K, Ramamoorty M (2020) New research trends in the processing and applications of iron-based nanoparticles as adsorbents in water remediation methods. Nanotechnol Environ Eng 5(2):12

Burdușel AC, Gherasim O, Grumezescu AM, Mogoantă L, Ficai A, Andronescu E (2018) Biomedical applications of silver nanoparticles: an up-to-date overview. Nanomaterials 8(9):681

Calderon-Jimenez B, Johnson ME, Montoro Bustos AR, Murphy KE, Winchester MR, Vega Baudrit JR (2017) Silver nanoparticles: technological advances, societal impacts, and metrological challenges. Front Chem 21(5):6

Campos EA, Pinto DV, Oliveira JI, Mattos ED, Dutra RD (2015) Synthesis, characterization and applications of iron oxide nanoparticles-a short review. J Aerosp Technol Manag 7(3):267–276

Canivet J, Fateeva A, Guo Y, Coasne B, Farrusseng D (2014) Water adsorption in MOFs: fundamentals and applications. Chem Soc Rev 43(16):5594–5617

CDC NA (2013) Current intelligence bulletin 65: occupational exposure to carbon nanotubes and nanofibers

Cervantes-Aviles P, Huang Y, Keller AA (2019) Incidence and persistence of silver nanoparticles throughout the wastewater treatment process. Water Res 1(156):188–198

Chae SR, Hotze EM, Wiesner MR (2014) Possible applications of fullerene nanomaterials in water treatment and reuse. In: Nanotechnology applications for clean water, pp 329–338. William Andrew Publishing

Cheng YF, Zhang Q, Li GF, Xue Y, Zheng XP, Cai S, Zhang ZZ, Jin RC (2019) Long-term effects of copper nanoparticles on granule-based denitrification systems: performance, microbial communities, functional genes and sludge properties. Biores Technol 289:121707

Cruz-rosa S, Perez-reyes O (2023) Titanium oxide nanoparticles as emerging aquatic pollutants: an evaluation of the nanotoxicity in the freshwater shrimp larvae Atyalanipes. Ecologies. 4(1):141–151

Desa AF, Habidin NF, Hibadullah SN, Fuzi NM, Zamri FI (2013) Occupational Safety and Health Administration (OSHA) practices and OSHA performance in Malaysian automotive industry. J Stud Soc Sci 4(1)

Dhull P, Saini N, Aamir M, Parveen S, Husain S (2024) Function of nanomaterials in the treatment of emerging pollutants in wastewater. In: Detection and treatment of emerging contaminants in wastewater, pp 93–112

Donaldson K, Aitken R, Tran L, Stone V, Duffin R, Forrest G, Alexander A (2006) Carbon nanotubes: a review of their properties in relation to pulmonary toxicology and workplace safety. Toxicol Sci 92(1):5–22

Drabik M, Grzeczkowicz A, Bącal P, Kwiatkowska A, Strawski M, Antosiak-Iwańska M, Kazimierczak B, Godlewska E, Granicka LH (2021) Nanocomposite membrane scaffolds for cell function maintaining for biomedical purposes. Nanomaterials 11(5):1094

Droepenu EK, Wee BS, Chin SF, Kok KY, Maligan MF (2022) Zinc oxide nanoparticles synthesis methods and its effect on morphology: a review. Biointerface Res Appl Chem 12:4261–4292

Dube E, Okuthe GE (2023) Engineered nanoparticles in aquatic systems: Toxicity and mechanism of toxicity in fish. Emerging Contaminants 9(2):100212

Duncan R, Izzo L (2005) Dendrimer biocompatibility and toxicity. Adv Drug Deliv Rev 57(15):2215–2237

Duncan TV (2015) Release of engineered nanomaterials from polymer nanocomposites: the effect of matrix degradation. ACS Appl Mater Interfaces 7(1):20–39

Elsehly EM, Chechenin NG (2024) The performance of functionalized multi-walled carbon nanotube-based filters for water treatment applications

Environmental Protection Agency, U. S. (2004) Water quality criteria for zinc. https://www.epa.gov/sites/default/files/2015-06/documents/nrwqc-2004.pdf

Fadeel B, Bussy C, Merino S, Vázquez E, Flahaut E, Mouchet F, Evariste L, Gauthier L, Koivisto AJ, Vogel U, Martín C (2018) Safety assessment of graphene-based materials: focus on human health and the environment. ACS Nano 12(11):10582–10620

Farahbakhsh J, Vatanpour V, Ganjali MR, Saeb MR (2021) Magnetic nanoparticles in wastewater treatment. In: Magnetic nanoparticle-based hybrid materials, pp 547–589. Woodhead Publishing

Fiordaliso F, Bigini P, Salmona M, Diomede L (2022) Toxicological impact of titanium dioxide nanoparticles and food-grade titanium dioxide (E171) on human and environmental health. Environ Sci Nano 9(4):1199–1211

Flahaut E (2011) Toxicity and environmental impact of carbon nanotubes. In: Carbon nanotubes for biomedical applications, pp 211–219

Franklin NM, Rogers NJ, Apte SC, Batley GE, Gadd GE, Casey PS (2007) Comparative toxicity of nanoparticulate ZnO, bulk ZnO, and $ZnCl_2$ to a freshwater microalga (Pseudokirchneriella subcapitata): the importance of particle solubility. Environ Sci Technol 41(24):8484–8490

Freixa A, Acuna V, Gutierrez M, Sanchis J, Santos LH, Rodriguez-Mozaz S, Farre M, Barcelo D, Sabater S (2018a) Fullerenes influence the toxicity of organic micro-contaminants to river biofilms. Front Microbiol 9:1426

Freixa A, Acuña V, Sanchis J, Farre M, Barcelo D, Sabater S (2018b) Ecotoxicological effects of carbon-based nanomaterials in aquatic organisms. Sci Total Environ 619:328–337

Gangwar J, Pratap Singh A, Marimuthu N, Kadanthottu Sebastian J. Environmentally sustainable zinc oxide nanoparticles for improved hazardous textile dye removal from water bodies. AQUA—Water Infrastruct Ecosyst Soc 72(7):1198–1210

Gardner J, Dhai A (2014) Nanotechnology and water: ethical and regulatory considerations. In: Application of nanotechnology in water research, pp 1–20

Ghulam AN, Dos Santos OA, Hazeem L, Pizzorno Backx B, Bououdina M, Bellucci S (2022) Graphene oxide (GO) materials—spplications and toxicity on living organisms and environment. J Funct Biomater 13(2):77

Guerranti C, Renzi M (2015) Ecotoxicity of nanoparticles in aquatic environments: a review based on multivariate statistics of meta-data. J Environ Anal Chem 2(04):1000149

Guo H, Ma F, Feng X, Lou X, Tade MO (2013) Application of iron oxide-based nanomaterials (NMs) in magnetic assisted chemical separation (MACS) processes for water/wastewater treatment. Adv Mater Res 610:1242–1251

Gupta AK, Gupta M (2005) Synthesis and surface engineering of iron oxide nanoparticles for biomedical applications. Biomaterials 26(18):3995–4021

Haider AJ, Jameel ZN, Al-Hussaini IH (2019) Review on: titanium dioxide applications. Energy Procedia 157:17–29

Haidri I, Shahid M, Hussain S, Shahzad T, Mahmood F, Hassan MU, Al-Khayri JM, Aldaej MI, Sattar MN, Rezk AA, Almaghasla MI (2023) Efficacy of biogenic zinc oxide nanoparticles in treating wastewater for sustainable wheat cultivation. Plants 12(17):3058

Canada H (2005) Guidelines for Canadian drinking water quality: guideline technical document—Silver. Health Canada, Ottawa

Henshaw DL, O'Carroll MJ (2009) Scientific committee on emerging and newly identified health risks (SCENIHR). European Commission, Brussels

Hou W, Gao Y, Wang J, Blackwood DJ, Teo S (2020) Recent advances and future perspectives for graphene oxide reinforced epoxy resins. Mater Today Commun 23:100883

Howarth AJ, Liu Y, Li P, Li Z, Wang TC, Hupp JT, Farha OK (2016) Chemical, thermal and mechanical stabilities of metal–organic frameworks. Nat Rev Mater 1(3):1–5

Hu L, Zhong H, He Z (2021) Toxicity evaluation of cadmium-containing quantum dots: a review of optimizing physicochemical properties to diminish toxicity. Colloids Surf, B 200:111609

Hyseni S (2016) Toxicological effects of nanomaterials on aqueous and terrestrial ecosystems. Center Dev Strat 2016(1)

Iijima S (1991) Helical microtubules of graphitic carbon. Nature 354(6348):56–58

Iravani S, Korbekandi H, Mirmohammadi SV, Zolfaghari B (2014) Synthesis of silver nanoparticles: chemical, physical and biological methods. Res. Pharm. Sci. 9:385–406

Iravani S, Varma RS (2022) Nanosponges for water treatment: progress and challenges. Appl Sci 12(9):4182

Isibor PO, Kayode-Edwards II, Oyewole OA, Agbontaen DO, Onwaeze OO, Paul E, Oyegbade SA, Ilyasu US, Chimbekujwo KI (2024) Risk assessment and management in nanotoxicology. In: Environmental nanotoxicology: combatting the minute contaminants, pp 267–293. Springer Nature Switzerland, Cham

Jackson P, Jacobsen NR, Baun A, Birkedal R, Kühnel D, Jensen KA, Vogel U, Wallin H (2013) Bioaccumulation and ecotoxicity of carbon nanotubes. Chem Cent J 7:1–21

Jamil A, Anwar-ul-Haq M, Ahmed R, Rizwan A, Umair M, Zia-ur-Rehman M (2024) Potential applications of carbon nanotubes for the removal of organic and inorganic pollutants from wastewater. In: Water treatment using engineered carbon nanotubes, pp 307–329. Elsevier

Kadhom M, Deng B (2018) Metal-organic frameworks (MOFs) in water filtration membranes for desalination and other applications. Appl Mater Today 11:219–230

Kausar A (2024) Fullerene in water remediation nanocomposite membranes—cutting edge advancements. Charact Appl Nanomater 7(2):4945

Kaya H, Aydın F, Gürkan M, Yılmaz S, Ates M, Demir V, Arslan Z (2015) Effects of zinc oxide nanoparticles on bioaccumulation and oxidative stress in different organs of tilapia (Oreochromis niloticus). Environ Toxicol Pharmacol 40(3):936–947

Keat CL, Aziz A, Eid AM, Elmarzugi NA (2015) Biosynthesis of nanoparticles and silver nanoparticles. Bioresour Bioprocess 2:1–1

Kolosnjaj J, Szwarc H, Moussa F (2007) Toxicity studies of carbon nanotubes. In: Bio-applications of nanoparticles, pp 181–204

Kornberg TG, Stueckle TA, Antonini JM, Rojanasakul Y, Castranova V, Yang Y, Rojanasakul LW (2017) Potential toxicity and underlying mechanisms associated with pulmonary exposure to iron oxide nanoparticles: conflicting literature and unclear risk. Nanomaterials 7(10):307

Kroto HW, Heath JR, O'Brien SC, Curl RF, Smalley RE (1985) C60: buckminsterfullerene. Nature 318(6042):162–163

Krysanov EY, Pavlov DS, Demidova TB, Dgebuadze YY (2010) Effect of nanoparticles on aquatic organisms. Biology Bulletin 37:406–412

Lahir Y (2017) Impacts of fullerene on biological systems. Clin Immunol Endocrine Metab Drugs (Discontinued) 4(1):47–58

Lam CW, James JT, McCluskey R, Hunter RL (2004) Pulmonary toxicity of single-wall carbon nanotubes in mice 7 and 90 days after intratracheal instillation. Toxicol Sci 77(1):126–134

Laurent S, Forge D, Port M, Roch A, Robic C, Vander Elst L, Muller RN (2008) Magnetic iron oxide nanoparticles: synthesis, stabilization, vectorization, physicochemical characterizations, and biological applications. Chem Rev 108(6):2064–2110

Lei M, Wang N, Zhu L, Xie C, Tang H (2014) A peculiar mechanism for the photocatalytic reduction of decabromodiphenyl ether over reduced graphene oxide–TiO_2 photocatalyst. Chem Eng J 241:207–215

Leonel AG, Mansur AA, Mansur HS (2022) Nanotoxicity and environmental risks of magnetic iron oxide nanoparticles and nanohybrids. In: Handbook of magnetic hybrid nanoalloys and their nanocomposites 2022, pp 1–26. Springer International Publishing, Cham

Levchuk I, Sillanpää M (2020) Titanium dioxide–based nanomaterials for photocatalytic water treatment. In: Advanced water treatment 2020, pp 1–56. Elsevier

Lewinski N, Graczyk H, Riediker M (2013) Human inhalation exposure to iron oxide particles. Bionanomaterials 14(1–2):5–23

Li L, Guo W, Zhang S, Guo R, Zhang L (2023) Electro spun nanofiber membrane: an efficient and environmentally friendly material for the removal of metals and dyes. Molecules 28(8):3288

Liu J, Legros S, Ma G, Veinot JG, Von der Kammer F, Hofmann T (2012) Influence of surface functionalization and particle size on the aggregation kinetics of engineered nanoparticles. Chemosphere 87(8):918–924

Liu X, Shan Y, Zhang S, Kong Q, Pang H (2023) Application of metal organic framework in wastewater treatment. Green Energy Environ 8(3):698–721

Liu Y (2017) Application of graphene oxide in water treatment. IOP Conf Ser: Earth Environ Sci 94(1):012060

Lopez-Covarrubias JG, Soto-Muñoz L, Iglesias AL, Villarreal-Gómez LJ (2019) Electrospun nanofibers applied to dye solar sensitive cells: a review. Materials 12(19):3190

Ma H, Williams PL, Diamond SA (2013) Ecotoxicity of manufactured ZnO nanoparticles–a review. Environ Pollut 172:76–85

Malara A (2024) Environmental concerns on the use of the electrospinning technique for the production of polymeric micro/nanofibers. Scient Reports 14(1):8293

Malik N, Evagorou EG, Duncan R (1999) Dendrimer-platinate: a novel approach to cancer chemotherapy. Anticancer Drugs 10(8):767–776

Marval J, Tronville P (2022) Ultrafine particles: a review about their health effects, presence, generation, and measurement in indoor environments. Build Environ 216:108992

Mat Lazim Z, Salmiati S, Marpongahtun M, Arman NZ, Mohd Haniffah MR, Azman S, Yong EL, Salim MR (2023) Distribution of silver (Ag) and silver nanoparticles (AgNPs) in aquatic environment. Water 15(7):1349

Michaels D, Wagner GR (2020) Occupational Safety and Health Administration (OSHA) and worker safety during the COVID-19 pandemic. JAMA 324(14):1389–1390

Mishra AK (ed) (2014) Nanocomposites in wastewater treatment. CRC Press

Mishra S, Sundaram B (2023) Efficacy and challenges of carbon nanotube in wastewater and water treatment. Environ Nanotechnol Monit Manage 19:100764

Mishra Y, Mishra V, Chattaraj A, Aljabali AA, El-Tanani M, Farani MR, Huh YS, Serrano-Aroca Ã, Tambuwala MM (2023) Carbon nanotube-wastewater treatment nexus: where are we heading to? Environ Res:117088

Mittal M, Tripathi S, Shin DK (2024) Biopolymeric nanocomposites for wastewater remediation: an overview on recent progress and challenges. Polymers 16(2):294

Moezzi A, McDonagh AM, Cortie MB (2012) Zinc oxide particles: Synthesis, properties and applications. Chem Eng J 185:1–22

Mohajerani A, Burnett L, Smith JV, Kurmus H, Milas J, Arulrajah A, Horpibulsuk S, Abdul KA (2019) Nanoparticles in construction materials and other applications, and implications of nanoparticle use. Materials 12(19):3052

Mohamed WA, Abd El-Gawad H, Mekkey S, Galal H, Handal H, Mousa H, Labib A (2021) Quantum dots synthetization and future prospect applications. Nanotechnol Rev 10(1):1926–1940

Monse C, Raulf M, Jettkant B, van Kampen V, Kendzia B, Schurmeyer L, Seifert CE, Marek EM, Westphal G, Rosenkranz N, Merget R (2021) Health effects after inhalation of micro-and nano-sized zinc oxide particles in human volunteers. Arch Toxicol 95:53–65

Mordkovich VZ, Khaskov MA, De VV, Naumova VA, Karaeva AR (2024) Thermo-oxidative degradation of carbon nanotubes and related nanostructures: role of acidic environment and chloride ions. Full Nanotubes, Carbon Nanostruct 32(7):682–689

More PR, Pandit S, Filippis AD, Franci G, Mijakovic I, Galdiero M (2023) Silver nanoparticles: bactericidal and mechanistic approach against drug resistant pathogens. Microorganisms 11(2):369

Mori T, Takada H, Ito S, Matsubayashi K, Miwa N, Sawaguchi T (2006) Preclinical studies on safety of fullerene upon acute oral administration and evaluation for no mutagenesis. Toxicology 225(1):48–54

Navratil JD (1999) Wastewater treatment technology based on iron oxides. Natural Microporous Mater Environ Technol 31:417–424

Nayl AA, Abd-Elhamid AI, Awwad NS, Abdelgawad MA, Wu J, Mo X, Gomha SM, Aly AA, Brase S (2022) Review of the recent advances in electrospun nanofibers applications in water purification. Polymers 14(8):1594

Negrescu AM, Killian MS, Raghu SN, Schmuki P, Mazare A, Cimpean A (2022) Metal oxide nanoparticles: review of synthesis, characterization and biological effects. J Funct Biomater 13(4):274

Niculescu AG, Mihaiescu B, Mihaiescu DE, Hadibarata T, Grumezescu AM (2024) An updated overview of magnetic composites for water decontamination. Polymers 16(5):709

Nikic J, Watson M, Tubic A, Solic M, Agbaba J (2024) Recent trends in the application of magnetic nanocomposites for heavy metals removal from water: a review. Sep Sci Technol 59(2):293–331

NIOSH (2011) Occupational exposure to titanium dioxide. Curr Intell Bull 63:1–111

Noga M, Milan J, Frydrych A, Jurowski K (2023) Toxicological aspects, safety assessment, and green toxicology of silver nanoparticles (AgNPs)—critical review: state of the art. Int J Mol Sci 24(6):5133

Novoselov KS, Geim AK, Morozov SV, Jiang DE, Zhang Y, Dubonos SV, Grigorieva IV, Firsov AA (2004) Electric field effect in atomically thin carbon films. Science 306(5696):666–669

Oberdörster E (2004) Manufactured nanomaterials (fullerenes, C60) induce oxidative stress in the brain of juvenile largemouth bass. Environ Health Perspect 112(10):1058–1062

Oghyanous FA (2021) Nanoparticles in wastewater treatment. In: 2021 Water conservation—inevitable strategy, 21 Sept. https://doi.org/10.5772/intechopen.100239

Olawade DB, Wada OZ, Fapohunda O, Egbewole BI, Ajisafe O, Ige AO (2024) Nanoparticles for microbial control in water: mechanisms, applications, and ecological implications. Front Nanotechnol 6:1427843

Ou L, Song B, Liang H, Liu J, Feng X, Deng B, Sun T, Shao L (2016) Toxicity of graphene-family nanoparticles: a general review of the origins and mechanisms. Part Fibre Toxicol 13:1–24

Papillomaviruses H (2011) IARC monographs on the evaluation of carcinogenic risks to humans. IARC, Lyon, France

Perera AA, Madhushani KA, Kumar A, Gupta RK (2023) Metal-organic frameworks for wastewater treatment: recent developments, challenges, and future prospects. Chemosphere 5:139713

Perez-Ferreiro M, M. Abelairas A, Criado A, Gomez IJ, Mosquera J (2023) Dendrimers: exploring their wide structural variety and applications. Polymers 15(22):4369

Petersen EJ, Huang Q, Weber WJ Jr (2008) Bioaccumulation of radio-labeled carbon nanotubes by Eisenia foetida. Environ Sci Technol 42(8):3090–3095

Phenrat T, Saleh N, Sirk K, Tilton RD, Lowry GV (2007) Aggregation and sedimentation of aqueous nanoscale zerovalent iron dispersions. Environ Sci Technol 41(1):284–290

Radoor S, Karayil J, Jayakumar A, Siengchin S (2024) Efficient removal of dyes, heavy metals and oil-water from wastewater using electrospun nanofiber membranes: a review. J Water Process Eng 1(59):104983

Rashid MM, Forte Tavčer P, Tomsic B (2021) Influence of titanium dioxide nanoparticles on human health and the environment. Nanomaterials 11(9):2354

Rath S, Harshavardhan M, Srivastava P (2024) A concise overview of effect of nanomaterials in soil and associated microbiota. J Environ Nanotechnol 13(2):183–193

Roberts JC, Bhalgat MK, Zera RT (1996) Preliminary biological evaluation of polyamidoamine (PAMAM) StarburstTM dendrimers. J Biomed Mater Res 30(1):53–65
Rocha JM, Sousa RP, Fangueiro R, Ferreira DP (2024) The potential of electrospun membranes in the treatment of textile wastewater: a review. Polymers 16(6):801
Sahu A, Dosi R, Kwiatkowski C, Schmal S, Poler JC (2023) Advanced polymeric nanocomposite membranes for water and wastewater treatment: a comprehensive review. Polymers 15(3):540
Samadi-Maybodi A, Sadeghi-Maleki MR (2016) In-situ synthesis of high stable CdS quantum dots and their application for photocatalytic degradation of dyes. Spectrochim Acta Part A Mol Biomol Spectrosc 152:156–164
Samberg ME, Monteiro-Riviere NA, Bhushan B (201) In vitro and in vivo toxicity of silver nanoparticles. Encyclopedia Nanotechnol 10:1069–1077
Santhosh C, Malathi A, Dhaneshvar E, Bhatnagar A, Grace AN, Madhavan J (2019) Iron oxide nanomaterials for water purification. In: Nanoscale materials in water purification, pp 431–446. Elsevier
Sayes CM, Fortner JD, Guo W, Lyon D, Boyd AM, Ausman KD, Tao YJ, Sitharaman B, Wilson LJ, Hughes JB, West JL (2004) The differential cytotoxicity of water-soluble fullerenes. Nano Lett 4(10):1881–1887
Seabra AB, Paula AJ, de Lima R, Alves OL, Durán N (2014) Nanotoxicity of graphene and graphene oxide. Chem Res Toxicol 27(2):159–168
Senthil R, Sumathi V, Tamilselvi A, Kavukcu SB, Aruni AW (2022) Functionalized electrospun nanofibers for high efficiency removal of particulate matter. Sci Rep 12(1):8411
Shah SN, Shah Z, Hussain M, Khan M (2017) Hazardous effects of titanium dioxide nanoparticles in ecosystem. Bioinorg Chem Appl 2017(1):4101735
Sharma VK (2009) Aggregation and toxicity of titanium dioxide nanoparticles in aquatic environment—a review. J Environ Sci Health Part A 44(14):1485–1495
Sharma H, Mondal S (2020) Functionalized graphene oxide for chemotherapeutic drug delivery and cancer treatment: a promising material in nanomedicine. Int J Mol Sci 21(17):6280
Sharma R, Pathak K (2011) Polymeric nanosponges as an alternative carrier for improved retention of econazole nitrate onto the skin through topical hydrogel formulation. Pharm Dev Technol 16(4):367–376
Sharma YC, Srivastava V, Upadhyay SN, Weng CH (2008) Alumina nanoparticles for the removal of Ni (II) from aqueous solutions. Ind Eng Chem Res 47(21):8095–8100
Sherje AP, Dravyakar BR, Kadam D, Jadhav M (2017) Cyclodextrin-based nanosponges: a critical review. Carbohyd Polym 173:37–49
Singh D, Thapa S, Singh KR, Verma R, Singh RP, Singh J (2023) Cadmium selenide quantum dots and its biomedical applications. Materials Letters: X 18:100200
Skocaj M, Filipic M, Petkovic J, Novak S (2011) Titanium dioxide in our everyday life; is it safe? Radiol Oncol 45(4):227–247
Sondi I, Salopek-Sondi B (2004) Silver nanoparticles as antimicrobial agent: a case study on E. coli as a model for Gram-negative bacteria. J Coll Interface Sci 275(1):177–182
Spoiala A, Ilie CI, Trușca RD, Oprea OC, Surdu VA, Vasile BȘ, Ficai A, Ficai D, Andronescu E, Dițu LM (2021) Zinc oxide nanoparticles for water purification. Materials 14(16):4747
Sun C, Lee JS, Zhang M (2008) Magnetic nanoparticles in MR imaging and drug delivery. Adv Drug Deliv Rev 60(11):1252–1265
Swaminathan S, Pastero L, Serpe L, Trotta F, Vavia P, Aquilano D, Trotta M, Zara G, Cavalli R (2010) Cyclodextrin-based nanosponges encapsulating camptothecin: physicochemical characterization, stability and cytotoxicity. Eur J Pharm Biopharm 74(2):193–201
Tate JA, Ogden JA, Strawbridge RR, Pierce ZE, Hoopes PJ (2009) Toxicity and biodistribution of activated and non-activated intravenous iron oxide nanoparticles. In: Energy-based treatment of tissue and assessment, vol 7181, pp 211–218. SPIE
Theodore L, Kunz RG (2005) Nanotechnology: environmental implications and solutions. John Wiley & Sons

Tripathy J, Mishra A, Pandey M, Thakur RR, Chand S, Rout PR, Shahid MK (2024) Advances in nanoparticles and nanocomposites for water and wastewater treatment: a review. Water 16(11):1481
Tshangana CS, Muleja AA, Kuvarega AT, Malefetse TJ, Mamba BB (2021) The applications of graphene oxide quantum dots in the removal of emerging pollutants in water: an overview. J Water Process Eng 1(43):102249
Vandebriel RJ, De Jong WH (2012) A review of mammalian toxicity of ZnO nanoparticles. Nanotechnol Sci Appl 15:61–71
Waghmode MS, Gunjal AB, Mulla JA, Patil NN, Nawani NN (2019) Studies on the titanium dioxide nanoparticles: biosynthesis, applications and remediation. SN Appl Sci 1(4):310
Walser T, Limbach LK, Brogioli R, Erismann E, Flamigni L, Hattendorf B, Juchli M, Krumeich F, Ludwig C, Prikopsky K, Rossier M (2012) Persistence of engineered nanoparticles in a municipal solid-waste incineration plant. Nat Nanotechnol 7(8):520–524
Wang Z, Chen J, Li X, Shao J, Peijnenburg WJ (2012) Aquatic toxicity of nanosilver colloids to different trophic organisms: contributions of particles and free silver ion. Environ Toxicol Chem 31(10):2408–2413
Wang S, Liu Z, Wang W, You H (2017) Fate and transformation of nanoparticles (NPs) in municipal wastewater treatment systems and effects of NPs on the biological treatment of wastewater: a review. RSC Adv 7(59):37065–37075
Wang X, He K, Hu Y, Tang M (2022) A review of pulmonary toxicity of different types of quantum dots in environmental and biological systems. Chem Biol Interact 368:110247
Wen Y, Yuan J, Ma X, Wang S, Liu Y (2019) Polymeric nanocomposite membranes for water treatment: a review. Environ Chem Lett 17(4):1539–1551
Xie Y, Zhang T, Wang B, Wang W (2024) The application of metal-organic frameworks in water treatment and their large-scale preparation: a review. Materials 17(9):1972
Xu P, Zeng GM, Huang DL, Feng CL, Hu S, Zhao MH, Lai C, Wei Z, Huang C, Xie GX, Liu ZF (2012) Use of iron oxide nanomaterials in wastewater treatment: a review. Sci Total Environ 1(424):1
Xu X, Lv H, Zhang M, Wang M, Zhou Y, Liu Y, Yu DG (2023) Recent progress in electrospun nanofibers and their applications in heavy metal wastewater treatment. Front Chem Sci Eng 17(3):249–275
Yan A, Chen Z (2019) Impacts of silver nanoparticles on plants: a focus on the phytotoxicity and underlying mechanism. Int J Mol Sci 20(5):1003
Yehia HM, Said SM (2023) Silica nanoparticles for water purification and monitoring in point-of-use water supply systems. Am J Water Resour 11(3):98–102
Yu Y, Zhou Z, Huang G, Cheng H, Han L, Zhao S, Chen Y, Meng F (2022) Purifying water with silver nanoparticles (AgNPs)-incorporated membranes: recent advancements and critical challenges. Water Res 15(222):118901
Yusuf VF, Malek NI, Kailasa SK (2022) Review on metal–organic framework classification, synthetic approaches, and influencing factors: applications in energy, drug delivery, and wastewater treatment. ACS Omega 7(49):44507–44531
Zhang K, Song X, Liu M, Chen M, Li J, Han J (2023) Review on the use of magnetic nanoparticles in the detection of environmental pollutants. Water 15(17):3077
Zhao X, Striolo A, Cummings PT (2005) C60 binds to and deforms nucleotides. Biophys J 89(6):3856–3862
Zhao Y, Liu Y, Xu H, Fan Q, Zhu C, Liu J, Zhu M, Wang X, Niu A (2023) Preparation and application of magnetic composites using controllable assembly for use in water treatment: a review. Molecules 28(15):5799

Chapter 8
Technological Challenges and Future Prospects

Nimmy Srivastava, Jayeeta Chattopadhyay, and Ashish Ch. Singh

Abstract Industrial wastewater is mainly characterized by the high concentration of toxic pollutants and hazardous compounds that it contains. Generally, the traditional methods used for the treatment of wastewater have proved ineffective in cases of low concentration contaminants. The nanotechnology has therefore gained great importance in purification through enhancing contaminant removal efficiency with the use of nanoscale materials and devices. Some unique features of nanofibers, nanoadsorbents, and photocatalytic nanoparticles include high surface area and reactivity and selective adsorption. Such nanoscale materials are very efficient in targeting specific contaminants that are difficult to remove compared to traditional methods. Major challenges for the application of nanotechnology in industrial wastewater treatment are cost-effectiveness, scalability, environmental and health risks from nanomaterials, and lack of regulatory frameworks in place for safe deployment. Further research is required to improve material stability, optimize nanomaterial synthesis, and develop reusable and environmentally friendly nanomaterials to overcome these barriers. Future opportunities include combining nanotechnology with membrane filtration, advanced oxidation processes, and hybrid treatment systems to offer more effective, scalable, and sustainable solutions. If present technological as well as legal barriers are removed, nanotechnology may successfully revolutionize the safe and efficient treatment of industrial wastewater.

Keywords Nanomaterials · Technological challenges · Cost-effectiveness · Scalability

N. Srivastava
Amity Institute of Biotechnology, Amity University Jharkhand, Ranchi, India

J. Chattopadhyay (✉)
Amity Institute of Applied Sciences, Amity University Jharkhand, Ranchi, India
e-mail: jchattopadhyay@rnc.amity.edu

A. Ch. Singh
Chemistry Department, Birla Institute of Technology, Deoghar, Jharkhand, India

A. Mandpe et al. (eds.), *Nanotech Solutions for Industrial Wastewater*, Advances in Wastewater Research, https://doi.org/10.1007/978-981-96-6879-3_8

8.1 Introduction

It takes many compounds for living organisms to exist and thrive. But among all those compounds, the most important one is water. Since water covers over 70% of the earth's surface, Earth is referred to as a "blue planet." About 97.5% of water is thought to be saline, with the remaining 2.5% being considered fresh water, of which 68.9% is found in the form of glaciers, ice, and permanent snow. Moreover, only 0.3% of fresh water is readily accessible, making up 30.8% of the total (Definition of Freshwater Resources, n.d.). Clean, safe water is the need to establish a healthy economy and successful society (Kurniawan et al. 2012). However, there is a massive worldwide concern that the development of high-quality water supply resources is being severely eroded by rapid population growth, increasing industrialization, urbanization, and enormous agriculture activities (Olvera et al. 2017; Singh and Mishra 2020; Kamali et al. 2019). About 1.2 billion lack access to clean water for drinking, 2.6 billion suffer from accessing basic sanitation, and millions—especially children—die of diseases caused by water contamination or unsafe water supply (Kumar et al. 2014; Yamamura et al. 2011). The statistics show that about 1.8 million children die every year due to diarrhea caused by consumption of polluted water (Kumar et al. 2014; Qu et al. 2012).

These include: coagulation, flocculation, precipitation, adsorption, ion exchange, and membrane separation by electro-dialysis with aerobic, anoxic, or anaerobic oxidation processes. Physicochemical methods of treatment may depend on much more power-intensive and operatively loaded mechanical processes employing materials which include ion-exchange resins and their regenerants, aids for coagulation and filtration like ammonia, permanganate, alum, sodium hydroxide, hydrochloric acid, ozone, ferric salts, etc. These methods also demand high infrastructure and engineering capabilities. Besides, the conventional methods fail to remove heavy metals, phosphorus, nitrogen, and toxins from the contaminated wastewater. However, despite all these merits and demerits of each of the techniques, they are often expensive and labor-intensive although they help in mitigating various pollutants.

Nanotechnology is referred to essentially as the handling of matter at the atomic and molecular level to form new structures, devices, or systems with improved electrical, optical, magnetic, conductive, and mechanical properties. Theron et al. (2008), Humplik et al. (2011); Khan et al. (2019); Jain (2017); Jain (2018), mention that nanotechnology is a fast developing discipline with extraordinary potential, and it embraces wastewater treatment. The small size and high surface area of nanostructures, combined with easy functionalization, open broad prospects for developing more effective catalysts and redox-active media for purifying wastewater. It has already been demonstrated that nanomaterials are able to extract from wastewater a very large group of pollutants, from heavy metals and organic or inorganic solvents through dyes and biological toxins, to pathogens responsible for cholera and typhoid (Kumar et al. 2014).

8.2 Sources of Wastewater and Its Composition

Wastewater can be distinguished by its physical characteristics, chemical composition, and amount of microorganisms. Wastewater originates from different sources, which include sewages, commercial and industrial wastes, agricultural wastes, and many more. Toxic and destructive substances abound in wastewater (Ritter 2002). Wastewater is water contaminated with human activities and is often produced by natural biological activities (Rajasekhar et al. 2018; Rodriguez-Proteau and Grant 2005). Household, agricultural, industrial, and commercial wastes are major sources of wastewater (Ahmad et al. 2016). Almost all these major sources particularly the industries require pure water; however, at the expense of producing millions of polluted and contaminated water that are flushed into vast water bodies that contaminate them as well (Jassby et al. 2018; Deshpande et al. 2020).

Wastewater is a complex mixture, predominantly made up of 99% water and 0.1% organic matter. Organic matter includes suspended solids, dissolved biodegradable organics such as proteins, carbohydrates, and lipids, inorganic solids such as soil, salts, and metals, particulate matter with a chemical oxygen demand more than 500 mg/L, various microorganisms up to 10^9 per mL, heavy metals, micropollutants, and nutrients. According to several different studies, it is concluded that over 60% of phosphate compounds in wastewater are found in a soluble fraction (Warwick et al. 2013).

8.3 Different Nanomaterials Used for Removing Toxic Metals, Pathogens, Organic, and Inorganic Solutes from Wastewater

8.3.1 Nano Sorbents

Indeed, due to their high surface area and selective sorption abilities, are attracting much interest for water purification, remediation, and treatment applications. Because of their specific targeting of contaminants such as heavy metals, organic pollutants, and pathogens, they also add value in the removal of pollutants that conventional filtration may not catch or deal with very effectively.

Most commercially available nano sorbents used today are manufactured in the U.S. and Asia, but research is already tremendous in this field. More efforts have been directed toward tuning up the material for targeted removal of contaminants such as emerging contaminants that include pharmaceuticals and endocrine-disrupting compounds. We can look forward to production and commercialization increases in nano sorbents with growing interest in more efficient and better solutions for water treatment technologies.

They are also highly utilized in wastewater treatment, especially in removing organic contaminants. Because most contaminants lack magnetic properties, filtration aids should be used to absorb them and then magnetically separate. These magnetic nanosorbents are created by coating magnetic nanoparticles with specific ligands that have a targeted affinity (Apblett et al. 2001). More typically used to desorb the nanosorbents from a site of treatment when this may produce undesirable toxic side effects are magnetic forces, cleaning agents, and ion exchangers. Much more inexpensive, or more cost-effective regenerated nanosorbents, these are also more generally favored for application on commercial-scale.

8.3.2 *Nano Catalyst*

Catalytic nanoparticles also have several properties among them some are the surface areas and shape dependent that consequently make them to be notably reactive and therefore to work effectively in degrading the nature contaminants. These can degenerate environmentally harmful material especially polychlorinated biphenyls, dyes, halogenated aliphatic, herbicides, and halogenated herbicides among nitro aromatics as discussed by Zhao et al. (2011). Various laboratory tests affirm that catalytic nanoparticles work like good catalysts on the variety of contaminants. As most large-scale production of active catalysts requires quite a bit of hydrogen in redox reactions, reducing hydrogen usage and directly supporting the hydrogen economy by producing catalysts in metallic form is of prime importance.

The TiO_2-AGS composite is highly effective for the removal of Cr(VI) from wastewater due to modifications in the TiO_2 nanoparticles that shift their absorption from UV to natural light, enhancing degradation. Specific interactions between hydrogen and Pd-based nanoparticles have also been demonstrated (Zhang et al. 2012). Wastewater containing trace amounts of halogenated organic compounds (HOCs) can undergo selective biodegradation using advanced nanocatalysts. Here, HOCs are first biodegraded in treatment plants using nano-sized Pd catalysts. These nanocatalysts, having ferromagnetic properties, can easily be separated and therefore are recyclable. Hydrogen or formic acid could be used as reducers depending upon contamination levels (Hildebrand et al. 2008).

Another example is a silver and amidoxime fiber-based nanocatalyst stabilized through coordination interactions that can be repeatedly reactivated with tetrahydrofuran and is effective in the degradation of organic dyes (Wu et al. 2010). Nano-WO_3, prepared by the sol–gel method, was found to have enhanced photocatalytic activity in the disinfection of E. coli when used in combination with a 355 nm pulsed laser (Khalil et al. 2009). Research also shows that in-situ remediation by catalysis from Pd nanoparticles that reduce Cr(VI) to Cr(III) is feasible (Marcells et al. 2009). Another interesting method combines nanosorbents with catalysts to co-adsorb and co-degrade contaminants. It thus emphasizes nanocatalysis as a very potent tool for water treatment.

8.3.3 Nanostructured Catalytic Membranes

Nanostructured catalytic membranes have found wide applications in the treatment of water contamination with the advantages of uniform catalytic sites, optimization potential, controlled catalyst contact time, support for sequential reactions, and ease of scaling up for industrial applications. These membranes, particularly TiO_2 films, play several roles under UV and visible-light irradiation; these include decomposing organic pollutants, inactivating microorganisms, anti-biofouling action, and physical separation of contaminants (Choi et al. 2009). Moreover, N-doped "nut-like" ZnO nanostructured membranes enhance photodegradation under visible light, show antibacterial properties, and retain high water flow, making them very useful for water purification (Bai et al. 2012).

Several researchers have immobilized metallic nanoparticles in membranes such as cellulose acetate, polyvinylidene fluoride (PVDF), polysulfone, and chitosan for efficient degradation and dechlorination of toxic contaminants. Such immobilization has several advantages such as high reactivity, organic partitioning, prevention of nanoparticle leakage, resistance to agglomeration, and reduced surface passivation (Xu et al. 2009). More importantly, films of polyetherimide and palladium acetate nanocomposite were produced in which hydrogen-Pd nanoparticle interactions have provided good results for water treatment purposes. Metal nanoparticles were precipitated within the membrane matrix through annealing precursor films at different conditions by means of in-situ and ex-situ methods in order to materialize the production of materials with property customizability (Clémenson et al. 2010). Advancements in nanotechnology have even more enabled the synthesis of novel nanostructured catalytic membranes with enhanced permeability, selectivity, and fouling resistance, using techniques like bottom-up approaches and hybrid methods for multifunctional applications (Tarabara 2009). Studied with the help of various analytical instruments like High Resolution Transmission Electron Micrography (HRTEM), X-ray diffraction (XRD), and Energy Dispersive spectroscopy (EDS) Prakash et al. 2011).

MgO nanoparticles and cellulose acetate (CA) fibers with embedded Ag nanoparticles show high biocidal effect against Gram-positive bacteria, Gram-negative bacteria, and bacterial spores (Savage and Diallo 2005). Current and future nanotechnology-based approaches toward the detection of microbial pathogens will assist in the recognition of microbes and pathogens as well as diagnostics.

8.3.4 Bioactive Nanoparticles

Various infectious diseases are caused due to water pollution, while the pathogenic microorganisms responsible in most cases, are also antibiotic-resistant. To break this resistance also, recent alternatives in terms of nanoparticles with bio-activity are emerging from older chlorine-based biocides. Strong silver nanoparticles or AgNPs can be produced via extracellular action by the silver-resistant bacteria Bacillus cereus from a specific strain of exposed bacterium to varying concentration of silver.

8.3.5 Molecularly Imprinted Polymers (MIPs)

MIPs recently became of interest in the areas of biology, pharmacy, and environmental fields on grounds of their high selectivity. The basis for its selectivity lies in the synthesis, where it is attached to the template molecule by covalent, semi-covalent, or non-covalent bonds to the functional monomers. After removing the template, the place remaining becomes a specific binding place for the MIP. Essentially, molecular imprinting represents a free radical polymerization process between a functional monomer and a cross-linker that makes it feasible for MIPs to selectively act as absorbents with strong potential for detection and removal of water pollutants even at very low concentration levels (Caro et al. 2006). The same MIPs can further be formed into novel composite adsorbents or catalyst systems accompanied by an additional advantage beyond the conventional sorbent in terms of selective capability for extraction. Electrospinning can also be used to encapsulate MIP nanoparticles within nanofibers for various applications in pollution control, including water treatment (Kiparissides 2010). In addition, nano-MIPs, or molecularly imprinted nanospheres, have been engineered to specifically adsorb micropollutants from hospital wastewater. Preparing the nano-MIPs that range in size between 50 and 500 nm using the mini-emulsion polymerization technique is complex because it has to be prepared in one reaction chamber.

8.4 Application of Nanotechnology in Wastewater Management

Carbon-based nanomaterials, metal nanoparticles, polymeric nanoparticles, zeolites, self-assembled monolayers on mesoporous substrates, biopolymers, and more have been considered for applications in wastewater treatment. Various approaches are now used to treat wastewater including adsorption and biosorption, nanofiltration, photocatalysis, disinfection and disease control, sensing and monitoring, among others (Baruah et al. 2019).

8.4.1 Adsorption and Biosorption

Adsorption is the process of transferring a molecule, ion, or phase referred to as an adsorbate from a liquid or gaseous bulk onto a solid, or at times a liquid, surface

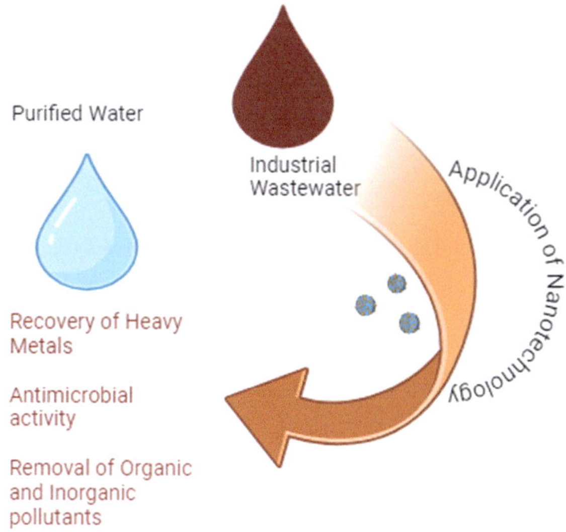

Fig. 8.1 Application of nanotechnology

known as an adsorbent. Under certain conditions, it forms a monomolecular layer on the surface through physicochemical or chemical interactions. Adsorption is a surface phenomenon and exothermic process (Gusain et al. 2020; Crawford and Quinn 2017; Alawady et al. 2020; Derco and Vrana 2018). Biosorption is a type of adsorption in which biological agents such as specific bacteria, algae, or fungi function as adsorbents through their inherent ability to interact with and accumulate heavy metals, even from very dilute aqueous solutions. This process occurs through spontaneous physicochemical uptake pathways that do not require ATP or metabolically mediated mechanisms. The key events in the biosorption process are microprecipitation, ion exchange, and cell-surface complexation (Kim et al. 2006; Yoon et al. 2006; Fard et al. 2011; Koyuncu et al. 2015) (Fig. 8.1).

8.4.2 *Nanofilters*

Water filtration is the removal or reduction of the concentration of particulate matter, including suspended particles, microorganisms, and other harmful biological and chemical contaminants from polluted or contaminated water for the production of safe and clean water for drinking, pharmaceutical, and medical purposes (Charcosset 2012). In the last two decades, interest in membrane technology has grown significantly with nanofiltration membranes becoming the latest breakthrough of this area. The prefix "nano" in nanofiltration membranes relates to its molecular weight cut-off for uncharged molecules at a nanometer size range (Nagy 2019). The novel application of NF membranes for the treatment of wastewater and drinking water is the favored method. Nanofiltration is a pressure-driven membrane process located between reverse osmosis and ultrafiltration. NF membranes are characterized with low rejection of monovalent ions, high rejection of divalent or higher valent ions, a flux of up to the hundredth value, with little energy consumption, and show better rejection performance as compared to RO and UF (Wu et al. 2017; Mulyanti and Susanto 2018; Abdel-Fatah 2018; Thanuttamavong et al. 2002; Malato et al. 2002); the pore sizes between 0.5–2.0 nm with an operating pressure that can be varied between 6–20 bars. More recently, nanofiltration, which is both aqueous and non-aqueous in nature, has been available. Its distinctive filtration mechanism and variety of membranes available have made NF one of the most crucial and widely used techniques in wastewater treatment (Timmer 2001).

8.4.3 Photocatalysis

There are two Greek words used in the term "photocatalysis." The terms "photo" and "catalysis" refer to substances that change the rate of reaction but are not a part of the chemical process. A light-induced reaction catalyzed and accelerated by light may thus be described as photocatalysis (Bethi et al. 2016). In other words, photocatalysis is the induction of a chemical reaction through the absorption of light (photons) by a solid catalyst called a photocatalyst (Rueda-Marquez et al. 2020). One of the Advanced Oxidation Processes (AOPs) is photocatalysis. This refers to the in-situ synthesis of very strong chemical oxidants in the presence of a catalyst, Fenton's reagent, hydrogen peroxide (H_2O_2), UV light, or ozone (O_3). The generated hydroxyl radicals ($^{\circ}OH$) are strong enough to oxidize strong organic molecules (Mehrjouei et al. 2015; McLain 2019).

Advanced Oxidation Processes are generally recognized methods for removing CECs from wastewater effluent. Among the methods, photocatalysis can oxidize a wide range of organic compounds: estrogens, dyes, organic acids, pesticides, crude oil, microbes, including viruses and chlorine-resistant organisms, as well as some inorganic molecules such as nitrous oxides. This makes it an area of research interest in the treatment of wastewater. Besides, in combination with filtration or precipitation, photocatalysis can also help in the removal of metals from wastewater, such as mercury (Gómez-Pastora et al. 2017; Ivanova et al. 2013). Nanoscale photocatalysts behave very differently compared to bulk photocatalysts due to superior surface properties, mechanical, chemical, electrical, magnetic, and optical capabilities along with quantum effects. Large surface areas and specific reactivity—properties that classical methods cannot match—are only some of the outstanding functionalities that nanomaterials confer on inactivating viruses in water (Zhang et al. 2010).

These nanoparticles deactivate through surface-based electrostatic interactions and photochemical reactions that cause cell walls to rupture, produce reactive oxygen species (ROS), and accurately apply disinfectants (Wang et al. 2015; Mauter et al. 2018; Moustafa 2017; Ahamed et al. 2014). The excellent surface properties and reactivity of these nanoparticles make it simple to regulate microorganisms in water (Li et al. 2008). Among them are Ag, CuO, ZnO, TiO_2 nanomaterials, polymeric nanoparticles, and carbon nanotubes (CNTs) that have been used in the studies of disinfection of wastewater (Mustapha et al. 2020; Liu et al. 2013; Elmi et al. 2014; Edwards-Jones 2009; Durán et al. 2005; Bhattacharyya et al. 2017).

8.4.4 *Sensing and Monitoring*

There are numerous living and non-living substances that have polluted the environment, ranging from disease-causing microorganisms to industrial and municipal waste, sewage discharge, animal faeces, and heavy metals (Farahi et al. 2012). The highly diluted concentration of pollutants, complexity, and variability of wastewater matrices make water quality monitoring difficult at both small and large scales (Graboski et al. 2020). In response to these challenges, it is essential to develop quick and effective techniques. Recently, interest has increased in the development of nanomaterial-based sensors for water quality monitoring. Nanosensors are an object or material that can sense response due to various stimuli such as thermal, chemical, and mechanical stress, volume changes, concentration, gravity, magnetic fields, and electrical forces. Nanosensors offer valuable information about nanoparticle behavior and characteristics between the nanoscale and the macroscopic level due to their remarkable property of excellent trace contaminant detection as well as rapid analysis (Scoville 2013; Riquelme et al. 2017). The three major components of the nanosensor include a transducer coupled with a recognizing component such as nanometals, nanotubes, and nanowires, among others, and a device for real-time monitoring a voltammetric, amperometric, conductometric, spectrophotometric, etc. transducer coupled together with a recognizing component, which could be nanometals, nanotubes, and nanowires, amongst others, and a show for real-time monitoring (Talari et al. 2021; Pardhi et al. 2019).

8.5 Challenges Associated with Nanotechnology Application in Wastewater Treatment and Future Prospects

Even though nanotechnology is very promising for wastewater treatment, there are several major hurdles that stand in the way of realizing the benefits. The most common barriers are those related to societal acceptability, cost-effectiveness, and nanomaterial toxicity. Some of the risks include water contamination, ecotoxicity of engineered nanomaterials, and transformation of nanomaterials (Raza et al. 2016).

8.5.1 *Toxicity of Nanomaterials*

Many past examples exist of wastewater treatment practices where the intention was opposite from the consequence. Another typical wastewater treatment process in historical times, chlorination, has even been supposed to promote more years in human lifespan overall worldwide (Sukhanova et al. 2018). Its generation of adverse by-products after its inception has instead established trihalomethanes and N-nitrosodimethylamine in this field. And finally, maybe the increasing threat is about the fact of wastewater treatment by using nanotechnology. The same factors which make nanoparticles useful will lead to the toxicity unleashed through them. The point at which the constituents turn toxic and their size is established at the molecular level in such a way that ultimately, the cellular uptake comes through the size. Due to their small size, NPs can penetrate endothelium and epithelial barriers into the lymphatic system, which then carries them into a range of organs and tissues, including the nervous system, brain, heart, liver, kidneys, and spleen (Fig. 8.2), as well as other body parts (Pan et al. 2020). Size and shape-related toxicity has been reported for Ag NPs, CNTs, and many other metal NPs. These nanoparticles, which have their size in between 1 to 100 nm can penetrate inside cells because they correlate in their dimensions, with that of the protein globules, DNA helices, and even the cell membrane's thickness, measured at 10 nm.

The small nanoparticles of gold, with sizes below 6 nm, easily penetrate into the nucleus. The bigger nanoparticles of size 10–16 nm can enter only the cell

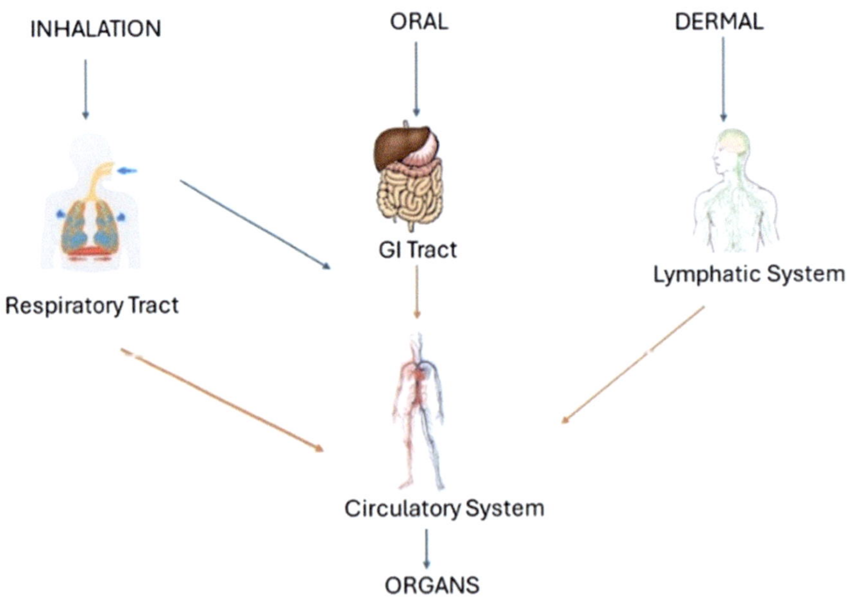

Fig. 8.2 Nanoparticles exposure and translocation potential inside human body

membrane and then get accumulated in the cytoplasm. TiO_2 nanoparticles have been reported to interfere with intracellular transport, cell division, and cell migration due to their ability to change the shape of tubulin and inhibit its polymerization. Nanoparticles come in different shapes, which include cubes, cylinders, spheres, ellipsoids, sheets, and rods. Nanotubes and nanofibers are less prone to endocytosis than spherical nanoparticles. It has also been observed that SWCNTs inhibit the calcium channels more potently than spherical NPs. Hydroxyapatite nanoparticles of varying shapes, including spherical, rod-like, plate-like, and needle-like forms, were evaluated for toxicity. The results were that the percentage of BEAS-2B lung epithelial cells killed was more in the case of plate-like and needle-like NPs than spherical and rod-like NPs. For the assessment of benefit-risk of nanoparticles, there exist a toxicological database including shared constituents and bulk equivalent. There exists no allotrope nanomaterial bulk equivalent in the available materials such as fullerenes and CNT, and thus evaluation is simply not possible for such chemicals, and again this throws open the requirements for extended toxicity research for all possible stages of nanomaterial life cycle and considered accordingly in an holistic perspective.

8.5.2 *Cost Effectiveness*

The acceptability of nanotechnology for wastewater treatment depends largely on its affordability and performance. While wastewater treatment in developed nations uses sophisticated methods to remove a wide range of contaminants, in underdeveloped nations it meets the most basic demands and is usually limited to something as simple as disinfectant. Both cases challenge the current state-of-the-art wastewater treatment models because the mixes that must be treated become much more complex to attain more stringent water quality at an economically feasible cost. Hurdle number three for a practical implementation is rather high but not insurmountable. Separation and purification are associated with the majority of the expense for producing nanomaterials. The cost of research-grade nanomaterials with high purity and consistent features has been fairly stable over the last two decades. Without greater demand and production scale-up, prices are unlikely to collapse dramatically. Additionally, with particles of lower purity, the rationality of their use in cleaning wastewater may be made stronger. For instance, amino-fullerene photocatalysts were produced by substituting ultrapure C60 with fullerene soot and resulted in a 95% cost saving without a discernible loss of efficiency (Westerhoff et al. 2013). Moreover, nanomaterials can be reused for long periods due to the repeated use that makes them economical. The two are multifunctional nanomaterials that can be reused several times because of magnetic separability and photocatalysts that can be recycled because of the regeneration of nanoadsorbents.

8.5.3 Risk of Nanomaterial Transformation in Water

Nanomaterial behaviors depend on interaction with the biotic and abiotic factors, such as dispersion and solubility of the nanoparticles. These nanoparticle particles settle slowly, compared to large-sized particle settling; however, on account of having large surfaces, they may adsorb more soil and sedimental particles. Carbon nanotubes and fullerenes are very soluble in water, and hence can easily be removed by the action of columns of water. Specific application demands engineered nanoparticles, sometimes called pristine nanomaterials. However, those materials may transform into the product-modified, product-weathered, and environmentally transferred nanomaterials. Photochemical changes can be induced by irradiation, and in some instances, oxidation reduction reactions can also occur. These changes would determine the behavior of interactions of nanomaterials with their environment, particularly impacting the adsorption or desorption of pollutants on water (Robichaud et al. 2009).

8.5.4 Environmental Hazards Associated with Nanomaterials

Ecotoxicity is the likelihood that biological, physical, or chemical stresses will disrupt the ecosystem. When products that have nanoparticles are manufactured, used, and disposed of, there is a potential for the nanomaterials to leach into the treated water. The new metal nanoparticles (NPs) such as Ag and TiO_2 are also being produced at much higher annual production rates (Kim et al. 2011). As indicated by a study in which the inhibitory potential of Ag NPs and TiO_2-NPs was determined on duckweed growth, an aquatic plant, these newly synthesized nanomaterials are thought to be of major concern to the aquatic environment.

The natural surface waters have estimated values of NP around nano/micro g/L. These concentrations rise along with the increase in production. Most developing countries are using silver, titanium, and zinc oxide NPs as disinfectants for filtration in water owing to its antibacterial property and replacing chemical disinfectants (Albukhari et al. 2019). Ag NP-doped ceramic filters are used. Water filtered through the nano Ag-coated sheets of the filter possessed an amount of detectable NPs Ag, far way below the WHO's recommendations for allowable limits. Virtually there were no reports found to evaluate leached NPs from the treated water regarding their toxic effects on any experimental animal. Investigations through rat that ingested contaminated water by TiO_2 NPs showed some degradation in the DNA content, along with the degenerative effects of key organs. Similarly, engineered nanoparticles have been found to induce a variety of health issues.

8.5.5 *Water Pollution*

Many laboratories investigate the toxicities of nanomaterial contaminated water. For instance, Cimbaluk et al. (2018) studied the toxic effects of multi-walled carbon nanotubes (MWCNTs) in two fish species, namely Astyanax altiparanae and Danio rerio. The researcher highlighted some potential concerns with the species including CNTs-DNA cross-linking, oxidative stress, and acute and subchronic neurotoxicity. Khan et al. have recently reported increased oxidative stress and genotoxicity in freshwater fish Labeo rohita exposed to Ag NP-treated water. The toxicity has also been incriminated for the dispersant used for the dispersion of nanomaterials themselves as nanomaterials are poorly soluble in water. Tetrahydrofuran, for example has been used as a potent dispersant for CNTs and C60 fullerenes, but its toxicity has been reported in the literature (Das et al. 2018).

There are various biological and ecological risks being investigated concerning the application of nanotechnology in wastewater treatment. After ruling out such a possibility, since research strategies of scientists are toward reducing the costs of nanomaterials, application of nanotechnology can succeed in this field as well. However, initial studies indicate that nanotechnology is capable of treating water with effectiveness, and further sophistication will enhance the safety factor.

8.6 Conclusions

For example, the requirement for pure, safe drinking water is on the increase globally. Other factors at play are an increase in population, increased industrialization, urbanization, and intensive agricultural techniques. Several techniques are employed to clean and decontaminate water. But because these methods are energy- and operation-intensive, and most often chemicals are used, a lot of infrastructure and engineering know-how are required. New wastewater treatment techniques must now be developed for the purpose of removing pollutants from wastewater. Nanotechnology is one possible approach in this regard for nanomaterial-based wastewater treatment. Special characteristics seen in nanostructure materials include high surface-to-volume ratios, high reactivity and sensitivity, a high capacity for adsorption, and ease of functionalization.

These characteristics make them capable of solving problems associated with traditional methods. However, there is another risk associated with nanotechnology. Researchers working on different applications and properties of nanomaterials have reported various challenges. Due to their small size, they can penetrate the bodies of human beings and other aquatic creatures, where they can cause harm. Degree of toxicity relies on several environmental factors such as pH, concentration, and exposure time. Even though scientists have made significant research concerning nanotechnology, much remains to be understood. Nanotechnology seems to be a very promising field with bright future prospects in wastewater treatment, but serious and committed efforts from government agencies and the scientific world are crucial.

References

Abdel-Fatah MA (2018) Nanofiltration systems and applications in wastewater treatment: review article. Ain Shams Eng J 9:3077–3092. https://doi.org/10.1016/j.asej.2018.08.001

Ahamed M, Alhadlaq H, Khan MAM, Karuppiah P, Al-Dhabi NA (2014) Synthesis, characterization, and antimicrobial activity of copper oxide nanoparticles. J Nanomater 2014:1–4. https://doi.org/10.1155/2014/637858

Ahmad HR, Aziz T, Zia-Ur-Rehman M, Sabir M, Khalid H, Hakeem KR, Akhtar J (2016) Sources and composition of wastewater: Threats to plants and soil health. In: Hakeem K, Akhtar J, Sabir M (eds) Soil science: agricultural and environmental prospectives. Springer, Cham, Switzerland, pp 349–370

Alawady RA, Alshahrani AA, Aouak T, Alandis MN (2020) Polysulfone membranes with CNTs/chitosan biopolymer nanocomposite as selective layer for remarkable heavy metal ions rejection capacity. Chem Eng J 15:124267

Albukhari SM, Ismail M, Akhtar K, Danish EY (2019) Catalytic reduction of nitrophenols and dyes using silver nanoparticles @ cellulose polymer paper for the resolution of waste water treatment challenges. Coll Surf A Physicochem Eng Asp. 577:548–561. https://doi.org/10.1016/j.colsurfa.2019.05.058

Apblett AW, Al-Fadul SM, Chehbouni M, Trad T (2001) Proc 8th Int Environ Petroleum Consortium

Bai H, Liu Z, Sun DD (2012) Hierarchical ZnO nanostructured membrane for multifunctional environmental applications. Colloids Surf A Physicochem Eng Asp 410:11–17

Baruah A, Chaudhary V, Malik R, Tomer VK (2019) Nanotechnology in water and wastewater treatment: theory and applications. In: Nanotechnology-based solutions for wastewater treatment. Elsevier, Amsterdam, The Netherlands, pp 337–368

Bethi B, Sonawane SH, Bhanvase BA, Gumfekar SP (2016) Nanomaterials-based advanced oxidation processes for wastewater treatment: a review. Chem Eng Process Process Intensif 109:178–189. https://doi.org/10.1016/j.cep.2016.08.016

Bhattacharyya S, Bennett J, Short LC, Theisen TS, Wichman MD, White JC et al (2017) Nanotechnology in the water industry, part 1: occurrence and risks. J Am Water Work Assoc 109:30–37. https://doi.org/10.5942/jawwa.109.0142

Caro E, Marcé RM, Borrull F, Cormack PAG, Sherrington DC (2006) Application of molecularly imprinted polymers to solid-phase extraction of compounds from environmental and biological samples. TrAC Trends Anal Chem 25(2):143–154

Charcosset C (2012) Membrane processes in biotechnology and pharmaceutics. Elsevier, Amsterdam, Netherlands. Some perspectives, pp 295–321

Choi H, Al-Abed SR, Dionysiou DD (2009) Nanostructured titanium oxide film and membrane based photocatalysis for water treatment. In: Nanotechnology applications for clean water, pp 39–46

Cimbaluk GV, Ramsdorf WA, Perussolo MC, Santos HKF, Da Silva De Assis HC, Schnitzler MC et al (2018) Evaluation of multiwalled carbon nanotubes toxicity in two fish species. Ecotoxicol Environ Saf 150:215–223. https://doi.org/10.1016/j.ecoenv..12.034

Clémenson S, Espuche E, David L, Léonard L (2010) Nanocomposite membranes of polyetherimide nanostructured with palladium particles: processing route, morphology and functional properties. J Membr Sci 361(1–2):167–175

Crawford CB, Quinn B (2017) Microplastic pollutants. The interactions of microplastics and chemical pollutants. Elsevier, Amsterdam, The Netherlands, pp 131–157

Das R, Leo BF, Murphy F (2018) The toxic truth about carbon nanotubes in water purification: a perspective view. Nanoscale Res Lett 13:1–10. https://doi.org/10.1186/s11671-018-2589-z

Definition of freshwater resources (n.d.) https://web.archive.org/web/20160411064155/. http://webworld.unesco.org/water/ihp/publications/waterway/webpc/definition.html

Derco J, Vrana B (2018) Biosorption. In: Introductory chapter: Biosorption. IntechOpen, London, UK

Deshpande B, Agrawal P, Yenkie M, Dhoble S (2020) Prospective of nanotechnology in degradation of wastewater: a new challenge. Nano-Struct Nano-Objects 22:100442. https://doi.org/10.1016/j.nanoso.2020.100442

Durán N, Marcato PD, Alves OL, Souza GIH, Esposito E (2005) Mechanistic aspects of biosynthesis of silver nanoparticles by several Fusarium oxysporum strains. J Nanobiotechnol 3:8. https://doi.org/10.1186/1477-3155-3-8

Edwards-Jones V (2009) The benefits of silver in hygiene, personal care and healthcare. Lett Appl Microbiol 49:147–152. https://doi.org/10.1111/j.1472-765X.2009.02648.x

Elmi F, Alinezhad H, Moulana Z, Salehian F, Tavakkoli SM, Asgharpour F et al (2014) The use of antibacterial activity of ZnO nanoparticles in the treatment of municipal wastewater. Water Sci Technol 70:763–770. https://doi.org/10.2166/wst.2014.232

Farahi RH, Passian A, Tetard L, Thundat T (2012) Critical issues in sensor science to aid food and water safety. ACS Nano 6:4548–4556. https://doi.org/10.1021/nn204999j

Fard RF, Azimi A, Bidhendi GN (2011) Batch kinetics and isotherms for biosorption of cadmium onto biosolids. Desalin Water Treat 28:69–74. https://doi.org/10.5004/dwt.2011.2203

Gómez-Pastora J, Dominguez S, Bringas E, Rivero MJ, Ortiz I, Dionysiou DD (2017) Review and perspectives on the use of magnetic nanophotocatalysts (MNPCs) in water treatment. Chem Eng J 310:407–427. https://doi.org/10.1016/j.cej.2016.04.140

Graboski AM, Martinazzo J, Ballen SC, Steffens J, Steffens C (2020) Nanotechnology in the beverage industry. Nanosensors for water quality control. Elsevier, Amsterdam, The Netherlands, pp 115–128

Gusain R, Kumar N, Ray SS (2020) Recent advances in carbon nanomaterial-based adsorbents for water purification. Coord Chem Rev 405:213111

Hildebrand H, Mackenzie K, Kopinke FD (2008) Novel nano-catalysts for wastewater treatment. Global NEST J 10(1):47–53

Humplik T, Lee J, O'Hern SC, Fellman BA, Baig MA, Hassan SF, Atieh MA, Rahman F, Laoui T, Karnik R et al (2011) Nanostructured materials for water desalination. Nanotechnology 22:292001. https://doi.org/10.1088/0957-4484/22/29/292001

Ivanova EP, Hasan J, Webb HK, Vervinskas G, Juodkazis S, Truong VK et al (2013) Bactericidal activity of black silicon. Nat Commun 4:2838. https://doi.org/10.1038/ncomms3838

Jain K (2017) Biopolymer-based composites: drug delivery and biomedical applications. Dendrimers: Smart nanoengineered polymers for bioinspired applications in drug delivery. Woodhead Publishing, Cambridge, UK, pp 169–220

Jain K (2018) Nanohybrids of dendrimers and carbon nanotubes: a benefaction or forfeit in drug delivery? Nanosci Nanotechnol Asia 9:21–29. https://doi.org/10.2174/2210681208666171204163622

Jassby D, Cath TY, Buisson H (2018) The role of nanotechnology in industrial water treatment. Nat Nanotechnol 13:670–672. https://doi.org/10.1038/s41565-018-0234-8

Kamali M, Persson KM, Costa ME, Capela I (2019) Sustainability criteria for assessing nanotechnology applicability in industrial wastewater treatment: current status and future outlook. Environ Int 125:261–276. https://doi.org/10.1016/j.envint.2019.01.055

Khalil A, Gondal MA, Dastageer MA (2009) Synthesis of nano-WO3 and its catalytic activity for enhanced antimicrobial process for water purification using laser induced photocatalysis. Catal Commun 11(3):214–219

Khan MS, Qureshi NA, Jabeen F (2017) Assessment of toxicity in fresh water fish Labeo rohita treated with silver nanoparticles. Appl Nanosci 7:167–179. https://doi.org/10.1007/s13204-017-0559-x

Khan I, Saeed K, Khan I (2019) Nanoparticles: properties, applications and toxicities. Arab J Chem 12:908–931. https://doi.org/10.1016/j.arabjc.2017.05.011

Kim E, Kim S-H, Kim H-C, Lee SG, Lee SJ, Jeong SW (2011) Growth inhibition of aquatic plant caused by silver and titanium oxide nanoparticles. Toxicol Environ Health Sci. 3:1–6. https://doi.org/10.1007/s13530-011-0071-8

Kim SJ, Cho JH, Kim TY, Choi SY (2006) New developments and applications in chemical reaction engineering. Biosorption of heavy metals and cyanide complexes on biomass. Elsevier, Amsterdam, The Netherlands, pp 141–144

Kiparissides C (2010) Nanotechnology meets water treatment. Dissemination workshop of the Nano4water cluster

Koyuncu I, Sengur R, Turken T, Guclu S, Pasaoglu M (2015) Advances in membrane technologies for water treatment. Advances in water treatment by microfiltration, ultrafiltration, and nanofiltration. Woodhead Publishing, Oxford, UK, pp 83–128

Kumar S, Ahlawat W, Bhanjana G, Heydarifard S, Nazhad MM, Dilbaghi N (2014) Nanotechnology-based water treatment strategies. J Nanosci Nanotechnol 14:1838–1858. https://doi.org/10.1166/jnn.2014.9050

Kurniawan TA, Sillanpää ME, Sillanpää M (2012) Nanoadsorbents for remediation of aquatic environment: local and practical solutions for global water pollution problems. Crit Rev Environ Sci Technol 42:1233–1295. https://doi.org/10.1080/10643389.2011.556553

Li Q, Mahendra S, Lyon DY, Brunet L, Liga MV, Li D et al (2008) Antimicrobial nanomaterials for water disinfection and microbial control: potential applications and implications. Water Res 42:4591–4602. https://doi.org/10.1016/j.watres.2008.08.015

Liu X, Wang M, Zhang S, Pan B (2013) Application potential of carbon nanotubes in water treatment: a review. J Environ Sci 25:1263–1280. https://doi.org/10.1016/S1001-0742(12)60161-2

Malato S, Blanco J, Vidal A, Richter C (2002) Photocatalysis with solar energy at a pilot-plant scale: an overview. Appl Catal B Environ 37:1–15. https://doi.org/10.1016/S0926-3373(01)00315-0

Marcells A, Omole IK, Owino O, Sadik OA (2009) Nanostructured materials for improving water quality: potentials and risks. In: Nanotechnology applications for clean water, pp 233–247

Mauter MS, Zucker I, Perreault F, Werber JR, Kim JH, Elimelech M (2018) The role of nanotechnology in tackling global water challenges. Nat Sustain 1:166–175. https://doi.org/10.1038/s41893-018-0046-8

McLain AA (2019) Photocatalytic properties of zinc oxide and graphene nanocomposites. Public Knowledge Project, Vancouver, BC, USA

Mehrjouei M, Müller S, Möller D (2015) A review on photocatalytic ozonation used for the treatment of water and wastewater. Chem Eng J 263:209–219. https://doi.org/10.1016/j.cej.2014.10.112

Moustafa MT (2017) Removal of pathogenic bacteria from wastewater using silver nanoparticles synthesized by two fungal species. Water Sci 31:164–176. https://doi.org/10.1016/j.wsj.2017.11.001

Mulyanti R, Susanto H (2018) Wastewater treatment by nanofiltration membranes. IOP Conf Ser Earth Environ Sci 142:012017. https://doi.org/10.1088/1755-1315/142/1/012017

Mustapha S, Ndamitso MM, Abdulkareem AS, Tijani JO, Shuaib DT, Ajala AO et al (2020) Application of TiO_2 and ZnO nanoparticles immobilized on clay in wastewater treatment: a review. Appl Water Sci 10:49. https://doi.org/10.1007/s13201-019-1138-y

Nagy E (2019) Basic equations of mass transport through a membrane layer. Nanofiltration. Elsevier, Amsterdam, Netherlands, pp 417–428

Olvera RC, Silva SL, Robles-Belmont E, Lau EZ (2017) Review of nanotechnology value chain for water treatment applications in Mexico. Resour Technol 3:1–11. https://doi.org/10.1016/j.reffit.2017.01.008

Pan Y, Liu X, Zhang W, Liu Z, Zeng G, Shao B et al (2020) Advances in photocatalysis based on fullerene C60 and its derivatives: properties, mechanism, synthesis, and applications. Appl Catal B Environ 265:118579. https://doi.org/10.1016/j.apcatb.2019.118579

Pardhi VP, Verma T, Flora S, Chandasana H, Shukla R (2019) Nanocrystals: an overview of fabrication, characterization and therapeutic applications in drug delivery. Curr Pharm des 24:5129–5146. https://doi.org/10.2174/1381612825666190215121148

Prakash A, Sharma S, Ahmad N, Ghosh A, Sinha P (2011) Synthesis of Agnps by Bacillus cereus bacteria and their antimicrobial potential. J Biomater Nanobiotechnol 2(2):155–161

Qu X, Brame J, Li Q, Alvarez PJ (2012) Nanotechnology for a safe and sustainable water supply: enabling integrated water treatment and reuse. Acc Chem Res 46:834–843. https://doi.org/10.1021/ar300029v

Rajasekhar B, Nambi IM, Govindarajan SK (2018) Human health risk assessment of groundwater contaminated with petroleum PAHs using Monte Carlo simulations: a case study of an Indian metropolitan city. J Environ Manag 205:183–191. https://doi.org/10.1016/j.jenvman.2017.09.078

Raza M, Kanwal Z, Rauf A, Sabri A, Riaz S, Naseem S (2016) Size- and shape-dependent antibacterial studies of silver nanoparticles synthesized by wet chemical routes. Nanomaterials 6:74. https://doi.org/10.3390/nano6040074

Riquelme MV, Leng W, Carzolio M, Pruden A, Vikesland P (2017) Stable oligonucleotide-functionalized gold nanosensors for environmental biocontaminant monitoring. J Environ Sci 62:49–59. https://doi.org/10.1016/j.jes.2017.08.005

Ritter KSL (2002) Sources, pathways, and relative risks of contaminants in surface water and groundwater: a perspective prepared for the Walkerton Inquiry. J Toxicol Environ Health A 65:1–142. https://doi.org/10.1080/152873902753338572

Robichaud CO, Uyar AE, Darby MR, Zucker LG, Wiesner MR (2009) Estimates of upper bounds and trends in nano-TiO_2 production as a basis for exposure assessment. Environ Sci Technol 43:4227–4233. https://doi.org/10.1021/es8032549

Rodriguez-Proteau R, Grant RL (2005) Toxicity evaluation and human health risk assessment of surface and groundwater contaminated by recycled hazardous waste materials. In: Kassim TA (ed) Water pollution: the handbook of environmental chemistry, vol 2. Springer, Berlin/Heidelberg, Germany, pp 133–189

Rueda-Marquez JJ, Levchuk I, Fernández Ibañez P, Sillanpää M (2020) A critical review on application of photocatalysis for toxicity reduction of real wastewaters. J Clean Prod 258:120694. https://doi.org/10.1016/j.jclepro.2020.120694

Savage N, Diallo MS (2005) Nanomaterials and water purification: opportunities and challenges. J Nanopart Res 7:331–342

Scoville S (2013) Nanotechnology safety. Implications of nanotechnology safety of sensors on homeland security industries. Elsevier, Amsterdam, The Netherlands, pp 175–194

Singh I, Mishra PK (2020) Nano-membrane filtration: a novel application of nanotechnology for wastewater treatment. Mater Today Proc. 29:327–332. https://doi.org/10.1016/j.matpr.2020.07.284

Sukhanova A, Bozrova S, Sokolov P, Berestovoy M, Karaulov A, Nabiev I (2018) Dependence of nanoparticle toxicity on their physical and chemical properties. Nanoscale Res Lett 13:1–21. https://doi.org/10.1186/s11671-018-2457-x

Talari FF, Bozorg A, Faridbod F, Vossoughi M (2021) A novel sensitive aptamer-based nanosensor using rGQDs and MWCNTs for rapid detection of diazinon pesticide. J Environ Chem Eng 9:104878. https://doi.org/10.1016/j.jece.2020.104878

Tarabara VV (2009) Multifunctional nanomaterial-enabled membranes for water treatment. In: Nanotechnology applications for clean water, pp 59–75

Thanuttamavong M, Yamamoto K, Oh JI, Choo KH, Choi SJ (2002) Rejection characteristics of organic and inorganic pollutants by ultra-low-pressure nanofiltration of surface water for drinking water treatment. Desalination 145:257–264. https://doi.org/10.1016/S0011-9164(02)00420-4

Theron J, Walker JA, Cloete TE (2008) Nanotechnology and water treatment: applications and emerging opportunities. Crit Rev Microbiol 34:43–69. https://doi.org/10.1080/10408410701710442

Timmer JMK (2001) Properties of nanofiltration membranes: model development and industrial application. [Phd Thesis 1 (Research TU/e/Graduation TU/e), Chemical engineering and chemistry], Technische Universiteit Eindhoven. https://doi.org/10.6100/IR545659

Wang H, Zhou Y, Jiang X, Sun B, Zhu Y, Wang H et al (2015) Simultaneous capture, detection, and inactivation of bacteria as enabled by a surface-enhanced Raman scattering multifunctional chip. Angew Chem Int Ed 54:5132–5136. https://doi.org/10.1002/anie.201412294

Warwick C, Guerreiro A, Soares A (2013) Sensing and analysis of soluble phosphates in environmental samples: a review. Biosens Bioelectron 41:1–11. https://doi.org/10.1016/j.bios.2012.07.012.21

Westerhoff PK, Kiser MA, Hristovski K (2013) Nanomaterial removal and transformation during biological wastewater treatment. Environ Eng Sci 30:109–117. https://doi.org/10.1089/ees.2012.0340

Wu ZC, Zhang Y, Tao TX, Zhang L, Fong H (2010) Silver nanoparticles on amidoxime fibers for photocatalytic degradation of organic dyes in wastewater. Appl Surf Sci 257(3):1092–1097

Wu L, Wang H, Xu TW, Xu ZL (2017) Membrane-based separations in metallurgy. Polymeric membranes. Elsevier, Amsterdam, Netherlands, pp 297–334

Xu J, Bachas L, Bhattacharyya D (2009) Synthesis of nanostructured bimetallic particles in poly ligand functionalized membranes for remediation applications. In: Nanotechnology applications for clean water, pp 311–335

Yamamura H, Hashino M, Kubota N (2011) Membrane filtration: principle and applications in water treatment. Sen-I Gakkaishi 67:3. https://doi.org/10.2115/fiber.67.P_81

Yoon J, Cao X, Zhou Q, Ma LQ (2006) Accumulation of Pb, Cu, and Zn in native plants growing on a contaminated Florida site. Sci Total Environ 368:456–464

Zhang D, Li G, Yu JC (2010) Inorganic materials for photocatalytic water disinfection. J Mater Chem 20:4529–4536. https://doi.org/10.1039/b925342d

Zhang QK, Kemp KC, Chandra V (2012) Homogeneous anchoring of TiO_2 nanoparticles on graphene sheets for wastewater treatment, pp 784–790

Zhao X, Lv L, Pan B, Zhang W, Zhang S, Zhang Q (2011) Polymer-supported nanocomposites for environmental application: a review. Chem Eng J 170(2–3):381–394

Chapter 9
Case Study: Nanofilters and Their Applications in Water and Wastewater Treatment

Vivek Yadav, Rahil Mathakia, Pritesh Bhatt, Amit Kumar Sharma, Arun K. Soni, and Viralkumar B. Mandaliya

Abstract Nanofilter (NF) and Nanofilter membranes are known to utilize in industrial water, wastewater, and solution filtrations. The current chapter is emphasized on types and property of nanomaterials used in nanofiltration. Further, it has highlighted on evolution of NF, methods and mechanism of fabricating NF, synthesis and production of NF, methods for NF membrane preparation, and types of NF membranes. The application of NF in various industrial applications has led to innovation in NF technology. The innovative technologies have driven for their protection under patent. The present article also emphasizes the patent scenario on NF for their year-wise trends in patent application and geographical distribution of NF patent application. Furthermore, International Patent Classification (IPC) vs Global Patent application trend is observed to analyze domain-specific weightage in patent application in broad regions. Also, top applicants and innovators who hold the patent portfolio in NF patent are identified. At the last, application of NF in various industrial filtration processes and IPC subclass C02F on technologies related to industrial water treatment and purification in multinational phase via PCT is discussed.

Keywords Nanofilter · Nanofilter membrane · Patent landscape · International patent classification · C02F

V. Yadav · R. Mathakia · V. B. Mandaliya (✉)
Department of Microbiology, Faculty of Science, Marwadi University, Rajkot, Gujarat, India
e-mail: viralmandaliya@gmail.com; viralkumar.mandaliya@marwadieducation.edu.in

P. Bhatt
Virus Research and Diagnostic Laboratory (VRDL), All India Institute of Medical Sciences, Rajkot, Gujarat, India

A. K. Sharma
Department of Bioscience, Saurashtra University, Rajkot, Gujarat, India

A. K. Soni
Ramanbhai Patel College of Pharmacy, Charotar University of Science and Technology, Changa, Gujarat, India

A. Mandpe et al. (eds.), *Nanotech Solutions for Industrial Wastewater*, Advances in Wastewater Research, https://doi.org/10.1007/978-981-96-6879-3_9

9.1 Introduction

Adventitious agents are a major concern in the manufacturing process of biological products, as they can cause contamination and pose a risk to patient safety. According to the provided research results, the four primary adventitious agents of concern are viruses, prions, mycoplasmas, and bacteria/fungi. To ensure patient safety, it is important to appropriately manage and remove these infectious agents from the manufacturing process and drug product. Among them viruses and prions are more critical due to lack of detection method complexity. Latent adventitious agents could be present in starting material like cell banks or raw materials from the animal origin used as a substrate in bioprocess. This latent adventitious agent may become active upon receiving favorable conditions.

The source of adventitious agents can come from various sources, including the source material, raw materials used in manufacturing, environment, personnel, and process equipment. Therefore, it is crucial to implement appropriate measures to prevent the introduction of adventitious agents into the manufacturing process. This can include using validated and controlled raw materials, implementing strict environmental controls, and regularly testing for the presence of adventitious agents and eliminating virus partials by terminal sterilization by filtration through nanofilters.

In early 1960s, an unbelievable growth was observed in membrane science and technology (Basile and Charcosset 2016). With this extraordinary growth in membrane science, interest of every other company drawn toward membrane filtration, particularly nanofiltration (NF). FilmTec was the one who introduced the term nanofiltration (Linder and Kedem 2021). With the growing interest, in 1980, nanofiltration membranes were introduced, and they received such nomenclature because their pore size is in nanometer (< 2 nm) (Bowen and Mukhtar 1996). These membranes have properties of both ultrafiltration (UF) and reverse osmosis (RO), and they were originally known as open RO, tight UF. These nanofiltration membranes are low-pressure RO membranes because they are pressure-driven and they operate between pressure 5 and 20 bar which has been widely used to separate molecules of different molecular weights (Basile and Charcosset 2016).

Previously, NF membranes were made up of cellulose acetate but due to various limitations, other membrane materials were introduced such as polyamides, polyether sulfones, chlorinated polyvinyl chloride, and PVDF (Linder and Kedem 2021). Major breakthrough came with noncellulosic composites invention, which was based on coating of UF, which supports with submicron selective barrier using interfacial polymerization methodology. Composites work was started in 1970s but NF membranes composites got huge popularity and demand in mid-1980s (Linder and Kedem 2021).

In 1973, first-generation nanofiltration membranes were commercially introduced in market which included gel cellophane (DuPont, Union Carbide), polyelectrolyte: anisotropic (Diaflo UM-3) (Amicon), anisotropic: Pellicon PSAC (Millipore). By the onset of 1990s and late 1980s, several patents were processed stating the production of NF membranes through the modifications in RO polyamide composites by using various reagents such as acids and bases. (Linder and Kedem 2021) For example, in the study of Strantz and Brehm, they shown the process of producing NF membranes by treating RO polyamide composites with permanganate salts which led to opening of selective barriers simultaneously to chemically stabilize the membrane they treated with hydrogen peroxide (Strantz and Brehm 1990). With the onset of 1990s, NF membranes having porous ceramic support coated with polymeric layer were introduced in the industries.

In early 1990s, nanofiltration technique was introduced in biopharmaceutical industries for the manufacturing of plasma-derived coagulation factors (Burnouf-Radosevich et al. 1994). Between 1997 and 2003, use of nanofiltration occupied huge strength and demand (Fig. 9.1) in manufacturing processes of various recombinant products such as monoclonal antibodies. Planova 35N filters of Asahi Kasei Corporation were the first nanofilters commercially available for virus filtration in late 1980s. In late 1990s, addition of nanofiltration during manufacturing process became important because it ensures the product safety against various enveloped and non-enveloped viruses (Inouye and Burnouf 2020). After that, various NF membranes came in the market which were used during the downstream processing of any biological product.

Since it was initially recognized in the late 1980s, nanofiltration (NF) membranes have progressed significantly. NF membranes, which have characteristics halfway between reverse osmosis (RO) and ultrafiltration (UF), have pores that are typically 1 nm in size, which is equivalent to a molecular weight cutoff (MWCO) of 300–500 Da (Mohammad et al. 2015). The number of publications published on NF membranes in numerous fields has significantly increased over the previous six years. A brief scan of the Scopus database revealed that 1642 papers on NF membranes have been published overall from 2008 to the present.

The development of membrane purification of active pharmaceutical ingredients (API) from common classes of genotoxic impurities (GTIs) has received a lot of attention lately (Székely et al. 2012). Martínez et al. (2012) investigated the potential of commercial NF270 (FilmTec, Dow) and homemade NF membranes with 25% and 27% PES for the cross-flow recovery of API, 1-(5-bromo-fur-2-il)-2-bromo-2-nitroethane (or G-1). Cross-flow NF has been used successfully, obtaining retentions of more than 80% for all of the examined membranes.

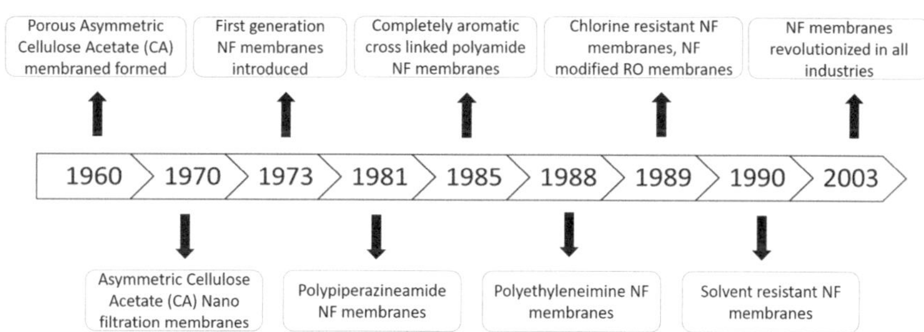

Fig. 9.1 Depiction of nanofilter technological evolution

Mohammad et al. (2015) published a review focusing on various application of NF in various sectors in which they have reported that NF looks to be a significant alternative to traditional methods of food processing, including those used in the production of beverages, dairy products, sugar, and vegetable oils. Recent studies show the use of membrane technology to separate chemicals like antioxidants and deacidification that are present in tiny concentrations in vegetable oils (Tres et al. 2010; Firman et al. 2013).

Today, nanofiltration membranes are produced in spiral wound, plate and frame, tubular, pillow-shaped, capillary, hollow fiber configurations (Giwa and Ogunribido 2012). From this, generally, two types of membranes can be used flat-sheet membrane and hollow fiber membrane (Du et al. 2022). At present, only flat membranes are in manufacturing process using interfacial polymerization strategy because existing composite production of membrane is only liable to flat membrane (Emonds et al. 2020). In future, flat-sheet membrane can be replaced by hollow fiber membrane due to its high surface area and filling density (Jye and Ismail 2016).

Nanofiltration membranes are made up of a variety of materials. Commonly used reacting monomers are trimethyl chloride, diaminobenzidine, phenylenediamine, piperazine, and acyl chloride (Saha and Joshi 2009), some polymers such as polyimide, polyacrylonitrile, polybenzimidazole, polyetherimide, polysulfone, polyethersulfone, polyvinylidene fluoride (Misdan et al. 2014), and some green organic materials like sodium alginate, tannic acid, bentonite clay (Yadav et al. 2022). Some inorganic materials modify organic membranes like mixed-matrix membranes (Aburabie et al. 2019). Inorganic membranes have high mechanical strength and long-term chemical and thermal stability (Yang et al. 2019), but the drawback is high cost and fragility (Mahalingam et al. 2019). Yang et al. (2019) reported that inorganic membranes mostly have metal oxide membranes (alumina, zirconia, titania) and carbon-based membranes (nanotubes and graphene).

9.2 Types of Nanomaterials Commonly Used in Nanofilters

Nanofiltration technology has emerged as a powerful tool for separating and purifying a wide range of mixtures. At the heart of this technology lie nanomaterials, the building blocks that endow nanofilters with their remarkable capabilities.

9.2.1 Nanoparticles: Tiny Titans with Big Potential

Nanoparticles, the workhorses of many nanofilters, are incredibly small particles with at least one dimension in the range of 1 to 100 nm (nm) (Bhuiyan et al. 2015). This minute size translates to a colossal surface area-to-volume ratio. Imagine a tennis ball shrunk down to the size of a marble—that is the dramatic increase in surface area nanoparticles offer compared to their larger counterparts. This expansive surface area is the key to a nanoparticle's prowess in filtration. Contaminants can become adsorbed or trapped on the vast surface, effectively being removed from the filtrate (the purified liquid).

The type of nanoparticle also plays a crucial role. Carbon nanotubes, for instance, possess a unique tubular structure that can act like efficient channels for selective permeation. Graphene, a single layer of carbon atoms arranged in a honeycomb lattice, offers exceptional strength and a highly customizable surface, making it a versatile candidate for nanofilters. Metal oxides, like titanium dioxide (TiO_2), zinc oxide (ZnO), and iron oxide (Fe_3O_4), bring their own set of advantages (Speranza 2021). These oxides often exhibit photocatalytic activity, which harnesses light to degrade organic contaminants. Silver nanoparticles, on the other hand, are renowned for their antimicrobial properties, making them ideal for applications requiring elimination of bacteria and other microbes (Bruna et al. 2021).

9.2.2 Nanofibers: Weaving a Web for Selective Separation

Nanofibers, another essential component of nanofilters, are essentially very thin fibers with diameters in the nanometer range. Polymeric nanofibers, constructed from polymers like polyacrylonitrile, polyvinyl alcohol, and polysulfone, are particularly attractive for filtration applications. These nanofibers boast an impressive porosity, meaning a significant volume within the fibers is filled with air or voids. This porosity allows the filtrate to pass through while capturing contaminants on the fiber surface or within the intricate pore network (Huang et al. 2003).

The interconnected nature of pores in nanofibers is another critical feature. Imagine a sponge with a well-connected network of channels—that is analogous to the pore structure of nanofibers. This interconnectedness ensures efficient flow of the filtrate while minimizing dead zones where contaminants might escape capture. Additionally, the size distribution of pores can be precisely controlled during nanofiber fabrication. This allows for tailoring the filter to selectively remove contaminants of specific sizes, achieving a superior level of separation (Xue et al. 2019).

9.2.3 Nanocomposites: The Combination of Strength and Performance

Nanocomposites offer the best of both worlds, combining the exceptional properties of nanoparticles and nanofibers. These materials are meticulously crafted by embedding nanoparticles or nanofibers within a polymeric matrix. This synergistic approach yields a composite material with enhanced mechanical strength, improved thermal stability, and, most importantly, superior filtration performance compared to its individual constituents (Rashid et al. 2024). The polymeric matrix in a nanocomposite acts as a supportive scaffold, providing structural integrity and preventing the nanoparticles or nanofibers from agglomerating. This translates to a more robust and durable nanofilter. Furthermore, the incorporation of nanoparticles can imbue the nanocomposite with functionalities not inherent to the polymer alone.

9.3 Property and Characteristics of Nanomaterials Commonly Used in Nanofilters

The Symphony of Properties: A Delicate Balance for Peak Performance. The properties and characteristics of nanomaterials in a nanofilter play a delicate yet crucial role in determining its effectiveness. Here is a closer look at how some key properties influence filtration performance:

Nanoparticle Size: As mentioned earlier, the miniscule size of nanoparticles translates to a high surface area, which is directly linked to the filter's adsorption capacity. The larger the surface area, the more contaminants can be adsorbed onto the nanoparticles (Altammar 2023).

Nanoparticle Shape: The shape of nanoparticles and nanofibers significantly impacts how densely they pack together within the filter. Spherical nanoparticles, for example, tend to pack more densely, potentially leading to smaller pores and increased resistance to flow. Conversely, rod-shaped nanoparticles or ribbon-like nanofibers may leave behind larger voids, creating a more open pore structure that facilitates better flow (Hoshyar et al. 2016).

Surface Chemistry: The chemical nature of a nanomaterial's surface, including its charge, functional groups, and wettability (hydrophilicity or hydrophobicity), governs how it interacts with contaminants. For instance, a positively charged surface might attract negatively charged contaminants, enhancing adsorption. Similarly, hydrophilic (water-loving) surfaces may be more effective in capturing polar contaminants, while hydrophobic (water-repelling) surfaces might be better suited for nonpolar contaminants (Mu et al. 2014).

Porosity and Pore Size Distribution: The porosity of a nanofilter, essentially the percentage of void space within the material, dictates the overall volume of filtrate that can pass through. A higher porosity allows for faster filtration rates. However, it is crucial to strike a balance. Excessive porosity can compromise the filter's ability to retain smaller contaminants. This is where pore size distribution comes into play. A well-defined distribution with a range of pore sizes ensures efficient capture of a wider spectrum of contaminants while maintaining a reasonable flow rate (Abdel-Fatah 2018).

The exploration of novel nanomaterials for nanofiltration is a continuous journey. Researchers are constantly striving to develop materials with even higher surface areas, more precisely tailored pore structures, and functionalities that address specific filtration challenges. The integration of advanced materials like metal–organic frameworks (MOFs) and covalent organic frameworks (COFs) holds immense promise for highly selective and efficient nanofiltration. MOFs and COFs possess highly ordered structures with tunable pore sizes and functionalities, making them ideal candidates for capturing specific contaminants (Ji et al. 2017).

As research delves deeper into the fascinating realm of nanomaterials, the capabilities of nanofiltration technology will undoubtedly continue to expand. With its potential for high-performance separation and purification, nanofiltration holds the key to addressing a multitude of challenges, from clean water production and wastewater treatment to food and beverage processing and pharmaceutical manufacturing. The future of nanofiltration is undeniably bright, driven by the ingenuity of material science and the relentless pursuit of cleaner and more sustainable solutions for our world (Joseph et al. 2023).

Nanomaterials commonly used in nanofilters include nanoparticles, nanofibers, and nanocomposites. Nanoparticles, such as carbon nanotubes, graphene, metal oxides (e.g., TiO_2, ZnO, Fe_3O_4), and silver nanoparticles, are widely employed in nanofilters due to their high surface area, unique physicochemical properties, and potential for surface functionalization. Nanofibers, especially polymeric nanofibers (e.g., polyacrylonitrile, polyvinyl alcohol, polysulfone), have been explored for their high porosity, interconnected pore structure, and tunable pore size distribution (Naseem and Durrani 2021). Nanocomposites, which combine nanoparticles or nanofibers with polymeric matrices, offer enhanced mechanical strength, thermal stability, and improved filtration performance compared to their individual counterparts (Fu et al. 2019).

The properties and characteristics of nanomaterials play a crucial role in their filtration performance. The size of nanoparticles, typically ranging from 1 to 100 nm, contributes to their high surface area-to-volume ratio, enhancing adsorption and catalytic properties. The shape of nanoparticles (e.g., spherical, tubular, irregular) and nanofibers (e.g., cylindrical, ribbon-like) influences their packing density and porosity in nanofilters. Surface chemistry, including surface charge, functional groups, and hydrophilicity/hydrophobicity, governs the interactions between nanomaterials and target contaminants, affecting adsorption and rejection capabilities. Porosity, pore size distribution, and interconnectivity of pores in nanofibers and nanocomposites determine the sieving and size-exclusion properties of nanofilters (Altammar 2023).

9.4 Various Methods for Fabricating Nanofilters

Nanofilters can be fabricated using a variety of techniques, each offering distinct advantages and limitations (Table 9.1). Here are some of the most common methods:

Table 9.1 Advantages and limitations of each fabrication technique

Technique	Advantages	Limitations
Electrospinning	High control over pore size and morphology, versatile material selection	Requires specialized equipment, limitations on scalability
Chemical vapor deposition (CVD)	Precise control over film thickness and composition, suitable for high-performance filters	Complex and expensive process, requires specialized equipment
Self-assembly	Potentially low-cost and scalable, simple process	Challenging to control, difficulty achieving precise pore size and uniformity

Source Amadi et al. (2022)

9.4.1 Electrospinning

This technique utilizes an electric field to draw a polymer solution into thin fibers, creating a nanofiber mesh that can function as a nanofilter. Electrospinning offers high control over pore size and morphology but may require specialized equipment and have limitations on scalability for large-scale production (Xue et al. 2019).

9.4.2 Chemical Vapor Deposition (CVD)

CVD involves depositing a thin film of material onto a substrate through a chemical reaction in the gas phase. This method allows for precise control over film thickness and composition, making it suitable for creating high-performance nanofilters. However, CVD can be a complex and expensive process requiring specialized equipment (Khlyustova et al. 2020).

9.4.3 Self-assembly

This approach utilizes the inherent properties of materials to organize themselves into specific structures at the nanoscale. Self-assembly offers a potentially low-cost and scalable approach for nanofilter fabrication. However, controlling the self-assembly process can be challenging, and achieving precise pore size and uniformity may be difficult.

9.5 Unveiling the Secrets: How Nanofilters Work Their Magic

Nanofilters achieve their remarkable purification prowess through a fascinating interplay of physical and chemical mechanisms. Let us delve into the scientific principles governing these mechanisms to understand how nanofilters function at the nanoscale.

9.5.1 Filtration Mechanisms: A Multifaceted Approach

Nanofilters do not rely on a single mechanism to capture contaminants. Instead, they employ a combination of techniques, each playing a crucial role in the overall filtration process. Here is a closer look at some of the most prominent mechanisms:

9.5.2 Sieving

This mechanism acts like a size-exclusion process. Imagine a colander separating large pasta pieces from smaller rice grains. In nanofiltration, the intricate pore network within the nanofilter acts as the colander. Contaminants larger than the pores are physically blocked from passing through, while the desired filtrate (the smaller component in the mixture) permeates the filter. The size distribution of pores in nanofibers and nanocomposites can be precisely controlled to achieve exceptional sieving efficiency for contaminants of specific sizes.

9.5.3 Electrostatic Attraction

Opposites attract, and this principle also applies in nanofiltration. The surface of some nanomaterials can be engineered to possess a specific electrical charge. This charge can attract contaminants with an opposite charge, causing them to adhere to the filter surface and be removed from the filtrate. For instance, positively charged nanofilters can effectively capture negatively charged pollutants like bacteria or certain heavy metal ions.

9.5.4 *Adsorption*

This mechanism leverages the high surface area of nanomaterials. As discussed earlier, nanoparticles possess a vast surface area due to their minute size. This extensive surface area acts like a magnet for contaminants, attracting and holding them onto the surface of the nanoparticles. The attractive forces between the contaminant and the nanomaterial can be due to various interactions, including van der Waals forces, hydrogen bonding, or even weak chemical interactions.

9.5.5 *Diffusion*

Diffusion is a natural phenomenon where molecules or particles tend to move from an area of high concentration to an area of low concentration. In nanofiltration, contaminants can diffuse through the pores of the nanofilter driven by this concentration gradient. However, the size-selective nature of the pores and the presence of other filtration mechanisms often limit the role of diffusion in contaminant removal (Mautner 2020).

9.6 The Synergy of Mechanisms: A Recipe for Success

In essence, nanofiltration is an orchestrated interplay of various mechanisms. Sieving forms the primary barrier, physically excluding large contaminants. Electrostatic attraction and adsorption come into play to capture contaminants that might squeeze through the pores due to their size. Diffusion, while not the dominant mechanism, can also contribute to the overall removal process. By understanding the theoretical principles governing these mechanisms, scientists and engineers can design nanofilters with tailored properties to achieve optimal performance for specific filtration challenges.

The combined effect of these mechanisms allows nanofilters to achieve exceptional separation efficiency for a wide range of contaminants, making them a valuable tool for various purification applications. As research continues to explore novel nanomaterials and refine filtration techniques, the potential of nanofiltration technology will undoubtedly continue to grow, shaping a cleaner and more sustainable future (Siddique et al. 2020).

9.7 Theoretical Principles at the Nanoscale

The effectiveness of nanofilters hinges on a few key theoretical principles that govern the filtration process at the nanoscale. Here are some of the most important concepts:

1. The Boltzmann Distribution: This principle describes the statistical distribution of particles based on their energy. In nanofiltration, smaller particles with higher thermal energy have a greater chance of squeezing through the nanopores compared to larger particles with lower thermal energy. This translates to a size-dependent separation process.
2. Hydrodynamic Interactions: As the filtrate flows through the nanofilter, it interacts with the nanomaterial surfaces. These interactions can create various forces, including drag forces and steric hindrance. Drag forces can slow down the movement of larger contaminants, further enhancing the sieving effect. Steric hindrance arises when the contaminant molecule is too bulky to fit comfortably within the pore due to the presence of functional groups or bulky side chains on the nanomaterial surface.
3. Surface Potential: The surface potential of a nanomaterial refers to the electrical charge distribution at its surface. This potential can influence the electrostatic interactions between the nanomaterial and contaminants, impacting adsorption and repulsion phenomena within the filter (Ryzhkov and Minakov 2016).

9.8 Gauging the Power: Evaluating Nanofilter Performance

Nanofilters hold immense promise for purification, but how do we assess their effectiveness? Scientists and engineers rely on a variety of methods to evaluate the performance of nanofilters. Here is a look at some key metrics:

9.8.1 Filtration Efficiency

This is the most fundamental measure of a nanofilter's success. It quantifies the filter's ability to remove target contaminants from the feed solution. Efficiency is typically expressed as a percentage and can be determined for specific contaminants or for a broader range of components. Different techniques, like measuring the concentration of contaminants in the filtrate compared to the feed, are used to calculate this value (Singh et al. 2020).

9.8.2 Rejection Rate

Closely linked to filtration efficiency, the rejection rate refers to the percentage of a specific contaminant that is retained by the filter. It provides a more granular understanding of the filter's performance for individual contaminants. For instance, a nanofilter might exhibit a 99% rejection rate for bacteria but a lower rejection rate for smaller organic molecules.

9.8.3 Permeability

This metric reflects the ease with which the desired filtrate flows through the nanofilter. It is typically expressed as the volume of filtrate permeating the filter per unit time per unit of applied pressure. A higher permeability translates to faster filtration rates, but it is crucial to strike a balance. Excessive permeability can compromise filtration efficiency if it allows some contaminants to pass through (Singh et al. 2020).

9.8.4 Pressure Drop

As the filtrate flows through the nanofilter, it encounters resistance due to the presence of pores and the nanomaterial itself. This resistance translates to a pressure drop across the filter. A higher pressure drop signifies greater resistance and can lead to increased energy consumption during operation. Ideally, nanofilters should achieve high filtration efficiency and permeability while maintaining a low pressure drop.

9.8.5 Durability

A nanofilter's effectiveness diminishes over time due to factors like fouling (accumulation of contaminants on the filter surface) and degradation of the nanomaterial. Evaluating the durability involves testing the filter's performance over extended periods and assessing how its efficiency, permeability, and pressure drop change. Understanding the long-term stability of a nanofilter is crucial for determining its cost-effectiveness and maintenance requirements (Heris et al. 2023).

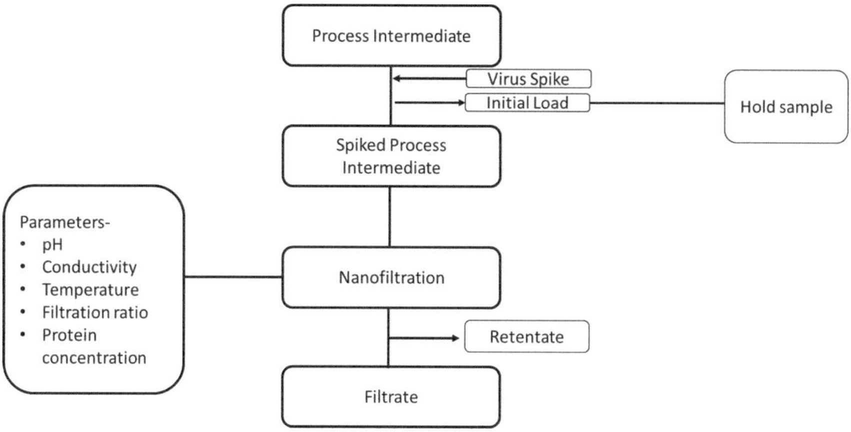

Fig. 9.2 Process of virus filtration with the help of nanofiltration membrane

9.9 Synthesis and Production of Nanofilter

Nanofiltration is a new class of membrane filtration technology, which possess properties in between both reverse osmosis (RO) membranes and ultrafiltration (UF) membranes. In this filtration process, the mechanism for separation involves both electrical effect and steric effect (Bowen and Mohammad 1998). To date, nanofiltration has emerged as great energy-saving and eco-friendly technology with properties such as pore size < 2 nm, molecular weight cutoff ranges from 100 to 2000 Da, and having charged surface (Cao et al. 2021). Along with these properties, these nanofiltration membranes also have charge due to dissociation of surface functional groups or because of adsorption of charged solutes (Oatley-Radcliffe et al. 2017). It is a process in which with the help of membranes having pore size in nanometers is used to filter protein solutions. This filtration method is exclusively dedicated for the separation of pathogen from any biological or other products intend to human use (Manabe 1996). Nanofilters are used for such separation. There are various companies in market which provide these nanofilters such as Asahi Kasei Corporation, Millipore Corporation, Sartorius AG, and Pall Corporation having different pore sizes and membrane materials (for nanofilter construction) depending upon the requirement (Caballero et al. 2014). Figure 9.2 shows the simple process of virus filtration with the help of nanofiltration membrane.

9.10 Methods for Nanofilter Membrane Preparation

There are various methods for the preparation of NF membranes such as phase inversion, layer by layer, and interfacial polymerization.

9.10.1 Phase Inversion

This has been widely used methodology for the preparation of NF membranes. There is separation of homogenous liquid polymer in this process by which membranes are created and then transformed into solid form. Mainly, this methodology involves either NIPS (non-solvent-induced phase separation) or TIPS (thermally induced phase separation) technologies (Nguyen Thi et al. 2020). Its main part is polymer transformation from liquid to solid. Immersion precipitation has most simple operation that is why it is mostly used in phase inversion. During this process polymer, solution is scraped on the support, immersed in a coagulation bath containing a non-solvent, and precipitated due to the exchange of the solvent and the non-solvent. As the entire process completes, either mass transfer or phase separation determines the structure of final polymer. Depending upon the required membrane properties, it can be modified and post-treated with solution (Yuchen et al. 2022). The chemical cross-linking of polymer chains improves the performance of the membrane by avoiding the reduction of membrane flux due to the compression of polymer chains or the swelling of the polymer matrix under the action of organic solvents (Yuchen et al. 2022).

9.10.2 Layer by Layer

It is an alternating method of electrostatic adsorption of cation and anion polyelectrolytes (PEs) on the surface of a UF membrane (Jye and Ismail 2016). During this whole preparative process of NF membrane, charged porous UF substrate is alternately immersed in a solution containing both cationic and anionic PEs, which reverse the surface charge, and simultaneous deposition can form a layered composite, the so-called self-assembly PE complex. Being advantageous over other methods, it too has some setback in engineering application, that is, it takes a lot of time to fabricate multilayer films (Yuchen et al. 2022).

9.10.3 Interfacial Polymerization

Interfacial polymerization is another way to prepare NF membrane. Using this method, ultra-thin selective PA layer having thickness in hundreds of nanometers on the microporous support membrane surface can be prepared. Hence, TFC membrane with good water flux and solute retention can be formed (Jye and Ismail 2016). The essence of interfacial polymerization method is to control the polymerization process to form a polymer network on the porous substrate at the interface of two immiscible phases (Kyriakou et al. 2020). Duong et al. used this methodology and polyetherimide (PEI) as the support layer to synthesize zwitterionic copolymer (ZCP) membrane.

9.11 Types of Nanofilter Membrane

Using the abovementioned methods, many researchers have polymerized NF membranes for several nanofilters to be used in different industrial sectors such as regenerated cellulose membrane, PVDF membranes, and polyethersulfone membrane. These membranes are discussed below.

9.11.1 Regenerated Cellulose Membrane

Commercially available regenerated cellulose fibers are also known as viscose rayon, lyocell fiber, and cuprammonium rayon. Non-solvent such as ketones can be used for the liquid–liquid phase separation in solution of cuprammonium cellulose (Ide 2022). Based on this information, porous regenerated cellulose membranes were made in a flat-sheet membrane configuration with mean pore diameter of 10–100 nm. To enable virus removal from any protein solution using membrane filtration, Asahi Kasei commercialized world's first hollow fiber membrane filter (Planova) which was made up of regenerated cuprammonium cellulose (Ide 2022). Kamide and Manabe observed that for membrane formation mechanism in phase separation methodology induced by non-solvents "particle growth concept" is quite important (Kamide and Manabe 1985). In the same study, they observed growth of primary and secondary particles during phase separation procedure during which cuprammonium cellulose solution was coagulated with ammonia, acetone, and aqueous solutions.

9.11.2 Polyethersulfone Membrane

Polyethersulfone (PES) is renowned as being one of the most significant polymeric materials used in the production of polymeric membranes. These membranes possess outstanding thermal, hydrolytic, and oxidative stability with excellent mechanical property. It is inherently hydrophobic and extremely prone to fouling. Various studies have been conducted to enhance the antifouling property and to improve surface hydrophilicity. It could be achieved through surface deposition, graft polymerization, and loading of inorganic nanofillers like oxides of aluminum, iron, silicon, titanium, and graphene into the membrane matrix (Zareei and Hosseini 2019). This PES has been extensively employed for wastewater treatment and treatment of drinking water.

9.11.3 Polyvinylidene Fluoride Membrane

Poly(vinylidene fluoride) (PVDF) has gained extensive attention as a membrane preparation material for its exceptional properties including superior hydrophobicity, thermal stability, high mechanical strength, and chemical resistance. PVDF membranes are broadly utilized in ultrafiltration and microfiltration membranes and later utilized in nanofiltration membranes (Oshima et al. 1996). Polypropylene and polytetrafluoroethylene membranes are prepared using the phase inversion process that have constraint in solvent choice, while PVDF membrane is comparatively more hydrophobic that gives broader utilization for membrane distillation and membrane contactor (Liu et al. 2011).

9.12 Patent Scenario on Nanofilter

Evaluating nanofilter performance is not a one-dimensional endeavor. A comprehensive assessment considers all these metrics in conjunction with the specific application. For instance, a nanofilter designed for desalination might prioritize high rejection rates for salt ions, even if it results in a slightly lower permeability. Conversely, a filter for food and beverage processing might emphasize high permeability to ensure efficient processing while maintaining an acceptable level of contaminant removal.

Nanofiltration is used in many interesting applications especially in water and wastewater treatment for water softening and removal of sulfate, micropollutant, and dissolved organic matter. Other applications include biotech or pharmaceuticals industries for antibody production and blood processing; in textile and dye industries for dye desalting, dye penetrate removal, dye concentration, optical brightening agent concentration, and desalination; in food industry for concentration and demineralization of lactose in dairy products and to produce beverages and fruit juices in beverage sector.

According to the research that has been done thus far, NF membranes continue to find new uses in a variety of industries, particularly in the treatment of water and wastewater, desalination of brackish water, pretreatment in seawater desalination, and separation of small organics from other species in pharmaceutical, biotechnology, and food applications. The industrial application of NF has boosted a roar of safeguarding the research and invention in a form of patent.

The present review aimed to conduct a comprehensive statistical analysis and classification of a dataset consisting of nanofilter patent applications. The dataset is obtained from the "Patentscope" database (https://www.wipo.int/patentscope). This database provides access to published patent applications in full text from participating national and regional offices. When Patentscope is searched on June 27, 2023 on nanofilter using following keywords "nanofilter*" OR "nano-filter*" OR "nano filter*," a total of 1427 applications are identified. The dataset is treated with inclusion and exclusion criteria (Fig. 9.3), and final dataset found consists of 1314 patent

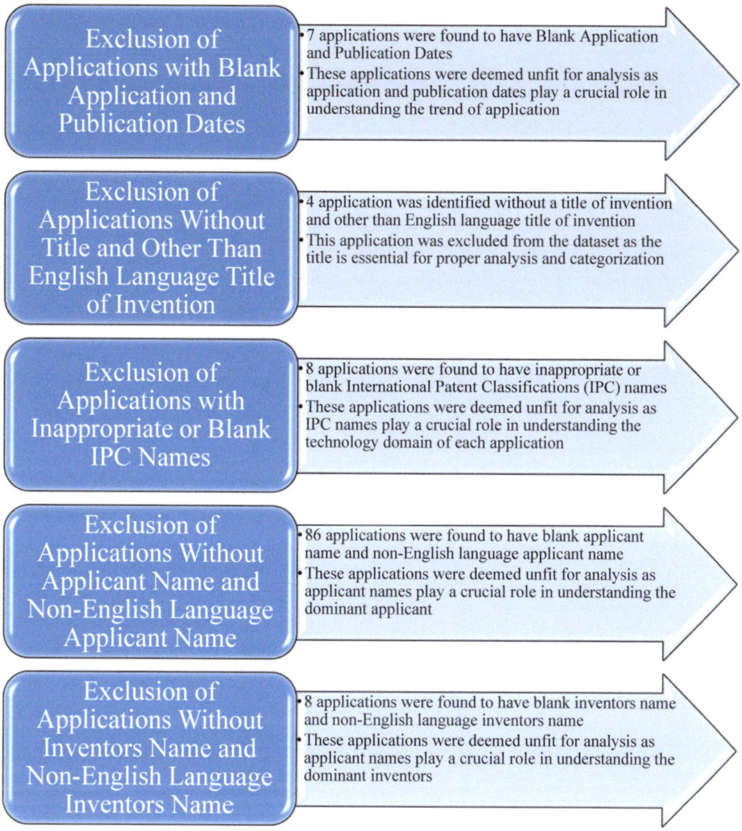

Fig. 9.3 Flowchart on treatment of dataset with inclusion and exclusion criteria to derive a final dataset of patent applications on nanofilter

applications. First title expression tree is derived from title on patent application on nanofilter (Fig. 9.4). This expression tree helps us to identify the multilevel usage of nanofilter in various applications like membrane, 3D, water, and instrument.

9.12.1 *Year-Wise Trends in Nanofilter Patent Applications*

The analysis of the patent applications related to nanofilters (Fig. 9.5) reveals interesting trends across last three decades. Though the usage and application of nanofilter is observed in 1960s (mentioned earlier in this review), the patent application is recorded first in 1991 from Canada, Patent No. 2038485 (Hadden et al. 1991). This was on "Nanofiltration process for making dextrose." It is observed that in the 1990s, the number of applications is relatively low, ranging from 1 to 23 per year. The highest count was observed in 1998 with 18 applications. The following decade, the 2000s,

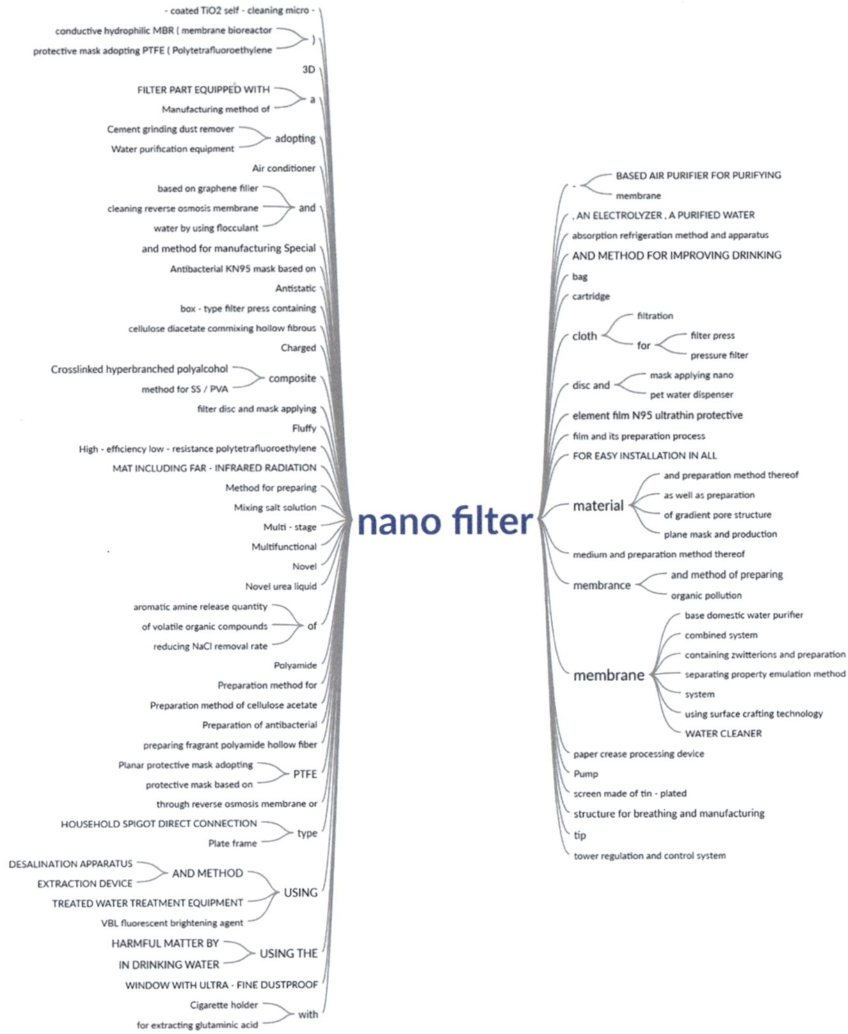

Fig. 9.4 Title expression tree on title on patent application on nanofilter

witnessed a notable increase in interest, with the count ranging from 23 to 59 applications per year. The peak was seen in 2005 with 59 applications. This upward trend continued into the 2010s, where the number of applications kept rising. In 2019, a peak of 104 applications is recorded, indicating sustained interest in nanofilter innovations. Throughout the 2010s, the counts ranged from 29 to 104 per year, with fluctuations but an overall positive trend. The 2020s began with a significant surge, with 119 applications in 2020, the highest count in the dataset. The following years, 2021 and 2022, maintained relatively high counts at 93 and 37 applications,

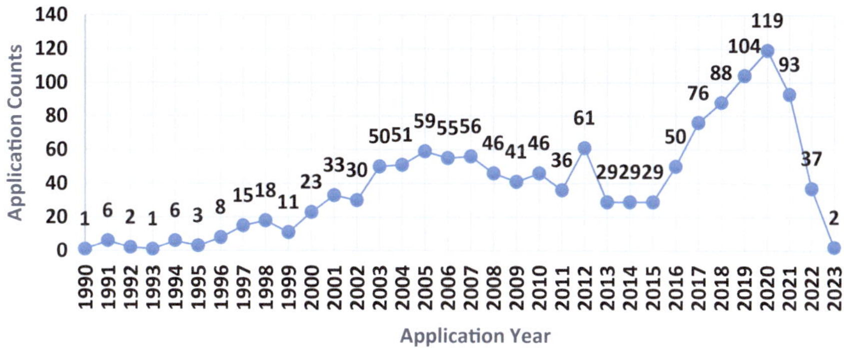

Fig. 9.5 Year-wise patent applications on nanofilters

respectively. However, it is important to note that the count for 2023, with only two applications, may not be representative due to incomplete data.

Comparing the trends, it is evident that the number of nanofilter patent applications has increased over time. Descriptive statistics reveal that the average number of applications across all years is 44.72, with a median of 41. The minimum count of 1 was observed in 1990s, while the maximum count of 119 occurred in 2020s. This indicates a wide range of variation in application counts. The standard deviation of 29.67 demonstrates a significant dispersion around the mean.

9.12.2 Geographical Distribution of Nanofilter Patent Applications

The geographical distribution of nanofilter patent applications reveals interesting insights into the global landscape of innovation. It is studied into two spectrums as distributed by Patentscope database: broad regions (Fig. 9.6) and country-wise (Fig. 9.7).

Among the broad regions, the majority of applications were concentrated in the "Asia–Europe" region, with a frequency of 1052, accounting for 80.1% of the total. This indicates a significant focus on nanofilter technologies in this region, possibly due to the presence of research institutions, technological advancements, and market opportunities.

The "Americas" region also demonstrated a considerable number of patent applications, with a frequency of 137, representing 10.4% of the total. This suggests a noteworthy level of innovation and interest in nanofilters within this region, which could be attributed to the presence of established industries, research centers, and technological advancements.

The "PCT" (Patent Cooperation Treaty) category is application at multinational phase processed by WIPO which is accounted for 106 applications, making up

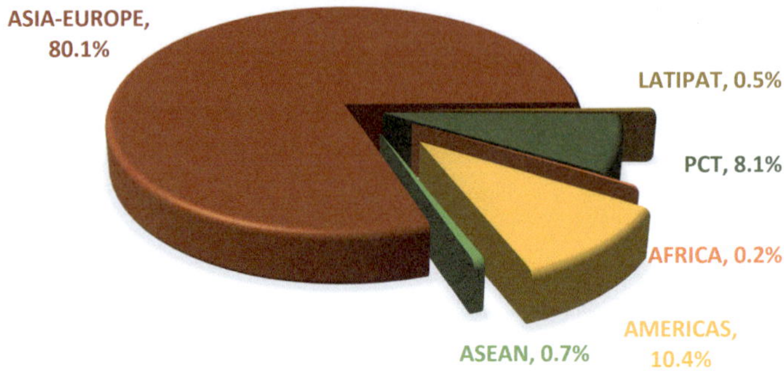

Fig. 9.6 Distribution of nanofilter patent applications by broad region

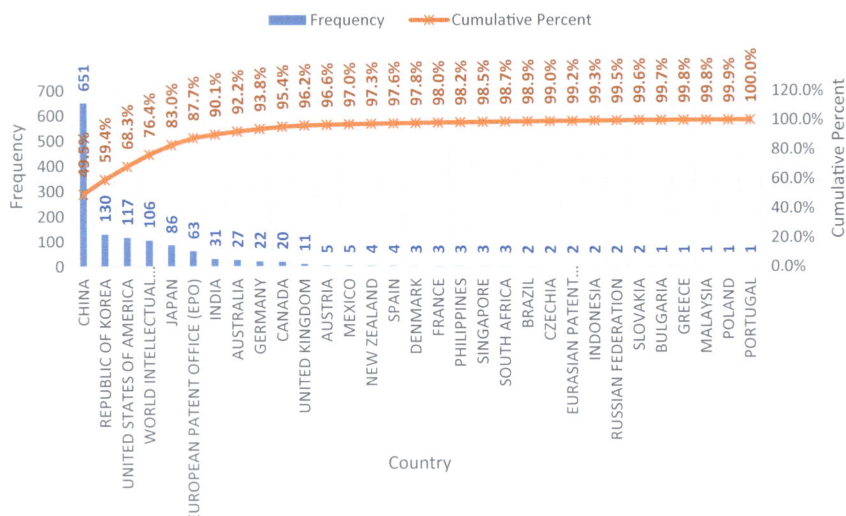

Fig. 9.7 Country-wise distribution of nanofilter patent applications

8.1% of the total. It indicates global perspective and potential interest in protecting nanofilter innovations on a broader scale.

Other regions, including “ASEAN” with nine applications (0.7%), “LATIPAT” with seven applications (0.5%), and “Africa” with three applications (0.2%), exhibited relatively lower frequencies. These regions may represent emerging markets or areas where nanofilter technologies are still in the early stages of development and adoption.

A country-wise analysis of nanofilter patent applications (Fig. 9.6) provides valuable insights into the global landscape of innovation and the involvement of different nations in this field. The analysis reveals the frequency and percentage distribution of patent applications across various countries.

China emerges as the leading contributor, with 651 patent applications, accounting for 49.5% of the total. This indicates a significant focus on nanofilter technologies within China and highlights its strong presence in this domain. Following China, the Republic of Korea demonstrates notable involvement with 130 patent applications, representing 9.9% of the total. These two Asian countries, China and the Republic of Korea, together account 59.4% hold on patent potential at global level.

The United States of America ranks third with 117 patent applications, comprising 8.9% of the total. This showcases the active participation and innovation in nanofilters within the United States. The World Intellectual Property Organization (WIPO) means PCT application as described earlier accounts for 8.1% of the total.

Other notable contributors include Japan with 86 patent applications (6.5%), the European Patent Office (EPO) with 63 patent applications (4.8%), and India with 31 patent applications (2.4%). These countries showcase their commitment to nanofilter research and development.

Several other countries exhibit a relatively lower number of patent applications, indicating varying levels of involvement. Australia and Germany contribute 27 and 22 patent applications, respectively, representing 2.1% and 1.7% of the total. Canada, the United Kingdom, and Austria demonstrate moderate engagement with 20, 11, and five patent applications, respectively. It is earlier observed that first patent application was from Canada but since then total number of patent application is very less.

The cumulative percentage provides insights into the concentration of patent applications across the contributing countries. The top five countries include China, the Republic of Korea, the United States of America, Japan, and India.

9.12.3 International Patent Classifications (IPC) Versus Global Patent Application Trend

The International Patent Classifications (IPC) provide insights into the technology domains covered by nanofilter patent applications. IPC-based classification may classify one patent into one or more than one IPC subclass code. Table 9.2 is prepared which shows the frequency and percentage of top 15 IPC subclass codes applied to patent application on nanofilter.

Among the top 15 IPC codes, B01D emerged as the most frequent, representing 521 responses, accounting for 36.8% of the total. This highlights the prevalence of separation and filtration processes in nanofilters. C02F followed closely with 352 responses, constituting 24.9% of the total, indicating a significant presence of water treatment and purification technologies. Additionally, A61K (72 responses, 5.1%) and F24F (30 responses, 2.1%) represented pharmaceutical compositions and air conditioning systems incorporating nanofilters, respectively. Other notable IPC codes included A23L (57 responses, 4.0%) for food compositions, C07K (43 responses, 3.0%) for peptides and proteins, and B01J (52 responses, 3.7%) for chemical or physical processes. These findings provide a valuable insight into the technological

Table 9.2 Frequencies and percentages of top 15 IPC subclass codes applied to patent application on nanofilter

Top 15 IPC subclass codes	N	Percent (%)
B01D (technologies related to separation and filtration processes)	521	36.80
C02F (technologies related to water treatment and purification)	352	24.90
A61K (technologies related to pharmaceutical compositions and drug delivery systems)	72	5.10
F24F (technologies related to air conditioning and ventilation systems)	30	2.10
A23L (technologies related to food compositions and additives)	57	4.00
C07K (technologies related to peptides and proteins)	43	3.00
B01J (technologies related to chemical or physical processes)	52	3.70
C07C (technologies related to organic compounds)	41	2.90
A61P (technologies related to therapeutic or diagnostic agents)	40	2.80
C12P (technologies related to biotechnological processes)	51	3.60
C01D (technologies related to inorganic compounds)	36	2.50
A41D (technologies related to personal protection equipment, such as clothing or helmets)	26	1.80
G01N (technologies related to investigative or analytical devices)	33	2.30
C13B (technologies related to macromolecular compounds)	27	1.90
C07H (technologies related to sugars, carbohydrates, or derivatives)	35	2.50
Total	1416	100.00

domains covered by the IPC subclass codes, revealing prominent areas of innovation and research within the field of nanofilters.

The minimum IPC count observed is 1, indicating that patent applications focus on a single technology domain related to nanofilters. On the other hand, the maximum IPC count is 23, suggesting certain applications cover a wide range of technology domains within the field of nanofilters (Table 9.3). Patent application no. US41583352 has one IPC count, i.e., B01D while patent application no. US38308491 has 23 IPC subclass count. Out of the 1314 patent applications included in the study, the average number of IPCs per application is 3.66, with a median value of 3. This indicates that, on average, each patent application is associated with minimum three IPC subclasses, which reflects the multiple application of nanofilter.

Table 9.3 Distribution of International Patent Classifications (IPC) applied to patent application on nanofilter

Descriptive statistics	Value
N	1314
Mean	3.66
Median	3
Minimum	1
Maximum	23
Sum	4803

In total, the sum of IPC counts across all patent applications amounts to 4803. It is needful to test statistical significance of IPC count among patent applications (Table 9.4). Here, a Kruskal–Wallis test is conducted to test the hypothesis about significant differences in IPC counts among the broad regions (Table 9.5). The test statistic, *H*, is 64.623, with five degrees of freedom. The asymptotic significance value is reported as 0.000, indicating that there are significant differences in the IPC counts among the broad regions.

The results of the Kruskal–Wallis test suggest that the IPC counts vary significantly across different broad regions. This finding highlights the diverse technological areas and innovation trends within the nanofilter patent landscape, as represented by the broad regions. The statistical analysis provides evidence for the need to consider regional factors and influences when studying nanofilter technologies and their patent applications.

Table 9.4 The number of IPC subclass counts among broad region

Broad region	*N*
Africa	3
Americas	137
ASEAN	9
Asia–Europe	1052
LATIPAT	7
PCT (WIPO)	106
Total	1314

Table 9.5 Kruskal–Wallis test results for IPC counts and broad region

Test statistics[a,b]	IPC counts
Kruskal–Wallis *H*	64.623
df	5
Asymp. Sig.	0

[a]Kruskal–Wallis test
[b]Grouping variable: broad region

Table 9.6 provides an insightful overview of the distribution of top IPC subclass codes in different regions. In the Americas region, the IPC code B01D stands out as the most frequent, with a count of 74 occurrences, representing 5.2% of the total. This suggests a significant emphasis on processes related to separation and filtration within the Americas. Following closely behind is the IPC code C02F, with 29 instances and accounting for 2.0% of the total. This code is associated with technologies for water treatment, indicating a focus on addressing water-related challenges in the region.

Moving to the Asia–Europe region, it observes a similar trend. The IPC code B01D appears as the dominant code, with a substantial count of 396 occurrences, making up 28.0% of the total. This reaffirms the importance of separation and filtration technologies in this region. Additionally, the IPC code C02F is also significant, with 307 instances, representing 21.7% of the total. These numbers highlight the attention given to water treatment processes and technologies in the Asia–Europe region.

Among other regions, there are also notable IPC codes that emerge. For instance, the IPC codes A61K, A23L, B01J, C07C, A61P, C12P, C01D, G01N, and others make appearances, although with varying frequencies. Each of these codes signifies distinct technological areas of interest within their respective regions.

Table 9.7 displays the distribution of IPC counts, including the mean, median, minimum, maximum, and sum of IPCs, for each broad region. Among the analyzed regions, the Americas exhibit the highest average IPC count, with a mean of 6.03 and a median of 5 IPCs per application. This suggests that nanofilter patent applications

Table 9.6 The cross-tabulation of top IPC subclass codes vs broad region

Top 15 IPC codes	Broad region						Total
	Africa	Americas	ASEAN	Asia–Europe	LATIPAT	PCT	
B01D	1	74	2	396	1	47	521
C02F	0	29	0	307	1	15	352
A61K	0	9	0	55	0	8	72
F24F	0	0	0	29	0	1	30
A23L	0	3	0	48	0	6	57
C07K	0	8	0	28	0	7	43
B01J	0	12	2	31	0	7	52
C07C	0	5	0	33	0	3	41
A61P	0	3	0	33	0	4	40
C12P	0	8	0	39	0	4	51
C01D	0	5	2	28	0	1	36
A41D	0	1	0	24	0	1	26
G01N	0	7	0	20	1	5	33
C13B	0	11	0	13	0	3	27
C07H	0	3	0	30	0	2	35
Total	1	178	6	1114	3	114	1416

Table 9.7 Distribution of IPC by broad region and country in patent applications on nanofilter

Broad region	Country full	IPC counts					
		N	Mean	Median	Minimum	Maximum	Sum
Africa	South Africa	3	1.33	1	1	2	4
	Total	3	1.33	1	1	2	4
Americas	Canada	20	3.4	3	1	8	68
	United States of America	117	6.48	6	1	23	758
	Total	137	6.03	5	1	23	826
Asian	Indonesia	2	1.5	1.5	1	2	3
	Malaysia	1	3	3	3	3	3
	Philippines	3	3	3	1	5	9
	Singapore	3	1.67	1	1	3	5
	Total	9	2.22	2	1	5	20
Asia–Europe	Australia	27	4.7	3	1	23	127
	Austria	5	7.2	8	4	9	36
	Bulgaria	1	1	1	1	1	1
	China	651	3.05	2	1	14	1986
	Czechia	2	1.5	1.5	1	2	3
	Denmark	3	4.67	4	3	7	14
	Eurasian Patent Organization (EAPO)	2	2	2	2	2	4
	European Patent Office (EPO)	63	4.17	3	1	11	263
	France	3	5	4	4	7	15
	Germany	22	2.95	3	1	6	65
	Greece	1	3	3	3	3	3

(continued)

Table 9.7 (continued)

Broad region	Country full	IPC counts					
		N	Mean	Median	Minimum	Maximum	Sum
	India	31	2.61	2	1	5	81
	Japan	86	6.31	5	1	23	543
	New Zealand	4	2.75	3	2	3	11
	Poland	1	2	2	2	2	2
	Portugal	1	2	2	2	2	2
	Republic of Korea	130	2.65	2	1	11	345
	Russian Federation	2	2.5	2.5	2	3	5
	Slovakia	2	2	2	1	3	4
	Spain	4	3.25	2.5	2	6	13
	United Kingdom	11	6.45	6	1	20	71
	Total	1052	3.42	3	1	23	3594
LATIPAT	Brazil	2	6	6	3	9	12
	Mexico	5	1	1	1	1	5
	Total	7	2.43	1	1	9	17
PCT	World Intellectual Property Organization (WIPO)	106	3.23	3	1	16	342
	Total	106	3.23	3	1	16	342

in the Americas cover a relatively broader range of technology domains compared to other regions. The maximum IPC count observed in the Americas is 23, indicating significant diversity in the technological aspects addressed.

Asia–Europe follows closely with a mean IPC count of 3.42 and a median of 3. This region exhibits a balanced distribution of IPCs across patent applications, indicating a moderate level of diversity in the technological domains related to nanofilters. The Asian region demonstrates a mean IPC count of 2.22, reflecting a narrower focus on specific technology domains within nanofilters. The maximum IPC count in this region is 5. LATIPAT and the PCT regions exhibit mean IPC counts of 2.43 and 3.23, respectively. These regions encompass a moderate range of technology domains in their nanofilter patent applications.

Moving on to the distribution of IPCs within countries, the Americas stand out with a mean IPC count of 6.03. Within this region, the United States of America demonstrates the highest mean IPC count of 6.48, indicating a significant diversity of IPCs within nanofilter patent applications. On the other hand, China, despite having a lower mean IPC count of 3.05, exhibits the highest IPC count of 651, suggesting a substantial number of nanofilter patents originating from China with a relatively focused narrow range of IPCs per application.

In terms of specific IPC codes and their associated countries, some notable examples can be observed. Australia displays a strong presence in IPC code A61K, indicating a focus on pharmaceuticals and medical technologies. The United States of America showcases significant activity in IPC codes C02F and A61K, highlighting advancements in water treatment and pharmaceuticals, respectively. Moreover, China demonstrates a dominant position in multiple IPC codes, notably A23L, C07K, and A61P, signifying its contributions to food technology, biotechnology, and pharmaceutical research. Overall, the analysis of IPC codes and their distribution across regions provides insights into the technological and scientific landscape of different countries. These findings can contribute to the understanding of research and innovation trends, as well as inform strategic decisions in various industries.

9.12.4 *Top Applicants and Inventors in Patent Application on Nanofilter*

The key players in patent application on nanofilter (Tables 9.8 and 9.9) highlights the prominent applicants and inventors who have made substantial contributions to the field. NFT Nanofiltertechnik GMBH emerges as the leading applicant, with 56 patent applications, showcasing their strong commitment to advancing nanofilter technologies. Toray Ind Inc follows closely behind with 21 patent applications, demonstrating their active participation and dedication to the field.

Table 9.8 Top five applicant in patent application on nanofilter

Sr. No.	Top five applicants	*N*
1	NFT Nanofiltertechnik GMBH	56
2	Toray Ind Inc.	21
3	Suh, Hee Dong	13
4	Hw Process Tech Inc.	11
5	Toyobo Co Ltd.	10

Table 9.9 Top five inventors in patent application on nanofilter

Sr. No.	Top five inventors	*N*
1	Hofmann Wilfried	51
2	Suh, Hee Dong	13
3	Charkoudian John	7
4	Fang Hui	7
5	Jocham Simon	7

Suh Hee Dong is applicant as well as inventor who makes a significant impact, with 13 patent applications that highlight their expertise and interest in nanofilter innovations. As an inventor, Hofmann Wilfried stands as top contributor with 51 patent applications. Other notable inventors include Charkoudian John, Fang Hui, and Jocham Simon, each with seven patent applications.

These inventors have made valuable contributions to nanofilter research and development, showcasing their innovative ideas and expertise in the field. The presence of these top applicants and inventors highlights their role in shaping the nanofilter industry. Their expertise, dedication, and contributions are instrumental in driving advancements and pushing the boundaries of nanofilter technologies.

9.13 Applications of Nanofilter

While reverse osmosis and ultrafiltration are two well-known separation techniques, nanofiltration membranes are a relatively recent development that fills the gap between them. The versatility enables the creation of specialized process applications across many industries. The creation of nanofiltration membranes to supplement the existing extensive selection of ultrafiltration and microfiltration membranes is a logical move as filtration is mostly used in process applications. Figure 9.8 depicts

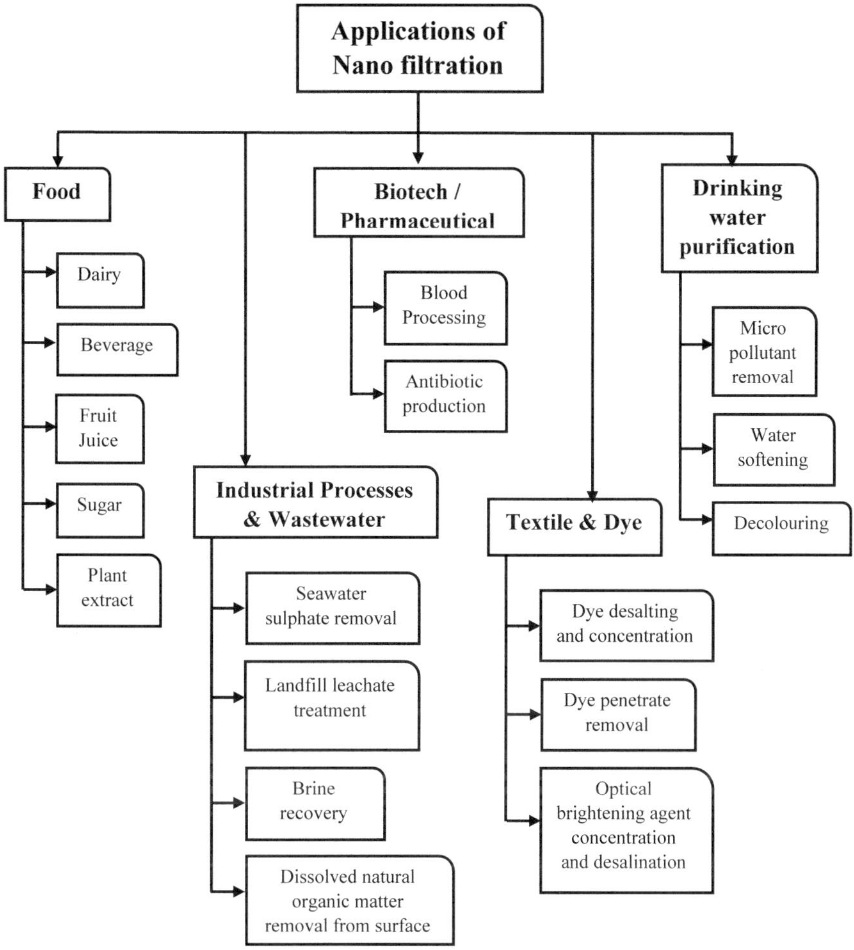

Fig. 9.8 Applications of NF membranes in different industrial sectors

industrial sectors-wise application of nanofilter, while Fig. 9.9 depicts a word cloud generated on patent application on nanofilter. This is direct indication of application of nanofilter in various categories. Nanofilter used in industrial processes is shown in Table 9.10, including food, environmental, chemical, textiles, metal finishing, pharmaceutical, and biotechnology.

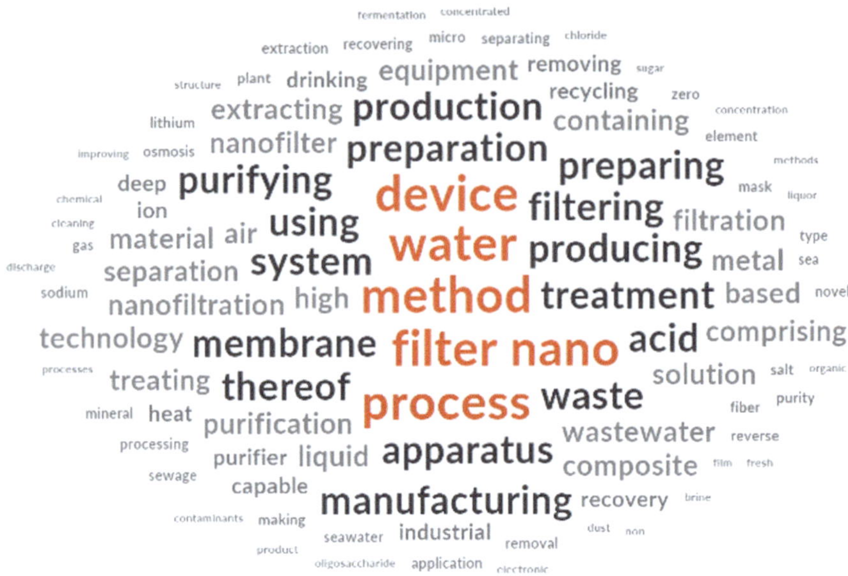

Fig. 9.9 Word cloud generated from patent application on nanofilter

C02F class represents "technologies related to water treatment and purification." PCT applications are those patent applications made to multinational phase for their broad-spectrum protection. As shown in Table 9.11, the patent application at multinational phase on IPC subclass C02F via PCT is 15. Table 9.11 shows this 15-patent application on nanofilter with their key features. The major patent is for method to purify, treat, or desalinate the drinking water, surface water, or seawater. The recent patent on method or device is to remove or separate oil–water to control the water pollution. This indicates that nanofilter has gain huge capital investment and protection in water and wastewater treatment.

Figure 9.10 summarizes the work carried out in the chapter.

Table 9.10 Applications of nanofilters in various industrial processes

Sr. No.	Applicability/Industry	Concentrate	Benefits	References
01	Plant, food, dairy, and beverage extracts	Lactose concentration and demineralization	Nanofiltering membranes may demineralize and concentrate lactose	Vellenga and Trägårdh (1998)
		Maple syrup concentration	Nanofiltration can be used to treat thick maple syrup more quickly and affordably than traditional boiling reduction procedures	
		Gibberellins	Plant hormones called gibberellins help growth and development By concentrating plant hormones like gibberellins using a process called nanofiltration, total product yields can be increased	
02	Textile and dyes	Desalting and concentrating dye	Enhancing dye strength, purity, and value through desalination and concentration uses nanofiltration. Nanofiltration membrane concentration capacity increases textile wastewater treatment by increasing product recovery and reuse	Abdel-Fatah (2018), Blocher et al. (Blöcher et al. 2002)
		Dye concentration	Nanofiltration allows salts and water to pass across an NF membrane, replacing textile dye concentration and desalination	
		Dye penetrate removal	Nanofiltration is widely used to recover dye and adhere to discharge constraints during penetration testing with fluorescent dyes	
		Optical brightening agent concentration and desalination	The appearance of colors can be improved using optical brightening agents. Operations costs can be reduced by using nanofiltration to increase the agent brightening concentration that can be reused	
03	Industrial processes and wastewater	Seawater sulfate removal	Scaling in the operation of oilfield waterfloods can be prevented by using nanofiltration membranes to specifically remove sulfate contained in seawater	Trebouet et al. (2001)
		Natural organic matter removal from surface water	Nanofiltration can remove natural organic matter from surface water to enrich it or use it in industrial process water	

(continued)

Table 9.10 (continued)

Sr. No.	Applicability/Industry	Concentrate	Benefits	References
		Landfill leachate treatment:	Nanofiltration can be used to process some landfill waste as a less energy-intensive option. Leachate by eliminating contaminants, lowering levels of TOC and COD, and reducing turbidity	
		Brine recovery	NF membranes can recover industrial brine by rejecting high sulfate concentrations and letting sodium chloride ions through	
04	Biotech/ Pharmaceutical	Antibiotics production	Nanofiltration membranes are attractive for hormone and antibiotic separation, concentration, and synthesis	Drioli et al. (2011)
		Blood serum	After separating blood serum and plasma, nanofiltration may eliminate fibrinogen and other clotting substances	
05	Recent applications	Systems water softening	Calcium, magnesium, and bicarbonate ions cause hardness. Hardness ions or minerals in potable water may cause scale in pipes, equipment, and systems that use water, such as heat exchangers. Softening units remove scale-causing ions to cleanse water	Abdel-Fatah (2018)
		Municipal wastewater treatment	To properly replace groundwater, a new municipal sewage treatment system was investigated. According to the data, NF may reject DOC and AOX in tertiary effluent at values below 2–3 mg/l and 20 lg/l. Biofouling is reduced by the NF membrane (Desal DK5)'s 1 m/s cross-flow velocity. A suitable pretreatment, such as progressive sand filtration, must be used. Using the concentration, ozonation experiments demonstrate that refractory DOC dissolves more quickly	Abdel-Fatah (2018)

(continued)

Table 9.10 (continued)

Sr. No.	Applicability/Industry	Concentrate	Benefits	References
		Water treatment	Food industries use membrane processing to treat wastewater. Food wastewater, especially under high pressure, fouls and concentrates quickly in the membrane layer. In whey production, a stream with a COD of 100,000 mg O_2 L^{-1} is filtered to reduce permeate with 2787 mg O_2 L^{-1}. This approach may lower COD better than RO/UF. About 88% of the protein stream is rejected	Abdel-Fatah (2018)
		Separation processes	With the use of NF, valuable compounds may be separated from unwanted or harmful substances in liquid streams, reducing costs and improving the environmental effect of industry. Exchange of solvents is possible using NF. The two key factors that can influence how effectively a process works are pH and material concentration. The important lactic acid is recovered utilizing organically made NF membrane systems (PES10, N30F, and MPF36). PES10 was used at a flow rate of 6.5 Lm 2 h 1 to get the best results. In the food business, NF can be used to separate various substances; however, further study is required to increase effectiveness and address anticipated issues	Abdel-Fatah (2018)

Table 9.11 The patent filed at multinational phase on IPC subclass C02F (technologies related to water treatment and purification) via PCT on nanofilter and their key features

Patent application Id	Patent application year	Key feature	References
WO1997049639	1997	A method for the treatment of polluted water	Maier (1997)
WO2010045502	2010	A transportable water disinfecting and purification device	O'Brien and Engel (2010)
WO2010138462	2010	An adapter kit	O'Brien (2010)
WO2010139773	2010	A method to desalinate water	Gaid and Coeytaux (2010)
WO2011046061	2011	A method for purifying drinking water	Yanagidaira and Ishibashi (2011)
WO2014057425	2014	A method to treat tannery wastes	Religa (2014)
WO2016040973	2016	A device to filter water	El Khiar et al. (2016)
WO2016066382	2016	A water purifier	Chatterjee et al. (2016)
WO2017132151	2017	An ion concentrator	Mertz et al. (2017)
WO2017199996	2017	A method for water treatment	Nakano and Oe (2017)
WO2018199357	2018	An oil–water separating vessel	Lee (2018)
WO2020249668	2020	A method to make drinkable water	Sauvignet et al. (2020)
WO2021234695	2021	A method and system of wastewater treatment for Zero liquid discharge	Banker and Zeevy Perez (2021)
WO2023026080	2023	A device to remove oil pollution form sea surface	Setayesh et al. (2023)
WO2023084298	2023	A method and device to separate oil water	Khodaparast and Khosropour (2023)

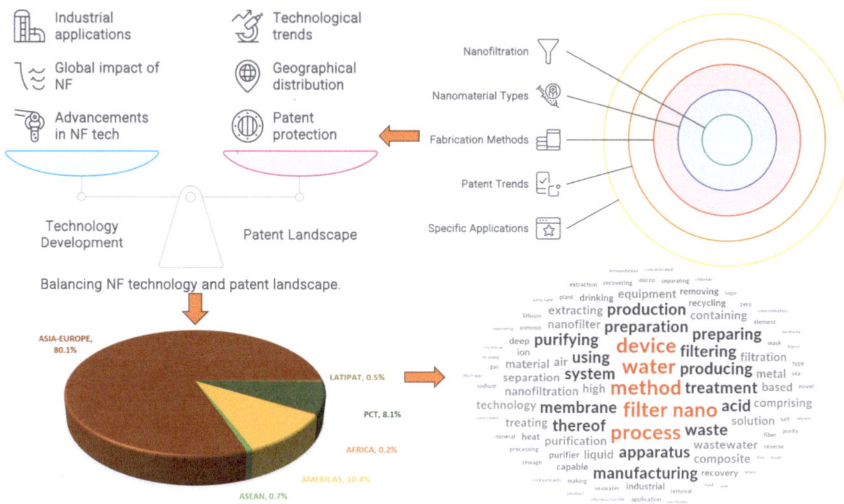

Fig. 9.10 Summary

9.14 Conclusion

The nanofilter has become known for their filter application since 1960. Synthesis and production of NF membrane especially for virus filtration and their various parameters is identified. Various nanomaterials commonly used and their relevant property in nanofiltration are discussed. Methods for NF membrane preparation like phase inversion, layer by layer, interfacial polymerization are outlined. Types of NF membrane like regenerated cellulose membrane, polyethersulfone membrane, and PVDF membrane are summarized. Applications in food, biotech pharma, drinking water filtration, and purification are derived. The patent scenario has identified year-wise and geographical distribution of patent applications. Among inventors, Hofmann Wilfried found as top innovator in NF technology. It is observed that NF has major innovation in C02F IPC subclass for technologies in industrial water treatment.

Acknowledgements The authors express their gratitude to the management of Marwadi University, Rajkot, Gujarat, for providing the necessary facilities for conducting this study, and equally grateful to Dr. Gaurav Sanghvi, (Head of the Department, Microbiology) for his encouragement in carrying out the investigation.

Declaration of Interest Conflict of interest: The authors declare that they do not have any conflict of interest.

Competing Interests: The authors have non-financial interests to disclose.

Funding: Not funding.

Ethical Approval: Not Required.

CRediT Authorship Contribution Statement #Mr. Vivek Yadav contributed to writing—original draft;

#Mr. Rahil Mathakia contributed to writing—original draft;

Dr. Pritesh Bhatt: Writing and editing draft;

Dr. Amit Kumar Sharma: Review and editing draft;

Dr. Arun K. Soni: Writing, review and editing draft;

Dr. Viralkumar B. Mandaliya: Writing, review, editing and final draft preparation;

At the last all authors have reviewed and approved the draft.

#Both have equally contributed.

References

Abdel-Fatah MA (2018) Nanofiltration systems and applications in wastewater treatment. Ain Shams Eng J 9(4):3077–3092. https://doi.org/10.1016/j.asej.2018.08.001

Aburabie J, Emwas AH, Peinemann KV (2019) Silane-crosslinked asymmetric polythiosemicarbazide membranes for organic solvent nanofiltration. Macromol Mater Eng 304(1):1800551. https://doi.org/10.1002/mame.201800551

Altammar KA (2023) A review on nanoparticles: characteristics, synthesis, applications, and challenges. Front Microbiol 14:1155622. https://doi.org/10.3389/fmicb.2023.1155622

Amadi EV, Venkataraman A, Papadopoulos C (2022) Nanoscale self-assembly: concepts, applications and challenges. Nanotechnology 33(13):132001. https://doi.org/10.1088/1361-6528/ac3f54

Banker R, Zeevy Perez Y (2021) ZLD (zero liquid discharge) wastewater treatment. PCT Application No. WO2021234695. WIPO, Geneva. Retrieved from https://patentscope.wipo.int/search/en/detail.jsf?docId=WO2021234695

Basile A, Charcosset C (2016) Ultrafiltration, microfiltration, nanofiltration and reverse osmosis in integrated membrane processes. In: Integrated membrane systems and processes, pp 1–22. ISBN: 978-1-118-73908-2

Bhuiyan MHU, Saidur R, Amalina MA, Mostafizur RM, Islam AKMS (2015) Effect of nanoparticles concentration and their sizes on surface tension of nanofluids. Procedia Eng 105:431–437. https://doi.org/10.1016/j.proeng.2015.05.030

Blöcher C, Noronha M, Fünfrocken L, Dorda J, Mavrov V, Janke HD, Chmiel H (2002) Recycling of spent process water in the food industry by an integrated process of biological treatment and membrane separation. Desalination 144(1–3):143–150. https://doi.org/10.1016/S0011-9164(02)00303-X

Bowen WR, Mohammad AW (1998) Characterization and prediction of nanofiltration membrane performance—a general assessment. Chem Eng Res des 76(8):885–893. https://doi.org/10.1205/026387698525685

Bowen WR, Mukhtar H (1996) Characterisation and prediction of separation performance of nanofiltration membranes. J Membr Sci 112(2):263–274. https://doi.org/10.1016/0376-7388(95)00302-9

Bruna T, Maldonado-Bravo F, Jara P, Caro N (2021) Silver nanoparticles and their antibacterial applications. Int J Mol Sci 22(13):7202. https://doi.org/10.3390/ijms22137202

Burnouf-Radosevich M, Appourchaux P, Huart JJ, Burnouf T (1994) Nanofiltration, a new specific virus elimination method applied to high-purity factor IX and factor XI concentrates. Vox Sang 67(2):132–138. https://doi.org/10.1111/j.1423-0410.1994.tb01647.x

Caballero S, Diez JM, Belda FJ, Otegui M, Herring S, Roth NJ, Douglas L, Rodrigo G, Jorquera JI (2014) Robustness of nanofiltration for increasing the viral safety margin of biological products. Biologicals 42(2):79–85. https://doi.org/10.1016/j.biologicals.2013.10.003

Cao Y, Chen G, Wan Y, Luo J (2021) Nanofiltration membrane for bio-separation: Process-oriented materials innovation. Eng Life Sci 21(6):405–416. https://doi.org/10.1002/elsc.202000100

Chatterjee J, Pratap S, Subbiah S (2016) A water purifier and a process of cleaning the membrane. PCT Application No. WO2016066382. WIPO, Geneva. Retrieved from https://patentscope.wipo.int/search/en/detail.jsf?docId=WO2016066382

Drioli E, Stankiewicz AI, Macedonio F (2011) Membrane engineering in process intensification—an overview. J Membr Sci 380(1–2):1–8. https://doi.org/10.1016/j.memsci.2011.06.043

Du Y, Pramanik BK, Zhang Y, Dumee L, Jegatheesan V (2022) Recent advances in the theory and application of nanofiltration: a review. Curr Poll Reports 8(1):51–80. https://doi.org/10.1007/s40726-021-00208-1

El Khiar H, Brandt V, Glavina M (2016) Transportable device for gravity filtering water. PCT Application No. WO2016040973. WIPO, Geneva. Retrieved from https://patentscope.wipo.int/search/en/detail.jsf?docId=WO2016040973

Emonds S, Roth H, Wessling M (2020) Chemistry in a spinneret–formation of hollow fiber membranes with a cross-linked polyelectrolyte separation layer. J Membr Sci 612:118325. https://doi.org/10.1016/j.memsci.2020.118325

Firman LR, Ochoa NA, Marchese J, Pagliero CL (2013) Deacidification and solvent recovery of soybean oil by nanofiltration membranes. J Membr Sci 431:187–196. https://doi.org/10.1016/j.memsci.2012.12.040

Fu S, Sun Z, Huang P, Li Y, Hu N (2019) Some basic aspects of polymer nanocomposites: a critical review. Nano Mater Sci 1(1):2–30. https://doi.org/10.1016/j.nanoms.2019.02.006

Gaid A, Coeytaux M (2010) Method for treating water in order to desalinate said water, including treating concentrates. PCT Application No. WO2010139773. WIPO, Geneva. Retrieved from https://patentscope.wipo.int/search/en/detail.jsf?docId=WO2010139773

Giwa A, Ogunribido A (2012) The applications of membrane operations in the textile industry: a review. Br J Appl Sci Technol 2(3):296. https://doi.org/10.9734/BJAST/2012/1520

Hadden DK, Inder TP, Sievers LJ (1991) Nanofiltration process for making dextrose. Canada Application No. CA2038485. Canada office. Retrieved from https://patentscope.wipo.int/search/en/detail.jsf?docId=CA2038485

Heris SZ, Ebadiyan H, Mousavi SB, Nami SH, Mohammadpourfard M (2023) The influence of nano filter elements on pressure drop and pollutant elimination efficiency in town border stations. Sci Rep 13(1):18793. https://doi.org/10.1038/s41598-023-46129-5

Hoshyar N, Gray S, Han H, Bao G (2016) The effect of nanoparticle size on in vivo pharmacokinetics and cellular interaction. Nanomedicine 11(6):673–692. https://doi.org/10.2217/nnm.16.5

Huang ZM, Zhang YZ, Kotaki M, Ramakrishna S (2003) A review on polymer nanofibers by electrospinning and their applications in nanocomposites. Compos Sci Technol 63(15):2223–2253. https://doi.org/10.1016/S0266-3538(03)00178-7

Ide S (2022) Filter made of cuprammonium regenerated cellulose for virus removal: a mini-review. Cellulose 29(5):2779–2793. https://doi.org/10.1007/s10570-021-04319-2

Inouye M, Burnouf T (2020) The role of nanofiltration in the pathogen safety of biologicals: an update. Curr Nanosci 16(3):413–424. https://doi.org/10.2174/1573413715666190328223130

Ji Y, Qian W, Yu Y, An Q, Liu L, Zhou Y, Gao C (2017) Recent developments in nanofiltration membranes based on nanomaterials. Chin J Chem Eng 25(11):1639–1652. https://doi.org/10.1016/j.cjche.2017.04.014

Joseph TM, Al-Hazmi HE, Śniatała B, Esmaeili A, Habibzadeh S (2023) Nanoparticles and nanofiltration for wastewater treatment: from polluted to fresh water. Environ Res:117114. https://doi.org/10.1016/j.envres.2023.117114

Jye LW, Ismail AF (2016) Nanofiltration membranes: synthesis, characterization, and applications. CRC Press. https://doi.org/10.1201/9781315181479

Kamide K, Manabe SI (1985) Role of microphase separation phenomena in the formation of porous polymeric membranes. ACS Symp Ser 269:197

Khlyustova A, Cheng Y, Yang R (2020) Vapor-deposited functional polymer thin films in biological applications. J Mater Chem B 8(31):6588–6609. https://doi.org/10.1039/d0tb00681e

Khodaparast B, Khosropour A (2023) Oil-water separator filter combining nanoparticles and soot impregnated on sponges, and method for preparing the same. PCT Application No. WO2023084298. WIPO, Geneva. Retrieved from https://patentscope.wipo.int/search/en/detail.jsf?docId=WO2023084298

Kyriakou N, Merlet RB, Willott JD, Nijmeijer A, Winnubst L, Pizzoccaro-Zilamy MA (2020) New method toward a robust covalently attached cross-linked nanofiltration membrane. ACS Appl Mater Interfaces 12(42):47948–47956. https://doi.org/10.1021/acsami.0c13339

Lee S (2018). Non-powered nano micro-mesh based real-time flow-type oil return vessel. PCT Application No. WO2018199357. WIPO, Geneva. Retrieved from https://patentscope.wipo.int/search/en/detail.jsf?docId=WO2018199357

Linder C, Kedem O (2021) History of nanofiltration membranes from 1960 to 1990. In: Nanofiltration: principles, applications, and new materials, vol 1. Wiley, pp 1–34. https://doi.org/10.1002/9783527824984.ch1

Liu F, Hashim NA, Liu Y, Abed MM, Li K (2011) Progress in the production and modification of PVDF membranes. J Membr Sci 375(1–2):1–27. https://doi.org/10.1016/j.memsci.2011.03.014

Mahalingam DK, Wang S, Nunes SP (2019) Stable graphene oxide cross-linked membranes for organic solvent nanofiltration. Ind Eng Chem Res 58(51):23106–23113. https://doi.org/10.1021/acs.iecr.9b05169

Maier J (1997) Process for water treatment in a closed circuit with a reactor. PCT Application No. WO1997049639. WIPO, Geneva. https://patentscope.wipo.int/search/en/detail.jsf?docId=WO1997049639

Manabe S (1996) Removal of virus through novel membrane filtration method. Dev Biol Standard 88:81–90. https://pubmed.ncbi.nlm.nih.gov/9119166/

Martínez MB, Van der Bruggen B, Negrin ZR, Alconero PL (2012) Separation of a high-value pharmaceutical compound from waste ethanol by nanofiltration. J Ind Eng Chem 18(5):1635–1641. https://doi.org/10.1016/j.jiec.2012.02.024

Mautner A (2020) Nanocellulose water treatment membranes and filters: a review. Polym Int 69(9):741–751. https://doi.org/10.1002/pi.5993

Mertz J, Simpson B, Luey J, Orme R, Carlson D (2017) System and method for manipulation of ion concentration to maximize efficiency of ion exchange. PCT Application No. WO2017132151. WIPO, Geneva. https://patentscope.wipo.int/search/en/detail.jsf?docId=WO2017132151

Misdan N, Lau WJ, Ismail AF, Matsuura T, Rana D (2014) Study on the thin film composite poly (piperazine-amide) nanofiltration membrane: impacts of physicochemical properties of substrate on interfacial polymerization formation. Desalination 344:198–205. https://doi.org/10.1016/j.desal.2014.03.036

Mohammad AW, Teow YH, Ang WL, Chung YT, Oatley-Radcliffe DL, Hilal N (2015) Nanofiltration membranes review: recent advances and future prospects. Desalination 356:226–254. https://doi.org/10.1016/j.desal.2014.10.043

Mu Q, Jiang G, Chen L, Zhou H, Fourches D, Tropsha A, Yan B (2014) Chemical basis of interactions between engineered nanoparticles and biological systems. Chem Rev 114(15):7740–7781. https://doi.org/10.1021/cr400295a

Nakano T, Oe T (2017) Water treatment method and device and regeneration method for ion exchange resin. PCT Application No. WO2017199996. WIPO, Geneva. https://patentscope.wipo.int/search/en/detail.jsf?docId=WO2017199996

Naseem T, Durrani T (2021) The role of some important metal oxide nanoparticles for wastewater and antibacterial applications: a review. Environ Chem Ecotoxicol 3:59–75. https://doi.org/10.1016/j.enceco.2020.12.001

Nguyen Thi HY, Nguyen BTD, Kim JF (2020) Sustainable fabrication of organic solvent nanofiltration membranes. Membranes 11(1):19. https://doi.org/10.3390/membranes11010019

Oatley-Radcliffe DL, Walters M, Ainscough TJ, Williams PM, Mohammad AW, Hilal N (2017) Nanofiltration membranes and processes: a review of research trends over the past decade. J Water Process Eng 19:164–171. https://doi.org/10.1016/j.jwpe.2017.07.026

O'Brien PW (2010) Water filtering and purification method and adapter kit. PCT Application No. WO2010138462. WIPO, Geneva. https://patentscope.wipo.int/search/en/detail.jsf?docId=WO2010138462
O'Brien PW, Engel M (2010) Portable drinking water purification device. PCT Application No. WO2010045502. WIPO, Geneva. https://patentscope.wipo.int/search/en/detail.jsf?docId=WO2010045502
Oshima KH, Evans-Strickfaden TT, Highsmith AK, Ades EW (1996) The use of a microporous polyvinylidene fluoride (PVDF) membrane filter to separate contaminating viral particles from biologically important proteins. Biologicals 24(2):137–145. https://doi.org/10.1006/biol.1996.0018
Rashid AB, Haque M, Islam SM, Labib KRU (2024) Nanotechnology-enhanced fiber-reinforced polymer composites: recent advancements on processing techniques and applications. Heliyon 10(2):e24692. https://doi.org/10.1016/j.heliyon.2024.e24692
Religa P (2014) Method of preliminary regeneration of tanning waste. PCT Application No. WO2014057425. WIPO, Geneva. https://patentscope.wipo.int/search/en/detail.jsf?docId=WO2014057425
Ryzhkov II, Minakov AV (2016) Theoretical study of electrolyte transport in nanofiltration membranes with constant surface potential/charge density. J Membr Sci 520:515–528. https://doi.org/10.1016/j.memsci.2016.08.004
Saha NK, Joshi SV (2009) Performance evaluation of thin film composite polyamide nanofiltration membrane with variation in monomer type. J Membr Sci 342(1–2):60–69. https://doi.org/10.1016/j.memsci.2009.06.025
Sauvignet P, Held HG, Sauer U, Gaid A (2020) Membrane method for making surface water drinkable without adding any sequestering agent. PCT Application No. WO2020249668. WIPO, Geneva. https://patentscope.wipo.int/search/en/detail.jsf?docId=WO2020249668
Setayesh B, Taheri S, Mousavizadeh Torbati E, Tamadi H (2023) Nano-filter device to remove oil pollution from sea surface. PCT Application No. WO2023026080. WIPO, Geneva. https://patentscope.wipo.int/search/en/detail.jsf?docId=WO2023026080
Siddique TA, Dutta NK, Roy Choudhury N (2020) Nanofiltration for arsenic removal: challenges, recent developments, and perspectives. Nanomaterials 10(7):1323. https://doi.org/10.3390/nano10071323
Singh R, Bhadouria R, Singh P, Kumar A, Pandey S, Singh VK (2020) Nanofiltration technology for removal of pathogens present in drinking water. In: Waterborne pathogens. Butterworth-Heinemann, pp 463–489. https://doi.org/10.1016/B978-0-12-818783-8.00021-9
Speranza G (2021) Carbon nanomaterials: synthesis, functionalization and sensing applications. Nanomaterials 11(4):967. https://doi.org/10.3390/nano11040967
Strantz J Jr, Brehm WJ (1990) Treatment for reverse osmosis membranes. US Application No. US4938872. United States Patent and Trademark Office. https://patentscope.wipo.int/search/en/detail.jsf?docId=US37965701
Székely G, Bandarra J, Heggie W, Sellergren B, Ferreira FC (2012) A hybrid approach to reach stringent low genotoxic impurity contents in active pharmaceutical ingredients: combining molecularly imprinted polymers and organic solvent nanofiltration for removal of 1,3-diisopropylurea. Sep Purif Technol 86:79–87. https://doi.org/10.1016/j.seppur.2011.10.023
Trebouet D, Schlumpf JP, Jaouen P, Quemeneur F (2001) Stabilized landfill leachate treatment by combined physicochemical–nanofiltration processes. Water Res 35(12):2935–2942. https://doi.org/10.1016/s0043-1354(01)00005-7
Tres MV, Ferraz HC, Dallago RM, Di Luccio M, Oliveira JV (2010) Characterization of polymeric membranes used in vegetable oil/organic solvents separation. J Membr Sci 362(1–2):495–500. https://doi.org/10.1016/j.memsci.2010.07.011
Vellenga E, Trägårdh G (1998) Nanofiltration of combined salt and sugar solutions: coupling between retentions. Desalination 120(3):211–220. https://doi.org/10.1016/S0011-9164(98)00219-7

Xue J, Wu T, Dai Y, Xia Y (2019) Electrospinning and electrospun nanofibers: methods, materials, and applications. Chem Rev 119(8):5298–5415. https://doi.org/10.1021/acs.chemrev.8b00593

Yadav D, Karki S, Ingole PG (2022) Current advances and opportunities in the development of nanofiltration (NF) membranes in the area of wastewater treatment, water desalination, biotechnological and pharmaceutical applications. J Environ Chem Eng. Advance online publication. https://doi.org/10.1016/j.jece.2022.108109

Yanagidaira S, Ishibashi S (2011) Method for purifying drinking water using nanofilter coated with nano-sized ferromagnetic ferrite, method for sterilizing harmful bacteria contained in drinking water using said nanofilter, and method for improving drinking water by sterilizing and removing harmful substances using said nanofilter. PCT Application No. WO2011046061. WIPO, Geneva. https://patentscope.wipo.int/search/en/detail.jsf?docId=WO2011046061

Yang Z, Zhou Y, Feng Z, Rui X, Zhang T, Zhang Z (2019) A review on reverse osmosis and nanofiltration membranes for water purification. Polymers 11(8):1252. https://doi.org/10.3390/polym11081252

Yuchen D, Pramanik BK, Zhang Y, Dumée L, Jegatheesan V (2022) Recent advances in the theory and application of nanofiltration: a review. Curr Poll Reports 8(1):51–80. https://doi.org/10.1007/s40726-021-00208-1

Zareei F, Hosseini SM (2019) A new type of polyethersulfone based composite nanofiltration membrane decorated by cobalt ferrite-copper oxide nanoparticles with enhanced performance and antifouling property. Sep Purif Technol 226:48–58. https://doi.org/10.1016/j.seppur.2019.05.077

Chapter 10
Economic and Societal Aspects

Jayeeta Chattopadhyay, Nimmy Srivastava, and Tara Sankar Pathak

Abstract This chapter explores the social and economic aspects of using nanotechnology to clean industrial effluent. The economic feasibility and cost-effectiveness of nanomaterials-based treatment technologies are assessed in the first section. The analysis illustrates possible cost reductions, long-term advantages, and the scalability of nanotechnology in industrial applications by contrasting these cutting-edge technologies with traditional techniques. The effectiveness and sustainability of nanomaterial applications are emphasized, and the conversation also touches on life cycle costs, maintenance needs, and the possibility of resource recovery. The chapter's second portion examines public opinion, society acceptance, and ethical issues related to the application of nanotechnology in wastewater treatment. It looks into things like safety issues, perceived benefits vs dangers, and transparency that affect public trust and acceptability. A comprehensive analysis is conducted of the ethical issues of using nanomaterials, including the effects on the environment and the fair distribution of technical advantages. The chapter seeks to provide a thorough understanding of the complex implications of nanotech solutions in industrial wastewater treatment by addressing these economic and societal factors. This will help to facilitate responsible innovation and well-informed decision-making.

Keyword Economic viability · Societal acceptance · Environmental impact · Nanomaterials

J. Chattopadhyay (✉)
Amity Institute of Applied Sciences, Amity University Jharkhand, Ranchi, India
e-mail: jchattopadhyay@rnc.amity.edu

N. Srivastava
Amity Institute of Biotechnology, Amity University Jharkhand, Ranchi, India

T. S. Pathak
Chemistry Department, Surendra Institute of Engineering and Management, Siliguri, WB, India

A. Mandpe et al. (eds.), *Nanotech Solutions for Industrial Wastewater*, Advances in Wastewater Research, https://doi.org/10.1007/978-981-96-6879-3_10

10.1 Introduction

Significant economic growth and progress have resulted from the last decades' fast industrialization and urbanization. Industrial wastewater pollution is one of the most urgent environmental problems that have arisen as a result of this progress. Aquatic ecosystems and human health are seriously threatened by industrial wastewater, which is known for its high concentration of contaminants, heavy metals, organic compounds, and other dangerous chemicals. While somewhat effective, conventional wastewater treatment techniques frequently fall short of meeting the volume and complexity of contemporary industrial effluents, which calls for the development of cutting-edge and more effective treatment approaches (Kamali et al. 2019).

Nanotechnology, the manipulation of matter at the atomic and molecular scale, has emerged as a groundbreaking approach in various fields, including wastewater treatment. Nanomaterials, with their unique physicochemical properties such as high surface area, reactivity, and tunable functionalities, offer remarkable potential for the removal of contaminants from wastewater (Freeda Gnana Rani et al. 2005; Ahmed et al. 2021; Kamali et al. 2019; Kato and Kansha 2024). The application of nanotechnology in this domain encompasses a variety of methodologies, including adsorption, catalysis, membrane filtration, and advanced oxidation processes, each leveraging the distinct advantages of nanomaterials to enhance treatment efficiency and effectiveness (Ali et al. 2021; Jones et al. 2021). Even while nanotechnology shows great promise in lab settings, solving important societal and economic issues will be necessary before it can be applied on a big scale in industry. This chapter attempts to give a thorough analysis of these factors, starting with an evaluation of the economic feasibility and cost-effectiveness of treatment solutions based on nanomaterials (Pratiwi et al. 2023).

Cost-effectiveness is a pivotal factor in the adoption and implementation of any new technology. For nanotechnology to be viable in the industrial wastewater treatment sector, it must not only demonstrate superior performance compared to existing methods but also be economically competitive. This entails a detailed examination of the initial outlay needed for the implementation of systems based on nanomaterials, as well as the ongoing expenses of operation and maintenance. The economic evaluation is further complicated by the possibility of resource recovery and reuse, which is a key benefit of some applications involving nanotechnology. This chapter aims to illustrate the financial benefits of nanomaterials-based technologies by contrasting them with traditional treatment methods. It also identifies areas where cost reduction measures can be applied to improve the technologies' economic appeal. In addition to economic factors, societal acceptance plays a crucial role in the widespread adoption of nanotechnology in wastewater treatment. Public perception and trust in nanotechnology are influenced by a myriad of factors, including the perceived benefits, safety concerns, and transparency in the technology's development and implementation. The chapter explores these dimensions, shedding light on the societal and ethical implications of utilizing nanomaterials in water purification processes.

Public perception of nanotechnology in wastewater treatment is shaped by several factors, including the visibility of its benefits, the perceived risks associated with its use, and the level of public understanding and awareness of the technology (Nagar and Pradeep 2020; Zahmatkesh et al. 2023). Transparent communication and effective educational strategies are essential to foster public trust and acceptance. By examining case studies of successful implementations as well as controversies and challenges faced in various regions, the chapter aims to provide a nuanced understanding of societal responses to nanotechnology in wastewater treatment. Ethical considerations are integral to the discourse on nanotechnology adoption. The potential environmental impacts of nanomaterials, including their production, use, and disposal, must be carefully evaluated to ensure that the benefits of nanotechnology do not come at the expense of environmental sustainability. This includes addressing concerns about secondary pollution caused by the release of nanoparticles into the environment and the long-term ecological effects of nanomaterial accumulation. Furthermore, the equitable distribution of the benefits of advanced wastewater treatment technologies is a crucial ethical consideration, particularly in ensuring that these innovations are accessible to diverse communities, including those in developing regions. The chapter also examines the current regulatory landscape governing the use of nanotechnology in wastewater treatment. This includes national and international standards and guidelines that aim to ensure the safety and efficacy of nanomaterials. Policy recommendations are provided to enhance regulatory frameworks, promote public engagement and transparency, and support the responsible development and deployment of nanotechnology in wastewater treatment. The future directions and research needs in the field of nanotechnology for wastewater treatment are also explored. Innovations in nanomaterials and treatment technologies hold the potential to further enhance the efficiency and cost-effectiveness of wastewater treatment processes.

However, bridging the economic and societal gaps remains a critical challenge. Multidisciplinary approaches that integrate technical, economic, and social perspectives are essential to develop sustainable solutions that can be widely adopted. By addressing these economic and societal aspects, this chapter aims to provide a holistic understanding of the implications of integrating nanotechnology into industrial wastewater treatment systems. The goal is to inform stakeholders, including policymakers, industry leaders, researchers, and the public, enabling them to make informed decisions and foster responsible innovation. Through this comprehensive examination, the chapter underscores the importance of balancing technological advancements with economic feasibility and societal well-being, ultimately contributing to sustainable and effective wastewater management solutions.

The introduction of nanotechnology into the field of industrial wastewater treatment represents a significant leap forward in addressing one of the most critical environmental challenges of our time. Figure 10.1 represents the schematic diagram on nano catalysts utilized in industrial wastewater treatments. However, the successful adoption and implementation of these advanced technologies depend not only on their technical merits but also on their economic viability and societal acceptance. By

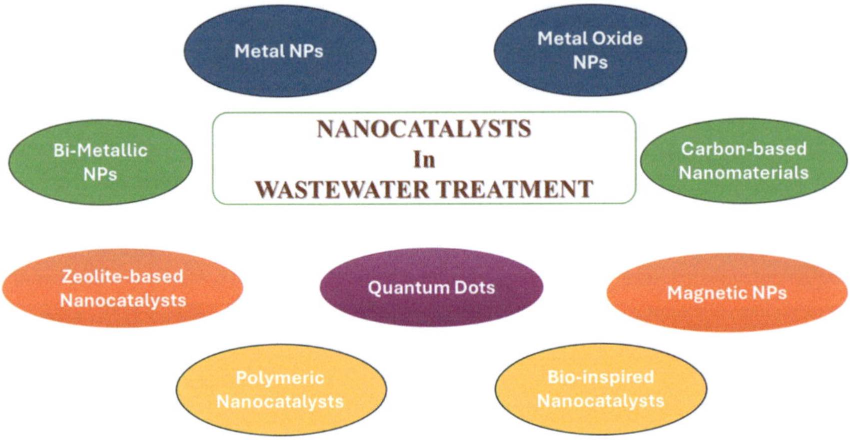

Fig. 10.1 Schematic diagram on nano catalysts used in industrial wastewater treatment

exploring the cost-effectiveness, public perception, ethical considerations, and regulatory frameworks associated with nanomaterials-based treatment technologies, this chapter provides a comprehensive framework for understanding and addressing the multifaceted challenges and opportunities presented by nanotechnology in wastewater treatment. Through this balanced approach, the chapter aims to pave the way for informed decision-making and responsible innovation, ultimately contributing to a cleaner and more sustainable future (Savage and Diallo 2005).

10.2 Economic Aspects of Nanomaterials-Based Wastewater Treatment Technologies

The economic viability of nanomaterials-based wastewater treatment technologies is crucial for their adoption and scalability in industrial settings. This analysis covers various facets including initial investment, operational and maintenance costs, long-term economic benefits, efficiency and performance metrics, potential savings, return on investment, life cycle cost assessment, and the environmental and economic sustainability (Yaqoob et al. 2020; Xu et al. 2021). Each of these aspects plays a significant role in determining the overall cost-effectiveness and feasibility of deploying nanotechnology for wastewater treatment.

10.3 Cost-Effectiveness Analysis

Nanomaterials-based wastewater treatment technologies offer promising improvements over traditional methods in terms of efficiency and effectiveness. However, their cost-effectiveness must be thoroughly evaluated to ensure that these benefits justify the financial investment. This involves analyzing the costs associated with the development, implementation, and operation of these technologies, and comparing them to the savings and economic gains they provide.

Initial Investment and Capital Costs: The initial investment in nanomaterials-based wastewater treatment systems includes expenses related to research and development (R&D), production, and installation of the treatment infrastructure.

Research and Development Costs: Developing novel nanomaterials tailored for specific wastewater treatment applications involves substantial R&D efforts. This includes laboratory research, pilot-scale testing, and optimization processes. The costs associated with R&D are often high, but they are crucial for ensuring the effectiveness and safety of the nanomaterials used.

Production Costs: The production of nanomaterials, such as carbon nanotubes, metal oxides, and quantum dots, can be expensive due to the sophisticated techniques required. However, advancements in production technologies and scaling up manufacturing processes can help reduce these costs over time. Bulk production and economies of scale are essential to making these materials more affordable.

Installation Costs: Integrating nanomaterials-based treatment systems into existing wastewater treatment facilities or setting up new infrastructure involves significant installation costs. This may require modifications to current treatment processes or the addition of specialized equipment, such as nanofiltration membranes or adsorption columns. Initial capital outlay can be a barrier to adoption, particularly for small to medium-sized enterprises.

Operational and Maintenance Costs: Once installed, the operational and maintenance (O&M) costs of nanomaterials-based treatment systems become critical in determining their overall cost-effectiveness.

Operational Costs: These include the expenses related to running the treatment systems, such as energy consumption, chemical usage, and labor. Nanomaterials-based methods, like photocatalysis or advanced oxidation processes, may require substantial energy inputs, which can increase operating expenses. Conversely, certain nanomaterials, such as efficient adsorbents, can reduce overall energy consumption by minimizing the need for intensive mechanical or chemical treatments.

Maintenance Costs: Regular maintenance involves inspections, cleaning, and repair of the treatment systems. Nanomaterials, while effective, can degrade or foul over time, necessitating periodic replacement or regeneration. For example, nanofiltration membranes may require frequent cleaning to prevent clogging, and adsorbents might need regeneration to restore their capacity. These maintenance activities add to the ongoing expenses and must be factored into the overall cost analysis.

Durability and Lifespan: The stability and longevity of nanomaterials are crucial. Materials that demonstrate high durability can reduce the frequency of replacement and associated costs. Research aims to enhance the lifespan of nanomaterials, thereby improving their cost-effectiveness over time.

10.4 Long-Term Economic Benefits

The long-term economic benefits of nanomaterials-based wastewater treatment technologies extend beyond immediate cost savings. These benefits include improved treatment efficiency, compliance with regulatory standards, and potential revenue generation from by-products.

Nanomaterials-based wastewater treatment technologies offer significant long-term economic advantages, driven by enhanced operational efficiency, compliance with environmental regulations, and resource recovery potential. These technologies improve treatment efficiency, which leads to reduced operational costs over time by minimizing energy and chemical consumption (Crini and Lichtfouse 2019). For example, Das et al. (2020) highlight that carbon nanotubes and graphene-based adsorbents enable rapid contaminant removal, lowering the need for prolonged treatment stages and associated costs. Moreover, reduced sludge generation from nanofiltration processes translates into lower disposal expenses and environmental liabilities (Zhu et al. 2022; Mohan et al. 2018). Additionally, compliance with stringent environmental regulations is crucial for industrial operations, as failure to meet wastewater discharge standards can result in substantial fines. According to Kamali et al. (2019), companies adopting advanced nanomaterials-based treatments not only avoid penalties but also enhance their corporate reputation, improving competitiveness and attracting sustainable investment. Another key economic advantage is the ability of nanomaterials to facilitate the recovery of valuable by-products, such as heavy metals or treated water, which can be reused within industrial processes. Ma et al. (2019) demonstrates that metal oxide nanomaterials, such as Fe_3O_4 and TiO_2, effectively capture metals like copper and nickel from industrial effluents, offering industries opportunities to offset treatment costs through resource reuse. Additionally, the recyclability of certain nanomaterials ensures lower replacement costs over time, further enhancing economic sustainability (Gao and Feng 2022). In the long run, the operational savings, resource recovery, and compliance benefits create favorable return on investment (ROI), making nanomaterials-based technologies economically attractive solutions for wastewater management (Zahmatkesh et al. 2023).

Improved Treatment Efficiency: Nanomaterials offer enhanced treatment efficiency, capable of removing a wide range of contaminants more effectively than conventional methods. This can lead to significant cost savings by reducing the need for secondary or tertiary treatment processes. Nanomaterials can achieve higher removal rates for heavy metals, organic pollutants, and pathogens, resulting in cleaner effluent that meets stringent regulatory standards.

Regulatory Compliance: Compliance with environmental regulations is critical for industrial operations. Failure to meet wastewater discharge standards can result in hefty fines and penalties. Nanomaterials-based technologies can help industries achieve and maintain compliance, thereby avoiding financial penalties. Companies that adopt advanced treatment technologies may benefit from a positive public perception and enhanced corporate reputation, potentially leading to competitive advantages and market differentiation.

Revenue Generation from By-Products: Certain nanomaterials-based processes offer opportunities for resource recovery and reuse, which can generate additional revenue streams. For instance, nanomaterials can facilitate the recovery of valuable metals from industrial effluents, which can be sold or reused in manufacturing processes. Treated water can be recycled for industrial use, reducing overall water consumption and associated costs.

10.5 Efficiency and Performance Metrics

Efficiency and performance metrics are critical in evaluating the effectiveness of nanomaterials-based wastewater treatment technologies. These metrics provide a quantifiable means to compare these technologies against conventional methods and assess their real-world applicability.

Nanomaterials-based wastewater treatment technologies exhibit superior efficiency and performance compared to conventional methods, driven by their high surface area, reactivity, and selectivity for specific contaminants. Removal efficiency is a key metric, with nanomaterials such as titanium dioxide (TiO_2) and zinc oxide (ZnO) demonstrating exceptional photocatalytic degradation of organic pollutants and microbial contaminants under UV and visible light conditions (Chen and Mao 2019; Yaqoob et al. 2020). Similarly, carbon-based nanomaterials like graphene oxide (GO) and carbon nanotubes (CNTs) excel in adsorbing heavy metals and organic pollutants, achieving removal efficiencies above 90% for contaminants like lead and cadmium (Das et al. 2020; Ma et al. 2019). Another critical metric is energy efficiency, as many nanomaterials-based processes, such as adsorption and photocatalysis, consume less energy compared to conventional aeration or chemical oxidation methods (Gao and Feng 2022).

Scalability and adaptability are also essential, with modular nanomaterials-based systems showing flexibility in handling varying volumes of wastewater and suitability for decentralized applications (Keller and Lazareva 2018). Nanomaterials demonstrate excellent throughput, maintaining stable performance even under high contaminant loads, which enhances operational reliability (Mohan et al. 2018). Moreover, these technologies often produce minimal sludge, reducing the need for post-treatment sludge handling and disposal, which is a major drawback of conventional wastewater treatment (Kamali et al. 2019). These efficiency metrics collectively contribute to the economic feasibility of nanotechnology-based systems, making them viable solutions for both industrial and municipal wastewater treatment.

Removal Efficiency: The primary metric for evaluating wastewater treatment technologies is their contaminant removal efficiency. Nanomaterials can achieve high removal rates for various pollutants, including heavy metals, organic compounds, and pathogens. This metric is crucial for determining the overall effectiveness of the treatment process and its ability to meet regulatory standards.

Energy Efficiency: Energy consumption is a significant operational cost. Nanomaterials-based technologies that reduce energy usage compared to conventional methods can offer substantial cost savings. For instance, nano-adsorbents can quickly and efficiently remove contaminants, reducing the need for prolonged and energy-intensive treatment stages.

Throughput and Capacity: The ability of a treatment system to handle large volumes of wastewater is another important performance metric. Nanomaterials-based systems need to demonstrate high throughput and capacity to be viable for industrial applications. This metric assesses the scalability of the technology and its suitability for large-scale operations.

Longevity and Stability: The lifespan and stability of nanomaterials are critical factors in determining their long-term cost-effectiveness. Materials that maintain their performance over extended periods without significant degradation can reduce maintenance and replacement costs, enhancing overall efficiency.

10.6 Potential Savings and Return on Investment

The potential savings and return on investment (ROI) of nanomaterials-based wastewater treatment systems are key indicators of their financial viability. These savings are derived from various sources, including reduced operational costs, enhanced treatment efficiency, and revenue from resource recovery.

Nanomaterials-based wastewater treatment systems offer compelling potential savings and favorable return on investment (ROI) by enhancing operational efficiency and reducing long-term costs. These systems reduce energy and chemical consumption compared to conventional treatments, directly lowering operating expenses (Ali and Jain 2017). For example, nanomaterial-based adsorbents, such as zeolites and metal oxides, enable faster contaminant removal, minimizing the need for secondary treatment and chemical inputs (Shen et al. 2020). Kang et al. (2019) emphasize that the reduced energy requirements of processes like nanofiltration or photocatalysis further improve cost-effectiveness. Additionally, Omar (2018) highlights the financial benefits from reduced sludge generation, which cuts down on waste handling and disposal costs, typically a significant operational burden for traditional systems.

Revenue generation through resource recovery further improves ROI. Nanomaterials facilitate the extraction of valuable elements such as gold, copper, and rare earth metals from wastewater, which can be reused or sold, offsetting costs (Thompson and Tiwari 2021). Additionally, treated water can be recycled within industrial processes, reducing both water procurement and discharge fees (Tang et al. 2023). The ability to comply with environmental regulations without incurring fines also provides financial benefits, as companies can avoid penalties while enhancing their market reputation (Neale et al. 2021a, b). Over time, these combined savings and revenue streams contribute to an attractive ROI, making nanomaterials-based technologies a sound investment. As the cost of nanomaterials continues to decrease with advances in production, the economic appeal of these systems will further improve (Scott 2019).

Energy Savings: One significant source of potential savings is the reduction in energy consumption. Nanomaterials-based processes that are more energy-efficient compared to traditional methods can lower operating expenses over time. For example, nano-adsorbents that quickly remove contaminants can reduce the need for energy-intensive treatment stages, leading to significant energy savings.

Chemical Savings: Enhanced treatment efficiency also contributes to cost savings by reducing the consumption of chemicals and other treatment agents. Nanomaterials that achieve higher removal rates can minimize the need for additional chemical treatments, lowering both operational costs and environmental impact.

Revenue from Resource Recovery: The ability to recover valuable materials from wastewater can offset initial investment and operational costs. For instance, recovering precious metals or rare earth elements using nanomaterials can provide a substantial revenue stream. Similarly, recycling treated water within industrial processes can reduce water procurement costs, adding to overall savings.

Return of Investment (ROI) Calculation: Calculating ROI involves considering both tangible and intangible benefits. Tangible benefits include direct cost savings and revenue generation, while intangible benefits encompass improved regulatory compliance, enhanced corporate reputation, and potential market advantages. A comprehensive ROI analysis helps stakeholders make informed decisions regarding the adoption of nanotechnology in wastewater treatment.

10.7 Life Cycle Cost Assessment

A life cycle cost (LCC) assessment provides a holistic view of the economic sustainability of nanomaterials-based wastewater treatment technologies by evaluating all costs associated with the system over its entire lifespan. It offers a comprehensive view of the economic sustainability of nanomaterials-based wastewater treatment technologies by evaluating all associated costs across the system's entire lifespan, including capital investment, operational expenses, maintenance, and end-of-life disposal. LCC ensures that decision-makers consider both short-term and long-term financial implications, helping industries make informed choices. Initial capital expenditures can be high due to research and development, production, and infrastructure setup; however, these costs are offset over time by the enhanced efficiency and reduced operational expenses of nanotechnology (Shen et al. 2020). For example, Xu et al. (2021) highlight that nanomaterials-based systems require less frequent maintenance, as many materials, such as graphene-based membranes, exhibit superior durability and fouling resistance, reducing downtime and replacement costs.

Operational costs, such as energy and chemical consumption, are critical components of LCC. Ali et al. (2019) emphasize that nanomaterials, like nano-adsorbents and photocatalysts, can substantially lower energy usage compared to conventional aeration or oxidation processes. Gao and Feng (2022) further argue that resource recovery facilitated by nanomaterials—such as metal extraction or water reuse—generates additional economic value, contributing to long-term sustainability. LCC also accounts for environmental costs associated with the production and disposal of nanomaterials, which can influence overall sustainability if not properly managed (Neale et al. 2021a, b). Additionally, incorporating circular economy principles—such as recycling spent nanomaterials—into LCC assessments can reduce waste and improve financial outcomes (Tang et al. 2023). Ultimately, an LCC framework ensures that the economic sustainability of nanomaterials-based technologies is assessed holistically, balancing financial, operational, and environmental considerations over the entire lifecycle (Thompson and Tiwari 2021).

Initial and Ongoing Costs: LCC assessment includes the initial capital investment, operational and maintenance costs, and end-of-life disposal or recycling costs. By considering these factors, stakeholders can understand the total cost of ownership and make informed decisions about the financial feasibility of nanomaterials-based systems.

Environmental Costs: LCC assessment also considers environmental costs, such as the potential environmental impact of nanomaterial production and disposal. This includes evaluating the ecological footprint of manufacturing processes and the potential for secondary pollution from nanomaterials released into the environment.

Economic Sustainability: Assessing the long-term economic sustainability of nanomaterials-based technologies involves evaluating their ability to provide consistent cost savings and revenue generation over their lifespan. This includes considering the durability and stability of nanomaterials, as well as their effectiveness in maintaining high treatment efficiency over time.

Cost–Benefit Analysis: LCC assessment incorporates cost–benefit analysis to compare the economic benefits of nanomaterials-based systems against their total costs. This analysis helps identify the most cost-effective solutions and informs decision-making processes for stakeholders.

10.8 Environmental and Economic Sustainability

The sustainability of nanomaterials-based wastewater treatment technologies encompasses both environmental and economic dimensions.

Environmental Sustainability: The environmental sustainability of nanomaterials-based technologies involves assessing their ecological footprint, including the production, use, and disposal of nanomaterials. This includes evaluating the potential for secondary pollution, such as the release of nanoparticles into the environment, and their long-term ecological impacts.

Economic Sustainability: Economic sustainability involves ensuring that nanomaterials-based technologies provide consistent cost savings and revenue generation over their lifespan. This includes evaluating the durability and stability of nanomaterials, as well as their ability to maintain high treatment efficiency over time.

Resource Efficiency: Nanomaterials-based technologies that enhance resource efficiency, such as reducing energy consumption and minimizing chemical usage, contribute to both environmental and economic sustainability. These technologies help lower operational costs while reducing environmental impact, creating a win–win scenario.

Sustainable Production Practices: The adoption of sustainable production practices for nanomaterials, such as green synthesis methods and recycling of waste materials, can further enhance the sustainability of these technologies. Sustainable production practices reduce the environmental impact of nanomaterial manufacturing and contribute to overall sustainability goals.

10.9 Resource Recovery and Reuse Potential

The potential for resource recovery and reuse is a significant advantage of nanomaterials-based wastewater treatment technologies.

Recovery of Valuable Materials: Nanomaterials can facilitate the recovery of valuable metals and other resources from industrial effluents. For example, certain nanomaterials can selectively adsorb precious metals like gold, silver, and platinum, which can be recovered and reused in various industrial processes. This recovery not only offsets treatment costs but also provides a revenue stream.

Water Reuse: Treated water from nanomaterials-based systems can be recycled for industrial use, reducing overall water consumption and associated costs. This is particularly important in water-scarce regions where water reuse can significantly alleviate water shortages and enhance sustainability.

Circular Economy: The ability to recover and reuse resources aligns with the principles of a circular economy, where waste materials are converted into valuable resources, minimizing waste and reducing the environmental impact. Nanomaterials-based technologies that support resource recovery and reuse contribute to the development of sustainable industrial practices.

Economic Incentives: Governments and regulatory bodies may offer economic incentives for adopting technologies that promote resource recovery and reuse. These incentives can further enhance the economic viability of nanomaterials-based wastewater treatment systems, encouraging their adoption and integration into industrial processes.

The economic aspects of nanomaterials-based wastewater treatment technologies encompass a comprehensive analysis of initial investment, operational and maintenance costs, long-term economic benefits, efficiency and performance metrics, potential savings, return on investment, life cycle cost assessment, and sustainability. By addressing these factors, stakeholders can make informed decisions about the feasibility and benefits of integrating nanotechnology into wastewater management systems, ultimately contributing to more sustainable and cost-effective wastewater treatment solutions.

10.10 Comparative Analysis with Conventional Treatment Methods

Nanomaterials-based wastewater treatment technologies present significant advancements over conventional treatment methods. This comparative analysis evaluates the differences between these approaches by examining conventional wastewater treatment technologies, their limitations, and the advantages offered by nanomaterials-based methods. We will also delve into performance and economic comparisons to provide a comprehensive understanding of how these technologies stack up against each other.

10.11 Conventional Wastewater Treatment Technologies

Conventional wastewater treatment typically involves a combination of physical, chemical, and biological processes designed to remove contaminants from industrial effluents. The primary stages of conventional treatment include various treatment methods.

Preliminary Treatment: This stage involves the removal of large solids and debris through screening and grit removal. It prepares the wastewater for subsequent treatment processes.

Primary Treatment: Physical processes such as sedimentation are used to remove suspended solids. The wastewater is allowed to sit in large tanks, where heavier solids settle at the bottom, forming sludge, while lighter materials float to the surface and are skimmed off.

Secondary Treatment: Biological processes are employed to degrade organic matter using microorganisms. This stage typically involves activated sludge processes, trickling filters, or biofilm reactors, where bacteria break down organic pollutants.

Tertiary Treatment: Advanced chemical and physical processes are used to remove remaining contaminants, including nutrients (e.g., nitrogen and phosphorus), pathogens, and trace organic compounds. Common tertiary treatments include filtration, disinfection (chlorination or UV treatment), and chemical precipitation.

10.12 Limitations

Despite their widespread use and effectiveness in many applications, conventional wastewater treatment technologies have several limitations.

Incomplete Contaminant Removal: Conventional methods often struggle to remove certain contaminants, such as heavy metals, pharmaceuticals, and other emerging pollutants. These residual contaminants can pose environmental and health risks.

Energy and Chemical Intensive: Many conventional treatment processes, especially secondary and tertiary treatments, are energy and chemical intensive. This increases operational costs and environmental impact.

Sludge Production: The sedimentation and biological treatment stages generate large quantities of sludge, which requires further treatment and disposal. Managing sludge is a significant challenge and adds to the overall cost.

Infrastructure Requirements: Conventional treatment plants require substantial infrastructure and space, which can be a limitation in urban areas or regions with land constraints.

Operational Complexity: Managing and operating conventional treatment systems requires skilled personnel and continuous monitoring to ensure optimal performance. This complexity can be a barrier for small-scale or resource-limited facilities.

10.13 Advantages of Nanomaterials-Based Approaches

Nanomaterials-based wastewater treatment technologies offer several advantages over conventional methods.

Enhanced Contaminant Removal: Nanomaterials, due to their high surface area, reactivity, and tunable properties, can effectively remove a wide range of contaminants, including heavy metals, organic pollutants, and pathogens. This leads to cleaner effluent and reduced environmental impact.

Energy Efficiency: Many nanomaterials-based processes, such as adsorption and photocatalysis, require less energy compared to conventional methods. This can result in significant operational cost savings.

Reduced Chemical Use: Nanomaterials can achieve high contaminant removal efficiency with minimal use of chemicals. This reduces the overall chemical consumption and associated costs.

Lower Sludge Production: Certain nanomaterials-based treatments, such as membrane filtration and advanced oxidation processes, produce less sludge compared to conventional biological and physical methods. This simplifies sludge management and disposal.

Compact and Modular Systems: Nanomaterials-based technologies often involve compact and modular systems that require less space. This makes them suitable for urban areas and decentralized treatment applications.

Scalability and Flexibility: Nanomaterials-based treatments can be scaled up or down to suit different wastewater volumes and treatment requirements. They offer flexibility in terms of application and integration with existing infrastructure.

Potential for Resource Recovery: Nanomaterials can facilitate the recovery of valuable resources, such as metals and water, from wastewater. This aligns with circular economy principles and provides additional economic benefits.

10.14 Performance Comparisons

10.14.1 Contaminant Removal Efficiency

Nanomaterials-based technologies typically demonstrate superior performance in contaminant removal compared to conventional methods.

Heavy Metals: Nanomaterials like metal oxides and carbon nanotubes exhibit high adsorption capacities for heavy metals such as lead, mercury, and arsenic, achieving removal efficiencies exceeding 90%.

Organic Pollutants: Advanced oxidation processes (AOPs) using nanocatalysts can degrade persistent organic pollutants, including pharmaceuticals and pesticides, more effectively than conventional chemical treatments.

Pathogens: Nanomaterials with antimicrobial properties, such as silver nanoparticles, can effectively inactivate pathogens, providing enhanced disinfection compared to traditional chlorination or UV treatment.

10.14.2 Operational Efficiency

Nanomaterials-based treatments often operate more efficiently by various ways.

Energy Consumption: Processes like adsorption and membrane filtration using nanomaterials typically require less energy compared to conventional aeration and chemical oxidation methods.

Treatment Speed: Nanomaterials can accelerate treatment processes, reducing the overall time required for contaminant removal. For example, photocatalytic degradation of pollutants can occur rapidly under appropriate light conditions.

10.14.3 Scalability and Adaptability

Nanomaterials-based technologies are highly scalable and adaptable.

Modular Systems: Many nanomaterials-based treatments are designed as modular systems that can be easily scaled to match varying wastewater volumes and treatment needs.

Decentralized Treatment: Compact nanomaterials-based systems are suitable for decentralized treatment applications, making them ideal for remote or small-scale installations where conventional infrastructure is impractical.

10.14.4 Economic Comparisons

The initial capital investment for nanomaterials-based treatment systems can be higher than conventional systems due to the costs associated with R&D, production, and installation of specialized equipment. However, these costs are expected to decrease as production technologies improve and economies of scale are realized.

Research and Development: Significant investment in R&D is required to develop effective nanomaterials tailored for specific treatment applications. This includes costs for laboratory research, pilot-scale testing, and optimization.

Production and Installation: The production of nanomaterials, such as carbon nanotubes and metal oxides, can be expensive due to the sophisticated manufacturing processes involved. Installation costs may also be higher initially due to the need for specialized equipment.

10.14.5 Operational and Maintenance Costs

Nanomaterials-based treatment systems often have lower operational and maintenance costs compared to conventional methods.

Energy Costs: Many nanomaterials-based processes are more energy-efficient, leading to reduced energy consumption and lower operational costs. For example, adsorption and photocatalytic processes typically require less energy than conventional aeration and chemical oxidation.

Chemical Costs: The reduced need for chemicals in nanomaterials-based treatments results in lower chemical costs. High contaminant removal efficiency with minimal chemical use is a significant advantage.

Maintenance Costs: Nanomaterials with high stability and durability require less frequent replacement and maintenance. Processes that produce less sludge also reduce the costs associated with sludge handling and disposal.

10.15 Long-Term Economic Benefits

The long-term economic benefits of nanomaterials-based wastewater treatment technologies include:

Compliance with Regulatory Standards: Enhanced contaminant removal efficiency helps industries comply with stringent environmental regulations, avoiding fines and penalties. This compliance can also improve corporate reputation and market competitiveness.

Resource Recovery and Reuse: The ability to recover valuable materials, such as metals and treated water, provides additional revenue streams. This resource recovery aligns with circular economy principles and contributes to overall economic sustainability.

Operational Savings: Lower energy and chemical costs, combined with reduced maintenance expenses, result in significant operational savings over the long-term. These savings can offset the higher initial investment.

Return on Investment (ROI): The ROI for nanomaterials-based treatment systems can be attractive when considering the total cost of ownership. Operational savings, compliance benefits, and revenue from resource recovery contribute to a favorable ROI.

Nanomaterials-based wastewater treatment technologies offer significant advantages over conventional methods in terms of contaminant removal efficiency, operational efficiency, scalability, and long-term economic benefits. While the initial investment for nanomaterials-based systems may be higher, the lower operational and maintenance costs, combined with the potential for resource recovery and compliance with regulatory standards, make them economically attractive in the long run. The enhanced performance and sustainability of nanomaterials-based approaches provide compelling reasons for their adoption in industrial wastewater treatment, contributing to cleaner effluents, reduced environmental impact, and overall economic viability.

10.16 Societal Acceptance of Nanotechnology in Wastewater Treatment

The integration of nanotechnology into wastewater treatment offers substantial promise for enhancing efficiency and effectiveness. However, societal acceptance of this technology hinges on various factors including public perception, trust, and transparent communication. This section explores the public perception of nanotechnology and the factors influencing trust and acceptance, critical elements for the successful implementation of nanotechnology in wastewater treatment. Table 10.1 represents various social aspects of nanomaterials used in industrial wastewater treatment.

10.17 Public Perception of Nanotechnology

Awareness and Understanding: Public perception of nanotechnology is significantly influenced by the level of awareness and understanding. Many people have limited knowledge about what nanotechnology entails and its potential applications. This lack of understanding can lead to fear and skepticism, as people tend to be wary of technologies they do not comprehend.

Educational Outreach: Increasing public awareness through educational campaigns, workshops, and informational resources can help demystify nanotechnology. Simplifying complex scientific concepts and highlighting practical applications can make nanotechnology more accessible to the general public.

Table 10.1 Social aspects of nanomaterials used in industrial wastewater treatment

Social aspect	Description
Public health	Improved water quality leading to reduced exposure to harmful contaminants and improved community health
Environmental protection	Reduced pollution and environmental contamination, contributing to the preservation of ecosystems
Economic impact	Potential for cost savings in wastewater treatment processes and creation of new job opportunities in nanotechnology sectors
Regulatory safety concern	Need for stringent regulations to ensure safe handling, usage, and disposal of nanomaterials to protect workers and the public
Access to clean water	Enhanced ability to provide clean water to underserved and remote communities, improving overall quality of life
Public perception and acceptance	Public awareness and acceptance of nanotechnology applications in water treatment, influenced by perceived risks and benefits
Technological advancement	Encouragement of further research and innovation in nanotechnology, leading to advancements in other fields
Ethical consideration	Addressing ethical concerns related to the environmental and health impacts of nanomaterial usage
Educational opportunities	Increased demand for education and training programs in nanotechnology and environmental sciences
Global collaboration	Promoting international cooperation and knowledge exchange in the development and implementation of nanotechnology solutions

Media Influence: The role of media in shaping public perception cannot be overstated. Balanced and accurate reporting can help mitigate fears and present nanotechnology as a beneficial tool for environmental sustainability. Conversely, sensationalized media coverage focusing on potential risks without contextual balance can exacerbate public fears.

10.18 Perceived Benefits and Risks

The perception of benefits versus risks plays a crucial role in public acceptance of nanotechnology in wastewater treatment.

Tangible Benefits: When the public perceives clear and tangible benefits, such as improved water quality, cost savings, and environmental sustainability, acceptance tends to increase. Real-world examples and case studies demonstrating the successful application of nanotechnology in wastewater treatment can enhance perceived benefits.

Health and Environmental Concerns: Concerns about the potential health and environmental risks of nanomaterials are common. The public is often wary of unknown long-term impacts, such as nanoparticle toxicity, bioaccumulation, and ecological disruption. Addressing these concerns through transparent communication about safety protocols, regulatory oversight, and ongoing research is essential.

10.19 Historical Context and Analogies

Public perception of new technologies is often influenced by historical context and analogies to previous technological developments.

Comparisons to Past Innovations: Drawing parallels to past successful technological advancements that have become widely accepted can help ease public apprehensions. For instance, comparing the initial skepticism and eventual acceptance of water chlorination or vaccination programs can provide a relatable context for understanding the potential of nanotechnology.

10.20 Factors Influencing Trust and Acceptance

Transparency and Communication:

Trust in nanotechnology for wastewater treatment is largely built on transparent communication from scientists, policymakers, and industry stakeholders.

Clear and Honest Communication: Providing clear, accurate, and honest information about the development, benefits, and risks of nanotechnology is crucial. This includes communicating uncertainties and limitations openly rather than glossing over potential drawbacks.

Engaging Stakeholders: Involving a broad range of stakeholders, including local communities, in discussions and decision-making processes fosters a sense of ownership and trust. Public consultations, forums, and participatory approaches allow for direct communication and feedback.

10.21 Regulatory Oversight

The presence of robust regulatory frameworks that ensure the safety and efficacy of nanotechnology applications is vital for public trust.

Stringent Safety Standards: Implementing stringent safety standards and protocols for the production, use, and disposal of nanomaterials reassures the public that potential risks are being managed effectively. Regulatory bodies must be transparent about the criteria and processes used to evaluate safety.

Independent Assessments: Independent assessments and peer-reviewed studies on the safety and environmental impact of nanotechnology can provide credible evidence to support its use. Transparency in the regulatory review process and the publication of findings further build public confidence.

10.22 Scientific Credibility and Integrity

Trust in the scientific community and research institutions developing and advocating for nanotechnology is essential.

Reputable Institutions: Involvement of reputable research institutions and collaboration with trusted organizations lend credibility to nanotechnology projects. The public is more likely to trust initiatives backed by well-known and respected entities.

Peer-Reviewed Research: Publishing research findings in peer-reviewed journals and making data publicly accessible enhances scientific credibility. It allows independent verification of results and fosters an environment of openness and accountability.

10.23 Risk Communication and Management

Effective communication of risks associated with nanotechnology and the measures taken to mitigate these risks is crucial.

Proactive Risk Management: Proactively addressing potential risks through rigorous testing, continuous monitoring, and adaptive management strategies shows a commitment to safety and responsibility. Clearly outlining these measures helps alleviate public concerns.

Responsive and Adaptive Approaches: Being responsive to emerging risks and adaptive to new information demonstrates a commitment to public safety. Regular updates on safety assessments and adjustments to management practices in light of new evidence are important for maintaining trust.

10.24 Ethical Considerations and Social Responsibility

Ethical considerations and social responsibility play significant roles in influencing public trust and acceptance.

Equitable Access: Ensuring that the benefits of nanotechnology in wastewater treatment are equitably distributed and accessible to all segments of society is essential. Addressing potential disparities and ensuring that marginalized communities are not disproportionately affected by potential risks fosters social acceptance.

Long-Term Commitment: Demonstrating a long-term commitment to sustainable development and environmental stewardship underscores the ethical foundations of nanotechnology applications. Publicizing efforts to reduce environmental footprints and enhance sustainability can build trust.

10.25 Future Directions and Research Needs

As nanotechnology continues to advance, its potential applications in wastewater treatment are becoming increasingly promising. However, to fully realize this potential, future research must address key areas such as innovations in nanomaterials and technologies, bridging economic and societal gaps, and adopting multi-disciplinary approaches to sustainable solutions. This section explores these future directions and research needs in detail.

10.26 Innovations in Nanomaterials and Technologies

10.26.1 Advanced Nanomaterials

Future research should focus on developing advanced nanomaterials with enhanced properties tailored for wastewater treatment.

Functionalization and Modification: Tailoring the surface properties of nanomaterials to enhance their reactivity and selectivity for specific contaminants is crucial. Functionalization techniques, such as adding functional groups or coatings, can improve the performance of nanomaterials in adsorption, catalysis, and filtration processes.

Composite Materials: Combining different types of nanomaterials to form composites can synergistically enhance their properties. For example, hybrid nanocomposites that integrate carbon-based materials with metal or metal oxide nanoparticles can offer superior performance in removing a wide range of pollutants.

Nano-catalyst types	Applications
Metal nanoparticles	
– Gold (Au)	Catalytic degradation of organic pollutants, reduction of heavy metals
– Silver (Ag)	Antimicrobial activity, degradation of dyes and organic contaminants
– Platinum (Pt)	Oxidation of volatile organic compounds (VOCs), reduction of nitrites
– Palladium (Pd)	Hydrogenation reactions, removal of organic chlorides
Metal oxide nanoparticles	
– Titanium dioxide (TiO_2)	Photocatalytic degradation of organic pollutants, disinfection of pathogens
– Zinc oxide (ZnO)	Photocatalytic degradation of dyes and organic contaminants, antimicrobial activity
– Iron oxide (Fe_3O_4, Fe_2O_3)	Removal of heavy metals, magnetic separation of contaminants
– Cerium oxide (CeO_2)	Oxidation of organic pollutants, removal of heavy metals
Bimetallic nanoparticles	
– Au–Pd	Synergistic catalytic degradation of organic pollutants
– Au–Ag	Enhanced photocatalytic and antimicrobial activities
– Pt–Pd	Improved hydrogenation and oxidation reactions
Carbon-based nanoparticles	
– Carbon nanotubes (CNTs)	Adsorption of organic pollutants, catalytic degradation of dyes and pharmaceuticals
– Graphene oxide (GO)	Adsorption and catalytic degradation of heavy metals and organic contaminants
– Fullerenes	Photocatalytic degradation of pollutants, reactive oxygen species generation
Zeolite-based nanocatalyst	
– Zeolite-Y	Adsorption of heavy metals and organic pollutants, catalytic degradation of VOCs
– Zeolite-A	Adsorption and ion exchange for removal of heavy metals and ammonium
Magnetic nanoparticles	
– Magnetite (Fe_3O_4)	Removal and magnetic separation of heavy metals, organic pollutants
– Cobalt ferrite ($CoFe_2O_4$)	Adsorption and magnetic separation of dyes and heavy metals
Nanocomposites	
– TiO_2–CNT	Enhanced photocatalytic degradation of organic pollutants
– Ag–ZnO	Synergistic antimicrobial and photocatalytic degradation activities

(continued)

(continued)

Nano-catalyst types	Applications
– Fe_3O_4–GO	Magnetic separation and catalytic degradation of heavy metals and organic contaminants
Polymeric nanoparticles	
– Polymer-coated metal	Enhanced stability and reusability for catalytic degradation of pollutants
– Polymer Nanocomposites	Combined adsorption and catalytic degradation of dyes and organic contaminants
Quantum dots	
– CdS quantum dots	Photocatalytic degradation of organic pollutants, visible light activation
– ZnS quantum dots	Photocatalytic and antimicrobial activities under UV light
Bio-inspired nanocatalysts	
– Enzyme-mimetic	Catalytic degradation of specific organic pollutants, mimicking natural enzymatic processes
– Biomolecule-coated	Targeted removal and degradation of specific contaminants, enhanced selectivity and biocompatibility

10.26.2 Nanotechnology-Enhanced Filtration

Innovations in nanotechnology-enhanced filtration systems can significantly improve the efficiency and cost-effectiveness of wastewater treatment.

Membrane Technologies: Developing nano-engineered membranes with high permeability and selectivity can lead to more efficient filtration processes. Research into antifouling coatings and self-cleaning properties can extend membrane lifespan and reduce maintenance costs.

Nanofibers and Aerogels: Exploring the use of nanofibers and aerogels in filtration can provide high surface area and porosity, leading to effective removal of contaminants. These materials can be integrated into existing treatment systems or used in standalone applications.

10.26.3 Photocatalytic and Electrocatalytic Processes

Enhancing the efficiency of photocatalytic and electrocatalytic processes using nanotechnology is a promising area for future research.

Nanophotocatalysts: Developing nanophotocatalysts with high photoactivity under visible light can improve the degradation of organic pollutants. Innovations in doping

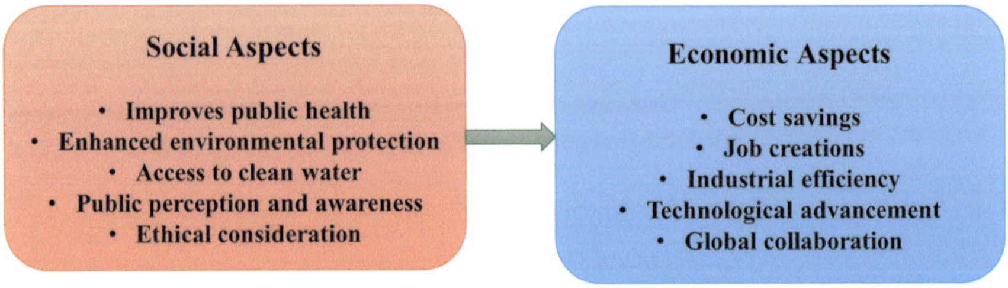

Fig. 10.2 Social and economic aspects

and heterojunction formation can enhance the photocatalytic efficiency of materials like titanium dioxide (TiO_2) and zinc oxide (ZnO).

Electrocatalysts: Research into nanostructured electrocatalysts for electrochemical treatment processes can enhance the degradation of contaminants and energy efficiency. Advances in materials like graphene and metal–organic frameworks (MOFs) can lead to more effective and sustainable electrocatalytic systems.

10.26.4 Smart and Responsive Nanomaterials

The development of smart and responsive nanomaterials that can adapt to changing environmental conditions or contaminant levels is a key future direction.

Stimuli-Responsive Materials: Designing nanomaterials that respond to external stimuli such as pH, temperature, or light can enable dynamic and targeted treatment processes. These materials can be used to release or capture contaminants based on specific triggers.

Self-Healing Materials: Research into self-healing nanomaterials that can repair themselves after damage can improve the durability and lifespan of treatment systems, reducing the need for frequent replacements (Fig. 10.2).

10.27 Conclusions

The application of nanomaterials in industrial wastewater treatment holds significant promise for addressing both social and economic challenges, marking a transformative shift in environmental management practices. On the social front, the deployment of nanomaterials can substantially enhance public health by effectively removing a wide array of contaminants, including heavy metals, organic pollutants, and pathogens, from industrial effluents. This leads to the provision of cleaner water, reducing the incidence of waterborne diseases and contributing to overall community well-being. In regions where access to clean water is limited, the adoption of nanotechnology can bridge critical gaps, ensuring safer drinking water for underserved populations and thus improving their quality of life. The environmental benefits are equally compelling; by mitigating pollution and promoting the recovery of valuable resources from wastewater, nanomaterials contribute to the preservation of ecosystems and biodiversity. Economically, the integration of nanomaterials in wastewater treatment processes can result in significant cost savings. Traditional water treatment methods often involve high operational and maintenance costs, whereas nanomaterials, with their high reactivity and efficiency, can streamline these processes, reducing energy and chemical consumption. This efficiency translates into lower treatment costs and operational expenses for industries, enhancing their economic viability. Furthermore, the burgeoning field of nanotechnology spurs economic growth through the creation of new jobs and industries centered around the development, production, and application of nanomaterials. The growth of these sectors not only drives innovation but also attracts investments, fostering a robust economic ecosystem.

However, the widespread adoption of nanomaterials is not without challenges. Regulatory frameworks need to evolve to address the unique safety and environmental concerns associated with nanomaterials. Ensuring safe handling, usage, and disposal is paramount to protecting both workers and the public from potential risks. This necessitates comprehensive regulations and standards that can keep pace with the rapid advancements in nanotechnology. Public perception also plays a crucial role in the acceptance and success of these technologies. While the benefits are substantial, there are apprehensions about the long-term impacts of nanomaterials on health and the environment. Therefore, transparent communication and public engagement are essential to build trust and acceptance. Educating communities about the advantages and addressing their concerns can pave the way for broader acceptance and support. The ethical considerations surrounding the use of nanomaterials must also be addressed, particularly in terms of environmental justice and equitable access. Ensuring that the benefits of these technologies reach all segments of society, especially marginalized and vulnerable communities, is critical.

Technological advancements in nanomaterials open up further opportunities for international collaboration and knowledge exchange. By sharing research findings and technological innovations, countries can collectively enhance their wastewater treatment capabilities, promoting global environmental sustainability. This collaborative approach can accelerate the development of more efficient and cost-effective solutions, benefiting countries with varying levels of technological advancement. Additionally, the emphasis on research and education in nanotechnology can stimulate academic and professional growth, preparing a skilled workforce equipped to tackle future challenges in environmental management. In conclusion, the social and economic implications of utilizing nanomaterials in industrial wastewater treatment are profound and multifaceted. While the benefits in terms of public health, environmental protection, and economic efficiency are clear, addressing the accompanying regulatory, ethical, and public perception challenges is crucial for the sustainable and equitable deployment of these technologies. The path forward lies in fostering innovation, ensuring robust safety standards, and engaging in transparent public dialogue, all while promoting global cooperation and education to maximize the positive impact of nanotechnology on society and the economy.

References

Ahmed J, Thakur A, Goyal A (2021) Industrial wastewater and its toxic effects. In: Shah MP (ed) Biological treatment of industrial wastewater, 1st edn. The Royal Society of Chemistry, London, pp 1–14

Ali I, Jain CK (2017) Advances in adsorption technology for water treatment. Crit Rev Environ Sci Technol 47(23):2416–2448. https://doi.org/10.1080/10643389.2017.1393262

Ali I et al (2019) Advances in nanotechnology for sustainable wastewater treatment. Crit Rev Environ Sci Technol 49(8):733–777. https://doi.org/10.1080/10643389.2018.1549244

Ali MN, Fouad HA, Meky MM, Elhefny RM (2021) Pilot-scale study based on integrated fixed-film activated sludge process for cement industrial wastewater treatment. J Degrad Min Lands Manag 9:3073–3081. https://doi.org/10.15243/jdmlm.2021.091.3073

Chen X, Mao SS (2019) Titanium dioxide nanomaterials: synthesis, properties, modifications, and applications. Chem Rev 107(7):2891–2959. https://doi.org/10.1021/cr050039k

Crini G, Lichtfouse E (2019) Advantages and disadvantages of techniques used for wastewater treatment. Environ Chem Lett 17:145–155. https://doi.org/10.1007/s10311-018-0785-9

Das R et al (2020) Carbon nanotubes and graphene for emerging contaminant removal technologies. Chem Eng J 401:126135. https://doi.org/10.1016/j.cej.2020.126135

Freeda Gnana Rani D, Arunkumar K, Sivakumar SR (2005) Physicochemical analysis of waste water from cement units. J Ind Pollut Control 21:337–340

Gao W, Feng X (2022) Nanotechnology-enhanced membranes for water purification. Chem Soc Rev 51(11):2838–2859. https://doi.org/10.1039/D2CS00012G

Jones ER, van Vliet MTH, Qadir M, Bierkens MFP (2021) Country-level and gridded estimates of wastewater production, collection, treatment and reuse. Earth Syst Sci Data. 13:237–254. https://doi.org/10.5194/essd-13-237-2021

Kamali M, Persson KM, Costa ME, Capela I (2019) Sustainability criteria for assessing nanotechnology applicability in industrial wastewater treatment: current status and future outlook. Environ Int 125:261–276. https://doi.org/10.1016/j.envint.2019.01.055

Kang S et al (2019) Antimicrobial properties of nanomaterials in water disinfection. Water Res 43(3):1051–1058. https://doi.org/10.1016/j.watres.2018.12.017

Kato S, Kansha Y (2024) Comprehensive review of industrial wastewater treatment techniques. Environ Sci Pollut Res 31:51064–51097

Keller AA, Lazareva A (2018) Global life cycle releases of engineered nanomaterials into the environment. Environ Sci Technol 47(5):2513–2520. https://doi.org/10.1007/s11051-013-1692-4

Ma W et al (2019) Nanomaterials for heavy metal removal from wastewater. Environ Pollut 255:113196. https://doi.org/10.1016/j.envpol.2019.113196

Mohan D, Pittman CU, Steele PH (2018) Emerging green technologies for wastewater treatment. J Hazard Mater 361:120–135. https://doi.org/10.1016/j.jhazmat.2018.07.113

Nagar A, Pradeep T (2020) Clean water through nanotechnology: needs, gaps, and fulfillment. ACS Nano 14(6):6420–6435. https://doi.org/10.1021/acsnano.9b01730

Neale PA et al (2021) Ecotoxicology of engineered nanoparticles: Pathways to sustainability. Sci Total Environ 756:143983. https://doi.org/10.1016/j.scitotenv.2020.143983

Neale PA et al (2021) Nanoparticle ecotoxicology in freshwater systems. Sci Total Environ 701:134934. https://doi.org/10.1016/j.scitotenv.2019.134934

Omar HH (2018) Nanofiltration for water desalination and wastewater recycling. Desalination 455:100–109. https://doi.org/10.1016/j.desal.2018.02.012

Pratiwi D, Sumiarsa D, Oktavia D, Sunardi S (2023) Water quality influences self-purification in the Cihawuk and Majalaya segments upstream of the Citarum river, West Java, Indonesia. Water 15:2998. https://doi.org/10.3390/w15162998

Savage N, Diallo MS (2005) Nanomaterials and water purification: opportunities and challenges. J Nanopart Res 7(4):331–342. https://doi.org/10.1007/s11051-005-7523-5

Scott NR (2019) Nanotechnology and the environment: ethical issues. Environ Eng Sci 36(1):5–15. https://doi.org/10.1089/ees.2018.0342

Shen L et al (2020) Performance of nanomaterials in water treatment. Sep Purif Technol 232:115881. https://doi.org/10.1016/j.seppur.2019.115881

Tang C et al (2023) Nanotechnology in industrial wastewater recycling. J Environ Chem Eng 9(1):104386. https://doi.org/10.1016/j.jece.2022.104386

Thompson S, Tiwari D (2021) Advances in nanomaterials for environmental remediation. Chemosphere 281:130812. https://doi.org/10.1016/j.chemosphere.2021.130812

Xu L, Li S, Yang Y (2021) Nanotechnology's role in meeting sustainable development goals for water management. Nano Res 14(5):1237–1250. https://doi.org/10.1007/s12274-021-3346-1

Yaqoob A, Parveen T, Umar K, Ibrahim NM (2020) Role of nanomaterials in the treatment of wastewater: a review. Water 12(2):495. https://doi.org/10.3390/w12020495

Zahmatkesh S, Hajiaghaei-Keshteli M, Bokhari A, Sundaramurthy S, Panneerselvam B (2023) Wastewater treatment with nanomaterials for the future: a state-of-the-art review. Environ Res 216:114652. https://doi.org/10.1016/j.envres.2022.114652

Zhu X, Yang J, Huang Q, Liu T (2022) A review on pollution treatment in cement industrial areas: from prevention techniques to python-based monitoring and controlling models. Processes 10. https://doi.org/10.3390/pr10122682

Chapter 11
Metal Oxide Nanocomposites for Industrial Wastewater Remediation

Alaa El Din Mahmoud and Ghada Ashour

Abstract Due to the escalating impacts of the industrial wastewater on the environment, especially water pollution, we highlight and discuss the role of metal oxide nanocomposites as promising materials for industrial wastewater remediation. We summarize their chemical and green synthesis processes as well as their applications including adsorption and photocatalytic degradation revealing their efficiency in eliminating a wide spectrum of pollutants. In addition to the limitations, and future implications for those metal oxide nanocomposites in industrial wastewater remediation. Several synthesis strategies are reviewed, including sol–gel, hydrothermal, and precipitation procedures, green synthesis using plant and algae extracts emphasizing their capacity to alter the structure, and characteristics of nanocomposites for specific remedial needs. Furthermore, the underlying methods of adsorption and photocatalytic degradation routes are investigated, offering insights into the fundamental processes that drive pollution removal. Despite their immense potential, concerns including scalability, cost-effectiveness, and environmental impact must be addressed carefully. Nevertheless, with ongoing advancements in nanotechnology, materials science, and environmental engineering, the prospects for metal oxide nanocomposites in industrial wastewater treatment appear promising. Through interdisciplinary collaboration and sustained research, this type of nanomaterials offer a viable pathway toward sustainable solutions for water pollution mitigation and environmental conservation.

Keywords Nanotechnology · Chemical synthesis · Green synthesis · Nanomaterials · Wastewater treatment · Adsorption · Photocatalysis

A. E. D. Mahmoud (✉)
Environmental Sciences Department, Faculty of Science, Alexandria University, Alexandria, Egypt
e-mail: alaa-mahmoud@alexu.edu.eg

Green Technology Group, Faculty of Science, Alexandria University, Alexandria, Egypt

G. Ashour
Environmental Science and Industrial Development Department, Faculty of Postgraduate Studies for Advanced Sciences, Beni-Suef University, Beni Suef, Egypt

A. Mandpe et al. (eds.), *Nanotech Solutions for Industrial Wastewater*, Advances in Wastewater Research, https://doi.org/10.1007/978-981-96-6879-3_11

11.1 Introduction

The release of industrial effluents into the environment is a source of tremendous worry, especially in developing countries. In some cases, discharges of wastewater from industrial facilities are either untreated or insufficiently treated before release, posing a harm to the environment and human health (Mahmoud 2020a; Quintelas et al. 2021). Industrialization has resulted in the discharge of industrial pollution into rivers and other bodies of water. This has led to variations in the water environmental condition of river basins throughout various regions. The spatial pattern and evolution of transport capability of pollutants from industries are influenced by the level of industrial development, industrial sites and conditions, and water environmental constraints (Chen and Chen 2021). Gaur et al. (2020) highlights the three of the most significant industries: the food, paper, and pulp industries, as well as agriculture, all of which are growing quickly. Figure 11.1 summarizes the potential fate of industrial wastes.

In the context of global water consumption, the industrial sector accounts for approximately 22% of usage and around 80% of the wastewater generated by these industries is discharged directly into waterways, posing significant risks to aquatic ecosystems and contributing to environmental pollution (Dutta et al. 2021). The ecosystem and human health are both harmed by the toxic substances and chemicals found in wastewater such as volatile organic contaminants, persistent organic pollutants, and toxic organic substances (Ansari et al. 2024; Saravanakumar et al. 2022; Guozhu et al. 2022). According to the sector, wastewater can have a variety of contents, including paper, leather, and wool are examples of biodegradable materials, while non-biodegradable waste includes heavy metals, pesticides, and plastic. Industrial wastewater may be flammable, reactive, toxic, or carcinogenic. As a result, dumping rubbish into bodies of water can have severe consequences for the environment and human health if proper treatment and management methods are not implemented.

Wastewater from industry is produced as a result of human activity and is carelessly discharged into the environment and water bodies without being treated, includes impurities such as heavy metals, lubricants, antibiotics, poisonous colors, and supplementary compounds (Mahmoud and Mostafa 2023; Mahmoud and Krasucka 2023). Toxins in industrial wastewater are determined by the type of plant and industry. Mining, steel, and iron production are examples of industries that produce wastewater, power plants, industrial laundry facilities, fracking for oil and gas, food and beverage industry, as well as metal finishers. Pesticides, oils, chemical substances, heavy metals, silt, pharmaceuticals, and other industrial wastes are among the many contaminants commonly found in industrial water outflows (Ahmed et al. 2021; Abdelhaleem et al. 2024).

The efficacy of metal oxides and their nanocomposites materials in the process of treating wastewater is being investigated, the chemical and ecological ways of creating metal oxides and nanocomposites of it, in addition to the methods and characteristics of their characterization (Ani and Egbosiuba 2023). To elaborate the role

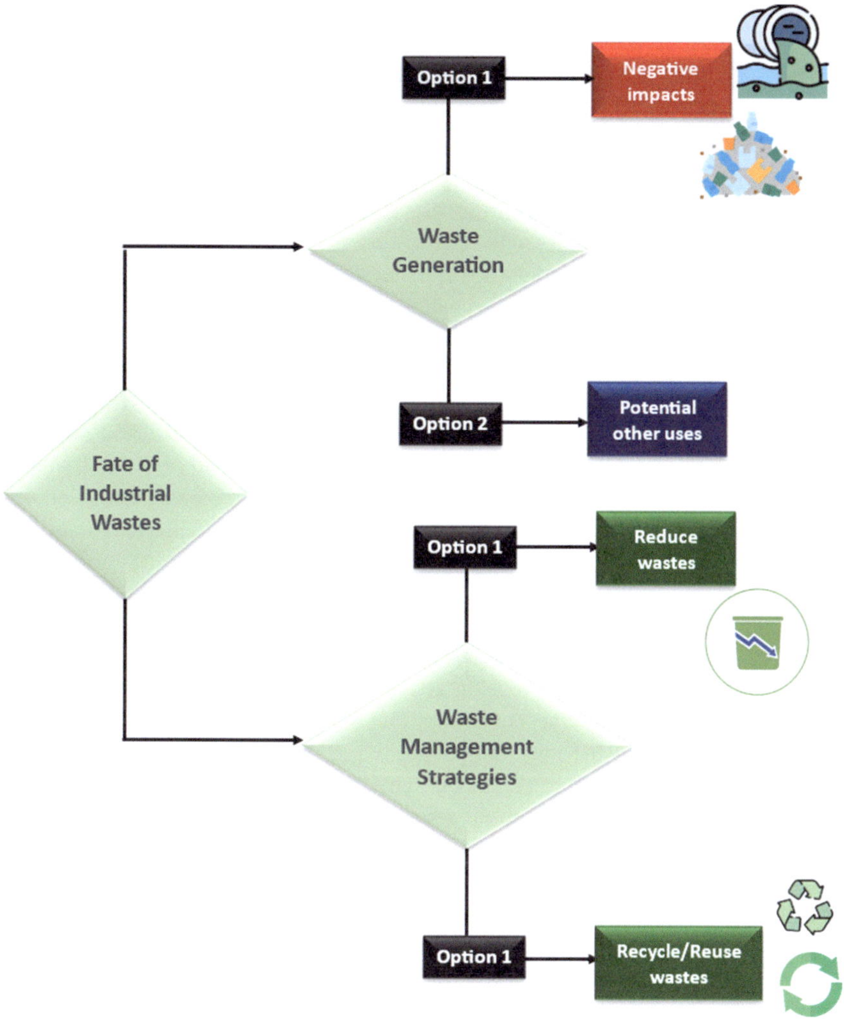

Fig. 11.1 The potential fate of industrial wastes

of metal oxide nanocomposites, we start with defining nanocomposites. Nanocomposites are solid materials that made up of many phases, at least one of which has one, two, or three nanometer-sized dimensions. The goal of establishing synergy between different substances is accomplished by nanoscale phase process. Phase boundaries are created when material diameters are lowered to the nanoscale levels, and these interfaces are critical for increasing the material's properties. Understanding the connection between structure and property is intimately tied to the surface area to volume ratio of the reinforced material used in the production of nanocomposites (Mahmoud 2020b; Omanović-Mikličanin et al. 2019).

Nanocomposites are categorized into ceramic-polymer, inorganic–organic polymer, polymer–layered nanocomposite of silicate, and inorganic–organic hybrid polymer nanocomposites (El-Kady et al. 2023). Those categories have numerous applications including environmental, biomedical, electronics, automotive, aerospace, building electromagnetic shielding, gas, and oil pipelines (Hassan et al. 2021).

Nanocomposites are likely a cutting-edge adsorbent for the effective and affordable heavy metal elimination from aquatic systems. Therefore, we highlight some examples of hybrid metal oxide/biochar materials. Methyl orange (MO) was effectively eliminated from one set of composites containing Ag and TiO_2 connected with biochar through the processes of calcination, mixing, and photodeposition. Due to the way that Ag, TiO_2, and biochar work together, it was discovered that all catalysts treated with Ag proved superior than solo TiO_2 in terms of photocatalytic decomposition performance; the greatest the decolorize efficiency and mineralized efficiency rates were 97.48% and 85.38%, respectively (Shan et al. 2020).

Another example, as a powerful adsorbent, a solvent-free ball milling process was used to prepare the nanostructures of Ag/MgO/biochar. Challenges include the development of functional materials with sufficient adsorption to remove organic contaminants from water completely. Various characterization indicates that the presence of MgO nanoparticles dispersed evenly throughout charcoal surfaces in the biochar matrix. Powdered metal improved nanocomposites' macropores and mesopores during the breakdown of biochar and compression of MgO. Modified biochar nanocomposites may be synthesized, according to XPS analysis. When enough MgO was added to charcoal, the nanocomposites' capacity to eliminate methylene blue (MB) increased (Venkatesh et al. 2022). Venkatesh et al. (2022) shows that the use of ball milling technique offers more adaptability and effectiveness in a molecular and structural formulation that may be strategically utilized to remove organic and inorganic contaminants from the environment.

Many studies have focused on the developing nanoparticles for wastewater treatment. Metal oxide compounds are garnering greater interest in the water and wastewater treatment because they have better qualities than their separate counter elements (Shirsat et al. 2022). Compared with their bulk counterparts, the nanocomposites of metal oxides show noticeably better surface properties, including a large number of desirable surface functions, active sites, improved surface energy, high surface area to volume ratio, nano structural flaws, and homogeneous metallic oxide dispersion nanoparticles. These attributes create metal oxide nanocomposites attractive candidates for surface-driven applications like catalysis and adsorption (Joshi et al. 2020). Metal oxide nanocomposites' large surface area and multiple active sites promote adsorption of molecules of gas and liquid-based contaminants onto their surface. This is a crucial prerequisite for their effective application in wastewater remediation processes such as adsorptive removal, photocatalysis, electrocatalysis, and others (Gusain et al. 2019; Kathi and Mahmoud 2024).

11.2 Synthesis of Metal Oxide Nanocomposites

It is crucial to synthesize uniformly sized nanocomposite materials because, depending on their size and dimensions, these materials can have optical, magnetic, electrical, and biological properties. Gas phase synthesis, vapor phase synthesis, and solution-based synthesis are the three classes that are commonly used to categorize synthetic techniques. An alternative strategy would be to categorize these synthetic procedures into two main groups: (a) top-down approaches, which comprise physical techniques, and (b) bottom-up approaches, which include wet methods (Mahmoud 2020b). The desirable properties needed in nanoparticle composites, such as size, shape, crystal structure, determine which synthetic technique is best (Lateef and Nazir 2017).

11.2.1 Chemical Production of Metal Oxide Nanocomposites

11.2.1.1 Chemical Vapor Deposition (CVD)

Among various methods, Chemical Vapor Deposition (CVD) procedures are well-known routes to oxide-based systems since they entail the regulated molecular precursors are converted from the vapor phase into the target compounds. High flexibility, rapid growth, reaction environment control, superb covering conformal steps, and being scalable to large-scale processing are the primary benefits of CVD (Bekermann et al. 2012).

One method of chemical deposition is chemical vapor deposition (CVD) wherein a volatile fluid or gaseous precursor reacts chemically with heat to generate thin coatings on an overheated surface. As an outcome of numerous adjustments made to traditional CVD techniques to address the precursor volatility issues, new branches of CVD have developed. The advanced CVD techniques are categorized as ultrahigh vacuum CVD (UHVCVD) at pressures below 10−8 torr and a low-pressure CVD (LPCVD) at sub-atmosphere pressures based on the operating conditions. Low pressure minimizes undesired gas-phase reactions and produces a coating film of superior quality. Additionally, the CVD techniques are divided into two categories: direct liquid injection CVD (DLICVD) and aerosol-assisted CVD (AACVD) depending on the vapor's physical properties. Direct liquid injection CVD (DLICVD) and aerosol-assisted CVD (AACVD) depending on the vapor's physical properties (Ehsan et al. 2021) as illustrated in Fig. 11.2

On Al_2O_3 substrates, CuO-TiO_2-Au nanocomposites were created using a multi-step chemical vapor deposition (CVD) process (Bharti and Sadhu 2022). TiO_2 was put on the copper oxide layer after CuO was first deposited on Al_2O_3. Finally, employing RF sputtering, gold was deposited. The precursor for CuO, Cu(hfa)2 · TMEDA

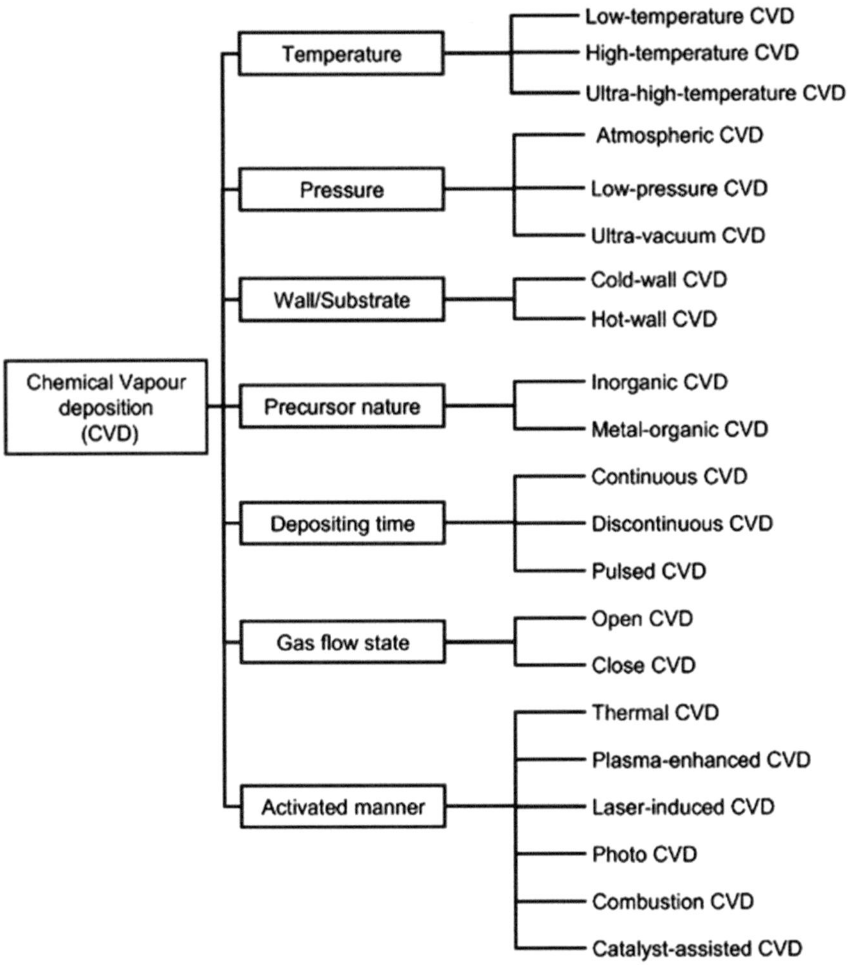

Fig. 11.2 CVD method classification. Adapted from Saeed et al. (2020)

where (hfa = 1,1,1,5,5,5-hexafluoro-2,4-pentanedionate; TMEDA = N,N,N′,N′-tetramethylethylenediamine), was vaporized at 60 °C and moved by O_2 flow toward Al_2O_3 substrates that were utilized as growth surfaces (Barreca et al. 2011).

The same was true for $Ti(O\text{-}iPr)_2(dpm)_2$ this chemical compound is composed of titanium (Ti) coupled with two isopropoxy (O-iPr) groups and two molecules containing ligand 2,2,6,6-tetramethyl-3,5-heptanedionate (dpm). The dpm ligand is widely utilized in coordination chemistry because of its propensity for creating stable complexes with metal ions. The substance evaporates at a comparatively low temperature of 80 °C. This shows that it transitions from a solid to a vapor at this level of temperature. When the vapor comes into touch with a surface in which there is a flow of H_2O (water) and O_2 (oxygen), it settles there. This procedure is frequently used in thin film deposition techniques when fine control over material deposition is needed. Radio frequency (RF) sputtering involves bombarding a target material (in this example, gold) with high-energy ions, which are normally created from a plasma, to expel atoms or molecules off the target surface. Argon gas is often utilized as a sputtering gas due to its inertness and ability to generate a stable plasma. The introduction of argon flow into the sputtering chamber helps to sustain the plasma and facilitates the sputtering process. Gold atoms are discharged from the target and deposit onto the substrate (in this example, $CuO \cdot TiO_2$). It is vital to keep in mind that during this procedure, the material's temperature, pressure, and radio frequency power are carefully managed to produce the required deposition properties while protecting the integrity of the substrate material (Bharti and Sadhu 2022).

Another example to the process of Graphene-based nanocomposites, Graphene is made via a variety of CVD techniques, including thermal CVD, CVD with hot/cold walls, and CVD boosted by plasma. To produce graphene, hydrocarbon precursors like ethylene, acetylene, or methane are typically heated up to extreme temperatures. In addition to gaseous hydrocarbons, liquid precursors like pentane or hexane have found widespread use. Additionally, various resources including food, insects, and garbage can be utilized in the technique of CVD to make different graphene materials as shown in Fig. 11.3.

In addition, the synthesis of MoS_2/C nanocomposites, for sensing and electrochemical applications, a uniform MoS_2 deposition is required, the CVD procedure is typically used to synthesize MoS_2 thin films. It's an adaptable technique that allows you to deposit any kind of chemical over another material. High purity of the sample is obtained, and it is also cost-effective (Ahmaruzzaman and Gadore 2021).

11.2.1.2 Sol–Gel Method

Similar to other solution-based approaches, sol–gel technologies are inexpensive and offer a number of benefits, including low temperatures, excellent homogeneity and purity, accurate composition control, and thin film, fiber production, monolith, and flexible powder. Figure 11.4 illustrates the general procedures for thin film deposition using sol–gel. Precursors for metals are combined with a solvent to create a sol. This sol is applied via spray coating, spin or dip, or to a substrate. There is a thin gel layer created after polymerization. The gel film then undergoes low-temperature thermal treatment, turning it into an amorphous film. A dense, crystalline film is produced by high-temperature thermal processing. All it takes to make thicker films is to repeat the coating and heat-treatment procedures (Fei et al. 2013).

Because of the sluggish reaction kinetics, which enable excellent structural design of the finished product, the sol–gel method is also very suitable for producing nanostructures with multiple components. The reactions are carried out at room temperature or at low temperatures, which provides an additional benefit. Through a sequence of chemical reactions involving inorganic precursors, a three-dimensional molecule structure is formed during the sol–gel process. One of the most frequent ways is the condensation and hydrolysis of metal alkoxides to produce larger metal oxide

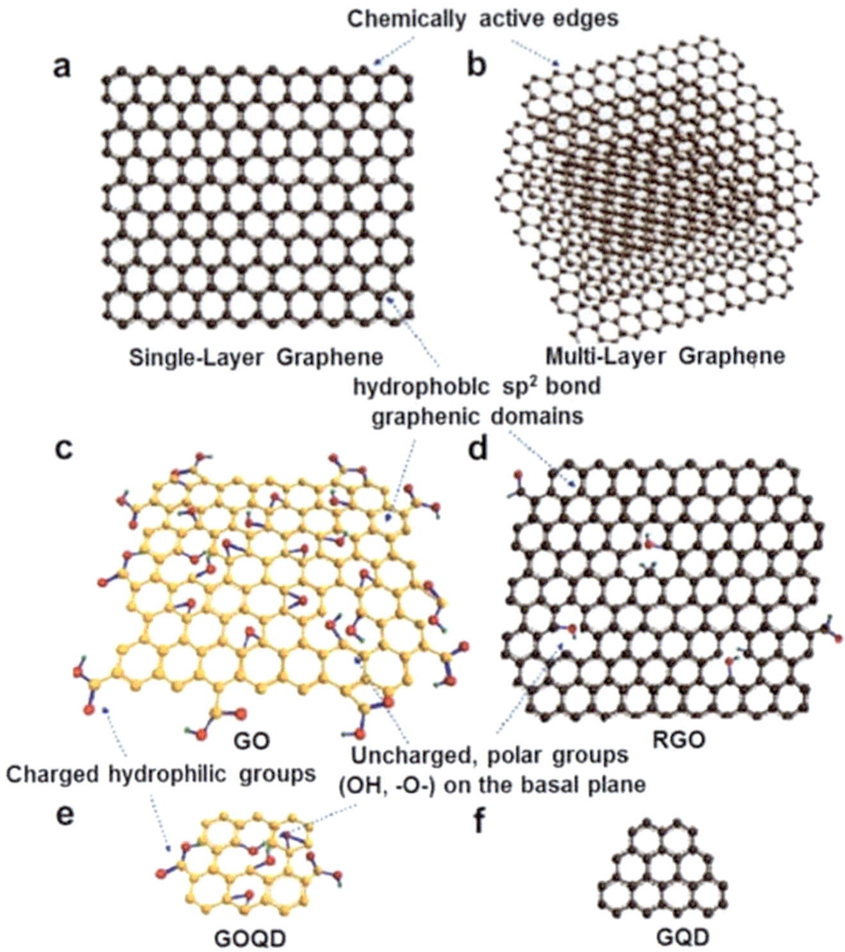

Fig. 11.3 Graphene's structure and different forms. Adapted from Yang et al. (2021)

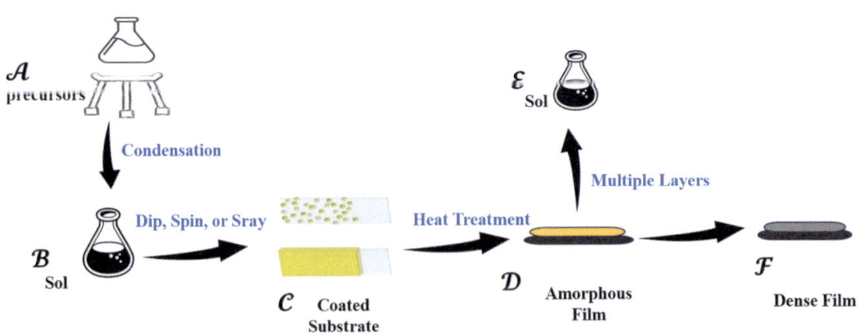

Fig. 11.4 Schematic representation of an example of film deposition using the sol–gel process

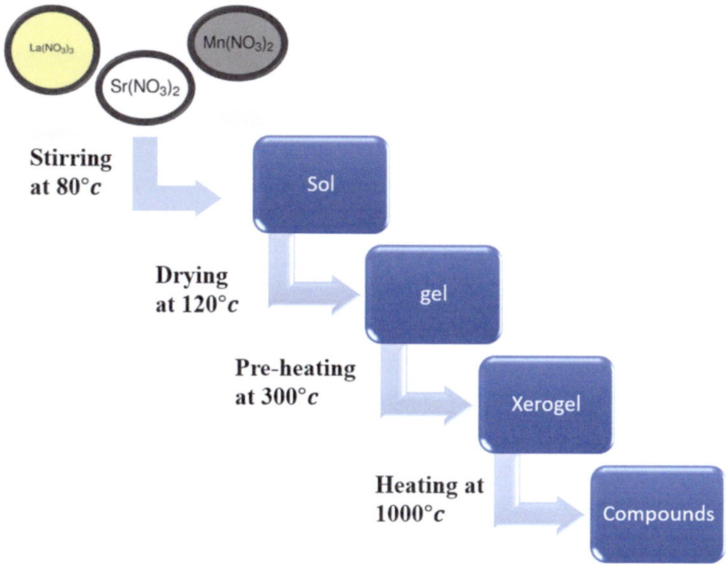

Fig. 11.5 Sol–gel technique schematic

molecules, which subsequently polymerize to form the coating as shown in Fig. 11.5 which adapted from Lateef and Nazir (2017).

Many research groups have used the sol–gel technique up to this point to generate high-quality thin films. Regarding structure, transport, and epitaxy quality, a comparison between sol–gel prepared epitaxial $La_{0.67}Ca_{0.33}MnO_3$ (LCMO) films atop $LaAlO_3$ (LAO) and PLD-grown films shows that the post-annealed LCMO films obtained using a straightforward sol–gel process are similar to PLD films (Fei et al. 2013). In sol–gel processing, synthetic control over nanoparticle size has been and continues to be a challenge. Convenient pathways to monodisperse nanocomposites are still required, even if control over the host matrix pores' size provides some control over the nanoparticles' size. Because the length and temperature of heat treatment affect the average nanoparticle size and particle size distribution, less energy-intensive synthetic techniques are required to produce highly monodisperse nanocomposites (Kwiatkowski and Lukehart 2000).

To give an example, we have sol–gel silica matrix was used in the production of the Au-NiO nanocomposite material. Similarly, titanium (IV) butoxide and acetylacetone in an ethanol–water combination were used to create the Au-TiO_2-NiO films. The sol–gel process was utilized to generate the final TiO_2-NiO nanocomposites, and the same methodology was employed to introduce gold into the nanocomposites (Bharti and Sadhu 2022). AuNPs were deposited on porosity sol–gel metal oxide films as part of a synthetic approach published for nanocomposite films of SiO_2, NiO, and Au. The bifunctional ligand 3-(2 aminoethylamino) propyltrimethoxysilane was used to prepare the NiO precursor solution. The amine groups in this ligand connected with the metal ions and helped bind using the silica matrix that contains gold salt

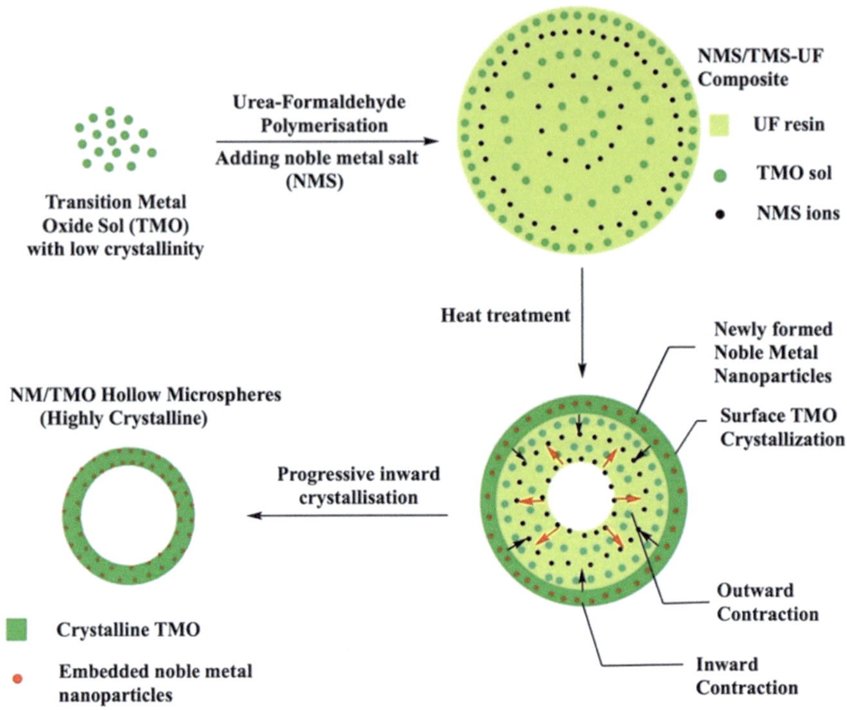

Fig. 11.6 Sol–gel method as a synthetic approach for hollow Au-TiO_2. Adapted from Bharti and Sadhu (2022)

that was synthesized in ethanol. As shown in Fig. 11.6, Lin et al. devised a synthetic approach regarding the hollow Au-TiO_2 and $AuCe_xTi1\text{-}xO_2$ nanocomposites (Bharti and Sadhu 2022).

11.2.1.3 Hydrothermal Synthesis

Many different metal oxide nanostructures are generated using hydrothermal synthesis. This approach is regarded as one of the easiest and most practical methods. It doesn't need complex preparatory conditions or specialized equipment. By altering reaction parameters such as temperature of reaction, reactant concentration, and reaction time, it regulates the chemical makeup, particle dimensions and the shape of the produced products (Nandagudi et al. 2022). A heterogeneous chemical reaction that takes place throughout a closed system at an ambient pressure more than 1 atm and above room temperature in an aquatic or non-aqueous fluid is what the hydrothermal technique entails (Fig. 11.7) (Byrappa and Adschiri 2007). Utilizing surfactants, capping agents, and mineralizers is a standard procedure to alter the size and characteristics (Lateef and Nazir 2017).

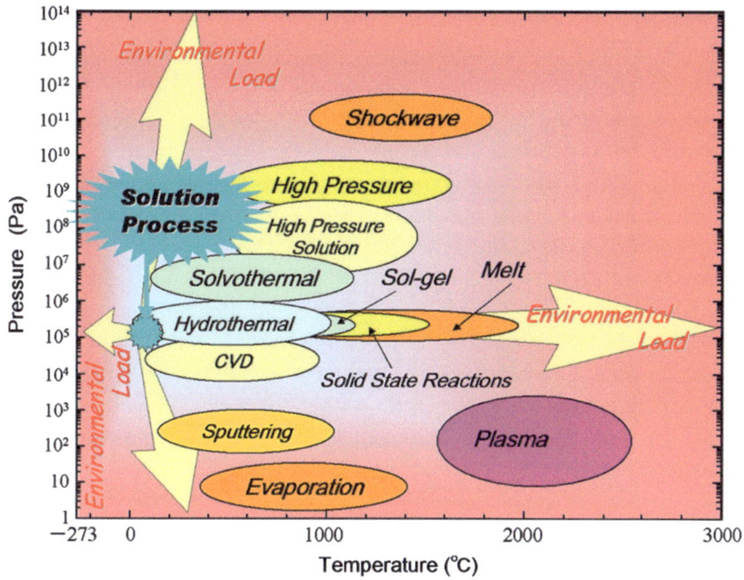

Fig. 11.7 Pressure–temperature map of methods used in material processing. Adapted from Byrappa and Adschiri (2007)

To reduce particle size, an in-situ, surfactant-free simple hydrothermal technique was utilized to generate the nanoparticles of Ag, Sn, and Ag-Sn (Pandiyan et al. 2022). Additionally, the study of Mapossa et al. (2021) generated cobalt ferrite by apply of the hydrothermal process. With 86 emu/g of saturation magnetization (Ms), the resulting material included crystallites of 70 nm.

Applications of hydrothermal also include creation of nanocomposites with mixed metal oxides. A hydrothermal method was employed to produce a $CuO/NiO/ZrO_2$ nanocomposite using $Cu(NO_3)$ with $3H_2O$, sodium hydroxide and $Ni(NO_3)2 \cdot 6H_2O$, and $ZrOCl_2 \cdot 8H_2O$. To the aforementioned solution, mix 3 M sodium hydroxide with copper and nickel solutions (0.5 M Cu $(NO_3)_2 \cdot 3H_2O$, $Ni(NO_3)_2 \cdot 6H_2O$, and 0.5 M $ZrOCl_2 \cdot 8H_2O$ in 50 mL deionized water) were added. For three hours, the combined solution was constantly swirled. After stirring, the mixture was transferred to a heated bottle. In a hot air oven, at 140 °C, the hydrothermal treatment was conducted for 12 h. After that, the precipitate was dried in a hot air oven (Kannan et al. 2023).

11.2.1.4 Co-precipitation Method

A popular method for creating nanoparticles of metal oxide and metal/ceramic nanocomposites is co-precipitation. The starting reagent concentration, pH, and heating action are only a few of the variables that greatly affect nanoparticles' shape and size (Ahmad et al. 2021). In co-precipitation, the purpose is to make multicomponent materials by producing intermediate precipitates, often oxalates or hydrous oxides, which cause a close mixing of components during precipitation and preserve chemical homogeneity during calcination. Depending on the precipitating agent used in the reaction, the co-precipitation technique has the benefit of creating crystalline sizes in the small range when compared to other synthesis techniques. Additionally, the generated material's crystallite shape and size may be controlled using capping agents by this process (Bajaj and Joshi 2021).

This method is used to create various forms of metal nanocomposites, such as oxide-matrix, metal–metal oxide, and oxide-oxide. Precipitating agents such as NaOH, NH_3 or NH_4OH, Na_2CO_3, are frequently used. In order to prevent agglomeration, which also affects particle sizes of the composites made using this method, surfactants are frequently used. The method's benefits include low cost, ease of use, water-based reaction, flexibility, mild response circumstances, and control over size (Lateef and Nazir 2017).

We have an application of the production of magnetic nanocomposites of iron oxide and bacterial cellulose co-precipitated, according to Chanthiwong et al. (2020) one of the most commonly utilized processes for generating BC/iron oxide nanocomposites is co-precipitation. This approach can produce an excellent dispersion of magnetite nanoparticles and is simple. Managing the sizes and their distribution is also made possible by this method by modifying the proper processing parameters (Fig. 11.8) (Chanthiwong et al. 2020). The co-precipitation of iron oxide (Fe_3O_4) requires Fe^{2+} ions (e.g., $FeCl_2$, $FeSO_4$) as well as Fe^{3+} ions (e.g., $FeCl_3$, $FeNO_3$). Common alkaline agents for converting Fe(II) and Fe(III) salts into Fe_3O_4 nanoparticles include sodium hydroxide or ammonia. The most popular beginning reactants for making BC/Fe_3O_4 nanocomposites are $FeCl_2$ and $FeCl_3$.

In Kong et al. (2022) and Mohammed et al. (2022) study, ZnO/Co_3O_4 nanocomposites are a novel magnetic semiconductor that can be applied to get rid of ciprofloxacin and azithromycin. First, a straightforward co-precipitation technique is utilized to generate the nanocomposite. The ZnO/Co_3O_4 cubic structure in the pure nanocomposites is clearly defined, these peaks have an acceptable intensity and position ratio in comparison to the reference patterns for ZnO (80-0075) and Co_3O_4 (JCPDS = 43-1003), respectively.

Also, using the co-precipitation approach, $CuMn_2O_4$/CuMnO nanocomposites were created and utilized to eliminate dye contamination from water-based solution (Sobhani and Alinavaz 2023). In low fields, the nanocomposites exhibited ferromagnetic behavior (the coercivity is around ~ 1082.55 Oe), however, with high fields, they showed antiferromagnetic behavior. Additionally, exposed to visible light, their photocatalytic abilities for the breakdown of Eriochrome black T were studied for the initial instance. The tests were conducted again with sucrose added as a capping agent. Two samples were prepared for Eriochrome black T breakdown in either the usage and without a capping agent. Their structure, morphology, and photocatalytic properties of capabilities were analyzed and compared. The sample that was made without the use of a capping agent had superior morphology and photocatalytic

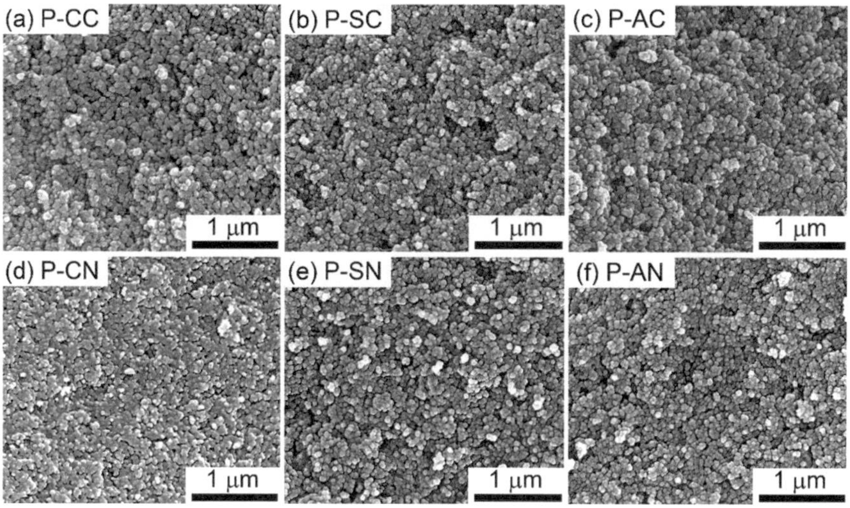

Fig. 11.8 Typical scanning electron microscopy pictures of Fe_3O_4 nanoparticles made by co-precipitation with various initial reactants: P-CC, P-SC, P-AC, P-CN, P-SN, and P-AN (**a**, **b**, **c**, **d**, **e**, **f**). Adapted from Chanthiwong et al. (2020)

performance when it came to removing Eriochrome black T from water-based solution, according to the data. Using a 0.03 g nanocomposite, 76% of the photodegradation of an Eriochrome black T solution at an amount of 10 ppm was achieved (Fig. 11.9). Compared to other pHs, the rate of Eriochrome black T degradation at acidic pH was greater, taking 90 min to reach 80.45% (Sobhani and Alinavaz 2023).

11.3 Green Synthesis of Metal Oxide Nanocomposites

11.3.1 Utilizing Extracts from Plants

Green synthesis techniques are becoming more and more important as viable avenues for the environmentally friendly and less harmful production of nanoparticles. Green synthesis methods can provide nanomaterials with lower production costs and energy inputs than conventional synthesis methods, which is encouraging for large-scale manufacturing (Mahmoud 2020c). In addition, the use of plants or plant extracts to generate metal and metal-based hybrid nanoparticles is a revolutionary green synthesis approach that has piqued scientists' interest (Jeevanandam et al. 2022). Nowadays, using green synthesis is recommended since it reduces pollution and byproducts, prevents or minimizes waste, and uses cleaner, non-toxic solvents. Plant extracts and plant-derived components are the most extensively researched green

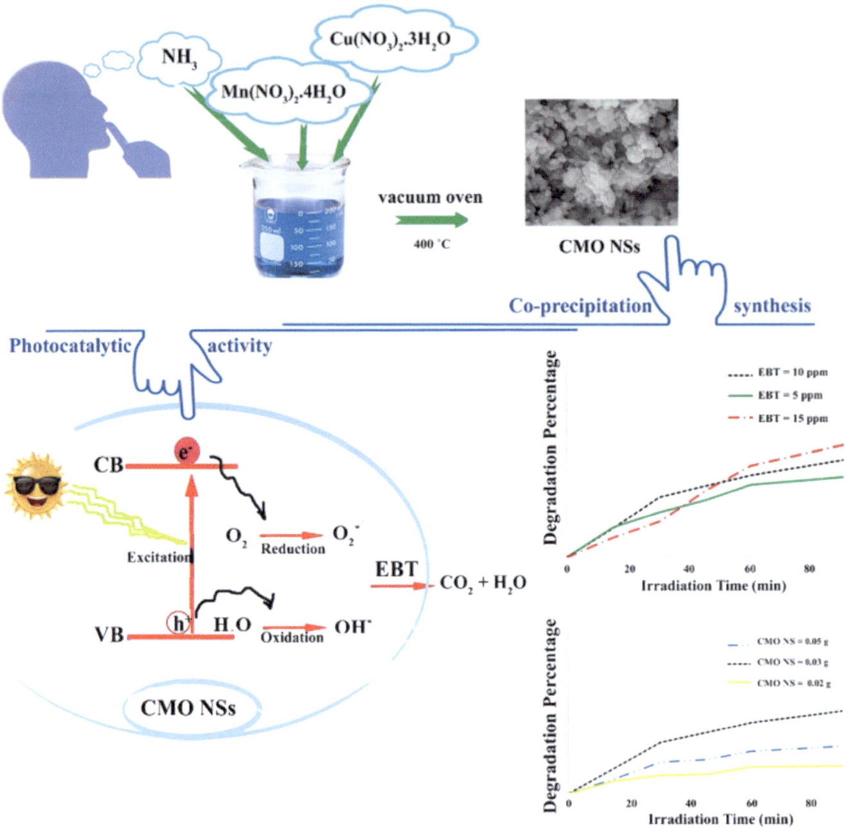

Fig. 11.9 $CuMn_2O_4$/CuMnO nanocomposites were synthesized by co-precipitation without the need of a capping agent and exploration of their potential uses in eliminating contaminants coming from wastewater via biomass conversion and biorefinery. Adapted form Sobhani and Alinavaz (2023)

production methods for metal oxide nanoparticles (NPs). It has a straightforward construction procedure that can be repeated on a larger scale (Radulescu et al. 2023).

Plant extracts from *Stachys pilifera Benth* (SB) and *Thymus daenensis* (TD) were employed as aqueous carbon-rich sources and as green-reducing agents during the formation of a carbon nanofiber/zinc oxide nanocomposite (ZnO-CNFs) due to their antibacterial and photocatalytic qualities for wastewater treatment (Arabkhani et al. 2023). In this process ZnO-CNFSB ZnO-CNFTD and were the labels given to the two varieties of ZnO-CNFs that were produced utilizing aqueous plant extracts from *Stachys pilifera Benth* (SB) and *T. daenensis* (TD). To sum up, 6.2 g zinc acetate dehydrate were applied to every aqueous extract produced and agitated at room temperature for 30 min, or until the zinc acetate dehydrate was completely dissolved. The above-prepared solutions were then mixed with a specific volume of the sodium hydroxide solution (0.1 mol L^{-1}) dropwise until pH of the mixture reached 11. The mixtures were then agitated for two hours at 50 °C till precipitation became apparent. The resulting precipitations were gathered, cleaned in ultrapure water, and then left to dry overnight at 110 °C in a hot air oven. Ultimately, light-yellow powders were prepared by calcining using a microwave to dry the pale yellow precipitates for three minutes at 800 W. In contrast, the same procedure was used to create the pristine ZnO nanoparticles (ZnO-NPs) in an ultrapure water environment without the need for aqueous extracts. Figure 11.10 shows a schematic depiction showing the synthesizing techniques (Arabkhani et al. 2023).

Since plant phytoconstituents serve as stabilizing agents for reduction and capping, using plant biomolecules is shown to be the most efficient method of creating stable nanomaterials. Bio-based nanocomposites have been shown through experimentation to have enormous potential for the deterioration of environmental contaminants. Therefore, research into the production of innovative metal oxide nanocomposites derived from plants and their potential applications in water treatment industry is imperative (Mishra and Devi 2023).

(a) **Using Algae Extracts**:

Algae come in two types: microalgae and macroalgae. Microalgae are phytoplankton that survive floating in the water column. Microalgae are photosynthetic microorganisms that are unicellular, form colonies, and have filamentous structures. Except for bacterial cyanobacteria, all microalgae contain a nucleus surrounded by a membrane, which makes them eukaryotic. Microalgae have emerged as significant and inexpensive cell factories that manufacture commercial products such as fatty acids and carotenoids. Macroalgae are occasionally known as seaweed. They constitute a varied class of multicellular, chlorophyll-containing, photosynthetic, marine eukaryotes. They lack genuine roots, stalks, and leaves and proliferate using rudimentary reproductive systems (Omran 2020).

Algae have traditionally been utilized in agriculture and manufacturing to make food, feed, cosmetics, medications, and fertilizers. However, more research is being conducted on using algae to create green nanomaterials. Algae are an abundant source of bioactive chemicals, easy to grow, scalable, and rapidly reproducible, which is why this movement is gaining momentum every day. The natural substance found in algae serves as a capping agent and stabilizing during the transformation of metal compounds into metal, metal oxides, or bimetallic nanoparticles. Because of their ability to manufacture stable, manageable nanoparticles without the need for cell maintenance, algae are also referred to as bionanofactories (El-Sherif et al. 2024). Algae are naturally occurring plants that provide a large supply of phytochemicals that are utilized to make metallic nanomaterials (Sampath et al. 2022).

As reported in Sadek et al. (2019), *Sargassum muticum* (brown algae) extract was effectively used to create nanocomposite of CuO and ZnO nanoparticles, which was subsequently disseminated in polyethylene oxide polymer. The *S. muticum* species Brown algae is used to make green nanoparticles because it includes phenolic chemicals, fucoidan, terpenoids, steroids, and flavonoids, all of which act as reducing and stabilizing agents. When compared to chemical synthesis, the biological approach is less hazardous and more cost-effective. Furthermore, they exhibit active groups of

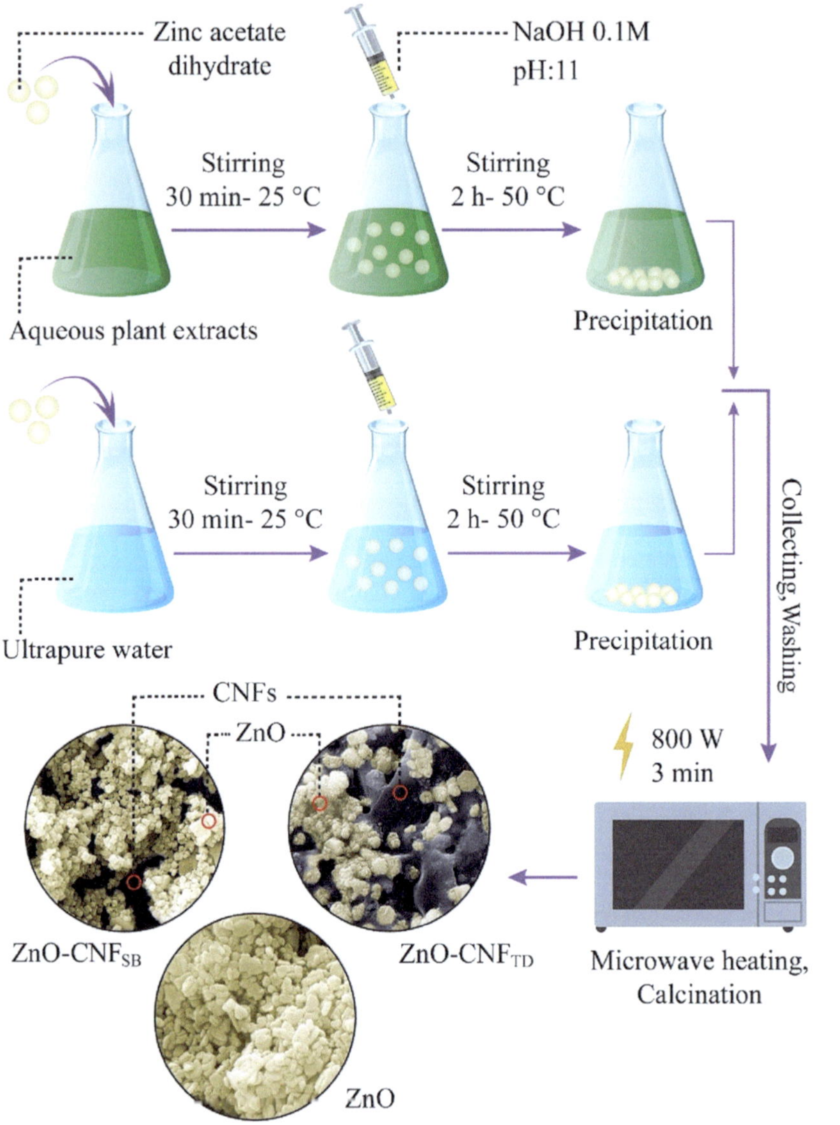

ZnO-CNF$_{TD}$: Synthesized ZnO-CNFs in presence of *Thymus daenesis celak* plant extract.

ZnO-CNF$_{SB}$: Synthesized ZnO-CNFs in presence of S*tachys pilifera Benth* plant extact.

Fig. 11.10 Diagram showing the fundamental phases in manufacturing of samples of ZnO-NPs and ZnO-CNFs. Adapted from Arabkhani et al. (2023)

chemicals (e.g., NH_2, CO, and CHO) that are adsorbed on the surface of the metal to promote protection. Nanocomposite demonstrated a powerful inhibitor of biofilm formation as well as antibacterial activity against gram-positive and gram-negative bacteria. It demonstrated an anti-hydrophobicity effect on biocorrosive bacteria, which boosted its effectiveness as a biocorrosion inhibitor (Sadek et al. 2019).

Also, after reviewing the study of Sharma et al. (2018) identified that Titanium dioxide (TiO_2) nanoparticles are synthesized utilizing green algae *Chlorella pyrenoidosa* (Fig. 11.11). Furthermore, TiO_2-GO nanocomposite was created by placing TiO_2 nanoparticles on GO sheets. Algae can be thought of as "bionanofactories" because of their ability to produce nanomaterials with high stability because to the existence of phytochemicals like carboxyl, amino, and hydroxyl groups, which serve as capping and reducing agents. Given this, a species of green microalgae, *C. pyrenoidosa*, was employed in the green manufacture of TiO_2 nanoparticles (Sharma et al. 2018).

Heavy metals and metalloids, such as Ba, Co, Cu, Mo, Mn, and Zn, can be consumed by microalgae and used for cell metabolism as well as for intracellular and extracellular enzyme processes. Heavy metals at appropriate concentrations have been demonstrated to boost the development and metabolism of the microalgae Nanochloropsis. Due to their stress tolerance mechanism, blue-green algae like Spirogyra, Oscillatoria, Anabaena, and Nostoc can thrive in heavy metal-contaminated wastewater. Microalgae also contain highly reactive groups with active heavy metal binding sites that form complexes with them, resulting in flocculation and a reduction in total suspended solids (TSS) and total dissolved solids. Microalgae can endure harsh environmental circumstances, including heavy metal toxicity, by employing specific self-protective mechanisms such as gene regulation,

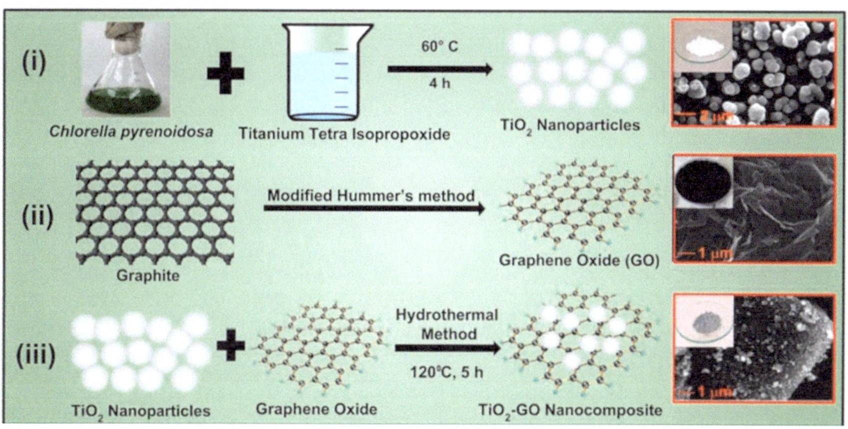

Fig. 11.11 Synthesis of (i) TiO_2 nanoparticles derived from *Chlorella pyrenoidosa* green algae, (ii) GO using an adapted version of Hummer's approach and (iii) TiO_2-GO nanocomposite made with a hydrothermal technique (Sharma et al. 2018)

chelation, antioxidant activity, the presence of enzymatic reducing agents, and self-immobilization. Microalgae use protein molecules to form metal compounds that do not alter their activities and maintain the concentration of heavy metal ions in the cytoplasm. Furthermore, the presence of heavy metals in microalgae stimulates phytochelatins' (PCs) synthesis and lowers heavy metal stress (Sharma et al. 2021).

11.4 Treatment of Industrial Wastewater with Metal Oxide Nanocomposites

We highlight some examples of metal nanocomposites used to remediate heavy metals and dyes such as:

(i) Silica-gold (SiO_2@Au) spherical core–shell nanocomposites:

Because of their greater efficiency than their bulk counterparts, core–shell nanocomposites and other nanomaterials such Au, Ag, and Cu can now be produced using straightforward methods and may eventually become more affordable thanks to nanotechnology. Reactive plasmonic nanoparticles of this kind have been used to the elimination of harmful metallic ions from water samples as well as the identification of heavy metallic ions such as Pb^{2+}, Cd^{2+}, Hg^{2+}, and As^{3+}, also created silica-gold (SiO_2@Au) spherical core–shell nanocomposites utilizing ~ 30 nm gold nanoparticles and ~ 420 nm silica cores, which function as a great sensor for trace-level metal ion detection (Thatai et al. 2014).

(ii) Ag@Fe_3O_4 nanocomposite (NC):

Hasan et al. (2023) examines the possibility of treating oily wastewater with Ag@Fe_3O_4 nanocomposite (NC), which is made biosynthesized with leaf extract from *Laurus nobilis* in aqueous form. Ag@Fe_3O_4 NC also demonstrated strong antibacterial qualities and good biocompatibility. These results aid in the creation of practical and affordable techniques for environmental cleanup and wastewater treatment.

(iii) $CaFe_2O_4$-based nanocomposite:

Ferrites made of metal, such as cobalt, zinc, copper, nickel, and other metals, are quite common and are employed as catalysts in many procedures. In contrast, the most common non-toxic alkali metal ferrite is $CaFe_2O_4$. Because of its magnetic characteristics, $CaFe_2O_4$ is a superparamagnetic substance that is simply retrieved from the reaction medium, extensively employed in a broad set of uses, including drug administration, gas sensing, photocatalytic water splitting, dye degradation, heavy metal ion removal, transesterification process, etc. Prospects for creating innovative, long-lasting $CaFe_2O_4$-based nanocomposite are discussed in Yadav and Ahmaruzzaman (2023).

(iv) Nanocomposites based on graphene:

The fabrication of graphene-based nanocomposites involved encasing Calcium alginate (CA) contains magnetite graphene (G-Fe_3O_4). The resulting G-Fe_3O_4/CA composite hydrogel was employed as an adsorbent to extract MB from aqueous solutions. The adsorption capability of G-Fe_3O_4/CA toward MB was good. The MB adsorption on G-Fe_3O_4/CA was essentially pH-independent. Because of the new G-Fe_3O_4/CA's quick and effective adsorption process and easy to use magnetic separation, it has a lot of potential as an affordable and ecologically acceptable adsorbent for dye removal from water-based solutions (Song et al. 2015).

Herein, we give further details about the use of metallic nanocomposites to remove heavy metals or dyes.

11.5 Removal of Heavy Metals

Insecure metals with densities higher than 5 g cm^{-3} are more widespread in ordinary life and are considered harmful even at ppb levels. Metals such as chromium (Cr), lead (Pb), cadmium (Cd), mercury (Hg), and arsenic (As) pose potential threats to human well-being, particularly in the brain, kidney, liver, and immune systems. The elimination of heavy metals has still limitations such as low adsorption capacity, sludge generation, and instability (Mahmoud et al. 2016). Practically, the adsorption approach has garnered clear interest due to its affordable, straightforward operation, reuse and regenerative abilities, increased HMs extraction in complexed matrix, and a wide range of pH values (Mahmoud et al. 2024; Natarajan et al. 2023).

The most regularly examined nanoscale metal oxides (NMOs) for heavy metal adsorption from aquatic systems are iron oxides, titanium oxides, manganese oxides, and aluminum oxides. They exist in various forms, particles, tubes, and other structures are included. The shape and size of NMOs have a crucial role in their sorption performance. Efficient synthetic approaches for producing shaped with control, extremely stable, and monodisperse metallic oxide nanoparticles have been extensively studied throughout the last decade. In general, synthesis techniques can be divided into two categories: (1) physical strategies, which include inert gas condensation, severe plastic deformation, high-energy ball milling, ultrasound shot peening, and (2) chemical strategies, which include opposite micelle (or microemulsion), controlled chemical co-precipitation, chemical vapor condensation, pulse electrode position, liquid flame spray, liquid-phase reduction, and gas-phase reduction, etc. (Hua et al. 2012).

Natural clay (NC) and Co_3O_4 nanoparticles (NPs) are co-precipitated to make super NC@Co_3O_4 nanocomposites with outstanding salt resistance, environmental stability, and reusability, as well as the elimination of heavy metal contamination such as hazardous Pb(II) and Cd(II) ions. The NC@Co_3O_4 adsorbent has the advantages of being easy to synthesize and biocompatible. Furthermore, NC@Co_3O_4 can maintain its high adsorption capability by taking into consideration numerous environmental characteristics like the pH solution, NC@Co_3O_4 dosage, adsorption process duration, and the starting heavy metal concentration. PSO kinetic and isotherm of

Langmuir modeling accurately described the kinetic and stability adsorption of lead and cadmium ions from water onto NC@Co_3O_4 nanocomposites. The Monte Carlo computer studies and the experimental findings were in good agreement. The experimental results revealed that metal ions (cadmium and lead) have significant negative adsorption energies. Overall, findings indicated that NC@Co_3O_4 nanocomposites are promising solutions for wastewater treatment and environmental remediation (Fig. 11.12) (Mouden et al. 2023).

11.6 Dyes Removal and Degradation

The release of organic pollutants like organic dye pollution into natural water bodies has been a significant difficulty in supplying and providing clean water to a huge percentage of the population (Mahmoud et al. 2023). As the population grows, underground water alone will not be enough to meet everyone's water needs. Purification and reuse of wastewater is the only option to meet the necessity for water. Nanotechnology has been utilized to remediate wastewater very well by photocatalytic destruction of dye molecules. Over the last few years, numerous studies have been conducted to improve metal oxide semiconductors' photocatalytic activity in water treatment (Yadav et al. 2022). Because of graphene's unique properties, such as its broad range of light absorption, nanocomposites have been used as efficient photocatalysts, charge separation, and high capacity of adsorption of dye pollutants (Mahmoud et al. 2020; Mahmoud et al. 2022a, b).

Nanocomposites based on metallic oxides that have improved surface characteristics, stability, and controlled structural and surface qualities are viable for photochemical destruction of organic contaminants. These materials have the potential for use in the organic dyes' photochemical breakdown due of their significant photodegradation abilities, regulated bandgap, and efficient synthesis techniques. The metal/oxide/graphene oxide, polymer/metal oxide, metal/metal oxide, metal oxide/silica, metal oxide/metal oxide, and metal oxide/biomaterials group of nanocomposites (Table 11.1), have efficiently photodegraded several colors from aqueous solutions (Mahmoud and Fawzy 2023; Joshi et al. 2021).

TiO_2/graphene nanocomposites were created by chemically exfoliating graphite is included in titanium alkoxide. The graphene sheets were covered with a TiO_2 nano-layer, extending the absorption edge into the visible light area while inhibiting electron–hole pair recombination. The novel TiO_2/graphene nanocomposite outperformed metal/metal oxide, metal/silica, metal oxide/metal oxide, metal oxide/ biomaterials, and conventional TiO_2/rGO and TiO_2-P25 polymer/metal oxide in photodegrading various dyes in aqueous solutions (Joshi et al. 2021).

The UiO-66(Ti) nanocomposite was utilized as a photocatalyst for methylene blue (MB) dye degradation in the availability of sunshine (Wang et al. 2016). Adding Ti to UiO-66 did not change the framework because the XRD patterns of UiO-66 and

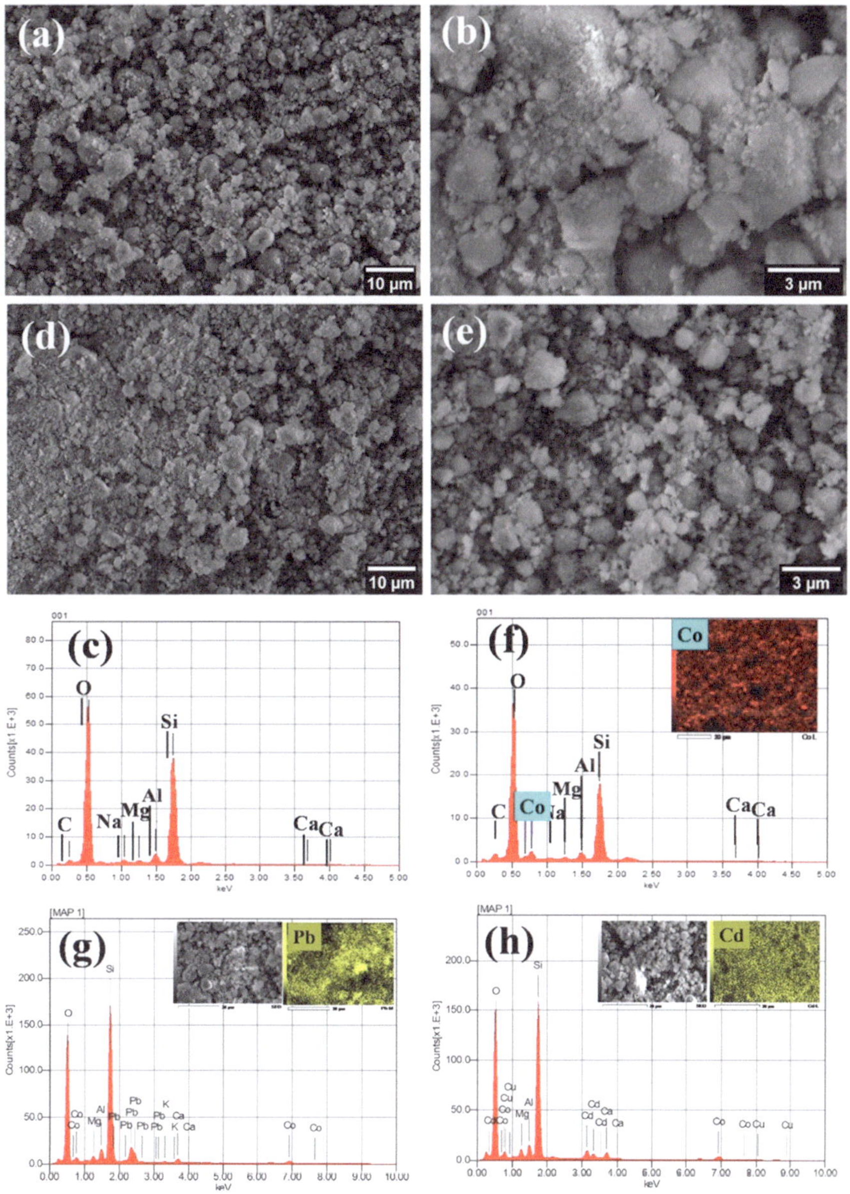

Fig. 11.12 Images from the SEM and accompanying EDS spectra for **a–c** NC and **d–f** NC@Co_3O_4. EDS spectra of NC@Co_3O_4 following the sorption of Cd(II) and Pb(II) in **h** and **g**. Adapted from Mouden et al. (2023)

Table 11.1 Metal oxide nanocomposites applied for dyes removal from water

Metal oxide nanocomposites	Dyes	Time (min)	Removal (%)	References
Gamma mixed iron and aluminum oxide (Fe_2O_3–Al_2O_3) nanocomposites	Congo red	15	~ 100	Mahapatra et al. (2013)
Graphene oxide-metal oxide (rGO/TiO_2/Fe_3O_4) nanocomposites	Methylene blue	9	94	Banerjee et al. (2018)
Ni/MnO/chitosan nanocomposites	Malachite green	90	89	Sobhani (2024)
ZnO-Al_2O_3 nanocomposite	Rose Bengal	90	80	Wani et al. (2024)
	Congo red	60	90	

UiO-66(Ti) were identical. However, the addition of Ti improved photodegradation process when compared to UiO-66. UiO-66(1.25Ti) had the most efficient MB deterioration at 82.2% (Wang et al. 2016).

The performance of metallic oxide nanocomposites in organic pollution degradation depends on many factors. In just 90 min of sunlight, the GO-V2O5 nanocomposite destroyed 97.95% of the VB dye. However, degradation of VB dye under visible conditions (96.93%) and ultraviolet (UV) radiation (78.35%) sources requires 180 and 210 min, respectively. The as-synthesized GO-V_2O_5 nanocomposite demonstrated higher activity of photocatalytic reactions than pure V_2O_5 because of its efficient charge transfer abilities, which prevent electron–hole pairs from recombining. Also, upon being under direct visible light, the ZnO-Fe_3O_4/rGO nanocomposite sample removed 97% of the MB dye from a water-based solution in 150 min, whereas Fe_3O_4/GO only decomposed 45% of the entire MB content. When exposed to light visible in the ZnO-Fe_3O_4/rGO system, electrons in VB of ZnO can be driven into CB of ZnO and successfully transported to Fe^{3+}/Fe^{2+} redox pair in Fe_3O_4 via the graphene interlayer (Kumari et al. 2023).

The catalysts of the second generation consist of binary metal oxides, including WO_3–ZnO, PbO–Al_2O_3, Fe_3O_4–CeO_2, and ZnO–SnO_2. Recently, various semiconductor-based nanomaterials have attracted interest for dye degradation in wastewater. TiO_2, CuO, ZnO, Co_3O_4, CdS, MgO, Fe_2O_3, Mn_2O_3, V_2O_5, and WO_3 are some of the most frequently investigated photocatalysts. Nevertheless, agglomeration, the quick recombination of photo excited electron hole pairs, and broadband gap energy place limitations on these materials. Several procedures have been investigated to solve these issues, including the doping of metallic and nonmetallic materials utilizing various synthetic techniques (Lanjwani et al. 2024).

11.7 Adsorption Process

According to Zhan et al. (2014)'s research, metal oxide nanomaterials were used to remove As(V) and methylene blue. Nano-Al_2O_3, -CuO, -TiO_2, and -ZnO had the maximal adsorption capacities for methylene blue were 229.09, 220.52, 206.60, and 182.39 mg g^{-1}, while for As(V) they were 36.81, 36.39, 29.24, and 18.21 mg g^{-1} based on Langmuir models. As(V) adsorption on metal oxide nanocomposites was pH dependent, with higher adsorbed rates observed at lower pH levels. When the pH was reduced from 9 to 3, nano-TiO_2, -Al_2O_3, -CuO, and -ZnO showed an increase in adsorption capabilities of 40, 34, 25, and 29%. Adsorption and photodegradation both enhanced methylene blue degradation rates, with the nano-Al2O3 adsorption experiment achieving a maximum rate of 90.5%.

The dye is removed from genuine textile manufacturing effluent using adsorbents include iron oxide nanorods and copper oxide nanoparticles. The study of Rather and Sundarapandian (2024) looked at the effects of adsorbent dose, adsorbent-adsorbate contact time, and wastewater pH on dye adsorption %. These findings indicate that increasing the adsorbent dosage and contact duration leads in a higher percentage of dye removal. Notably, copper oxide nanoparticles outperform iron oxide nanorods at pH 12 in terms of dye removal. The study achieved an excellent process efficiency of 95.24% for copper oxide nanoparticles and 62.5% for iron oxide nanorods.

Chemical adsorption creates ionic or covalent connections via chemical processes. Nonetheless, compared to chemisorption, physical adsorption is less selective but reversible, which is irrevocable but more particular. Biosorption refers to the phenomenon of adsorption that occurs across biological systems. Biosorption is a method that integrates the removal and recovery of metal. Biosorption is useful because adsorbents are inexpensive and easy to regenerate. Bacteria, fungus, algae, waste from industries, agricultural waste, natural leftovers, and other biological components are commonly used to extract heavy metals from wastewater. Adsorption of heavy metal ions on bioadsorbents can occur via pore adsorption, simple diffusion, intra-particle diffusion, chemical adsorption, physical adsorption, electrostatic interactions, precipitation, ion exchange, complexation, hydrogen bonding, and redox interactions (Damiri et al. 2022).

11.8 Photocatalytic Degradation Pathways

Photocatalysis refers to raising the kinetics of a photoreaction by using a catalyst. Photocatalysis is a photochemical mechanism that is in the advanced oxidative reactions group and produces highly oxidizing free radicals that breakdown organic compounds in water. The processes of photocatalysis can be homogeneous or heterogeneous. The heterogeneous photocatalytic approach varies from the homogeneous one in that it uses solid-state catalysts. As a result, incoming light causes electron/hole (e^-/h^+) pairs to interact with the medium, resulting in radical formation. This radiation is necessary for the homolytic breakdown of the oxidant to occur (Araújo et al. 2023).

In metal oxide-based systems, an artificial or natural light stimulus starts photocatalysis. Photogenerated water molecules on metal oxide surfaces produce $HO^{\cdot}$ and $O_2^{\cdot -}$ species, which drive redox processes that degrade contaminants. An oxide-based metal photocatalyst's activity in a water-based environment is summarized below (Mahmoud 2022; Zhu and Zhou 2019). Initially, the semiconductor is exposed to a light source with sufficient energy to activate it; as a result, greater energy photons are absorbed when they cross the bandgap. This process transfers a single electron (e^-) from the valence to the conduction band, resulting in a hole (h^+) within the valence band. CB's electrons (e^-) interact with the water absorbed molecules' oxygen at the oxide surface, creating superoxide (O_2^-) and/or hydroperoxide (OH^-) free radicals. Simultaneously, as water is adsorbed on the metal oxide surface, holes (h^+) produce HO free radicals (Fig. 11.13) (Araújo et al. 2023).

Photocatalysis has been shown to be the most efficient technology for treating wastewater because of its excellent photocatalytic elimination efficiency, quick oxidation process, cheaper costs, and non-detrimental. For ecological applications, the perfect photocatalyst must be stable, affordable, safe, effective, and light-responsive. Furthermore, photocatalysts must have high photoactivity to create ROS by the oxidation of surface H_2O/hydroxyl groups and O_2 reduction (Lee et al. 2023). We will discuss an example from the study of Majumder et al. (2022), Carbamazepine (CBZ) breakdown via photocatalysis can occur via four distinct mechanisms. The olefinic double bond of CBZ molecule has an extremely high Fukui index.

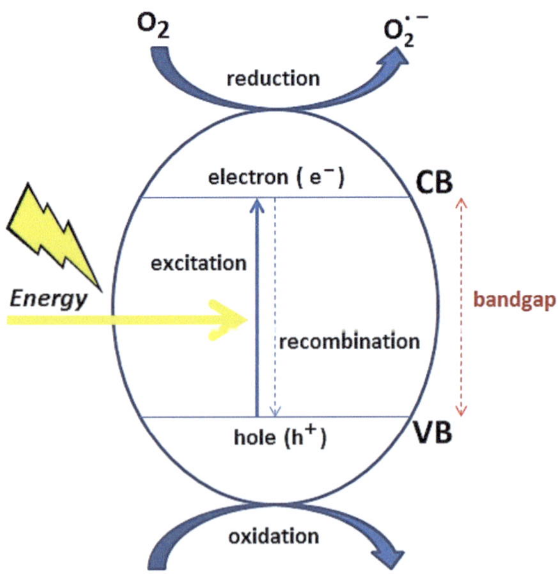

Fig. 11.13 An illustration of heterogeneous photocatalysis based on metal oxides (Araújo et al. 2023)

According to the degrading process (I), CBZ's olefinic bond is likely to be attacked by either $O_2^{\cdot}$, $OH^{\cdot}$, or both $O_2^{\cdot}$ and $OH^{\cdot}$, producing IP-A. IP-A may have extra hydroxylation events to produce IP-B. The following destruction route (II) may include $OH^{\cdot}$ attacking the core heterocyclic ring's olefinic double bond while simultaneously removing the NH_2-group. This produces IP-C (acridine-9-aldehyde) can be further oxidized to form IP-D (acridine). IP-D may lose its nitrogen atom, forming IP-H. According to the degradation process (III), to create its isomer, the molecule could undergo internal cyclization (IP-E) after breaking the olefinic double bond (Zhao et al. 2020). According to the breakdown route IV, IP-F (Acridone-*N*-carbaldehyde) can be generated through simple ring contraction. IP-G (acridone) can be created by further oxidizing IP-F's carbonyl group. $O_2^{\cdot}$ and $OH^{\cdot}$ can further oxidize the produced IPs (IP-B, IP-H, IP-E, and IP-G) to produce CO_2, H_2O, and other inactive species (Majumder et al. 2022).

11.9 Challenges and Future Perspectives

(a) **Scalability and Cost-Effectiveness**:

Nanomaterial cost-efficiency and suitability are the most common impediments to nanocomposites throughout the treatment of water. There is a vital necessity for the production of nanomaterials that are extremely efficient, simple to handle, and environmentally friendly. It is also necessary to be cost-effective in order to facilitate commercialization of wastewater treatment. The various applications of nanomaterials can show to be a benefit for delivering pure water for drinking to the entire planet (Sharma et al. 2021).

The manufacturing of MONPs involves chemical and physical synthesis techniques, (such as, bottom-up processes), which create an efficient amount of nanoparticles but have the drawback of being more costly (Negrescu et al. 2022). The ability to scale up the production of metal oxide nanocomposites to suit needs in industry can be difficult because of the complexity of the process of synthesis and the necessity for precise monitoring of nanoparticle characteristics. But there is mention of some prosperities, metal oxides are an excellent material in new-generation photovoltaics due to their extraordinary flexibility and the viability of modest, low cost, and easily scalable processes (Mahdy et al. 2022). And an example talks about automatic mechanochemical technique as economically viable and sustainable strategy for the synthesis of CuO NPs, based on the article of Mahdy et al. (2022).

In addition, the costs of manufacturing, especially at scale, might be prohibitively costly, particularly when using high-purity metal precursors and advanced synthesis equipment (Dikshit et al. 2021). And we can list future prospects as, optimizing synthesis processes to increase efficiency and cost-effectiveness, for as employing environmentally friendly synthesis or continuous flow reactors. Investigating alternate metal sources or precursors that are more plentiful and affordable. And cooperating with industry partners to improve manufacturing processes and lower costs.

(b) **Stability and Reusability**:

Designing environmentally friendly nanomaterials with specified qualities such as cheap preparatory costs, exact selective, high stability, rapid recovery, and remarkable recyclable by adjusting their size and shape is a critically required challenge regarding practical applications (AlGarni and Al-Mohaimeed 2022). Nanocomposites of metal and metal oxide have gained popularity as an alternate to traditional antibacterial agents due to their unique form, excellent selectivity, activity, chemical and physical stability, and surface-to-volume ratio (Basavegowda and Baek 2023).

The reusability of nanocomposites was tested by Rani and Shanker (2018), A reusability test of the utilized nanocomposite during the reaction was performed. It is possible to reuse the nanocomposite up to ten times without significantly changing its structure, according to XRD measurements. The data were determined to be consistent in terms of dye degradation. This demonstrated the catalyst's recyclability and reusability, with approximately 99% degradation in the first cycle and 96% in the tenth run.

Nano-sized cerium oxide (CeO_2) is a kind of rare earth oxide that has been used in a range of applications, including photocatalysis, sensing, UV blocking, water purification, and so on. It has been demonstrated that cerium oxide's bulk density, surface area, crystalline size, and porosity all significantly affect the material's dispersion behavior, stability, activity, and selectivity, changing the outcome of the dangerous metals deduction (Ali et al. 2023).

Fe_2O_3 Magnetite is regarded a promising nanomaterial for water purification due to its large surface-to-volume ratio, magnetic characteristics, and ease of surface modification and reuse. Because of the aforementioned benefits, it has significant potential for ecological applications as a sorbent, photocatalyst, and modifier. Its magnetic qualities are especially significant, as they assist to the development of more successful water purification schemes through the combination of adsorption with magnetic separation, which is becoming increasingly common in the field of water treatment. Furthermore, magnetite has a comparatively low cost, excellent sorption capacity, and stability (Dontsova et al. 2019).

Renewal of metal oxide nanocomposites, sorbent reusable nature is a significant aspect that determines their feasibility in massive scale uses for water purification, and this feature must satisfy the sorbent qualities. Reusing sorbents in the sorption process is significant from an economic and wastewater management perspective, with pH playing a vital function in the regeneration of sorbent materials using acids, alkalis, or different chemicals (Aigbe and Osibote 2022).

And from Li et al. (2020)'s research, we have an experiment for Fe_3O_4 and MnO_2 nanocomposites recycling. The Fe_3O_4/MnO_2 composite materials sample was originally immersed for 24 h in wastewater including heavy metals such Cu^{2+}, Cd^{2+}, Zn^{2+}, and Pb^{2+} (pH 6.5) with a dose of sorbent of 1 g L^{-1}. The sorbents of Fe_3O_4/MnO_2 composites containing heavy metals, using a magnet to gather the heavy metals, the sorbent was rinsed in 0.1 M HCl for one hour in order to remove the

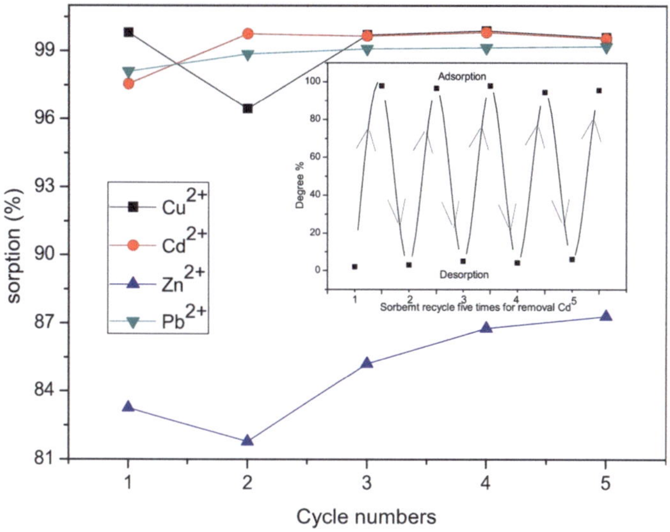

Fig. 11.14 Adsorption and desorption of Fe_3O_4/MnO_2 composites across five cycles (Li et al. 2020)

heavy metals. Unloaded sorbents were repeatedly employed in wastewater treatment to remove heavy metals. After five cycles of removing Cu^{2+}, Cd^{2+}, Zn^{2+}, and Pb^{2+} from wastewater, the Fe_3O_4/MnO_2 composites showed a significant capacity for adsorption of heavy metal ions (Fig. 11.14).

Nanocomposite materials with certain properties including targeted selectivity, efficient recovery, low preparation rate, outstanding stability, and exceptional recyclability through shape and size modification are some of the instantly important jobs for practical applications (Ali et al. 2023). But some of the challenges facing stability and reusability, the long-term stability of metal oxide nanocomposites in difficult industrial wastewater conditions, such as increased temperatures, pH fluctuations, and the existence of corrosive chemicals, can be an issue. Improving the reusability of nanocomposites across several cycles of wastewater treatment without considerable loss of ability to adsorb or structural integrity is necessary yet difficult.

Prospects for the Future, looking into cutting-edge surface alterations or stabilization approaches for enhancing the stability of metal oxide nanoparticles in a range of environmental circumstances, creating regeneration procedures to increase the lifetime and adsorption capability of nanocomposites and incorporating nanocomposites into resilient membrane systems or support matrices to increase their mechanical stability and make recovery and reuse simpler.

(c) **Integration with Existing Treatment Technologies**:

The selection of acceptable technology is dependent on its efficiency in permanently eliminating heavy metal ions (Saleh et al. 2022). Integration with existing treatment techniques is an important feature of using metal oxide nanocomposites for industrial wastewater purification. It entails merging the distinct features of nanocomposites with the established processes of traditional treatment methods to improve overall efficiency and pollutant removal capacities. For example, metal oxide nanocomposites are able to given into the activated sludge process, in two ways as a pre-treatment step or straight into the biological reactor (Zabermawi and El-Bestawy 2023), metal oxide nanomaterials have the potential to be precipitated on membrane surfaces or embedded in membrane structures to improve fouling resistance and the effectiveness with which pollutants are removed (Tran et al. 2020) and metal oxide nanocomposites may be employed in batch as adsorbents or continuous flow adsorption procedures to eliminate a variety of contaminants, such as heavy metals, dyes, and emerging contaminants (Ghosh et al. 2022). And with talking about challenges may face integrating metal oxide nanocomposites with existing wastewater treatment technologies, compatibility, efficiency, and operational feasibility were all carefully considered. Perspectives for tomorrow, developing composite treatment systems that combine the benefits of nanocomposites and traditional approaches to improve pollution reduction and process efficiency and performing extensive pilot-scale experiments to determine the feasibility and efficacy of incorporating metal oxide nanocomposites into current wastewater treatment systems.

(d) **Emerging Trends and Research Directions**:

Wastewater extraction and treatment are becoming increasingly important in the twenty-first century as cities and industries expand. To address the problem of increasing environmental dangers, a variety of treatment for wastewater techniques are applied, including biological, chemical, and physical (methods of treatment range from primary to tertiary). Various treatments strategies utilized carry the possibility of creating secondary contaminants (Younas et al. 2021). Most new contaminants lacked clear criteria, but even at low quantities, they might affect humans and marine organisms. To tackle the aforementioned ecological problems, there have been substantial attempts made to improve the efficiency of restoration methods or create new methods for sample identification, quantification, and efficiency (Khan et al. 2022).

Around one hundred techniques for treating inorganic and organic streams of wastewater have been reported; numerous of these innovations have emerged, and these vary from chemical, physical, and biological processes (Armah et al. 2021). As a result, implementing individual physical, biological, or chemical remediation strategies is frequently insufficient to adequately remediate polluted environments. As a result, for more cost-effective and efficient environmental cleanup, a mix of treatments meeting the efficiency and treatment time requirements is advised (Ugrina and Jurić 2023).

Membrane technology (MT) is the technical and scientific approaches for transferring components, species, or substances across or through membranes. This technique is commonly applied to illustrate the mechanical procedures involved in separating gas or liquid streams. Membranes are classed as a light layer of obstacle for size

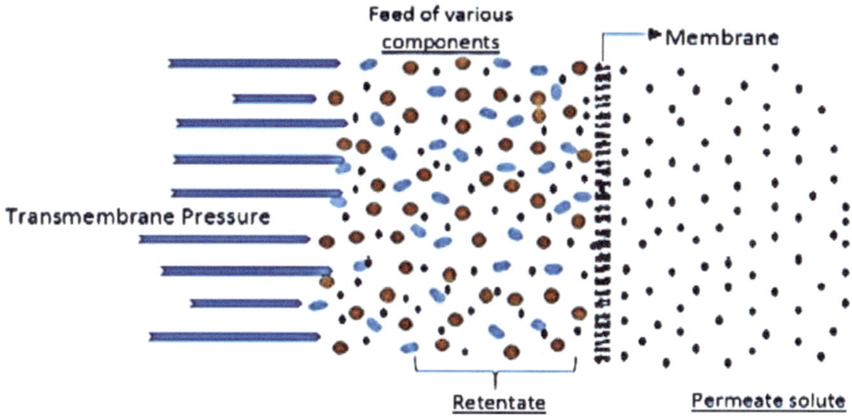

Fig. 11.15 Membrane selective permeability for a range of solutes (Armah et al. 2021)

variance separation, and they are typically used in conjunction with involving biological and chemical processes or functioning independently in secondary wastewater treatment. A semi-permeable barrier is a common driving force in a standard membrane mechanism that regulates the rate of substance flow by fractional penetration and decreases through pores of various diameters. The size of the membrane pore and chemical affinity dictate permeation and selective rejection, which results in a product stream free of target components (Fig. 11.15) (Armah et al. 2021).

Advanced oxidation processes (AOPs) are thought to be a promising procedure for removing refractory organic chemicals from wastewater through oxidation reactions using powerful, nonselective hydroxyl radicals. The AOP technique is a chemical filtration process that is used to remove unwanted organic waste in addition to dissolve and breakdown chemicals that are challenging or impossible to completely degrade using oxidation and standard biological approaches. The efficiency of AOPs depends upon the formation of reactive free radicals, one of the most important is the hydroxyl radical (Chakraborty et al. 2024). This type of radical is a potent oxidant capable change most organic compounds into CO_2, mineral acids, and water. It's vital to employ advanced oxidation procedures and choose the most efficient sewage remediation method. High quantities of harmful organic contaminants, which are routinely identified while treating wastewater from homes, municipalities, or industries, have piqued scientific attention around the world due to their significant risk to human health. Some advantages of integrating the Hydrodynamic cavitation (HC) approach in comparison to other AOPs standard wastewater treatment strategies include increased degrading efficiency, reduced sludge production, less chemical consumption, and a lower environmental effect (Chakraborty et al. 2024).

To remove solid particles, bacteria, and even nanomaterials from water, nanocomposites can be used as materials or filtration membranes. Their nanoscale small pores enable excellent filtration. Because of their catalytic properties, some nanocomposites can degrade organic pollutants utilizing advanced oxidation techniques such as processes of the Fenton type or photocatalysis. These substances can help complex pollutants break down into less dangerous forms. Polymer nanocomposites are highly useful materials used in wastewater treatment systems (Thamarai et al. 2024).

Efforts to increase water treatment with polymer nanocomposites performance have concentrated on nanoscale fillers as well as polymeric matrices. Among the many advantages of synthetic-polymer nanocomposites are their extended longevity as well as their strong chemical and mechanical resistance. However, they can harm the ecosystem through secondary sources. Natural polymer-based nanocomposites are thought to be valuable biodegradable material alternatives for removing contaminants and hazardous chemicals from drinking water and wastewater (Thamarai et al. 2024).

Magnetic nanoparticles have become known as viable techniques for removing heavy metals due to their distinct characteristics. One of the major drawbacks of magnetic nanoparticles is their tendency to cluster in solution, reducing their efficacy and restricting their use in water treatment. In order to maintain magnetic nanoparticles, researchers have begun to alter their surfaces includes a range of supporting resources. Aside from carbon-based materials, natural and artificial polymers and other compounds, metal organic frameworks, and layered double hydroxides have all been used as carriers in new magnetic nanocomposites (Nikić et al. 2024).

Taking into consideration the challenges that we can identify, such as keeping up with fast changing studies and emerging contaminants in industrial wastewater, including microplastics, pharmaceuticals, and developing pollutants, addressing key knowledge gaps, such as the mechanisms controlling nanoparticle behavior in complicated wastewater matrices and their long-term environmental fate.

Future perspectives Developing cross-disciplinary research partnerships to address emerging challenges at the intersection of nanotechnology and environmental science, metal oxide nanocomposites are being investigated for novel uses for instance photocatalytic pollutant degradation, antimicrobial treatment, and selective removal of certain contaminants and using advanced analytical techniques and modeling methodologies to better grasp the intricate interactions between nanocomposites and contaminants in industrial wastewater and improve methods for treatment.

11.10 Conclusion and the Future Expectations

Metal oxide nanocomposites in wastewater treatment offers a promising solution to the escalating problem of water pollution. Various synthesis approaches have been developed to tailor the properties and functionalities of these nanomaterials, enabling the precise targeting of specific pollutants commonly found in industrial effluents. Their versatility in treatment processes—ranging from adsorption to photocatalytic degradation—demonstrates their effectiveness against a wide spectrum of contaminants. Understanding the adsorption mechanisms provides insight into how pollutants interact with metal oxide nanocomposites, laying the groundwork for the development and optimization of more efficient remediation strategies. Additionally, exploring the pathways of photocatalytic degradation deepens our knowledge of the processes responsible for pollutant breakdown, supporting the creation of more effective and sustainable treatment technologies. Despite notable progress in this field, several challenges remain. These include issues related to scalability, cost-effectiveness, and the environmental impact of synthesizing and applying metal oxide nanocomposites. Overcoming these barriers requires interdisciplinary collaboration and continued research focused on improving material performance, enhancing treatment efficiency, and minimizing potential ecological risks. Looking ahead, the future of metal oxide nanocomposites in industrial wastewater treatment appears promising. Advances in nanotechnology, materials science, and environmental engineering are expected to unlock new opportunities and address current limitations. By harnessing the synergistic power of innovative research, technological progress, and supportive regulatory frameworks, we can maximize the potential of these nanomaterials to safeguard water resources and support sustainable industrial practices.

References

Abdelhaleem A, Aziz F, Alalm MG, Mahmoud AED (2024) 2 Hybrid technologies treatment for the removal of pharmaceuticals in wastewater. Environmental Nexus approach: management of water, waste, and soil, 20

Ahmad A, Ahmad A, Ramzan S, Kiyani MZ, Dubal DP, Mubarak NM (2021) Nanomaterial synthesis protocols. Elsevier eBooks, pp 73–85. https://doi.org/10.1016/b978-0-12-820773-4.00010-x

Ahmaruzzaman M, Gadore V (2021) MoS2 based nanocomposites: an excellent material for energy and environmental applications. J Environ Chem Eng (Print) 9(5):105836. https://doi.org/10.1016/j.jece.2021.105836

Ahmed J, Thakur A, Goyal A (2021) Industrial wastewater and its toxic effects. The Royal Society of Chemistry eBooks, pp 1–14. https://doi.org/10.1039/9781839165399-00001

Aigbe UO, Osibote OA (2022) Fluoride ions sorption using functionalized magnetic metal oxides nanocomposites: a review. Environ Sci Pollut Res Int 29(7):9640–9684. https://doi.org/10.1007/s11356-021-17571-7

AlGarni TS, Al-Mohaimeed AM (2022) Water purification by adsorption of pigments or pollutants via metaloxide. J King Saud Univ Sci 34(8):102339. https://doi.org/10.1016/j.jksus.2022.102339

Ali Z, Ullah R, Tüzen M, Ullah S, Rahim A, Saleh TA (2023) Colorimetric sensing of heavy metals on metal doped metal oxide nanocomposites: a review. Trends Environ Anal Chem 37:e00187. https://doi.org/10.1016/j.teac.2022.e00187

Ani IJ, Egbosiuba TC (2023) Metal oxides and their nanocomposites in wastewater treatment. Elsevier eBooks, pp 185–203. https://doi.org/10.1016/b978-0-443-18456-7.00009-2

Ansari I, El-Kady MM, Mahmoud AED, Arora C, Verma A, Rajarathinam RK, Singh P, Verma DK, Mittal J (2024) Persistent pesticides: accumulation, health risk assessment, management and remediation: an overview. Desalination Water Treat 100274

Arabkhani P, Saeedi N, Sadeghi H, Nouripour-Sisakht S, Gharaghani M, Asfaram A (2023) Plant extracts-mediated green synthesis of zinc oxide/carbon nanofiber nanocomposites with highly efficient photocatalytic and antimicrobial properties for wastewater treatment. J Water Process Eng 54:104020. https://doi.org/10.1016/j.jwpe.2023.104020

Araújo ES, Pereira MFG, Da Silva GMG, Tavares GF, Oliveira CYB, Faia P (2023) A review on the use of metal oxide-based nanocomposites for the remediation of organics-contaminated water via photocatalysis: fundamentals, bibliometric study and recent advances. Toxics 11(8):658. https://doi.org/10.3390/toxics11080658

Armah EK, Chetty M, Adedeji J, Kukwa DT, Mutsvene B, Shabangu KP, Bakare BF (2021) Emerging trends in wastewater treatment technologies: the current perspective. IntechOpen eBooks. https://doi.org/10.5772/intechopen.93898

Bajaj NS, Joshi RA (2021) Energy materials: synthesis and characterization techniques. Elsevier eBooks, pp 61–82. https://doi.org/10.1016/b978-0-12-823710-6.00019-4

Banerjee S, Benjwal P, Singh M, Kar KK (2018) Graphene oxide (rGO)-metal oxide (TiO2/Fe3O4) based nanocomposites for the removal of methylene blue. Appl Surf Sci 439:560–568. https://doi.org/10.1016/j.apsusc.2018.01.085

Barreca D, Carraro G, Comini E, Gasparotto A, Maccato C, Sada C, Tondello E (2011) Novel synthesis and gas sensing performances of CuO–TiO_2 nanocomposites functionalized with Au nanoparticles. J Phys Chem C 115(21):10510–10517. https://doi.org/10.1021/jp202449k

Basavegowda N, Baek KH (2023) Advances of metal and metal oxide nanocomposites: synthesis, characterization and biomedical applications. MDPI-Multidisciplinary Digital Publishing Institute, p 206. https://doi.org/10.3390/books978-3-0365

Bekermann D, Barreca D, Gasparotto A, Maccato C (2012) Multi-component oxide nanosystems by chemical vapor deposition and related routes: challenges and perspectives. CrystEngComm 14(20):6347. https://doi.org/10.1039/c2ce25624j

Bharti K, Sadhu KK (2022) Syntheses of metal oxide-gold nanocomposites for biological applications. Results Chem 4:100288. https://doi.org/10.1016/j.rechem.2022.100288

Byrappa K, Adschiri T (2007) Hydrothermal technology for nanotechnology. Prog Cryst Growth Charact Mater 53(2):117–166. https://doi.org/10.1016/j.pcrysgrow.2007.04.001

Chakraborty A, Chowdhury BR, Nandy S, Samanta S, Bhunia S, Roy N (2024) AOPs for municipal and tannery based industrial wastewater treatment. Elsevier eBooks, pp 47–73. https://doi.org/10.1016/b978-0-323-95656-7.00005-1

Chanthiwong M, Mongkolthanaruk W, Eichhorn SJ, Pinitsoontorn S (2020) Controlling the processing of co-precipitated magnetic bacterial cellulose/iron oxide nanocomposites. Mater des 196:109148. https://doi.org/10.1016/j.matdes.2020.109148

Chausali N, Saxena J, Prasad R (2021) Nanobiochar and biochar based nanocomposites: advances and applications. J Agric Food Res 5:100191. https://doi.org/10.1016/j.jafr.2021.100191

Chen M, Chen H (2021) Spatiotemporal coupling measurement of industrial wastewater discharge and industrial economy in China. Environ Sci Pollut Res Int 28(34):46319–46333. https://doi.org/10.1007/s11356-021-14743-3

Damiri F, Andra S, Kommineni N, Balu SK, Bulusu R, Boseila AA, Akamo DO, Ahmad Z, Khan F, Rahman H, Berrada MS, Cavalu S (2022) Recent advances in adsorptive nanocomposite membranes for heavy metals ion removal from contaminated water: a comprehensive review. Materials (Basel) 15(15):5392. https://doi.org/10.3390/ma15155392

Dikshit PK, Kumar J, Das AK, Sadhu S, Sharma S, Singh VP, Gupta PK, Kim BS (2021) Green synthesis of metallic nanoparticles: applications and limitations. Catalysts 11(8):902. https://doi.org/10.3390/catal11080902

Dontsova T, Nahirniak S, Астрелін IM (2019) Metaloxide nanomaterials and nanocomposites of ecological purpose. J Nanomater 2019:1–31. https://doi.org/10.1155/2019/5942194

Dutta D, Arya S, Kumar S (2021) Industrial wastewater treatment: current trends, bottlenecks, and best practices. Chemosphere 285:131245. https://doi.org/10.1016/j.chemosphere.2021.131245

Ehsan MA, Shah SS, Basha SI, Hakeem AS, Aziz MA (2021) Recent advances in processing and applications of heterobimetallic oxide thin films by aerosol-assisted chemical vapor deposition. Chem Record (New York, N.Y. Online 22(7). https://doi.org/10.1002/tcr.202100278

El-Kady MM, Ansari I, Arora C, Rai N, Soni S, Verma DK, Singh P, Mahmoud AED (2023) Nanomaterials: a comprehensive review of applications, toxicity, impact, and fate to environment. J Mol Liq 370:121046

El-Sherif DM, Mahmoud AED, Saber AN (2024) The application of the algal approach in the technological, socioeconomic, and wastewater treatment domains. In: Algae as a natural solution for challenges in water-food-energy nexus: toward carbon neutrality. Springer Nature, Singapore, pp 261–295

Fei L, Naeemi M, Zou G, Luo H (2013) Chemical solution deposition of epitaxial metal-oxide nanocomposite thin films. Chem Rec 13(1):85–101. https://doi.org/10.1002/tcr.201200022

Gaur VK, Sharma P, Sirohi R, Awasthi MK, Dussap C, Pandey A (2020) Assessing the impact of industrial waste on environment and mitigation strategies: a comprehensive review. J Hazard Mater (Print) 398:123019. https://doi.org/10.1016/j.jhazmat.2020.123019

Ghosh N, Das S, Biswas G, Haldar PK (2022) Review on some metal oxide nanoparticles as effective adsorbent in wastewater treatment. Water Sci Technol 85(12):3370–3395. https://doi.org/10.2166/wst.2022.153

Guozhu M, Han Y, Liu X, Crittenden JC, Huang N, Ahmad UM (2022) Technology status and trends of industrial wastewater treatment: a patent analysis. Chemosphere 288:132483. https://doi.org/10.1016/j.chemosphere.2021.132483

Gusain R, Gupta K, Joshi P, Khatri OP (2019) Adsorptive removal and photocatalytic degradation of organic pollutants using metal oxides and their composites: a comprehensive review. Adv Coll Interface Sci 272:102009. https://doi.org/10.1016/j.cis.2019.102009

Hasan GG, Laouini SE, Khelef A, Mohammed HA, Althamthami M, Meneceur S, Alharthi FA, Alshareef SA, Menaa F (2023) Efficient treatment of oily wastewater, antibacterial activity, and photodegradation of organic dyes using biosynthesized Ag@Fe_3O_4 nanocomposite. Bioprocess Biosyst Eng. https://doi.org/10.1007/s00449-023-02946-6

Hassan T, Salam A, Khan A, Khan SU, Khanzada H, Wasim M, Khan MQ, Kim IS (2021) Functional nanocomposites and their potential applications: a review. J Polym Res 28(2). https://doi.org/10.1007/s10965-021-02408-1

Hua M, Zhang S, Pan B, Zhang W, Lv L, Zhang Q (2012) Heavy metal removal from water/wastewater by nanosized metal oxides: a review. J Hazard Mater (Print) 211–212:317–331. https://doi.org/10.1016/j.jhazmat.2011.10.016

Jeevanandam J, Kiew SF, Ansah SB, Lau SY, Barhoum A, Danquah MK, Rodrigues JMM (2022) Green approaches for the synthesis of metal and metal oxide nanoparticles using microbial and plant extracts. Nanoscale 14(7):2534–2571. https://doi.org/10.1039/d1nr08144f

Joshi P, Gupta K, Gusain R, Khatri OP (2020) Metal oxide nanocomposites for wastewater treatment. Metal oxide nanocomposites: synthesis and applications. https://doi.org/10.1002/9781119364726.ch11

Joshi NC, Gururani P, Gairola S (2021) Metal oxide nanoparticles and their nanocomposite-based materials as photocatalysts in the degradation of dyes. Biointerface Res Appl Chem 12(5):6557–6579. https://doi.org/10.33263/briac125.65576579

Kannan K, Chanda D, Gautam J, Behera A, Meshesha MM, Jang SG, Yang BL (2023) Hydrothermally synthesized mixed metal oxide nanocomposites for electrochemical water splitting and photocatalytic hydrogen production. Int J Hydrogen Energy 48(93):36412–36426. https://doi.org/10.1016/j.ijhydene.2023.06.042

Kathi S, Mahmoud AED (2024) Trends in effective removal of emerging contaminants from wastewater: a comprehensive review. Desalination and water treatment, p 100258

Khan S, Naushad M, Govarthanan M, Iqbal J, Alfadul SM (2022) Emerging contaminants of high concern for the environment: current trends and future research. Environ Res 207:112609. https://doi.org/10.1016/j.envres.2021.112609

Kumari H, Sonia, Suman, Ranga R, Chahal S, Devi S, Sharma S, Kumar S, Kumar P, Kumar S, Kumar A, Parmar R (2023) A review on photocatalysis used for wastewater treatment: dye degradation. Water Air Soil Pollut (Print) 234(6). https://doi.org/10.1007/s11270-023-06359-9

Kwiatkowski KC, Lukehart CM (2000) Nanocomposites prepared by sol-gel methods. Synthesis and characterization. Elsevier eBooks, pp 387–421. https://doi.org/10.1016/b978-012513760-7/50011-3

Lanjwani MF, Tüzen M, Khuhawar MY, Saleh TA (2024) Trends in photocatalytic degradation of organic dye pollutants using nanoparticles: a review. Inorg Chem Commun 159:111613. https://doi.org/10.1016/j.inoche.2023.111613

Lateef A, Nazir R (2017) Metal nanocomposites: synthesis, characterization and their applications. *ResearchGate*. https://www.researchgate.net/publication/313634485_Metal_Nanocomposites_Synthesis_Characterization_and_their_Applications

Lee D, Kim M, Danish M, Jo W (2023) State-of-the-art review on photocatalysis for efficient wastewater treatment: attractive approach in photocatalyst design and parameters affecting the photocatalytic degradation. Catal Commun (Print) 183:106764. https://doi.org/10.1016/j.catcom.2023.106764

Li Q, Yang F, Zhang J, Cang Z (2020) Magnetic Fe_3O_4/MnO_2 core–shell nano-composite for removal of heavy metals from wastewater. SN Appl Sci/SN Appl Sci 2(8). https://doi.org/10.1007/s42452-020-3182-5

Mahapatra A, Mishra BG, Hota G (2013) Adsorptive removal of Congo red dye from wastewater by mixed iron oxide–alumina nanocomposites. Ceram Int 39(5):5443

Mahdy NK, El-Sayed M, Al-Mofty SE, Allaoui M, Karaly AH, El-Naggar ME, Nageh H, Sarhan WA, Azzazy HM (2022) Toward scaling up the production of metal oxide nanoparticles for application on washable antimicrobial cotton fabrics. ACS Omega 7(43):38942–38956. https://doi.org/10.1021/acsomega.2c04692

Mahmoud AED (2020) Graphene-based nanomaterials for the removal of organic pollutants: insights into linear versus nonlinear mathematical models. J Environ Manage 270:110911

Mahmoud AED (2020) Nanomaterials: green synthesis for water applications. In: Kharissova OV, Martínez LMT, Kharisov BI (eds) Handbook of nanomaterials and nanocomposites for energy and environmental applications. Springer International Publishing, Cham, pp 1–21

Mahmoud AED (2020) Eco-friendly reduction of graphene oxide via agricultural byproducts or aquatic macrophytes. Mater Chem Phys 253:123336

Mahmoud AED (2022) Recent advances of TiO_2 nanocomposites for photocatalytic degradation of water contaminants and rechargeable sodium ion batteries. In: Shalan AE, Hamdy Makhlouf AS, Lanceros-Méndez S (eds) Advances in nanocomposite materials for environmental and energy harvesting applications. Springer International Publishing, Cham, pp 757–770

Mahmoud AED, Krasucka P (2023) 1 Global water challenges and artificial intelligence and modeling for water sustainability: global challenges

Mahmoud AED, Fawzy M (2023) Decontamination of levofloxacin from water using a novel chitosan–walnut shells composite: linear, nonlinear, and optimization modeling. Appl Water Sci 13:244

Mahmoud AED, Mostafa E (2023) Nanofiltration membranes for the removal of heavy metals from aqueous solutions: preparations and applications. Membranes 13(9):789

Mahmoud AED, Fawzy M, Radwan A (2016) Optimization of cadmium (CD^{2+}) removal from aqueous solutions by novel biosorbent. Int J Phytorem 18(6):619–625

Mahmoud AED, Franke M, Stelter M, Braeutigam P (2020) Mechanochemical versus chemical routes for graphitic precursors and their performance in micropollutants removal in water. Powder Technol 366:629–640

Mahmoud AED, El-Maghrabi N, Hosny M, Fawzy M (2022) Biogenic synthesis of reduced graphene oxide from Ziziphus spina-christi (Christ's thorn jujube) extracts for catalytic, antimicrobial, and antioxidant potentialities. Environ Sci Pollut Res

Mahmoud AED, Franke M, Braeutigam P (2022) Experimental and modeling of fixed-bed column study for phenolic compounds removal by graphite oxide. J Water Process Eng 49:103085

Mahmoud AED, Fawzy M, Khan N (2023) Artificial intelligence and modeling for water sustainability: global challenges, 1st, Edition. CRC Press

Mahmoud AED, Ali R, Fawzy M (2024) Insights into levofloxacin adsorption with machine learning models using nano-composite hydrochars. Chemosphere 141746

Majumder A, Gupta AK, Sillanpää M (2022) Insights into kinetics of photocatalytic degradation of neurotoxic carbamazepine using magnetically separable mesoporous Fe_3O_4 modified Al-doped

ZnO: delineating the degradation pathway, toxicity analysis and application in real hospital wastewater. Colloids Surf, A 648:129250. https://doi.org/10.1016/j.colsurfa.2022.129250
Mapossa AB, Mhike W, Adalima JL, Tichapondwa SM (2021) Removal of organic dyes from water and wastewater using magnetic ferrite-based titanium oxide and zinc oxide nanocomposites: a review. Catalysts 11(12):1543. https://doi.org/10.3390/catal11121543
Mishra PM, Devi AP (2023) Current scenario on biogenic synthesis of metal oxide nanocomposites using plant specimens and their application towards treatment of wastewater. Environ Sci Pollut Res Int 30(50):108512–108524. https://doi.org/10.1007/s11356-023-29989-2
Mohammed T, Kadhim Alasedi K, Ruyid R, Abed Hussein S, Latif Jarallah A, Dahesh S, Sultan M, Salman Z, Bashar B, Kareem Obaid Aldulaimi A, Obaid M (2022) ZnO/Co_3O_4 Nanocomposites: novel preparation, characterization, and their performance toward removal of antibiotics from wastewater. J Nanostruct 12(3):503–509. https://doi.org/10.22052/JNS.2022.03.003
Mouden AE, Messaoudi NE, Guerraf AE, Bouich A, Mehmeti V, Lacherai A, Jada A, Sher F (2023) Multifunctional cobalt oxide nanocomposites for efficient removal of heavy metals from aqueous solutions. Chemosphere 317:137922. https://doi.org/10.1016/j.chemosphere.2023.137922
Nandagudi A, Nagarajarao SH, Santosh, Basavaraja BM, Malode SJ, Mascarenhas RJ, Shetti NP (2022) Hydrothermal synthesis of transition metal oxides, transition metal oxide/carbonaceous material nanocomposites for supercapacitor applications. Mater Today Sustain 19:100214. https://doi.org/10.1016/j.mtsust.2022.100214
Natarajan B, Kannan P, Rather JA, Sheikh RA (2023) Recent developments in metal nanoparticles functionalized nanocomposite adsorbents for heavy metals removal from wastewaters. J Taiwan Inst Chem Eng 147:104942. https://doi.org/10.1016/j.jtice.2023.104942
Negrescu AM, Killian MS, Raghu SNV, Schmuki P, Mazare A, Cîmpean A (2022) Metal oxide nanoparticles: review of synthesis, characterization and biological effects. J Funct Biomater 13(4):274. https://doi.org/10.3390/jfb13040274
Nikić J, Watson M, Tubić A, Šolić M, Agbaba J (2024) Recent trends in the application of magnetic nanocomposites for heavy metals removal from water: a review. Sep Sci Technol. https://doi.org/10.1080/01496395.2024.2315626
Kong L, Chu H, Xu M, Xu S, Pan Z, Zhao S, Li D (2022) Oxygen vacancy engineering of MOF-derived ZnO/Co_3O_4 nanocomposites for high harmonic mode-locking operation. J Mater Chem C 10(43):16564–16572. https://doi.org/10.1039/D2TC03503K
Omanović-Mikličanin E, Badnjević A, Kazlagić A, Hajlovac M (2019) Nanocomposites: a brief review. Heal Technol 10(1):51–59. https://doi.org/10.1007/s12553-019-00380-x
Omran BA (2020) Current trends in algae-mediated synthesis of metal and metal oxide nanoparticles (phyconanotechnology). In: Nanotechnology in the life sciences (Print), pp 111–143. https://doi.org/10.1007/978-3-030-46071-6_4
Pandiyan R, Dharmaraj S, Ayyaru S, Sugumaran A, Somasundaram J, Kazi AS, Samiappan SC, Ashokkumar V, Ngamcharussrivichai C (2022) Ameliorative photocatalytic dye degradation of hydrothermally synthesized bimetallic Ag-Sn hybrid nanocomposite treated upon domestic wastewater under visible light irradiation. J Hazard Mater 421:126734. https://doi.org/10.1016/j.jhazmat.2021.126734
Quintelas C, Mesquita DP, Ferreira EC (2021) Impact of industrial wastewater discharge on the environment and human health. The Royal Society of Chemistry eBooks, pp 15–39. https://doi.org/10.1039/9781839165399-00015
Radulescu D, Surdu V, Ficai A, Ficai D, Grumezescu AM, Andronescu E (2023) Green Synthesis of metal and metal oxide nanoparticles: a review of the principles and biomedical applications. Int J Molecul Sci (Online) 24(20):15397. https://doi.org/10.3390/ijms242015397
Rani M, Shanker U (2018) Sun-light driven rapid photocatalytic degradation of methylene blue by poly(methyl methacrylate)/metal oxide nanocomposites. Colloids Surf Physicochem Eng Aspects (Print) 559:136–147. https://doi.org/10.1016/j.colsurfa.2018.09.040
Rather MY, Sundarapandian S (2024) Utilization of eco-friendly copper oxide nanoparticles and iron oxide nanorods in dye removal from real textile industry effluent. Part Part Syst Charact. https://doi.org/10.1002/ppsc.202300223

Sadek RF, Farrag HA, Abdelsalam SM, Keiralla Z, Raafat AI, Araby E (2019) A powerful nanocomposite polymer prepared from metal oxide nanoparticles synthesized via brown algae as anti-corrosion and anti-biofilm. Front Mater (Lausanne). https://doi.org/10.3389/fmats.2019.00140

Saeed M, Alshammari Y, Majeed SA, Al-Nasrallah E (2020) Chemical vapour deposition of graphene—synthesis, characterisation, and applications: a review. Molecules 25(17):3856. https://doi.org/10.3390/molecules25173856

Saleh TA, Mustaqeem M, Khaled M (2022) Water treatment technologies in removing heavy metal ions from wastewater: a review. Environ Nanotechnol Monit Manag 17:100617. https://doi.org/10.1016/j.enmm.2021.100617

Sampath S, Yasasve M, Manjusha M, Sunderam V, Lawrance AV, Saravanan M (2022) A review on algal mediated synthesis of metal and metal oxide nanoparticles and their emerging biomedical potential. J Biotechnol 360:92–109. https://doi.org/10.1016/j.jbiotec.2022.10.009

Saravanakumar K, De Silva S, Santosh SS, Sathiyaseelan A, Ganeshalingam A, Jamla M, Sankaranarayanan A, Priya VV, MubarakAli D, Lee J, Thiripuranathar G, Wang M (2022) Impact of industrial effluents on the environment and human health and their remediation using MOFs-based hybrid membrane filtration techniques. Chemosphere 307:135593. https://doi.org/10.1016/j.chemosphere.2022.135593

Shan R, Lü L, Gu J, Zhang Y, Yuan H, Chen Y, Luo B (2020) Photocatalytic degradation of methyl orange by Ag/TiO_2/biochar composite catalysts in aqueous solutions. Mater Sci Semicond Process 114:105088. https://doi.org/10.1016/j.mssp.2020.105088

Sharma M, Behl K, Nigam S, Joshi M (2018) TiO_2-GO nanocomposite for photocatalysis and environmental applications: a green synthesis approach. Vacuum 156:434–439. https://doi.org/10.1016/j.vacuum.2018.08.009

Sharma P, Dutta D, Udayan A, Kumar S (2021) Industrial wastewater purification through metal pollution reduction employing microbes and magnetic nanocomposites. J Environ Chem Eng (Print) 9(6):106673. https://doi.org/10.1016/j.jece.2021.106673

Shirsat SD, Mane RS, Bauer J, Thorat ND (2022) Metal oxide nanocomposites in water and wastewater treatment. Elsevier eBooks, pp 479–522. https://doi.org/10.1016/b978-0-323-85705-5.00003-8

Sobhani A (2024) Ni/MnO/chitosan nanocomposites: synthesis, characterization and investigation of photocatalytic applications for degradation of malachite green. J Mater Sci. Mater Electron 35(9). https://doi.org/10.1007/s10854-024-12456-0

Sobhani A, Alinavaz S (2023) Co-precipitation synthesis of $CuMn_2O_4$/CuMnO nanocomposites without capping agent and investigation of their applications for removing pollutants from wastewater. Biomass Convers Biorefinery. https://doi.org/10.1007/s13399-022-03732-2

Song N, Wu X, Zhong S, Lin H, Chen J (2015) Biocompatible G-Fe_3O_4/CA nanocomposites for the removal of Methylene Blue. J Mol Liq (Print) 212:63–69. https://doi.org/10.1016/j.molliq.2015.08.059

Thamarai P, Raja K, Saravanan A, Swaminaathan P, Deivayanai V (2024) Emerging trends and promising prospects in nanotechnology for improved remediation of wastewater contaminants: present and future outlooks. Environ Nanotechnol Monit Manag. https://doi.org/10.1016/j.enmm.2024.100913

Thatai S, Khurana P, Boken J, Prasad S, Kumar D (2014) Nanoparticles and core–shell nanocomposite based new generation water remediation materials and analytical techniques: a review. Microchem J (Print) 116:62–76. https://doi.org/10.1016/j.microc.2014.04.001

Tran ML, Fu C, Chiang L, Hsieh C, Liu S, Juang R (2020) Immobilization of TiO_2 and TiO_2-GO hybrids onto the surface of acrylic acid-grafted polymeric membranes for pollutant removal: analysis of photocatalytic activity. J Environ Chem Eng (Print) 8(5):104422. https://doi.org/10.1016/j.jece.2020.104422

Ugrina M, Jurić A (2023) Current trends and future perspectives in the remediation of polluted water, soil and air—a review. Processes 11(12):3270. https://doi.org/10.3390/pr11123270

Venkatesh R, Sekaran PR, Udayakumar K, Jagadeesh D, Raju K, Bayu MB (2022) Adsorption and photocatalytic degradation properties of bimetallic AG/MGO/biochar nanocomposites. Adsorpt Sci Technol 2022:1–14. https://doi.org/10.1155/2022/3631584

Wang A, Zhou Y, Wang Z, Chen M, Sun L, Liu X (2016) Titanium incorporated with UiO-66(Zr)-type metal-organic framework (MOF) for photocatalytic application. RSC Adv 6(5):3671–3679. https://doi.org/10.1039/c5ra24135a

Wani AA, Rather RA, Shaari N, Khan U, Muhammad T, Hussain SM, Seddek LF, Abed AM (2024) Aspects of superior photocatalytic dye degradation and adsorption efficiency of reduced graphene oxide multiwalled carbon nanotubes with modified ZnO-Al_2O_3 nanocomposites. J Environ Chem Eng. https://doi.org/10.1016/j.jece.2024.112461

Yadav N, Ahmaruzzaman M (2023) Recent advancements in $CaFe_2O_4$-based composite: properties, synthesis, and multiple applications. Energy Environ 35(1):458–490. https://doi.org/10.1177/0958305x231155491

Yadav S, Shakya K, Gupta A, Singh D, Chandran AR, Aanappalli AV, Goyal K, Rani N, Saini K (2022) A review on degradation of organic dyes by using metal oxide semiconductors. Environ Sci Pollut Res Int 30(28):71912–71932. https://doi.org/10.1007/s11356-022-20818-6

Yang B, Gao L, Xue M, Wang H, Hou Y, Luo Y, Xiao H, Hu H, Cui C, Wang H, Zhang J, Li Y, Xie G, Tong X, Xie Y (2021) Experimental and simulation research on the preparation of carbon nano-materials by chemical vapor deposition. Materials 14(23):7356. https://doi.org/10.3390/ma14237356

Younas F, Mustafa A, Farooqi ZUR, Wang X, Younas S, Mohy-Ud-Din W, Hameed MA, Abrar M, Maitlo AA, Noreen S, Hussain MM (2021) Current and emerging adsorbent technologies for wastewater treatment: trends, limitations, and environmental implications. Water (Basel) 13(2):215. https://doi.org/10.3390/w13020215

Zabermawi NM, El-Bestawy E (2023) Effective treatment of petroleum oil–contaminated wastewater using activated sludge modified with magnetite/silicon nanocomposite. Environ Sci Pollut Res Int. https://doi.org/10.1007/s11356-023-26557-6

Zhan H, Jiang Y, Ma Q (2014) Determination of adsorption characteristics of metal oxide nanomaterials: application as adsorbents. Anal Lett 47(5):871–884. https://doi.org/10.1080/00032719.2013.850090

Zhao C, Liao Z, Li W, Liu F, Ye J, Liang J, Li Y (2020) Carbon quantum dots modified tubular g-C_3N_4 with enhanced photocatalytic activity for carbamazepine elimination: mechanisms, degradation pathway and DFT calculation. J Hazard Mater 381:120957. https://doi.org/10.1016/j.jhazmat.2019.120957

Zhu D, Zhou Q (2019) Action and mechanism of semiconductor photocatalysis on degradation of organic pollutants in water treatment: a review. Environ Nanotechnol Monit Manag 12:100255. https://doi.org/10.1016/j.enmm.2019.100255

Chapter 12
Conclusion and Future Outlook

Ashootosh Mandpe, Devaanshi Jagwani, and Maulin P. Shah

Abstract The journey of the book has brought forward the transformative role nanomaterials play in the treatment of industrial wastewater, showing not only enormous potential but also real-world and theoretical challenges posed when applied. From confronting all sorts of pollutants and their challenges to setting grounds for sustainable innovative treatment technologies, nanomaterials show that they can change wastewater treatment with higher efficiency, selectivity, and flexibility compared to traditional treatments. This final chapter encapsulates and summarizes core insights. It addresses bigger impacts by providing a more forward-looking vision for what the nanomaterial would bring to wastewater management roles.

Keywords Nanomaterials · Industrial wastewater treatment · Sustainable technologies · Pollutant removal

12.1 Detailed Summary of Core Insight

The key exploratory starting point behind the treatment of wastewater involves an appreciation of unique physical, chemical, and mechanical characteristics that uniquely characterize nanomaterials. Such tools have large surface areas and a high reactivity level; the material can be easily engineered for a specific application and thus has the ability to remove a wide variety of pollutants that are usually toxic and present in industrial wastewater. Chapter 2 provides several types of nanomaterials,

A. Mandpe (✉)
Department of Civil Engineering, Indian Institute of Technology Indore, Indore, Madhya Pradesh, India
e-mail: as_mandpe@iiti.ac.in

D. Jagwani
Civil Engineering Department, IPS Academy, Institute of Engineering and Science, Indore, India
e-mail: devaanshijagwani@ipsacademy.org

M. P. Shah
Microbiology Lab, Ankleshwar, Gujarat, India
e-mail: maulinshah1979@gmail.com

A. Mandpe et al. (eds.), *Nanotech Solutions for Industrial Wastewater*, Advances in Wastewater Research, https://doi.org/10.1007/978-981-96-6879-3_12

such as nanoparticles, nanocomposites, and nanogels that provide specific advantages with regard to their composition and structure, which are exploited to make targeted applications for removal of pollutants.

The synthesis and characterization procedures discussed in Chap. 3 have highlighted the necessity of preparing nanomaterials with specific shapes, sizes, and surface properties to enhance their effectiveness in wastewater environments. Scalable and environmentally friendly synthesis techniques would be essential for transitioning laboratory-scale successes into the industrial arena, while advanced TEM, SEM, XRD, and FTIR characterization techniques would prove to be absolutely indispensable in understanding how these materials interact with pollutants.

This understanding of interaction mechanisms, as presented in Chap. 4, is key to how nanomaterials work in pollutant adsorption, catalysis, and degradation processes. Case studies explored in Chap. 5 suggest potential selectivity and efficiency in removing certain contaminants with heavy metals, organic compounds, and dyes representing individual kinds of environmental hazards. The diversity of nanomaterials allows them to be used in flexible ways for treatment processes, depending on the type of industrial waste.

Chapter 6 maintained the theme of nanomaterials in advanced treatment processes, including advanced oxidation processes (AOPs) and membrane-based separations. Enhancing these processes through the incorporation of nanomaterials enables the development of more robust and comprehensive treatment systems that can more effectively address emerging contaminants and pathogenic threats than traditional approaches. Incorporation of nanomaterials into membranes, for example, has demonstrated promising results in overcoming fouling issues and increasing the durability of separation systems.

Chapter 7 discussed how the debate over the effectiveness of nanomaterials transcended into environmental implications. The balanced approach that must be adopted, including the benefits and risks of nanotechnology, calls for responsible practices and regulatory oversight of the entry of nanomaterials into natural ecosystems, the potential ecotoxicity, and uncertainties of their long-term environmental fate.

12.2 The Future of Wastewater Treatment

This has led nanomaterials to take the lead in innovation, at the forefront of revolutionizing effective, scalable, and sustainable treatment technologies, given that nanomaterials offer high performance in the removal of even trace contaminants and pollutants. Their application promises new avenues of meeting stringent regulatory requirements or reducing the impact of industrial processes on natural water bodies. Because their composition and functionality may vary significantly with the composition and structure of the substrate from which they are extracted, they are capable of undertaking a very broad scope of pollutants that are refractory to removal by other conventional methods, thereby opening pathways towards the development of adaptive water treatment systems based on the composition of wastewaters.

Future trends in this field are likely to involve incorporation with sources of renewable energy and some of the emerging technologies that encompass AI and machine learning capabilities. This integration can lead to real-time optimization of the use of nanomaterials, prediction of pollutant loads, and real-time adjustment of treatment parameters for maximum efficiency. Also, even more specialized nanomaterials may be found that can target specific contaminants with high precision and enable powerful yet environmentally friendly treatment systems.

12.3 Challenges and Opportunities for Research, Collaboration, and Innovation

Potential appears to be enormous; several challenges still restrict nanomaterials' full adaptation in wastewater treatment. Those challenges include scalability, cost-effectiveness, environmental, health risks, and compliance with regulations. The current methods of synthesis, effective within the lab settings, require cost-effectiveness and scalable to allow for a wider commercial application. Moreover, the ecotoxicological impact of nanomaterials must be examined so as not to inadvertently damage the environment through their application in wastewater treatment.

Interdisciplinary contributions from scientists, engineers, industry players, and policymakers will be essential for overcoming these barriers. Continuing interdisciplinary research will be an essential tool for innovation, at the laboratory-scale and also at the industrial scale. The expertise of materials science, environmental engineering, and regulatory bodies will establish guidelines for the safe and effective use of nanomaterials. It should also include the involvement of the public and education of communities on its benefits and risks to achieve societal acceptance and ethical application.

12.4 Economic, Societal, and Environmental Considerations

There is much significance that also relates to economic and social effects of the treatment using nanomaterials in water, and the problem has been discussed in Chap. 10. On one hand, various applications seem to have the potential for creating material or type-based economic efficiency. In certain types, savings are definitely great compared with conventional methods; long-term savings especially benefit if applied systems have an increased reutilization rate and/or a good recycling capability. Generally speaking, any investment into wastewater treatment made from nanotechnology can further strengthen wider society goals—save resources, improve the health conditions of populations, and protect environments.

However, societal acceptance and ethical considerations are critical factors that could influence the wider adoption of nanomaterials in wastewater treatment. The need to educate communities about the benefits of nanomaterial use as well as concerns regarding toxicity and environmental safety is an important step in building trust. Public perception and awareness will determine the extent to which these technologies are embraced and integrated into mainstream practices.

12.5 Towards a Greener Future

Nanomaterials are poised to change the face of wastewater treatment by offering high-performance adaptable systems to meet the increasing demands of a resource-scarce world. Over the next decade, advancements in hybrid nanomaterials, which will combine them with other treatment technologies, such as biological processes and renewable energy-driven systems, will continue to advance and offer even more effective yet sustainable and energy-efficient solutions.

Sustainable development goals (SDGs) center on having suitable clean water and sanitation as well, and nanomaterials therefore provide only one of the many possible ways that these ends should be fulfilled. Through such advancements in the aspects involving wastewater treatment that could be better, safer, and more economical, nanomaterials are to significantly play roles in water conservation and pollution-reduction ventures. As this furthers advancement in terms of industrial water treatment, the nanomaterials should lead the pack towards achieving this clean and healthy place.

12.6 Continued Innovation and Cooperative Efforts

The introduction and implementation of nanomaterials for wastewater treatment give both great promises and huge challenges. Therefore, it would be fundamental to advance the field forward by taking the path with sustainability, safety, and societal welfare at heart. Indeed, the great promise of nanotechnology will come about only by unrelenting research endeavors, openness, and multi-level cooperation.

Insights and case studies in this book provide a solid platform for a deeper understanding of the role of nanomaterials in the treatment of wastewater and spur continued innovation into the future. Water treatment technologies will be built upon interdisciplinary collaboration, respect for environmental integrity, and commitment to making clean water available for all, in the decades ahead. Embracing these principles, nanomaterials-based wastewater treatment will make long-term contributions towards safeguarding water security worldwide and safeguarding the environment (Fig. 12.1).

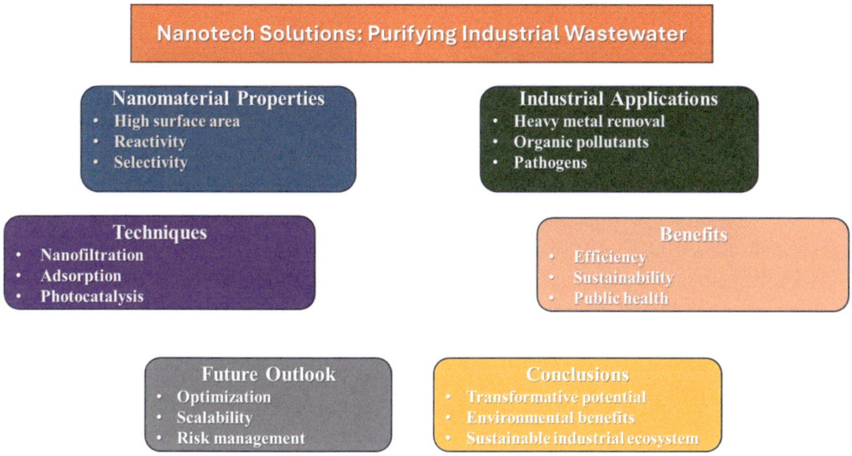

Fig. 12.1 Summary of book

12.7 Critical Elements

A. **Properties of Nanomaterials**

The nanomaterials display several properties that have led to them being suited to wastewater treatment:

- Extensive surface-to-volume ratio: High amounts of nanomaterial in a given volume lead to high surface area exposed which interacts with the pollutants more effectively.
- Reactiveness: Due to nanoscopic structure, these reactively engage in complex chemistry-related processes, which results from pollutant degradation
- Selection: Nanomaterial can be synthesized to favorly target certain contaminants enhancing higher efficiency in removing diversity-based types of pollutants.

These properties make nanomaterials versatile and efficient in various treatment processes to target the removal of specific industrial waste components.

B. **Industrial Applications**

Nanotechnology in wastewater treatment is very effective in dealing with several industrial pollutants, such as:

- Heavy Metal Removal: Nanomaterials can adsorb and neutralize toxic heavy metals, which reduces their impact on ecosystems and human health.
- Organic Pollutants: Carbon-based nanomaterials and other nano-enhanced compounds efficiently break down organic pollutants, including dyes and chemicals from industrial discharges.
- Pathogens: Nanomaterials possess some pathogens that destroy them, thus eliminating pathogens and making the water treated to be safe and pure in quality.

C. **Techniques**

These are various techniques put forward by nanotechnology employed during wastewater treatment process;

- Nanofiltration: This involves a method based on utilization of nanoporous membrane filters at the molecular level used for separating contaminants in the liquid. They are exceptionally excellent, especially in the case of suspended particles and small-size dissolved ions.
- Adsorption: These nanomaterials serve as adsorbents. It boasts high surface areas due to which it captures those particles and holds them in place.
- Photocatalysis: This is the process involving the degradation of pollutants through light-activated nanomaterials. Photocatalysts, such as titanium dioxide nanoparticles, break organic compounds when exposed to light, thus making it a friendly solution to the environment.

D. **Benefits**

Benefits of nanomaterials in wastewater treatment.

- High efficiency: Nanomaterials increase the rate and the extent of contaminant removal, thus making wastewater treatment faster and more effective.
- Sustainability: The efficiency of the treatment would be improved by reducing the total environmental impact of industrial wastewater treatment processes by using nanotechnology.
- Public Health: Improved removals of hazardous pollutants and pathogens from wastewater have direct outcomes on public health because these exposures to contaminated water are lessened.

E. **Future**

There is a huge potential future for nanotechnology in treating wastewater, given that most of the areas discussed remain open to new improvements, including:

- Optimization: A core focus of research nowadays, optimization of nanomaterials for increased selectivity, stability, and their reusability that result in more efficient as well as cost-effective use is emphasized.
- Scalability: Researchers are engaged to fabricate scalable manufacturing protocols allowing high-quality nanomaterials to be reached towards industrial applications so the uses can be made mass-compatible.
- Risk Management: Upon increasing usage, environment along with health risks by liberation of nanomaterial must be assessed and consequently taken measures to minimize or further prevent it from use leading to safer application protocols

12.8 Further Research and Innovation

Nanotechnology has really shown incredible promise in changing the face of industrial wastewater treatment, and yet, much is there to be explored for it to be both sustainable as well as commercially viable for the solutions. Future studies and innovations in this discipline will be directed towards improving upon the effectiveness, scalability, environmental compatibility, and nanomaterial safety of mainstream nanotechnology as a handy tool for addressing industrial pollution. Some of the key emphasis areas for the future are the following:

A. **Multifunctional Nanomaterials**

It will be multifunctionality as the next generation nanomaterials. With integrated functions, these nanomaterials can target multiresidue pollutants. For example, hybrid nanomaterials that combine adsorption with catalytic function to be used in treating various effluents containing the complexities of industrial effluents. Heavy metals and organic compounds are found in their combination. There is also a need for nanomaterials that can easily be tuned to accommodate other composition types of wastewater into one system.

B. **Environmentally Friendly Synthesis and Sustainability of Nanomaterials**

The methods that are used in the synthesis of nanomaterials are often quite hazardous chemicals or high-energy processes that can even reduce their environmental benefits. Future research focuses more on green synthesis techniques in the production of nanomaterials, such as plant extracts, bacteria, or fungi. These methods reduce the ecological footprint of the production of nanomaterials and ensure that produced materials have biocompatibility and are much less toxic, hence suitable for use in wastewater treatment.

C. **Increased Selectivity and Targeting**

Selective removal of specific contaminants is very crucial for industrial wastewater treatment since it has a complex mixture of pollutants to clean. Advanced research focuses on designing nanomaterials that express greater selectivity towards specific contaminants such as heavy metals, dyes, or pharmaceutical residues through surface modifications and functionalization. This would improve treatments with the potential to make the process more efficient, without the need for multistep or chemically based treatments.

D. **Reusability and regeneration of nanomaterials**

One of the biggest problems with nanomaterials is that they lose their effectiveness after one application. Future research is trying to develop reusable and regenerable nanomaterials that can withstand multiple treatment cycles without losing their performance. Self-cleaning nanomaterials, which can renew themselves through mechanisms like photocatalysis when exposed to light, are a promising area of research that could make nanomaterials more cost-effective and sustainable in the long run.

E. **Integration with New Technologies**

The integration of nanomaterials with emerging technologies such as artificial intelligence, machine learning, and Internet of Things has vast potential. Artificial intelligence and machine learning algorithms can predict pollutant loads and optimize nanomaterial deployment in real time, making the treatment process more adaptive and efficient. IoT-enabled sensors embedded in nanomaterial-based treatment systems can also provide continuous monitoring, ensuring precise pollutant removal and minimizing the need for frequent manual testing.

F. **Environmental Impact and Life Cycle Analysis**

As the application of nanomaterials in industrial wastewater treatment increases, there is an increasing demand to understand the long-term environmental impacts. LCAs and ecotoxicity studies will be required to evaluate the full environmental impact of nanomaterials from production to disposal. Future research should focus on developing safe degradation or retrieval methods for nanomaterials once their function is complete, thus reducing the risk of nanomaterial contamination in natural ecosystems.

G. **Economies of Scale and Cost Savings through Mass Production**

The primary challenge in nanomaterials being generally adopted for wastewater treatment purposes is the cost of these materials' production. It is therefore crucial that more research is conducted into the possibility of large-scale, inexpensive production of nanomaterials in order to enable nanotechnology to be accessible by industries worldwide. Innovation in material science may bring about dramatic reductions in costs by employing low-cost raw materials or by recycling industrial waste as sources for the production of nanomaterials. It also opens new directions in 3D printing and additive manufacturing, capable of producing nanomaterials with desired properties on scale.

H. **Hybrid Nanomaterials with Convention Treatments**

Nanomaterials alone will likely not be sufficient to fulfill the requirements of all industrial applications of wastewater. Thus, future developments would most probably focus on hybrid systems based on the association of nanomaterials with the more conventional approaches such as biological treatment, membrane filtration, and AOPs. Hybrid systems would exploit the relative strengths of both approaches through enhanced removal efficiencies of pollutants and increased ranges of contaminants being removed.

I. **Innovation in New Classes of Nanomaterial**

While carbon-based nanomaterials and metal nanoparticles are the most prominent classes of nanomaterials used in wastewater treatment, there is increasing attention to emerging classes of nanomaterials. For example, nanoclays, bio-inspired nanomaterials, and metalorganic frameworks (MOFs) display properties that might make them quite promising for particular treatment challenges. Future work will undoubtedly be directed towards such newer materials to enhance options to target challenging-to-treat pollutants.

J. **Regulatory and Safety Considerations**

When these nanomaterials come into view in increased and widespread applications in wastewater treatments, the corresponding rule-making systems would have to move forward to make sure they comply with safety.

Research studies done about safe handling and appropriate disposal of nanomaterials would help make and enforce policies that not only protect a human's health but protect and preserve the environment too. Cooperation between scientists, industries, and regulatory bodies will be absolutely necessary for the development and having of standardized guidelines for nanomaterial applications in wastewater treatments in the near future years.

12.9 Future Innovation Prospects

Future innovations in nanomaterials for industrial wastewater treatment will probably be better than the challenges that have been encountered in the past, very efficient, and sustainable. There are likely to be these three main places where innovations could make a great impact:

A. **Intelligent or Responsive Nanomaterials**

Self-regulating Nanomaterials: In relation to this category, such materials should respond real time to the contamination present in the water with adjustments accordingly. For instance, such "smart" nanomaterials would regulate themselves to change surface charges and structures to absorb specified pollution in different types of industrial wastewater.

Stimuli-responsive nanomaterials: create materials that change behavior for certain triggers, such as pH, temperature, or light. These may be most useful for the removal of contaminants that need selective and directed treatment.

B. **Advanced hybrid nanocomposites**

Functional multifunctional composites: Here, innovation would lie in the development of nanocomposites that allow for adsorption, catalysis, and filtration functionality in one material, hence, complex composition wastewater treated by one-step treatment.

Bio-nano Hybrid Materials: The combination of nanotechnology with bio-based materials, like algae or bacterial enzymes, may break down pollutants in a more environmentally friendly manner and permit the treatment of specific pollutants that are difficult for traditional materials.

C. **Nano-encapsulated Treatment Agents**

Encapsulation of Catalysts and Enzymes: The encapsulation of reactive agents, such as catalysts or enzymes, in nanoshells, may prevent degradation and extend their lifespan while providing a controlled release of reactive substances for sustained removal of pollutants.

Controlled Release Systems: Nanomaterials that have time-released properties can be used to deliver specific agents over an extended period of time, which may provide consistent treatment in long-term applications, especially in industrial effluents where contaminant levels fluctuate.

D. **3D-Printed Nano-enabled Filters and Membranes**

Customized 3D Printed Membranes: The use of 3D printing can enable the fabrication of nanomaterial-based filters with precise pore structures, which enhances contaminant capture. Customized filters for specific industrial processes can optimize treatment performance and reduce the need for additional filtration stages.

Nano-patterned membranes for flux enhancement and antifouling: It is also possible to pattern nano-sized surfaces on the membranes for improved flux and prevention of clogging, prolonging filtration system life with reduced maintenance costs and enhancing efficiency.

E. **Artificial Intelligence and Machine Learning to Optimize Nanomaterials**

It should further help predict the composition and optimal structure of nanomaterials for pollutant-type and concentration and properties, hence reducing the trial and error in the synthesis and possibly faster development cycles.

Real-time Monitoring and Adaptive Systems: AI-based monitoring systems may supplement nanomaterials to monitor contaminant levels in real-time, and treat them on-demand, adaptive, efficient, and resource-saving manner.

F. **Regenerative and Self-healing Nanomaterials**

Self-healing Nanostructures: Materials that self-heal after being damaged may prove to be a game-changer for applications that use aggressive industrial effluents. The materials may regain their active sites, thereby extending the operational lifetime and reducing the cost of replacement.

Photoregeneration and Electro-regeneration: The innovation could focus on materials that regenerate by using light or electrical currents to recover their adsorption or catalytic properties without requiring extra chemicals or complex procedures.

G. **Nano-bioremediation**

Nano-assisted Biodegradation: The integration of nanotechnology with bioremediation enhances the microbial processes through the delivery of nutrients or essential elements at the nanoscale, which boosts pollutant degradation.

Nano-augmented Microbial Filters: The combination of nanomaterials with microbial filters results in the effective breakdown of organic contaminants and even some inorganic compounds, providing highly effective nature-inspired treatment systems.

H. **Recyclable and Biodegradable Nanomaterials**

This can be biodegradable or natural nanomaterials, such as cellulose nanocrystals, in which the nanomaterial does not remain in the environment after use.

Recyclable Nanocatalysts: The recyclability and recoverability of nanocatalysts make it easy and thus lessen the cost as well as environmental impact. Example is a case where magnetic nanoparticles already have been examined for being retrieved using magnetic fields; more work soon will bring it into everyday occurrence.

Nano-enabled Reactive Barrier Systems

Nano-enabled Barriers: Nano-enabled barriers put in place in industrial premises can intercept the contaminated water at the entry point itself, thereby not letting it go underground. It could act as a nanomaterial-based target removal of some specific contaminants, acting like an in-situ treatment measure.

Passive Nano-reactive Systems: The passive treatment systems having incorporated nanomaterials can treat wastewater on an uninterrupted basis with a minimal amount of energy input. Therefore, the operating costs and impacts on the environment for sites in remote locations or ones without resources will also reduce.

I. **Carbon Capture and Water Treatment Hybrids**

Double-duty nanomaterials in water and carbon dioxide capture may solve, in a single step, the industrial problems of atmospheric and water pollution. Opportunities here will lead to multifunctional treatments that eliminate water pollutants and reduce emissions from the treatment process.

Two-in-one photocatalytic carbon reduction: photocatalysis of water pollutants using nanomaterials would not only break down wastewater pollutants but could potentially enable the capture and reduction of CO_2 in wastewater treatment, allowing for a more sustainable treatment process for the industries involved.

J. **Pathogen-targeted Nanomaterials for Water Disinfection**

Nano-coatings with Pathogen-specific Targeting Mechanism and Antibacterial Activity: the nano-coatings can be created as pathogen-specific for targeted mechanisms, thus without any harm to the positive bacteria, it is best used in wastewater having biological load.

Nanostructures for viruses or bacteria: nanostructures are designed to target industrial wastewater viruses or particular strains of bacteria present hence give highly effective disinfective action, especially used in pharmaceuticals, food industries, and other areas with similar applications.

K. **Energy Harvesting Nanomaterials**

Energy Generating Nanofilters: Nanofilters that generate energy either from the water stream or reaction of pollutants will make other treatment parts operational without need for external power supply in a remote area.

Photocatalytic Fuel Production: Photocatalytic nanomaterials that turn pollutants into fuel, for example, hydrogen, have the potential to revolutionize the way wastewater is treated and clean water produced into resources.

12.10 Concluding Remarks

Nanotechnology is rapidly replacing the conventional industrial methods of wastewater treatment and provides innovative solutions, making the purification more efficient and sustainable. Owing to the unique properties that nanomaterials present, it has emerged as a fundamental tool for removal of contaminants in wastewater streams across heavy metals, organic compounds, and pathogens.

This offers bright futures for nanomaterials in the industrial treatment of wastewater since these are constantly innovated and made more efficient, sustainable, and safe. Research into the challenges that face scaling, cost, and impact on the environment should pave the way for this science to change the landscape in the treatment of wastewater among industries. Continued interdisciplinary research and collaboration shall propel nanomaterials into the future where clean water will be accessible and sustainable for all industries.

Therefore, in summary, nanotechnology for wastewater treatment is transformative because it presents solutions that not only work well but also are environmentally friendly. This is because nanomaterials reduce the amount of pollutants and enhance water quality, thereby leading to a more sustainable industrial ecosystem. The future of nanotechnology in this area will therefore depend on further innovation and responsible implementation in collaboration between scientists, industry leaders, and policymakers in order to achieve long-term, sustainable impacts in the management of wastewater.

This brings forth a lot of future potential in nanotechnology and industrial wastewater treatment. Changes within the field of smart and responsive materials, greener synthesis techniques, AI-driven optimization, and multifunctional nanocomposites are predicted as the changes that will push the industry forward. Emphasis on sustainability, adaptability, and efficiency among these future innovations promises nanomaterial-based solutions to be an integral part of wastewater management and makes us step closer to sustainable and pollution-free future.

MIX
Papier aus verantwortungsvollen Quellen
Paper from responsible sources
FSC® C105338

If you have any concerns about our products,
you can contact us on
ProductSafety@springernature.com

In case Publisher is established outside the EU,
the EU authorized representative is:
Springer Nature Customer Service Center GmbH
Europaplatz 3, 69115 Heidelberg, Germany

Printed by Libri Plureos GmbH
in Hamburg, Germany